Springer Tracts in Modern Physics
Volume 206

Springer Tracts in Modern Physics

Springer Tracts in Modern Physics provides comprehensive and critical reviews of topics of current interest in physics. The following fields are emphasized: elementary particle physics, solid-state physics, complex systems, and fundamental astrophysics.
Suitable reviews of other fields can also be accepted. The editors encourage prospective authors to correspond with them in advance of submitting an article. For reviews of topics belonging to the above mentioned fields, they should address the responsible editor, otherwise the managing editor. See also springeronline.com

Matthias Wuttig   Xiangdong Liu

Ultrathin Metal Films

Magnetic and Structural Properties

With 234 Figures

Springer

Matthias Wuttig
Xiangdong Liu

I. Physikalisches Institut
RWTH Aachen
52056 Aachen, Germany
E-mail: wuttig@physik.rwth-aachen.de
xiang@physik.rwth-aachen.de

Library of Congress Control Number: 2004109593

Physics and Astronomy Classification Scheme (PACS):
75.70.-i,81.15.-z, 68.65.-k

ISSN print edition: 0081-3869
ISSN electronic edition: 1615-0430
ISBN 3-540-58359-9 Springer Berlin Heidelberg New York

Springer is a part of Springer Science+Business Media

springeronline.com

Printed in Germany

Typesetting: by the author using a Springer LaTeX macro package
Cover concept: eStudio Calamar Steinen
Cover production: *design &production* GmbH, Heidelberg

Printed on acid-free paper SPIN: 10477681 56/3141/jl 5 4 3 2 1 0

The search for truth seemed easy at first,
but ran into difficulty.
It is dark,
and a deep sea, with turbulent undercurrents,
and surmounting waves, stands between us.
Those in a calm harbor,
how can they know what we must go through?

Hafez (Persian poet), 1350

Preface

Without the help of many people, the work reported here could not have been done. It is with great pleasure that we acknowledge Professor Harald Ibach for fostering a dynamic research environment. His continuous support and unwavering interest in this project made this book possible.

The scope of this work is in part the result of the dedication shown by my students: Wolfgang Hoffmann, Johannes Thomassen, Benedikt Feldmann, Michael Hansen, Thomas Flores, Frank May, Rene Niebuhr, Jürgen Jandeleit, Holger Zilligen and Bernd Schirmer. We could always count on the help of Josef Larscheid who constructed and repaired many pieces of equipment. Hearthy acknowledgements also go to our crystallographer, Udo Linke who was always willing to cut and polish samples.

Dr. Colette Knight patiently reviewed the entire manuscript and provided thoughtful suggestions. Mrs. Thomassen efficiently turned our rough sketches into aesthetical drawings, while Mrs. Elbert patiently typed many pages.

What made science enjoyable from the start was the possibility to collaborate with and learn from many people. Twenty years ago, I took my first steps in surface science under the auspices of Dr. Rene Franchy who taught me more than just the running of an EELS spectrometer.

A substantial part of our research activities is focussed on structure determination using LEED. I learnt about the beauty and potential of this technique during a pleasant stay at the University of Californial at Berkeley in the group of Professor Gabor A. Somorjai and Dr. Michel Van Hove. Dr. Philip Rous and Dr. Adrian Wander were especially helpful in getting me acquainted with the newly developed Tensor LEED program. Professor W. Moritz from the University of München introduced me to an equally efficient automated optimization scheme. His student Stefan Pflanz showed me the potential of surface defect characterization using LEED. Over the last years, I had fruitful cooperations on structure determinations of alloy surface with Dr. Yves Gauthier from the University of Grenoble and Dr. Guy Treglia from Marseille. I also acknowledge pleasant interations with the group of Professor K. Heinz of the University of Erlangen. In addition, we have benifitted trenmendously from an exciting collaboration with Professor Dr. Stefan Blügel (IFF/KFA Jülich) and other former members of this Institute including Dr. Carlo Carbone, Dr. Elio Vescovo and Dr. Oliver Rader.

On a more personal note, we would like to give our appreciation to those dear to us: our parents for their encouragement and support, our friends and our relatives. In particular, we would like to thank our wives for their support and understanding even in times when work absorbed most of our time and minds.

Aachen,
April 2004

Matthias Wuttig
Xiangdong Liu

Contents

1 Introduction

The ability of mankind to produce and process materials has been an important aspect of human development. This is reflected in the names of certain historical periods, which mirror the ability of mankind at that time to manufacture or work different materials (e.g. the Bronze Age and Iron Age). From these examples, it is clear that for a long time the mechanical properties of matter were of key interest. This situation has changed tremendously in the last century, in which the main attention has been focused on the electronic properties of matter. Possibly, future historians will call this period the "Silicon Age" [1]. It is this material and its usage in semiconductor and information technology that have had one of the most pronounced influences on everyday life.

Over the last hundred years, the understanding of material properties has increased to the extent that we are now able to tailor materials to fulfill specific needs [2]. In ancient times, one would have wondered what options a given material offered. Today, one is often in the pleasant situation where one defines the desired properties and then asks a materials scientist to create a material that meets those requirements. This success is clearly related to an improved understanding of the relationship between structure and properties. The strategy of deriving properties from the structure of the building blocks is encountered not only in materials science but also in physics and chemistry. It can even describe the present approach towards an understanding of the properties of pharmaceutical substances.

In present times, however, the ability to manipulate materials is no longer just a luxury but unfortunately seems to be becoming a necessity for our survival. In a century that is realizing with pain that this word is finite and vulnerable, our dealings with resources have changed and need to change further. Thus, one now very often has to create devices to fulfill the required purpose with the boundary condition of using fewest resources possible (materials, energy, etc.). Clearly, these two driving forces, i.e., the needs to tailor materials and to save resources, are to a large extent responsible for the development and spread of thin-film technology. The greatest achievements of these techniques are in the realm of semiconductor heterostructures, notably devices based upon GaAs–GaAlAs, strained Si, etc. The high-technology applications of these structures include high-electron-mobility transistors and multiquantum-well lasers.

In this book, we shall not discuss semiconductor thin films, but concentrate on metallic films and, exclusively, on metallic substrates. This is partly justified by the fact that, so far, thin metal films have received less interest, even though the requirements on crystalline perfection are less stringent than for their semiconductor counterparts. Thus, a much wider range of different elements can be combined. These metallic films offer a large variety of present and potential applications employing, for example, their improved corrosion resistance or enhanced chemical activity. Their use as hard coatings has also received considerable attention. Furthermore, thin-film magnetoresistive heads are used for the readout of magnetic recording devices. Additionally, thin films with a unique coercive strength will find applications in the magnetic-materials industry [3], and perpendicular magnetic anisotropy is now being exploited in commercial products [4]. Another opportunity offered by thin-film growth is the possibility to grow metastable structures, which could possibly be magnetic. To use the full potential of these films, it is clearly necessary to expand and exploit our understanding of the relationship between the structure and the properties of thin films.

Figure 1.1 shows the typical dimensions of thin-film systems and compares them with the physical length scales of various "quasiparticles" (electrons, phonons and photons), as well as the magnetic coupling lengths. Whenever the film thickness or the period of a superlattice has dimensions comparable to the wavelength of the probing particle, characteristic film or superlattice effects are encountered.

It is the aim of this book to discuss the properties of ultrathin metallic films (0–20 Å thick). Associated with this small film thickness, a series of new physical phenomena can be observed in ultrathin films. First, new structures can be stabilized by pseudomorphic growth. These structures can be a metastable phase if the substrate's in-plane lattice happens to match that of the metastable phase. Another possibility is that they are only "forced structures", which are required to adopt the substrate's in-plane lattice within a limited thickness range owing to the interface interaction. No matter whether they are metastable or forced structures, these epitaxy-stabilized new structures provide a chance to observe properties which are unavailable from bulk materials. A comparison of the physical properties shown by ultrathin films with the same chemical composition but stabilized by different substrates and thus having different structures would add to our understanding of the correlation between structure and properties. Even for a given epitaxy-stabilized ultrathin film, structural relaxation, including the presence of various reconstructions, is observed with increasing film thickness. A typical example is iron films on a Cu(100) substrate. Structural relaxation with increasing thickness in epitaxy-stabilized ultrathin films again provides a good chance to study the correlation between structure and physical properties. In this

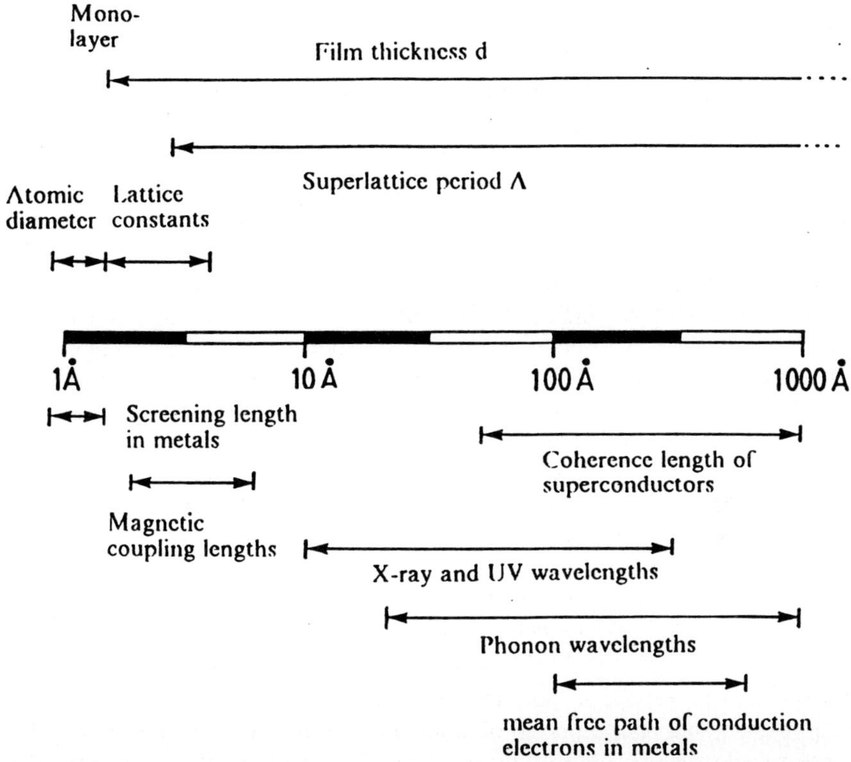

Fig. 1.1. Characteristic length scales of thin films. Typical dimensions of thin films and superlattices (*upper part* of the figure) are compared with physical length scales of phenomena such as Fermi screening in metals, the coherence length of superconductors, magnetic coupling, and the wavelengths of phonons and photons (x-ray and UV) (*lower part*). Characteristic film effects are encountered when the film thickness or superlattice period is comparable to one of the length scales. After [5]

book, an effort is undertaken to relate the structure to the properties of the film, mainly the magnetic properties of 3d transition metal films. It is also clear from Fig. 1.1 that for a film thickness up to 20 Å, the thickness is comparable to the electron wavelength and coincides with typical magnetic coupling lengths, giving rise to novel magnetic properties. For example, quantum size effects could become conspicuous, and new properties associated with the low dimensionality could be observed. These properties will be a second focus of this book.

In the following chapters, the growth (Chap. 2), structure (Chap. 3) and magnetism (Chap. 4) of thin films will be introduced first. In each of these chapters, a theoretical description is put forward first and then the relevant experimental techniques are discussed. In Chap. 5, various theoretically predicted metastable structures of 3d transition elements, as well as experimental results for the structure and magnetism of epitaxy-stabilized ultrathin 3d films, are summarized. In the following two chapters, an in-depth discussion of a small number of thin-film systems is provided. In these case studies, a large number of experimental techniques have been applied. This enables a detailed investigation of the relationship between structure and film properties, with a particular emphasis on magnetic properties. In Chap. 6, the correlation between the magnetism, structure and morphology of ultrathin iron films grown on Cu(100) is presented. Chapter 7 describes the growth and structure of manganese films on Cu(100) and Ni(100). Also, the mechanisms and atomic processes for the formation of MnCu/Cu(100) surface alloys are discussed. The outlook for the potential of ultrathin films and the associated research opportunities is considered in Chap. 8.

References

1. H. Queisser: *Kristallene Krisen*, Piper, München(1985)
2. T. Paxton: Physics world 5, Vol.11, 35(1992)
3. Artificially Structured Materials: High-Performance Building Blocks for Today and Tomorrow, published by Technical Insights Inc (1986)
4. D. Mergel and P. Hansen: *in Magnetismus von Festkörpern und Grenzflächen*, IFF/KFA Jülich (1993), ISBN 3-89336-110-3
5. W. Zinn: *in Magnetismus von Festkörpern und Grenzflächen*, IFF/KFA Jülich (1993), ISBN 3-89336-110-3

2 Growth of Ultrathin Metal Films

Driven by the requirement to tailor film properties for many technological applications, our understanding of the atomic processes involved in thin-film growth and the principles governing those processes has progressed tremendously in the last century. Growth of thin films has changed from an art to a science. For ultrathin metal films, the growth conditions are found to have important and sometimes crucial consequences for the structural and magnetic properties of the film. Therefore, manipulating the film growth by detailed control of the growth conditions is an effective pathway to exploring the ultimate potential of ultrathin magnetic films. In this chapter, the various growth modes are first introduced and described from the viewpoints of thermodynamics and kinetics. The growth parameters, such as growth temperature, growth rate and application of a surfactant, are discussed at some length as possible tools to modify the growth and thus the film properties. Experimental techniques for the analysis of the growth mode are described, with an emphasis on reflection high-energy electron diffraction (RHEED).

2.1 Growth Modes

In this book, we shall focus on ultrathin epitaxial films on metallic single-crystal substrates, where the films are prepared from the vapor phase under ultra high vacuum (UHV) conditions. The structure of such films can be described on the atomic level. This facilitates an investigation of the correlation between the film properties and the film structure and enables us to explore the limits of the modifications to the properties. The word "epitaxy" comes from the Greek words $\varepsilon\pi\iota =$"placed upon" and $\tau\alpha\xi\iota\zeta =$"arrangement". The term "epitaxy" was introduced by Royer [1] and describes the oriented growth of one crystal on another crystal. Sometimes a distinction is made between *homoepitaxy*, where the film and substrate are of the same material, and *heteroepitaxy*, where this is not the case. While homoepitaxial systems are usually studied as simple model systems with the goal of achieving a fundamental understanding of the growth kinetics, heteroepitaxial growth is of more practical relevance. For ultrathin magnetic films, the film and substrate need to differ from each other so that the properties of the film can be tailored by epitaxial growth. Therefore homoepitaxial growth is seldom used to produce ultrathin magnetic films.

Depending on the specific application, various different surface morphologies are required for the films. For example, in the production of magnetic storage devices, atomically smooth films are desired, whereas rough films with well-defined island sizes and densities are desired for heterogeneous catalysis. Manipulating the morphology of epitaxially grown films through detailed control of the growth conditions has always been a challenge. In reaching such a goal, the primary concern is to control the growth mode. The growth mode characterizes the nucleation and growth processes, and there is a direct correspondence between the growth mode and the film morphology. Three growth modes are frequently encountered. They are the Frank–van der Merwe mode, Stranski–Krastanov mode and Volmer–Weber mode (Fig. 2.1).

- *Frank–van der Merwe growth* [2] is defined by the sequential occupation of the layers. The growth of the $(n+1)$th layer starts only after the nth layer has been completed. Therefore it is also called *layer-by-layer growth*. Clearly, two-dimensional nucleation is a prerequisite for layer-by-layer growth. Frequently, imperfect layer-by-layer growth is found in which growth of the $(n+1)$th layer begins before the nth layer is completed. In cases where the aim is to produce smooth films, the layer-by-layer growth mode is most favorable.
- *Volmer–Weber growth* [3] is defined by the growth of three-dimensional islands on the substrate.
- *Stranski–Krastanov growth* [4] is encountered for systems that initially grow in a layer-by-layer fashion but then develop three-dimensional islands above a certain thickness. It can be described in simple terms as a layer-by-layer plus island mode.

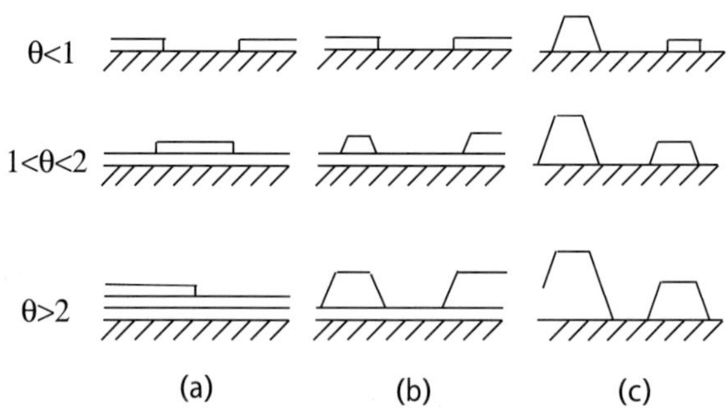

Fig. 2.1. Growth mode of thin films: morphology of a growing film for (**a**) Frank–van der Merwe growth, (**b**) Stranski–Krastanov growth and (**c**) Volmer–Weber growth, upon increasing coverage θ in monolayers (ML)

The three modes mentioned above describe phenomenologically the different growth behaviors observed on perfectly flat substrates. In fact, real substrates always contain steps and other defects. At a high temperature and low deposition rate, the mobility of the adatoms can be very high, so that on a real surface all the deposited adatoms reach the preexisting step edges and are captured there before they meet to form stable nuclei. In this case, the growth is characterized by the absence of nucleation and the advancement of steps over the lower terraces. This growth mode is called the *step-flow mode* [5]. For heteroepitaxy, elevated growth temperatures must usually be avoided owing to the risk of massive interdiffusion. Therefore we shall exclude step-flow growth from our further discussions.

In general, whether the epitaxial film undergoes layer-by-layer growth, island growth or layer-by-layer plus island growth is determined by both the surface energetics and kinetics. While the surface energetics predict the thermodynamic-equilibrium limit of the growth mode, the film growth is by definition a nonequilibrium kinetic process. The actual growth mode is determined by the growth conditions, such as growth rate and temperature, and the material parameters of the film/substrate combination. In the following, various contributions which influence or even control the growth mode will be discussed.

2.2 Thermodynamic Criterion for Growth Modes

Bauer has developed a thermodynamic criterion (Fig. 2.2) for the growth mode which neglects the strain energy of the film [6]. The criterion states that under equilibrium conditions, the growth mode is determined by the following energy difference:

$$\triangle\sigma = \sigma_f + \sigma_i - \sigma_s \ , \tag{2.1}$$

where σ_s is the surface free energy of the substrate σ_i the free energy of the interface, and σ_f the surface free energy of the film.

If $\triangle\sigma \leq 0$, the adatoms are more strongly bound to the substrate than to each other and the film tends to extend on (wet) the substrate to minimize the total energy. Under this condition, complete wetting of the substrate is favorable, and Frank–van der Merwe (layer-by-layer) growth should be

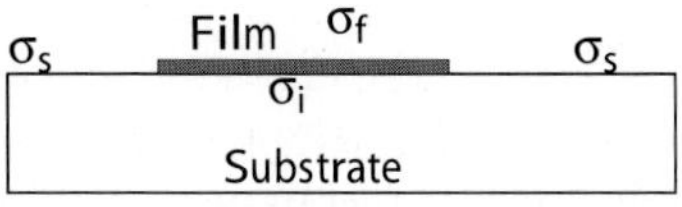

Fig. 2.2. Bauer's thermodynamic criterion for the growth mode: $\sigma_f + \sigma_i - \sigma_s < 0$ favors layer-by-layer growth. Island growth is observed if the film does not wet the substrate, i.e., $\sigma_f + \sigma_i - \sigma_s > 0$

observed. The inequality has the opposite sign when the atoms (or molecules) are more strongly bound to each other than to the substrate. In this case, one usually obtains Volmer–Weber growth (island growth), i.e., no wetting of the substrate. The film energy may have a contribution such as the strain energy, which increases linearly with increasing film thickness. After this energy is added to $\triangle\sigma$, it is possible for $\triangle\sigma$ to be smaller than zero until a certain thickness is reached and then larger than zero above this coverage. In this case, Stranski–Krastanov growth generally occurs.

2.2.1 Surface Free Energy and Interfacial Free Energy

In order to apply Bauer's criterion to any film–substrate pair, one needs to know the surface free energy and the interfacial energy of the materials considered. The surface free energy can be defined by the work W which is required to break a crystal so that these surfaces are left exposed. Let S be the area of the fracture section; then the surface free energy for a unit area, or the *surface tension*, is $\sigma = W/(2S)$. On the other hand, the interfacial free energy is defined by Dupre's relation [7]

$$\sigma_i = \sigma_s + \sigma_f - \beta \,, \tag{2.2}$$

where β is the adhesion energy, which is equal to the work required to separate the film from the substrate along the interface plane. Experimentally, it is rather difficult to determine surface free energies and interfacial energies precisely. Surface free energies of solid metals have been estimated from surface tension measurements on liquids [8] and directly measured by quantitative cleavage experiments [9], but the errors were reported to reach as much as 30% [10]. Therefore theoretical models have been proposed for calculating the surface free energies and interfacial energies from physical quantities that can be easily measured, such as the heat of sublimation [11, 12] and the internal free enthalpy of atomization [13]. An empirical relation [14] between the surface free energy per atom σ, the lattice constant a and the heat of vaporization $\triangle H$ (also expressed as an energy per atom) is frequently used to roughly estimate the surface free energy:

$$\sigma a/\triangle H \sim 0.3 - 0.4 \,. \tag{2.3}$$

Recently, a number of endeavors have been undertaken to develop reliable theoretical models to calculate the surface free energy. These attempts include the jellium model pioneered by Lang and Kohn [15, 16] and improved by Perdew and coworkers [17, 18, 19], the embedded-atom method developed by Daw and Baskes [20] and modified by Baskes and coworkers [21, 22, 23], and ab initio calculations using various techniques [25, 26, 27, 28]. However, the calculated values of surface free energies differ considerably from method to method [29] and even from group to group [30]. Data on surface free energies, both from experimental measurements and from theoretical calculations, are

Table 2.1. Surface energies σ (Jm^{-2}) for magnetic and nonmagnetic materials

Magnetic metal		Cr	Mn	Fe	Co	Ni	Gd
		2.3[1]	1.6[a]	2.48[a]	2.55[a]	2.45[a]	
		2.1[2]	1.4[b]	2.9[b]	2.7[b]	2.5[b]	0.9[b]
	bcc(110)	3.63[3]		3.09[c]			
		2.2[4]		2.4[d]			
	bcc(100)	2.3[d]		2.5[d]			
	bcc(111)	2.5[d]		2.7[d]			
	fcc(111)	3.09[c]	3.24[c]	3.28[c]	3.23[c]	2.63[c]	
	hcp(001)				3.18[c]		

Transition metal		Ti	V	Nb	Mo	Ru	Rh	Pd	Ta	W	Pt
		2.1[a]	2.55[a]	2.7[a]	3.0[a]	3.05[a]	2.7[a]	2.05[a]	2.8[a]	3.0[a]	
		2.6[b]	2.9[b]	3.0[b]	2.9[b]	3.4[b]	2.8[b]	2.0[b]	3.0[b]	3.5[b]	2.7[b]
	bcc(110)		2.02[c]	1.64[c]	3.18[c]						
			2.64[d]	2.49[d]	2.89[d]				2.78[d]	3.43[d]	
	bcc(100)		2.78[d]	2.72[d]	3.13[d]				3.04[d]	3.90[d]	
	bcc(111)		2.93[d]	2.92[d]	3.37[d]				3.25[d]	4.34[d]	
	fcc(111)	1.56[c]	2.55[c]	2.06[c]	2.5[c]	2.9[c]	2.78[c]	1.88[c]		2.20[5]	
	fcc(100)						2.9[c]	1.9[c]			
	hcp(001)	1.95[c]				3.32[c]					

Noble metal		Cu	Ag	Au
		1.83[a]	1.25[a]	
		1.9[b]	1.3[b]	1.6[b]
	fcc(111)	1.96[c]	1.12[c]	1.25[e]
		1.58[6]	1.21[f]	1.04[f]
	fcc(100)	2.09[c]	1.20[c]	
		1.71[f]	1.21[f]	1.33[f]
	fcc(110)	2.31[c]	1.29[c]	
		1.85[f]	1.26[f]	1.38[f]

[a] Derived from the surface tension of liquid metals. See [24].
[b] Calculated from the internal free enthalpies of atomization. See [13].
[c] Ab initio calculation results. See [25] and compare [26] for Fe, Co and Ni.
[d] Calculated by second-nearest-neighbor modified embedded-atom method. See [23].
[e] Calculation results from [27].
[f] Ab initio values from [28] and [33].

listed in Table 2.1 for the magnetic and nonmagnetic metals which are commonly involved in the preparation of ultrathin magnetic films. Magnetic materials always exhibit a relatively high surface energy, owing to their partially filled d shell, while noble-metal substrates have smaller surface free energies and insulating substrates have even smaller ones. The surface free energy depends not only on the material considered but also on the orientation of the surface. As a general rule, the surface free energy decreases with increasing atomic-layer spacing, i.e., increasing number of nearest neighbors of the surface atoms. For example, bcc crystals show the following sequence in the layer spacing: $d_{(110)} = (\sqrt{2}/2)a > d_{(100)} = (1/2)a > d_{(111)} = (\sqrt{3}/6)a$. Correspondingly, the surface free energy obeys the order $\sigma_{(110)} < \sigma_{(100)} < \sigma_{(111)}$. For fcc crystals, the spacing sequence is $d_{(111)} = (\sqrt{3}/3)a > d_{(100)} = (1/2)a > d_{(110)} = (\sqrt{2}/4)$, resulting in a different order in surface free energy, which is $\sigma_{(111)} < \sigma_{(100)} < \sigma_{(110)}$. This order of surface free energies is clearly seen in Table 2.1. The relationship between the surface free energy and the layer spacing explains why a crystal is easy to cleave parallel to the most densely packed crystal face, where the layer spacing is larger, since in this case the fewest bonds need to be broken.

If we conclude that the precise determination of the surface free energy is still a challenge, the situation for interface free energies is even worse. Interfaces, in essence, present the next step in complexity in comparison with the surface problem. The possible occurrence of interdiffusion and epitaxy-induced strain add further intricacy to the problem. At present, ab initio calculations of interface free energies have been performed only in a very limited number of cases, such as a metal–ceramic interface [31] and a semiconductor interface [32].

As we can see in Table 2.1, the experimental data on the surface free energy, either derived from the surface tension of liquid metals or calculated from the heat of vaporization, do not distinguish between different surface orientations, since the experimentally measured quantities from which the surface free energy is derived are not face-specific, at least in the way they are determined [34].

On the other hand, to test the validity and the precision of theoretical calculations, face-specific experimental data are required. Thanks to the advancements in microscopic techniques that enable one to monitor the shape of small crystals [35], the face dependence of surface free energy can be reliably determined by experiments with the help of Wulff's law. As was shown earlier by Wulff [36], the *equilibrium* shape of crystals reflects the orientational dependence of the surface free energy. More precisely, Wulff showed that at $T = 0$ K, the equilibrium shape of a crystal is a polyhedron with several, possibly different faces (see Fig. 2.3a). These faces have corresponding specific surface free energies σ_i and a minimal distance from the center of the crystal h_i such that

$$\frac{\sigma_i}{h_i} = A \,, \tag{2.4}$$

where A is a constant that is proportional to the chemical potential of the atoms in the equilibrium shape. If one face (k) of the crystal makes contact with the substrate B, then for this face (2.4) is replaced by

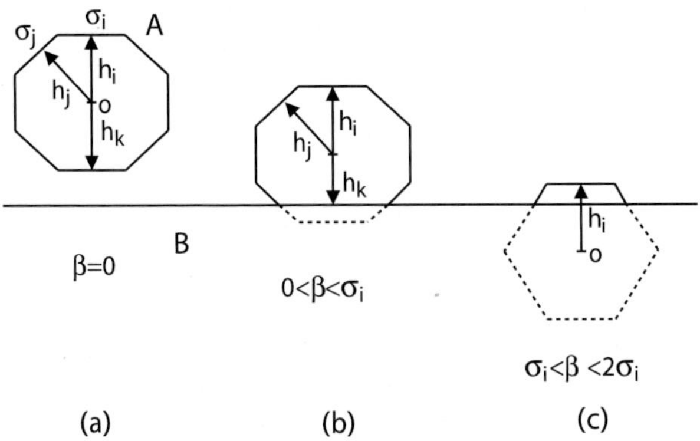

Fig. 2.3. Equilibrium shape of a crystal (A) on a substrate (B): dependence of the equilibrium form upon the specific adhesion energy β for three different values of β. h_i denotes the distance from the center of the crystal to a face which has a specific surface energy σ_i.(a) $\beta = 0$ corresponds to the case where the crystal and substrate are not in contact.(b) A finite value $(0 < \beta < \sigma_i)$ of the adhesion energy leads to $h_k < h_i$, i.e., a polyhedron truncated by the substrate. (c) When $\sigma_i < \beta < 2\sigma_i$, one has $h_i < 0$, signifying that the polyhedron becomes flat

$$\frac{\sigma_k - \beta_k}{h_k} = A \,, \tag{2.5}$$

where β_k is the specific free energy of adhesion. β describes the work per unit area necessary to separate the substrate and the contact plane k. This generalization of Wulff's theorem was first derived by Kaischew [38]. Consequences are depicted for three different values of β in Fig. 2.3 [45]. The equilibrium shape of monatomic-height islands (two-dimensional islands) on a surface, which is a two-dimensional analogue of the equilibrium crystal shape, is determined by the orientation-dependent step free energy δ. The corresponding equation for a two-dimensional island is [5]

$$\frac{\delta_i}{l_i} = B \,, \tag{2.6}$$

where l_i is the minimal distance from the center of the island to step i and B is a constant. Like (2.4), (2.6) is only valid at $T = 0$. However any experimental observation of the shape of crystals and islands occurs at finite temperature. Furthermore, to obtain the equilibrium shape experimentally, the measurement temperature is usually set to be close to the melting temperature [37].

At finite temperature, the faceted equilibrium shape may remain for three-dimensional crystals if the surface free energy is highly anisotropic, but more often the crystal shows a rounded equilibrium shape. As for two-dimensional islands, facets cannot be formed independently of the anisotropy of the step free energy. The absence of facets results from the fact that a step always has kinks at any finite temperature [40]. If a rounded equilibrium shape is observed experimentally, a plot of the relative surface free energy or of the step free energies can be obtained using an inverse Wulff plot [41]. Figure 2.4 shows how to construct an inverse Wulff plot for the three-dimensional case. The method of construction is similar for two-dimensional islands, where the tangential line to the equilibrium island contour is used in place of the tangential plane in Fig. 2.4.

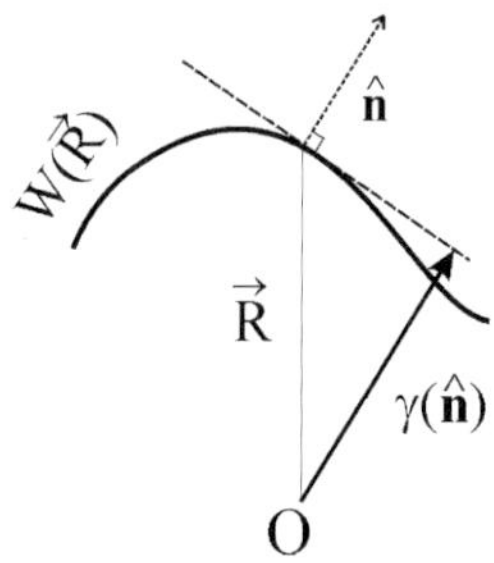

Fig. 2.4. Geometrical representation of the construction of the inverse Wulff plot : at each point $R(x_1, x_2, x_3)$ on the equilibrium crystal shape, denoted by $W(x_1, x_2, x_3) = 0$, a plane is drawn tangentially to the surface. The distance of the plane from the origin O is proportional to $\sigma(\boldsymbol{n})$, where $\boldsymbol{n}$ is the unit normal vector to the plane [39]

The inverse Wulff plot only allows one to derive the relative surface (or step) free energies from the equilibrium crystal shape, and not the absolute energy values. The precise experimental measurement of the absolute surface free energy and step free energy has been a challenge until recently. Several new methods have been proposed for this purpose, based on the equilibrium shapes of crystals and islands. These methods include measuring the temperature-dependent equilibrium island shapes [42] or the shape fluctuations of islands at a given temperature [43] to determine the absolute step free energies, and measuring the temperature dependence of the equilibrium size of a facet with a low index to determine its surface free energy [44]. In all these new methods, a common strategy is adopted in which k_BT is exploited as the energy scale for absolute energy measurements.

2.2.2 Strain Energy and Stranski–Krastanov Growth

In the above discussion of the thermodynamic criterion for the growth modes, we have ignored the effect of the strain energy upon the growth mode. Now we return to this point. In heteroepitaxy, the substrate and the deposited film generally have different equilibrium lattice constants. We say that there is a misfit between the film and the substrate. The misfit is defined as the relative difference between the equilibrium lattice constants of the substrate (a_s) and film (a_f), i.e.,

$$f = \frac{a_f - a_s}{a_f} \ . \tag{2.7}$$

The basic features of the role which a misfit plays can be described as follows. For a film–substrate system with $\triangle\sigma < 0$ (see (2.1)), the film grows in the layer-by-layer mode if there is no misfit. When the effect of misfit is added, the film is initially forced to be in register with the substrate lattice, resulting in a strain equal to the misfit f. With increasing film thickness, dislocation-free islands will form, because the strain energy can be reduced by the nonuniform strain field induced by an island array, compared with a uniformly stressed flat film [46]. For this reason, the growth mode changes to Stranski–Krastanov growth. If the island volume exceeds a certain critical size, the formation of a mismatch-strain-relieving edge dislocation is favorable. In fact, the detailed misfit effect is much more complicated for at least two reasons. First, the strain field and thus the total energy of the system are sensitive to the properties of the islands, such as island shape, island size and island density. Secondly even the surface free energy and the interface free energy are dependent on the film thickness, particularly for ultrathin films.

Grabow and Gilmer [49] have investigated the influence of strain upon film growth using molecular-dynamics simulations for a particular film–substrate system, where all particles in the film and the substrate interact with pair potentials such as the Lennard–Jones potential for molecular crystals and the Stillinger–Weber potential for silicon. Their results are shown in Fig. 2.5. The relative adlayer interaction strength W is defined as

$$W = \epsilon_{fs}/\epsilon_{ff} \ , \tag{2.8}$$

where ϵ_{fs} and ϵ_{ff} are the well depths of the potential between a film particle and a substrate particle and between two film particles, respectively. When

Table 2.2. Lattice parameters in Å of magnetic transition metals and of the substrates commonly used for their epitaxial growth [47, 48]. The number in parentheses denotes the nearest-neighbor spacing

	Cr (bcc)	Fe (bcc)	Co (bcc)	W (bcc)	Mo (bcc)	
$\sqrt{2}a$ (a)	4.080 (2.885)	4.053 (2.866)	3.99 (2.82)	3.165 (2.238)	3.15 (2.227)	
	Al (fcc)	Ag (fcc)	Au (fcc)	Pt (fcc)	Pd (fcc)	Rh (fcc)
a	4.041	4.086	4.078	3.924	3.890	3.804
	Fe (fcc)	Co (fcc)	Ni (fcc)	Cu (fcc)	Diamond	Cu_3Au
a	3.59	3.55	3.52	3.61	3.57	3.745
	Co (hcp)		Ru (hcp)		Re (hcp)	
a	2.507		2.698		2.755	

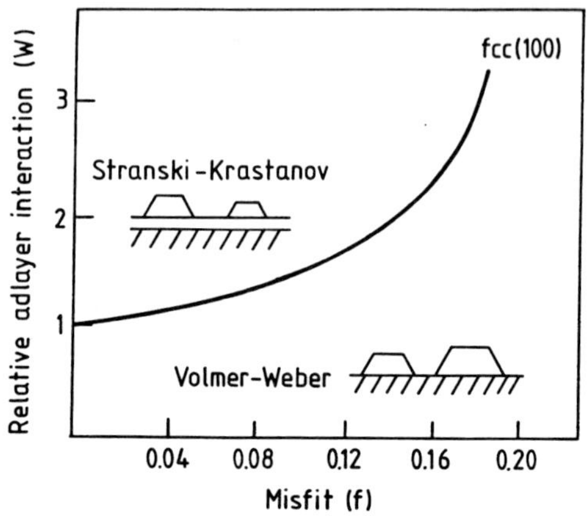

Fig. 2.5. Influence of misfit on the growth mode. The dependence of the equilibrium growth mode upon the misfit (f) and the relative adlayer interaction strength (W) is shown for epitaxial growth on an fcc (100) surface. Stranski–Krastanov growth can only be achieved for $W > 1$ (after [49])

W is smaller than one, Volmer–Weber growth is always observed. Only for a stronger film-substrate interaction, i.e., $W > 1$, can Stranski–Krastanov growth be observed. It should be noted that this transition shifts to higher values of W with increasing misfit.

Recently, Daruka and Barabási have investigated the equilibrium properties of strained wetting heteroepitaxial systems, incorporating the formation and growth of a wetting film, dislocation-free island formation, and ripening [50]. They determined a phase diagram for the growth mode (Fig. 2.6) as a function of coverage (Θ) and misfit (f) by minimizing the free energy of the system. In that work, linear elastic theory was used to calculate the nonuniform strain field and the strain energy [51]. The properties of the growth modes (phases) present in the coverage(Θ)–misfit (f) phase diagram are described below. The phase boundaries are determined by the material parameters for the specific system. In a subsequent work [52], Daruka and Barabási pointed out further that, for some specific systems, the phase boundaries may be located in different places and some phases may not appear in the phase diagram, but all growth modes can generally be described by four types of phase diagrams, depending on the film surface energy. The phase diagram shown in Fig. 2.6 is one of the four types of diagrams and it is the only case including all seven phases. In this diagram, seven phases are separated by four curved phase boundary lines ($\Theta_{c1}(f)$, $\Theta_{c2}(f)$, $\Theta_{c3}(f)$ and $\Theta_{c4}(f)$) and three vertical phase boundary lines ($f = f_1$, $f = f_2$ and $f = f_3$) (see the caption of Fig. 2.6):

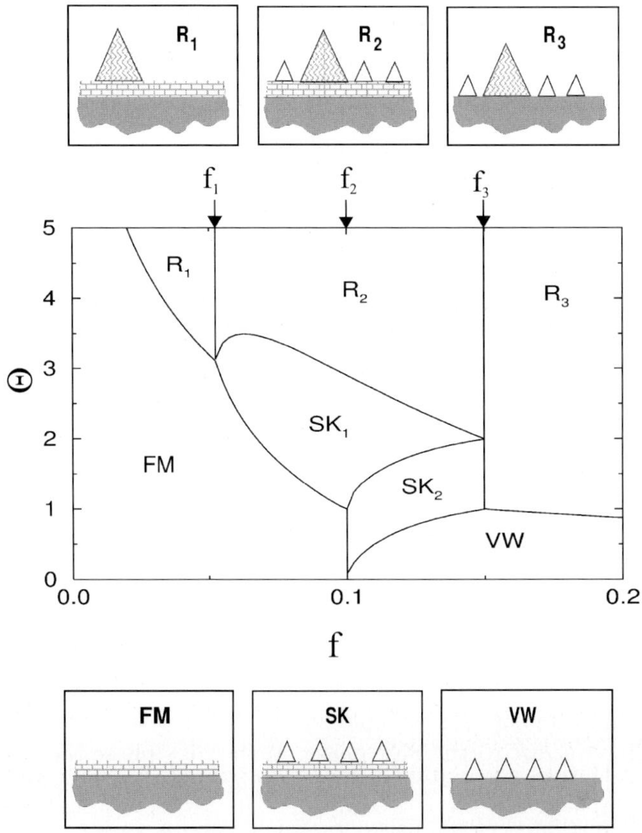

Fig. 2.6. The equilibrium phase diagram as a function of coverage Θ and misfit f. The small panels at the *top* and *bottom* illustrate the morphology of the surface for six different growth modes. The *small empty islands* indicate the presence of stable islands, while the *large shaded islands* refer to ripened islands. The phases are separated by the following phase boundary lines: $\Theta_{c1}(f)$, FM–R_1 and FM–SK_1; $\Theta_{c2}(f)$, SK_1–R_2; Θ_{c3}, SK_2–SK_1; Θ_{c4}, VW–SK_2 and VW–R_3. See text for details. From [50]

1. *FM phase.* If $\Theta < \Theta_{c1}(f)$ and $f < f_2$, the deposited material contributes to the pseudomorphic growth of the wetting film and islands are absent, reminiscent of the Frank–van der Merwe (FM) growth mode.
2. *R_1 phase.* When $\Theta > \Theta_{c1}(f)$ and $f < f_1$, the film first grows in the Frank–van der Merwe growth mode until $\Theta = \Theta_{c1}(f)$. Then the excess material deposited contributes to the formation of ripened islands. Since the free energy decreases monotonically with increasing island size, there is a tendency to accumulate the excess material $(\Theta - \Theta_{c1}(f))$ in islands as large as possible. These ripened islands, being infinitely large, have zero density.

3. *SK_1 phase.* Above $\Theta_{c1}(f)$, for $f_1 < f < f_2$, the deposited material (Θ) is distributed between a wetting layer and finite islands, similarly to the Stranski–Krastanov (SK) growth mode. At $\Theta_{c1}(f)$, the equilibrium island size jumps from zero (in the FM phase) to some finite value, while the island density is zero at $\Theta_{c1}(f)$ and increases with increasing Θ. Within the SK_1 phase, both the island size and the island density are continuous functions of Θ and f. The wetting layer continues to grow sublinearly with increasing Θ.
4. *R_2 phase.* In this phase, the deposited material is distributed between a wetting film, finite islands and ripened islands. The finite islands, which have formed in the SK_1 phase, are preserved and are stable with respect to ripening. Thus finite and ripened islands coexist.
5. *VW phase.* For large misfit ($f > f_2$) and small coverages ($\Theta < \Theta_{c4}(f)$), all the deposited material is accumulated in finite islands. Because of the large misfit, the wetting film is absent in this phase and the islands form directly on the substrate, similarly to the Volmer–Weber (VW) growth mode.
6. *SK_2 phase.* For $f_2 < f < f_3$, when we increase Θ above $\Theta_{c4}(f)$ we reach the SK_2 phase. The behavior of the system is different from the SK_1 growth mode: at the Θ_{c4} boundary, we have islands formed in the VW mode. In the SK_2 phase, the island density and the island size remain unchanged, and a wetting film starts forming. This process continues until a full monolayer is completed, at which point we enter the SK_1 phase. In contrast to the SK_1 phase, in the SK_2 phase the formation of further islands is suppressed until a wetting layer one monolayer thick is completed.
7. *R_3 phase.* In this phase, the formation of stable islands is suppressed, and all the material deposited after Θ_{c3} contributes only to the ripened islands, coexisting with the stable islands formed in the VW growth mode. However, in contrast to R_2, in R_3 a wetting film is absent.

2.3 Kinetic Aspects and Microscopic Models of Growth

So far, we have discussed the energetics of film growth, i.e., the equilibrium properties of the growing film. However, growth is by definition a non-equilibrium phenomenon. In most practical cases, growth inevitably occurs very far from equilibrium. In this case the growth is strongly influenced by kinetic processes. The final macroscopic state of the film is not necessarily the most stable but is the kinetically most favorable. Therefore a comprehensive understanding of film growth should cover both the thermodynamic principles and the atomistic processes. In this section, the kinetic aspects and microscopic models of growth will be discussed. Experimental parameters such as substrate temperature and growth rate, which are known to have a pronounced influence on film growth and film quality, will enter our discussion.

2.3.1 Atomistic Processes in Film Growth

The atomistic processes in film growth are schematically illustrated in Fig. 2.7. The growth of a film always starts with the arrival of atoms from the vapor. These atoms (adatoms) keep on migrating on the substrate surface until one of the processes described below occurs. These processes include re-evaporation into the vacuum, capture by existing steps and clusters, and nucleation into clusters. For metal film growth on a metal substrate, re-evaporation is negligible at the relevant growth temperatures. Before we go on to discuss nucleation and capture, let us first take a look at the migration properties of adatoms on a substrate surface.

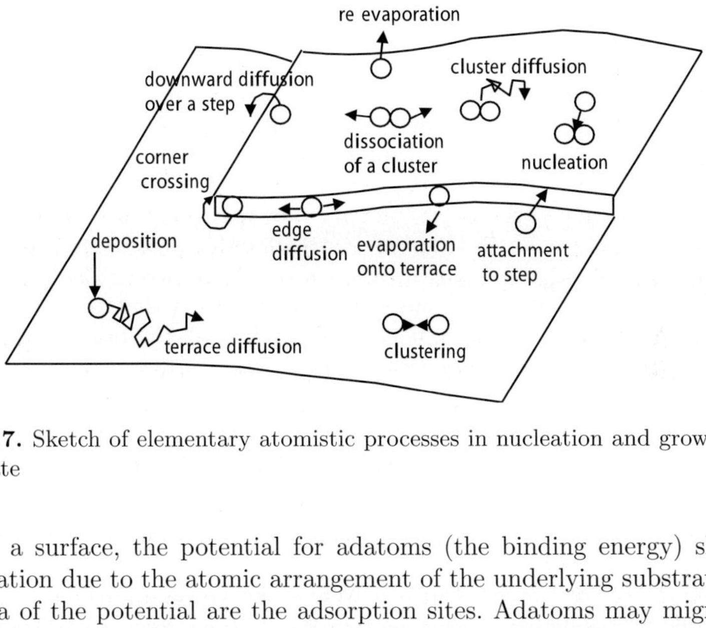

Fig. 2.7. Sketch of elementary atomistic processes in nucleation and growth on a substrate

On a surface, the potential for adatoms (the binding energy) shows a modulation due to the atomic arrangement of the underlying substrate. The minima of the potential are the adsorption sites. Adatoms may migrate on the surface by jumping from one adsorption site to a neighboring site. If the density of adatoms is low and there is no interaction between adatoms, the adatoms' migration can be characterized by an individual random walk. The statistical displacement $|\boldsymbol{\Lambda}|$ of an adatom after n random jumps is given by [53]

$$|\boldsymbol{\Lambda}|^2 = na^2 , \tag{2.9}$$

assuming that the adsorption sites form a two-dimensional square lattice with a lattice constant a. It follows that the statistical movement λ along a given direction, say the x direction, is given by

$$\lambda^2 = \frac{1}{4}|\Lambda^2| = \frac{1}{4}na^2 . \tag{2.10}$$

To describe the velocity of adatom migration, the diffusion coefficient D is defined by

$$\langle [r(t'+t) - r(t')]^2 \rangle = 2dDt , \tag{2.11}$$

where $r(t)$ is the position of an adatom as a function of time t, and d is the dimension of the diffusion path considered. Hence $d = 2$ for surface migration. If we compare (2.9) and (2.11), D should be proportional to the number of jumps performed per unit time. To move from one adsorption site to a neighboring one, the adatom must jump over an energy barrier E_d. According to the Gibbs–Boltzmann formula, the probability of an adatom in equilibrium overcoming this energy barrier is proportional to $\exp(-E_d/k_BT)$. Usually the frequency of successful jumps is expressed as

$$f = \nu_0 \exp(-E_d/k_BT) . \tag{2.12}$$

The prefactor ν_0 is often assumed to be of the order of a typical vibrational frequency of the atom, say $10^{13}\,\mathrm{s}^{-1}$, and depends only weakly on temperature. Since a theoretical determination of ν_0 is still a challenge, it is usually taken as an experimental fit parameter. If we apply (2.9), (2.11) and (2.12), the diffusion coefficient on a square surface lattice can be expressed as

$$D = \frac{a^2}{4}\nu_0 \exp\left(\frac{-E_d}{k_BT}\right) . \tag{2.13}$$

The diffusion barrier E_d at the surface is much smaller than in the bulk (see Fig. 2.23). Bulk diffusion takes places mainly through the motion of defects, especially vacancies and interstitial atoms. From (2.13), one can see that the migration properties of adatoms depend strongly on temperature. The diffusion coefficient defined above is called the tracer diffusion coefficient, to distinguish it from the Fick diffusion coefficient D_f. The latter is defined by Fick's law,

$$\frac{\mathrm{d}\rho(r,t)}{\mathrm{d}t} = D_f \nabla^2 \rho(r,t) , \tag{2.14}$$

where $\rho(r,t)$ is the particle density. While the tracer diffusion coefficient reflects purely the rapidness of the migration, Fick's diffusion coefficient also takes the mutual influence of migrating particles into account. A discussion of these two different diffusion coefficients can be found in [54].

The random-walk motion of an adatom will be disrupted when the adatom encounters a step along the diffusion path. When an adatom migrating over a terrace strikes an ascending step, it will be incorporated into it. It is even found that an adatom on the lower terrace is subject to a short-range attractive interaction when it approaches an ascending step. This adatom–step attraction has been observed by field ion microscopy on the Ir(111) surface [55], and is confirmed by the calculated smaller diffusion barriers near a step [56].

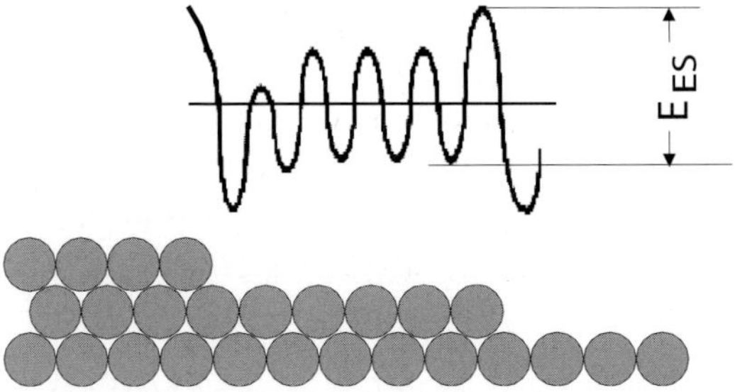

Fig. 2.8. Side view of a stepped fcc (100) surface and the corresponding schematic potential, showing a decreased barrier near an ascending step due to the step–adatom attraction and the Ehrlich–Schwoebel barrier at the descending step

In contrast, an adatom approaching a descending step has to overcome an additional potential barrier to roll over the edge before it can attach itself to the step, since the adatom will pass through a position in which it possesses fewer nearest neighbors (Fig. 2.8). The effects of this potential barrier on diffusion and crystal growth were first studied by Ehrlich and Hudda using field ion microscopy [57] and studied theoretically by Schwoebel [58]. Therefore this barrier is now called the Ehrlich–Schwoebel barrier. The Ehrlich–Schwoebel barrier plays an important role in film growth. It prevents the adatoms from diffusing to the lower terrace. With a large Ehrlich–Schwoebel barrier, nucleation of new islands on top of existing islands may occur before coalescence of the islands of the previous layer, resulting in island growth even though the energetics favor a layer-by-layer growth mode. Instead of the rolling-over mechanism, an alternative mechanism has been found for interlayer diffusion, namely, the exchange process (see Fig. 2.9). In this process, an adatom approaching a descending step can descend the step by pushing a terrace atom at the step edge further onto the lower terrace and sinking into the initial position of the terrace atom. The kinks at step edges and corners are believed to be the preferred sites for the exchange process. For the Al(111) surface, exchange diffusion is favored over direct hopping. The activation energy for interlayer diffusion is found to be dependent on the type and roughness of the

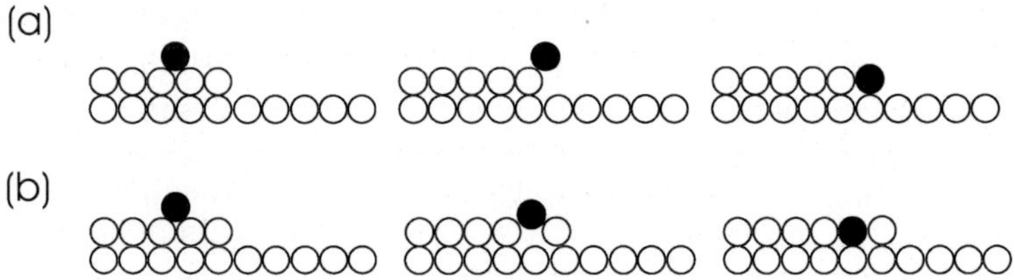

Fig. 2.9. Schematic side view of an adatom descending a step. (**a**) Adatom rolls over the step. (**b**) Adatom exchanges place with a step atom. From [61]

step [59, 60]. Thus the interlayer mass transport would be dependent on the island size, island shape and even island orientation if the exchange process were an efficient diffusion pathway.

Exchange processes have been observed at room temperature for Co on Pt(111) [61] using atomically resolved scanning tunneling microscopy. Exchange processes have also been proposed [62] to be responsible for the reentrant layer-by-layer growth at low temperatures observed in the Pt/Pt(111) system [63, 64], since at low temperatures the increased kink density of the smaller islands promotes interlayer diffusion owing to the smaller barrier. Recently, the Ehrlich–Schwoebel barrier has been found to be correlated with the occupation of surface states [65, 66]. By analyzing the ripening of multilayer islands on Cu(111), Giesen et al. have revealed that the Ehrlich–Schwoebel barrier is independent of the terrace width w until a critical value $w_c = 14 \pm 2\,\text{Å}$ is reached. When the terrace width is smaller than w_c, the Ehrlich–Schwoebel barrier vanishes abruptly [66]. The critical width w_c corresponds to the terrace width below which the surface state is pushed above the Fermi level owing to quantum confinement. The modification of the charge density in the surface state by quantum confinement leads to a reduction of the binding energy at steps and hence to the Ehrlich–Schwoebel barrier.

The migration of adatoms has for a long time been considered as the basic atomic process for mass transport. The migrating adatoms perform a random walk on a terrace. An adatom is reflected by a step or attached to it after overcoming the Ehrlich–Schwoebel barrier when it meets a descending step. When the adatom approaches an ascending step, it is incorporated into the step edge. On a simple metal surface, there is no obvious additional barrier to this incorporation. That is to say, the barrier for diffusion on the terrace should be the same as or larger than the barrier associated with this incorporation. However, the adatoms are not always the diffusing species. It has recently been found that the diffusion of Mn atoms incorporated into the Cu(100) surface is mediated by vacancies [67]. Furthermore, surface self-diffusion on Cu(100) has been shown to be mediated by single-atom vacancies on the terrace [68, 69]. In other words, vacancies are the prevailing mass-transport-carrying species on Cu(100). This conclusion was based upon the experimentally observed constant decay rate of small islands and its independence of the environment [68, 69], and has been confirmed by theoretical studies [56, 70]. Both the diffusion energy and the formation energy are calculated to be smaller for a vacancy than for an adatom (see Fig. 2.10). The mechanism for creation of a vacancy at a step has been discussed by Ibach et al. [71]. These vacancies and the atomic exchange processes in the vicinity of kinks are also responsible for the formation of surface alloys [67, 72, 73, 74, 75, 76]. This issue will be discussed in detail in Chap. 7.

An adatom striking a step will be attached to the step. The attached adatom will remain mobile along the step edge until it finds a kink position to rest in. The migration of an attached adatom along the periphery of a terrace

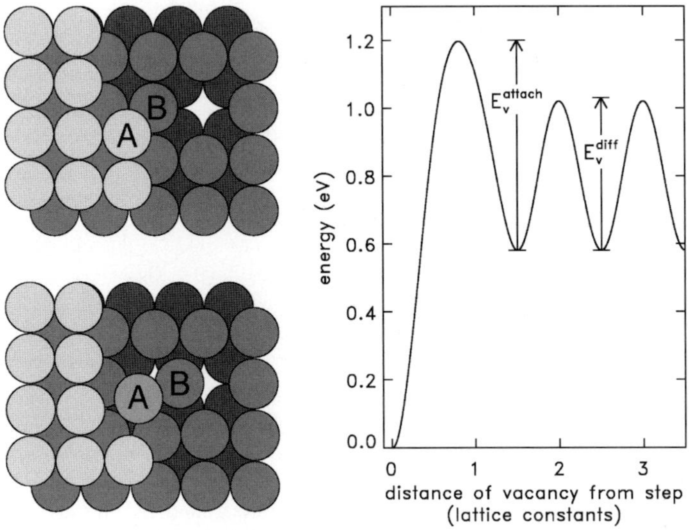

Fig. 2.10. Surface self-diffusion on Cu(100) mediated by single-atom vacancies rather than adatoms. The *left panel* shows a vacancy being incorporated into a step edge at a kink site: atom B moves into the vacancy and atom A drops down to replace B. The *right panel* shows the diffusion barriers experienced by a vacancy as it approaches a kink [68]

is called edge diffusion. The edge-diffusing adatoms still have a nonvanishing probability of reevaporating onto the terrace, but this probability is small for a metal/metal system. The energy barriers for an edge-diffusing atom to pass a corner (kinks and island corners) are usually larger than those on a smooth step edge. The edge diffusion and corner-crossing processes, which are generally dependent on the step direction, have a decisive effect on the evolution of the island shape [77, 78].

Up to now, we have discussed the migration of adatoms on a terrace and their motion across a step edge. On the other hand, the migrating adatoms may meet and form a cluster. These clusters may disappear again by decaying into single adatoms or, alternatively, develop into a stable nucleus. The stable nuclei will grow further by capturing single adatoms. Thus the atomic processes involved in the initial stage of film growth are quite complicated. The kinetic rate equations allow a quantitative description of the relevant processes. This is the subject that we shall discuss in the next subsection.

2.3.2 Kinetic Rate Equations for Nucleation and Growth

The atoms deposited from the vapor will diffuse on the substrate surface. The diffusion process will be terminated when the adatoms are captured by an existing step or by clusters, as we have discussed in the previous subsection.

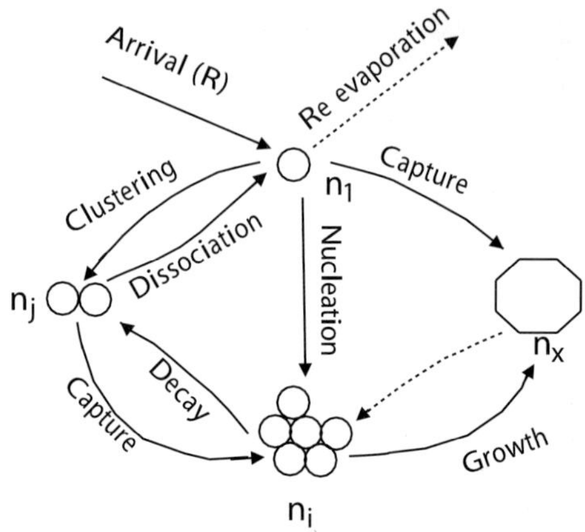

Fig. 2.11. Competition between nucleation and growth. The various processes that govern nucleation and growth are depicted. Unlikely processes are denoted by *dashed lines*. The single-adatom density n_1 is determined by the arrival rate R, and the characteristic times for nucleation (τ_n) and diffusion capture (τ_c) by stable clusters n_x. These stable clusters are formed from the addition of one more adatom to a critical cluster n_i. The capture of single adatoms by stable clusters leads to their growth. The subcritical clusters n_j may decay via dissociation or develop into stable nuclei by further capture of adatoms. After [79]

Alternatively, as the adatoms diffuse over the surface, they may encounter other adatoms, with which they can combine to form clusters. These clusters can also decompose, and by this process release adatoms. It is only above a critical cluster size (n_i) that growth becomes much more important than decay. Stable clusters grow by further capture of adatoms. The relationship between the atomistic processes included in nucleation and growth is illustrated in Fig. 2.11.

Let us consider the situation after atoms have been evaporated onto a step-free substrate at a constant rate R. Initially, the evaporation leads to a steady increase in the adatom density n_1. After some time, stable clusters are formed and their density (n_x) increases with time. Since the stable clusters act as adatom sinks, there exists a time τ_c, the capture time, above which the adatom density starts to decrease. τ_c is inversely proportional to the adatom diffusion coefficient D and the density of stable clusters n_x [80]. The total number of stable clusters (n_x) always increases up to the coverage at which cluster coalescence becomes important. Therefore one has time-dependent densities of adatoms and stable clusters, as shown in Fig. 2.12. Nucleation and growth are two competing processes which reduce the number of adatoms. Usually, nucleation dominates in the early stages and growth in the later stages of film formation.

To obtain a quantitative description of nucleation and growth, a set of kinetic equations incorporating the main atomistic processes has been devel-

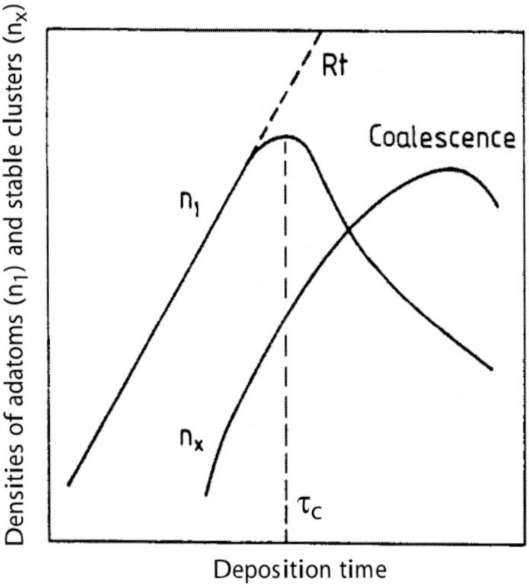

Fig. 2.12. Evolution of the density of single adatoms and stable clusters. The density of adatoms n_1 increases linearly with time until the capture time τ_c is reached. Then a pronounced increase in the density of stable clusters n_x is encountered. The number of stable clusters decreases only after longer deposition times, when coalescence becomes important. After [80]

oped by Zinsmeister [81], Logan [82], and Frankl and Venables [83]. These equations are called the kinetic rate equations or simply rate equations. Here we shall discuss the rate equations following the approach given by Venables et al. [80].

Assuming that only single atoms are mobile on the surface, the rate equations have the following general form:

$$\mathrm{d}n_1/\mathrm{d}t = R - n_1/\tau_a - 2U_1 - \sum_{j=2}^{\infty} U_j \ , \tag{2.15}$$

$$\mathrm{d}n_j/\mathrm{d}t = U_{j-1} - U_j \quad (j \geq 2) \ , \tag{2.16}$$

where n_1 and n_j are the surface densities (per unit area) of adatoms and of clusters consisting of j atoms, respectively, and U_j describes the net rate of capture of single atoms by the j-atom clusters. τ_a is the lifetime of an adatom, i.e., the time between its arrival and reevaporation. τ_a is related to the adatom's adsorption energy E_a and the substrate temperature by

$$\tau_a = \frac{1}{\nu_0} \exp\left(\frac{E_a}{k_B T}\right) \tag{2.17}$$

with ν_0 being a constant. The physical meaning of (2.15) and (2.16) is straightforward. The terms on the right side of (2.15) describe the processes of adsorption from the vapor at a deposition rate R, reevaporation, formation of two-atom clusters, and capture by other clusters, respectively. Accepting that a local equilibrium exists between the subcritical clusters leads to a simplification of the rate equations. In thermodynamic equilibrium, the subcritical clusters have a steady distribution, i.e.

$$\mathrm{d}n_j/\mathrm{d}t = 0 \quad (1 \le j \le i)\,. \tag{2.18}$$

The densities of the subcritical clusters have been derived from statistical principles and are given by the Walton relation [84]

$$(n_j/N_0) = (n_1/N_0)^j \sum_m c_j(m) \exp[E_j(m)/k_B T]\,, \tag{2.19}$$

where n_j is the density of j-atom clusters ($j < i$) and N_0 is the density of adsorption sites. $E_j(m)$ denotes the binding energy of a j-atom cluster in the configuration m and the coefficients $c_j(m)$ are statistical weights. In the case where the subcritical cluster is in local equilibrium, by summing all stable clusters via

$$n_x = \sum_{j=i+1}^{\infty} n_j\,, \tag{2.20}$$

the rate equations can be simplified to

$$\mathrm{d}n_1/\mathrm{d}t = R - n_1/\tau_a - \mathrm{d}(n_x w_x)/\mathrm{d}t\,, \tag{2.21}$$

$$\mathrm{d}n_j/\mathrm{d}t = 0 \quad (2 \le j \le i)\,, \tag{2.22}$$

$$\mathrm{d}n_x/\mathrm{d}t = U_i - U_c\,, \tag{2.23}$$

$$\mathrm{d}(n_x w_x)/\mathrm{d}t = (i+1)U_i + \sigma_x D n_1 n_x + RZ\,. \tag{2.24}$$

In (2.21), the last term represents the loss of adatoms to n_x stable clusters with an average of w atoms per cluster. The last term in (2.23) describes attempts to deal with coalescence. If stable clusters impinge on each other by growth with a rate U_c, then the number of stable clusters will be reduced. The first term in (2.24) is the nucleation rate, which can be expressed as

$$U_i = \sigma_i D n_1 n_i\,, \tag{2.25}$$

with D being the single-atom surface diffusion coefficient and σ_i the capture number of the critical clusters. The other two terms on the right side of (2.24) represent, respectively, the contributions from capture by surface diffusion and from direct impingement on growing clusters, which cover a fraction Z of the surface.

In general, the rate equations enable a numerical solution for the stable-cluster density n_x, once the associated parameters such as R, D, σ_i and σ_x are known. However, useful analytical formulae can also be deduced from an approximate treatment of the rate equations under specific conditions. For a metal/metal system, if the temperature is not too high, reevaporation does not take place. The single-atom concentration n_1 is hence limited only by the growth of clusters. If it is further assumed that cluster coalescence does not occur, and that the dominant term in (2.24) is the middle term, i.e., capture by the stable cluster, then the steady-state condition $dn_1/dt = 0$ gives

$$n_1 = \frac{R}{\sigma_x D n_x} \,. \tag{2.26}$$

Using (2.23), (2.25) and (2.26), as well as the Walton relation (2.19), one can obtain the following relationship:

$$n_x^{i+1}\,\mathrm{d}n_x = \frac{1}{N_0^{i-1}} \frac{\sigma_i}{\sigma_x^{i+1}} (\frac{R}{D})^i c_i \exp(E_i/k_B T) R\,\mathrm{d}t. \tag{2.27}$$

Since $\int R\,dt = \Theta$, integrating (2.27) gives

$$n_x \propto \frac{\sigma_i}{\sigma_x^{i+1}} \Theta^{1/(i+2)} \left(\frac{R}{D}\right)^{i/(i+2)} \exp\left\{E_i/[(i+2)k_B T]\right\} \,, \tag{2.28}$$

where Θ is the total coverage. Equation (2.28) contains several important results: *the island density depends on the two most important experimental parameters R and D via a scaling law $n_x \propto (R/D)^\chi$, and the scaling exponent is determined by the size of the critical cluster.* With (2.13), (2.28) can be rewritten as

$$n_x \propto \frac{\sigma_i}{\sigma_x^{i+1}} \Theta^{1/(i+2)} R^{i/(i+2)} \exp\left\{(E_i + iE_d)/[(i+2)k_B T]\right\} \,. \tag{2.29}$$

Equation (2.29) allows the determination of the energy term in the exponential function, i.e., $((E_i + iE_d)/(i+2))$, by experimentally measuring the density of stable clusters at a given coverage for different temperatures. With this knowledge, one can draw some conclusions about the magnitude of the activation energy for adatom diffusion and the binding energy of the critical nucleus. The situation is simplified when $i = 1$ and therefore $E_i = 0$. In this case the energy in the exponential term simply equals $E_d/3$.

Theory of nucleation and growth based on the rate equations are usually called atomistic theories, to distinguish them from classical theories, which are based on the continuum thermodynamic properties of clusters and on diffusion equations. Detailed discussions of classical theories of nucleation and growth can be found in the books by Lewis and Anderson [53] and Pimpinelli and Villain [54].

2.4 Experimental Techniques for Growth Mode Analysis

So far, we have discussed the basic aspects of the theory of film growth. Now we turn to the experimental aspects of epitaxial growth of ultrathin metal films. Let us start with the experimental techniques that are used to study the film growth and morphology. These methods include reflection high-energy electron diffraction (RHEED), low-energy electron diffraction (LEED), Auger electron spectroscopy (AES) and scanning tunneling microscopy (STM).

2.4.1 Reflection High-Energy Electron Diffraction (RHEED)

Reflection high-energy electron diffraction is one of the methods that have been most frequently employed to determine the growth mode experimentally. In a RHEED experiment, a monochromatic electron beam with a primary energy between 3 and 20 keV is reflected from a single-crystal surface at grazing incidence. The resulting diffraction pattern is detected on a fluorescent screen. Usually, for a given diffraction beam, conditions are chosen such that adjacent terraces interfere destructively. The advantage of RHEED is that it gives in-situ information on the film thickness and film morphology, owing to its excellent compatibility with molecular-beam epitaxy. The molecular (or atomic) beam is incident almost normal to the substrate, whereas the RHEED electrons hit the surface at grazing incidence so that the diffraction pattern can easily be observed during deposition.

Figure 2.13 shows the diffraction geometry of RHEED. Since the electron beam hits the sample surface at grazing incidence, the component of the incident wave vector in the direction of the surface normal is small. On the other hand, the cross sections for electron scattering off atoms are very large.

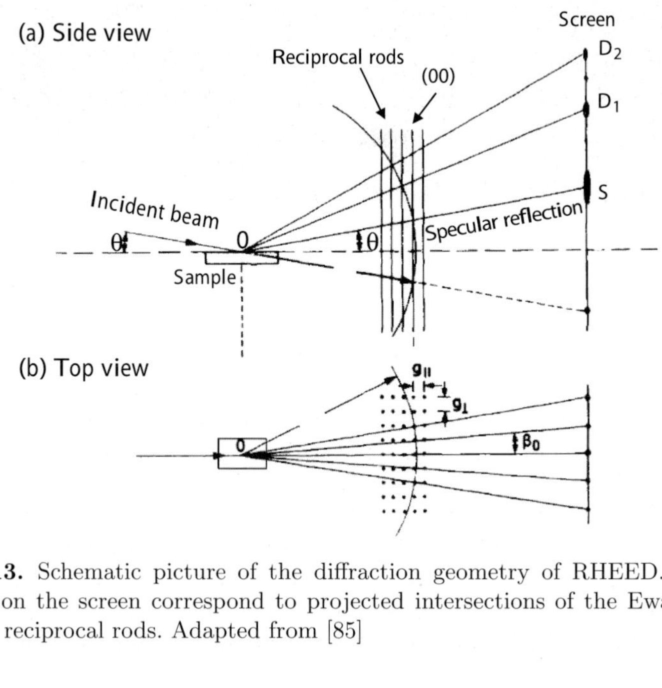

Fig. 2.13. Schematic picture of the diffraction geometry of RHEED. Intensity maxima on the screen correspond to projected intersections of the Ewald sphere with the reciprocal rods. Adapted from [85]

This means that the electron penetration depth in RHEED is very small, even though the attenuation length for electrons with energies between 3 keV and 10 keV lies between several atomic layers and several tens of atomic layers (see Fig. 2.21). Therefore RHEED is a surface-sensitive technique which detects only the surface atomic arrangement. In this case, the reciprocal lattice of the sample degenerates into an array of rods perpendicular to the sample surface (see Fig. 2.13). For a perfect surface, the reciprocal-lattice rods have vanishing size and are one-dimensional, so that the intersection of the reciprocal-lattice rods with the Ewald sphere leads to sharp diffraction spots. For an imperfect surface with ordered domains of diameter L, the reciprocal-lattice rods are no longer one-dimensional lines but form rods with a finite size inversely proportional to L. Since the electron energy in RHEED is high, as suggested by the name, the Ewald sphere is very large compared with the sample's reciprocal-lattice constant and cuts the reciprocal rods almost along their length. Therefore the RHEED diffraction pattern for an imperfect surface will consist of long streaks normal to the shadow edge of the sample. Figure 2.14 illustrates the typical streaky RHEED patterns of an ultrathin film of V grown on Fe(100). If there are small but high islands on the surface, the electron beam will go through them and will be diffracted by the three-dimensional atomic array. The RHEED pattern then shows a rectangular array of spots, instead of streaks or spots on semicircles.

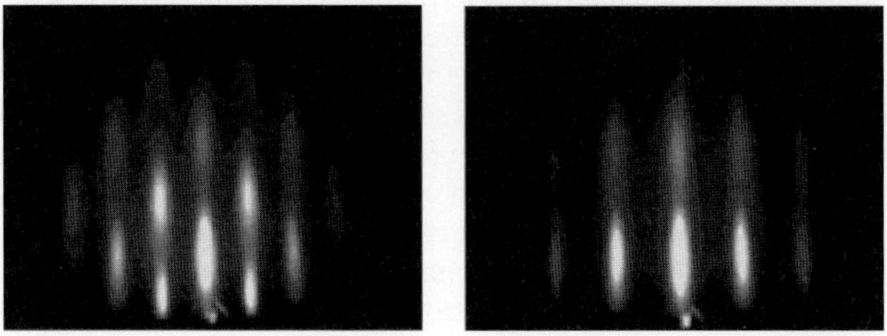

Fig. 2.14. RHEED patterns after 15 ML V has been grown on Fe(100) at 300 K. The pattern and the streak spacing vary with azimuthal angle. From [88]

The feature most commonly exploited for real-time monitoring is the temporal intensity of the diffracted beams, especially that of the specular beam. Layer-by-layer growth is characterized by regular intensity oscillations of the diffracted beams, where the oscillation period corresponds to the time necessary to complete a single monolayer [86, 87]. This is schematically depicted in Fig. 2.15, where the change in intensity of such a beam during film deposition is shown. The intensity of this beam decreases with increasing step density, leading to a minimum in intensity when the step density is largest. This minimum is often observed around a coverage of 0.5 ML. The intensity will recover to its initial value if the step density decreases to its original value

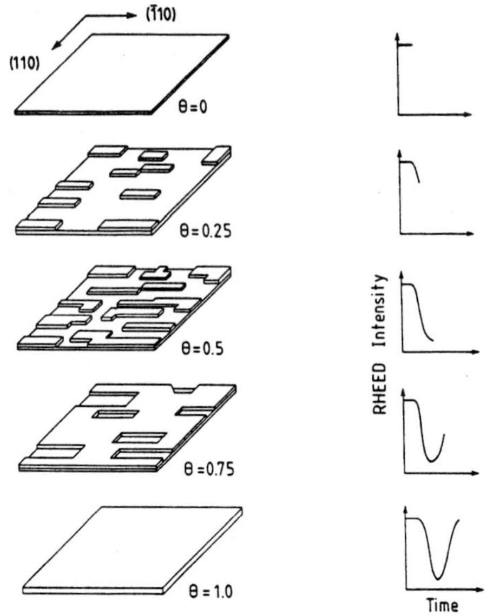

Fig. 2.15. Schematic pictures showing the relationship between the intensity oscillations of the RHEED beam and the film morphology for layer-by-layer growth

upon completion of a monolayer. Island growth necessarily leads to a continuous increase in step density and consequently a monotonic decrease in the intensity of the electron beam. When the Stranski–Kranstanov growth mode is adopted, the number of oscillations seen should correspond to the number of monolayers that grow in a layer-by-layer fashion. The subsequent growth of three-dimensional nuclei above the critical thickness manifests itself in a monotonic decrease in the intensity of the diffracted beam. Thus RHEED enables a precise characterization of the growth mode. In the case of Frank–van der Merwe and Stranski–Krastanov growth, it also allows a straightforward on-line determination of the film thickness simply by counting the number of oscillations.

Though it is generally accepted that the oscillating RHEED intensity is a manifestation of a layer-by-layer growth mode, the physical origin of the oscillations remains controversial [89, 90]. There have been two basic explanations. One is the two-level interference model [91], emphasizing interference effects in the electron beam reflected from different terraces. In this model, the changing exposure of different layers to the electron beam is responsible for the oscillations. However, this model fails to explain the fact that strong oscillations are still observed in experiments where the in-phase condition is satisfied and the interference of electrons reflected from different terraces is constructive [92]. The other explanation is based on the concept of step-edge diffuse scattering [86]. The diffuse scattering by steps and other defects decreases the intensity of the Bragg reflections, giving rise to the step density model. The step density varies with the stage of formation of the layer, resulting in an oscillation in the intensity of the RHEED beams, including the specular beam. The step density model is widely used to interpret RHEED data, though the diffuse-scattering mechanism has not been ruled out either. However, this model has recently been questioned by a number of authors [89, 90]. A better understanding of the origin of RHEED oscillations would be expected from full dynamical calculations of the diffracted intensity, but such calculations are still beyond the scope of present calculations. A systemic discussion about RHEED, including both experimental and theoretical aspects, can be found in Braun's monograph [93].

Two other diffraction techniques offer comparable potential for growth mode analysis. One is LEED, where a primary electron energy between 50 and 500 eV and a large angle of incidence between 45° and 90° are chosen (see Chap. 3). The second technique is thermal-atom scattering, notably using He atom beams [63]. However, all diffraction techniques provide information in reciprocal space only. To image the growth processes in real space, scanning tunneling microscopy is the most suitable tool.

2.4.2 Scanning Tunneling Microscopy (STM)

The invention of STM by Binnig and Rohrer [94] has provided another powerful technique that can be successfully applied to study thin film growth [95]. STM is used increasingly frequently to obtain a microscopic understanding of film growth. This method has the clear advantage that it directly images the film morphology with ultra high resolution. Thus it provides a wealth of information about the island size distribution, the average island spacing, correlations during growth, film and substrate defects, etc. The direct view of the surface provided by STM can be used to distinguish the different structural models that could otherwise be modeled in a diffraction study. Thus it is an important complementary technique to the diffraction approach. On the other hand, STM is mostly used for post-growth analysis and is seldom applied to monitor changes in morphology during film growth. The theory, experimental details and various applications of STM have been reviewed by a number of authors [96, 97, 98].

The functioning of the scanning tunneling microscope (also abbreviated to STM) is based upon the quantum mechanical mechanism of tunneling. If we consider two metal surfaces separated by a vacuum gap, the gap forms a potential barrier for electrons in both metals. However, the electron wave function does not drop to zero abruptly at the surface but extends into the barrier with an exponential tail. When the tails of the wave functions of both metals overlap, electrons can tunnel across the barrier. Thus, with an applied bias voltage, a tunneling current can flow from the occupied electronic states of one metal into the unoccupied states of the other metal. In the limit of low voltage, the tunneling-current density J in a planar junction has been given by Simmons [99] as

$$J = \frac{3}{4\pi}\frac{e^2}{h} V \frac{k_0}{s} \exp(-2k_0 s) , \tag{2.30}$$

where V is the applied bias voltage and s the barrier width. k_0 is the decay constant of the wave function in the barrier, which is related to the mean barrier height ϕ by $\hbar^2 k_0^2/2m = \phi$. One can see that the tunneling current decreases exponentially with increasing separation between the two metal surfaces. This feature enables the STM to act as a microscope with a resolution of less than 1 Å. Figure 2.16 shows the principle of operation of the STM. A sharp conducting tip is brought into close proximity to the sample surface within a distance of a few angstroms, resulting in a significant overlap of the electron wave functions of the tip and the sample. When a constant voltage is applied between the tip and the surface, a tunneling current is measured. A control unit regulates the z piezodrive to maintain a constant tunneling current while the tip is scanning over the sample surface, driven by P_X and P_Y. The recorded motion of the tip directly yields the topography of the surface.

In addition to the STM, another powerful technique was introduced in 1985 that is ideally suited to studying film growth [100]. This is low-energy electron microscopy (LEEM), which allows a spatial resolution of the order of 20–100 Å. The instrument itself is considerably more complex and expensive

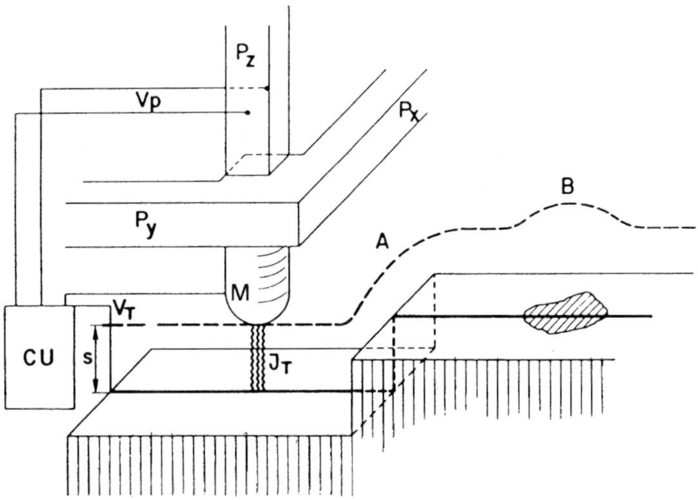

Fig. 2.16. Principle of operation of the STM. Note that distances and sizes are not drawn to scale. The piezodrives P_X and P_Y scan the metal tip M over the surface. The control unit (CU) applies the appropriate voltage V_p to the piezodrive P_Z to obtain a constant tunneling current J_T at constant tunneling voltage V_T. The *broken line* indicates the z displacement in a scan over a surface step (A) and a chemical inhomogeneity (B). From [94]

than an STM, which explains its less frequent application. On the other hand, its rapid data collection allows the study of kinetics over larger timescales. Since temperature stability is of no major concern either, a large temperature range is accessible. LEEM has been successfully used to observe shape oscillations in the growth of small crystals [101], and quantum wires in the growth of Ag on Si(100) [102].

2.4.3 Auger Electron Spectroscopy (AES)

Auger electron spectroscopy is another useful tool for studying film growth. AES is based on the Auger electron process. When an atom is bombarded by an energetic electron beam, electron holes can be created in an inner shell of the atom (see Fig. 2.17). This hole will be filled by an electron from a higher-energy outer shell. The released energy could be carried away by a photon, or alternatively transferred to another outer electron, which is ejected from the atom. The ejected electron is called an Auger electron. Clearly, the energy of an Auger electron depends only on the energy levels of the atom involved and not on the energy of the impinging particle creating the initial core hole, and thus the energy of the Auger electrons is characteristic of the atom. AES distinguishes the Auger electrons from the background of secondary electrons

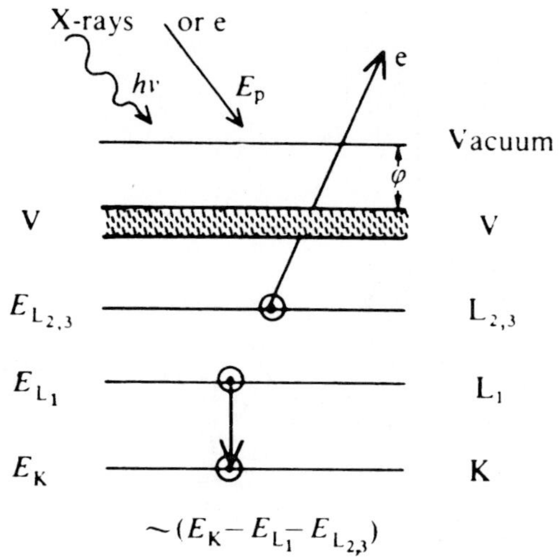

Fig. 2.17. Illustration of an Auger process. A hole is first created in the K level when the atom is ionized by an incident electron or photon. This hole is filled by an electron originally in the L_1 level, and an energy of $E_K - E_{L1}$ is released. A second electron in $L_{2,3}$ is ejected into the vacuum as an Auger electron by the released energy. The kinetic energy of the Auger electron equals $E_K - E_{L1} - E_{L2,3}$, and thus depends only on the atomic structure. The Auger electron produced in this way is referred to as a $KL_1L_{2,3}$ Auger electron

which are ejected directly by the primary electrons, and resolves the Auger electron energies by a modulation technique [104]. In AES, the primary electrons are usually chosen to have energies of 500–3000 eV, since the Auger process dominates over photon emission for low-energy electron bombardment (< 1000 eV). Therefore Auger electrons usually possess a low energy (10–500 eV), and thus can only come from the layers in the vicinity of the surface (see Fig. 2.18). Hence AES is both surface-sensitive and element-specific. It has been extensively used to check surface cleanliness and to determine the chemical composition of adsorbates. Furthermore, quantitative AES can also be used to analyze the film growth mode, especially in the initial stages [105]. The application of AES in this field has been reviewed by Argile and Rhead [106].

The capability of AES as a tool for growth mode analysis is based on the attenuation effect of the Auger electrons produced by the inner atoms (see

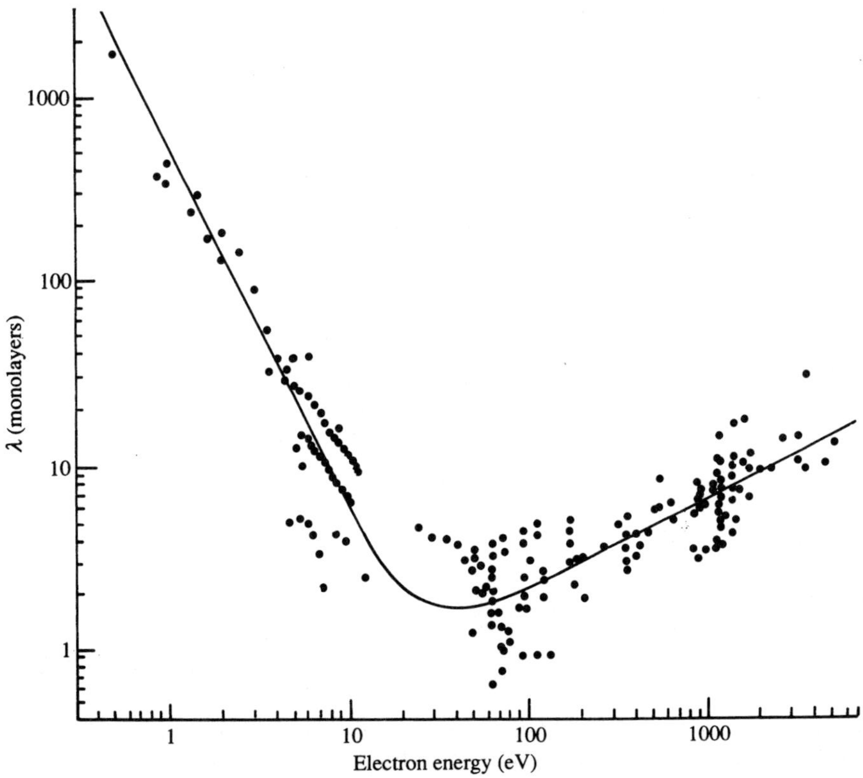

Fig. 2.18. The attenuation length plotted as a function of the electron kinetic energy. The *dots* correspond to the attenuation lengths measured in many different materials. The *curve* is drawn as a guide to the attenuation length of an escaping electron [103]

Fig. 2.20). The loss of part of the emission is due to inelastic scattering as the electrons escape into the vacuum. Let t_f represent the probability that an Auger electron emitted from a film atom will be transmitted through a layer of film atoms placed above it; for layer-by-layer growth, the Auger signal from n completed film-atom layers ($I_f^{(n)}$) is then given by

$$I_f^{(n)} = I_f^{(1)}(1 + t_f + t_f^2 + \cdots + t_f^{n-1}) = \frac{I_f^{(1)}(1 - t_f^n)}{1 - t_f} , \tag{2.31}$$

where $I_f^{(1)}$ is the signal from the first monolayer without attenuation. For the substrate, one has

$$I_s^{(n)} = t_s^n I_s^{(0)} , \tag{2.32}$$

where $I_s^{(0)}$ is the signal from the clean substrate and t_s is the transmission coefficient for the Auger emission from the substrate when it traverses a layer of film atoms. Suppose a layer-by-layer film growth proceeds to a point where n layers have been completed and a fraction Θ of the surface has been covered by the $(n+1)$th monolayer. The Auger signals are given by

$$I_f(\Theta) = \frac{I_f^{(1)}(1 - t_f^n)}{1 - t_f} + I_f^{(1)} t_f^n \Theta \tag{2.33}$$

for the film atoms and

$$I_s(\Theta) = I_s^{(0)} t_s^n + I_s^{(0)}(t_s^{n+1} - t_s^n)\Theta \tag{2.34}$$

for the substrate atoms. One can see that both the film signal and the substrate signal show a linear change with Θ during the growth of a particular layer, but the slopes change when a new layer starts to grow. With increasing film thickness, the absolute difference in the slope between neighboring layers becomes smaller and smaller. Assuming that the shape of the Auger peak remains constant, a plot of the absolute AES peak intensity of either the film or the substrate against the growth time should give linear segments separated by equally spaced breakpoints. These breakpoints appear upon completion of monolayers in layer-by-layer growth (Fig. 2.19).

In the case of Volmer–Weber growth, the Auger signals vary slowly and continuously with the deposited quantity. In the Stranski–Krastanov mode, linear segments are observed until a critical thickness is reached. Above this thickness, a continuous variation in the Auger signals is encountered.

To close this section, we present Fig. 2.20 to summarize the characteristic features of Frank-van der Merwe, Stranski–Krastanov and Volmer–Weber growth as seen by STM or LEEM, AES, and RHEED.

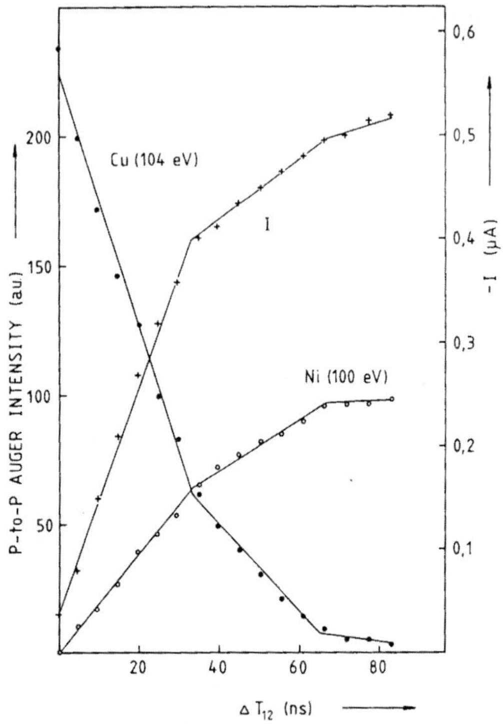

Fig. 2.19. The peak-to-peak intensities of the Auger lines of Cu and Ni as a function of the change in the period of quartz microbalance used to measure the thickness of the Ni film. The breakpoints of the segments were used for an absolute calibration of the microbalance. The sample current I shows the same segments as the Auger signal intensity when the energy and the flux of the primary electron beam are constant. From [107]

2.5 Manipulating the Growth

After we have gained a detailed understanding of the elementary processes involved in film growth and the principles governing these processes, we are capable of manipulating the film growth. To manipulate film growth requires us to choose the appropriate growth conditions and thus guide the nonequilibrium growth in the required direction. This goal can be reached by choosing the substrate temperature, tuning the growth rate and applying a surfactant. Ion beam bombardment has also been used to modify film growth.

2.5.1 Manipulation via Variation of Temperature and Deposition Rate

In typical growth experiments, it is very easy to vary the diffusion coefficient over several orders of magnitude by simply changing the substrate tempera-

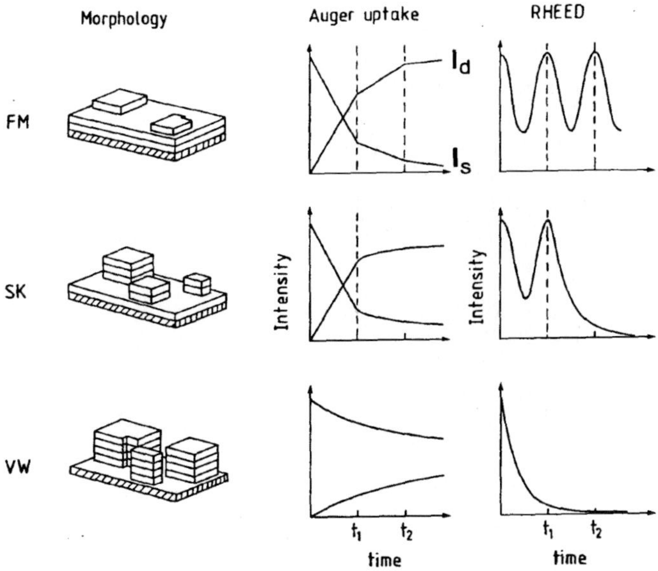

Fig. 2.20. Morphology, Auger uptake curve and RHEED intensity for the different growth modes. The growth modes depicted are layer-by-layer or Frank–van der Merwe (FM), Stranski–Krastanov (SK) growth and three-dimensional Volmer-Weber (VW) growth. The *left-hand side* shows the morphology after deposition of approximately 2.5 ML. STM and LEEM can be used to image the film topography. In an Auger uptake curve, for layer-by-layer growth, a sequence of linear segments is observed with breakpoints upon completion of full monolayers for both the deposit (I_d) and the substrate (I_s). A smaller, continuous slope is observed for VW growth. RHEED oscillations are observed for FM and SK growth. The latter growth type shows a subsequent exponential decay of intensity upon 3D nucleation. This decay is characteristic of the initial growth stage of VW growth as well

ture. The variation of the deposition flux can also span two or three orders of magnitude, though it is more difficult to modify than the diffusion coefficient. We shall see that the film growth can be manipulated effectively by changing the growth temperature and the growth rate.

In MBE growth, the growth mode usually changes from three-dimensional growth at low temperatures to layer-by-layer growth at higher temperatures. As we have seen in (2.13), all the diffusion processes show an exponential dependence on temperature, though the characteristic energies vary with the specific type of diffusion process. At higher temperatures, the adatoms become more mobile, so that they can jump over local energy barriers, including the barrier at step edges, the Ehrlich–Schwoebel barrier. This barrier can suppress interlayer transport at low temperatures. Hence the arriving adatoms nucleate on the top of clusters and give rise to the growth of three-dimensional islands with pyramid-like features. This kind of mode transition with tem-

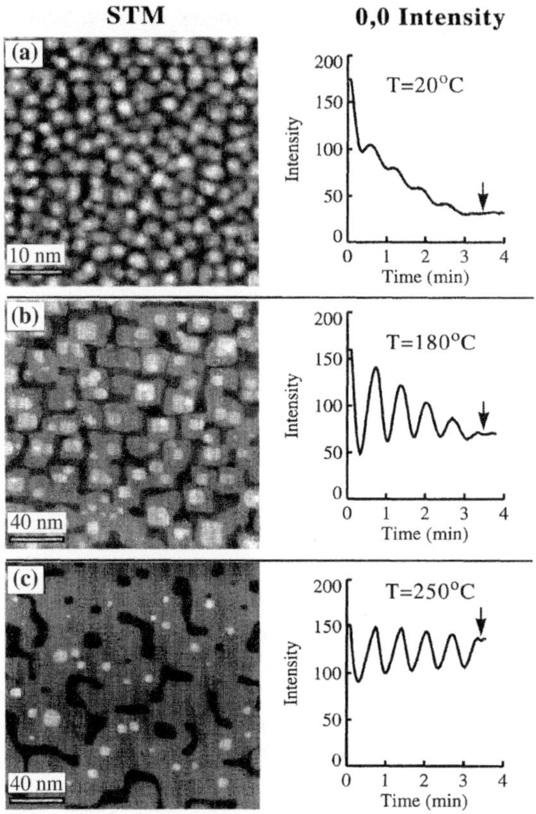

Fig. 2.21. Change of growth mode from layer-by-layer at high temperature to three-dimensional island growth at lower temperature for Fe growing on Fe(100) as seen by STM and RHEED. From [108]

perature has been clearly shown by Stroscio et al. for Fe growing on Fe(100) (Fig. 2.21) [108]. It should be pointed out that if the substrate temperature were to be increased further and a very low deposition rate or a substrate with high step density were used, the film growth could follow the step-flow mode, where the adatoms can diffuse to the nearest steps and are captured there. This is another mode for producing smooth films apart from layer-by-layer growth, but it is seldom exploited for heteroepitaxial film growth owing to the elevated interdiffusion between the substrate and film. Nevertheless, when high-melting-point substrates are used, such as W(110) or Mo(110), the step-flow mode provides an alternative pathway for manipulating the growth [109, 110]. A detailed discussion of interdiffusion will be given later in this subsection.

The role that temperature can play in manipulating film growth by regulating the adatom mobility has already been described above. Another less direct influence of temperature on growth lies in its capability to determine the density and size of nuclei. As shown by (2.28), the density of nuclei increases with the deposition flux and inversely with the diffusion coefficient in accordance with a power law. Therefore the density of nuclei is quite sensitive to temperature. On the other hand, a number of atomistic processes, for example inter-layer diffusion and edge diffusion, are strongly affected by the density and size of nuclei. Tersoff et al. pointed out that the onset of nucleation of the second layer on top of an island is associated with a critical island radius R_c [111], which is dependent on the Ehrlich–Schwoebel barrier at the step edge. This provides a simple criterion for layer-by-layer versus island growth: for a given island spacing L_n (determined by the density of nuclei), if $R_c < L_n$ the islands will nucleate a second layer before coalescence, resulting in multilayer growth. The existence of a critical island radius has been confirmed by experiments [63, 111, 112] and simulations [113]. This can be qualitatively understood by considering that for small islands, the adatoms have a high probability of reaching the step edge and thus have a high probability of reaching the lower terrace. The increased kink density for small islands could also contribute to interlayer diffusion [63, 64]. The existence of a critical island radius provides another means to obtain layer-by-layer growth, by increasing the density of nuclei. This can be realized by using a large deposition flux, by the application of a surfactant, and/or by decreasing the growth temperature. A possible phase diagram for growth, given by Tersoff et al., is shown in Fig. 2.22, which takes the effect of island density on interlayer diffusion into account. The reentrant layer-by-layer growth mode at low temperatures is of particular importance for systems with strong interdiffusion at high temperatures. This mode has been observed in a number of systems [63, 112], including ultrathin Co films on Cu(001) [114].

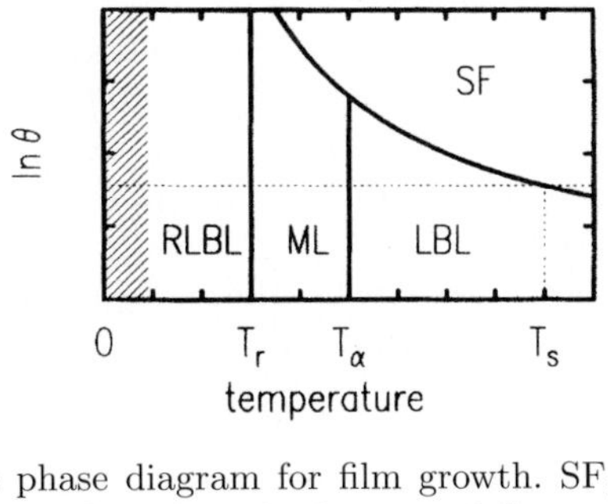

Fig. 2.22. A possible phase diagram for film growth. SF, LBL, ML and RLBL denote the regimes of step-flow, layer-by-layer, multilayer and reentrant layer-by-layer growth, respectively. θ is the angle of surface miscut, which determines the step density. T_s corresponds to the onset of step flow at a given angle of miscut. The boundary lines indicate a smooth crossover between regimes, not an abrupt transition. From [111]

Up to now, we have seen two conflicting effects of adatom mobility on the growth mode. To obtain layer-by-layer growth, on the one hand we wish the adatom diffusion to be small in the nucleation stage so that a large density of nuclei is obtained. In this case the island size could remain below the critical size for second-layer nucleation until coalescence is reached, and thus the island growth is suppressed even in the presence of an Ehrlich–Schwoebel barrier. On the other hand, to reach the same goal we wish to have a high adatom mobility in the growth stage so that the adatoms can roll over the steps effectively. Faced with this dilemma, Rosenfeld et al. proposed the "concept of two mobilities" to manipulate the film growth [115]. The basic idea is that the experimental parameters can be varied deliberately in the different stages so that a high density of nuclei is obtained in the nucleation stage and a high mobility is employed in the growth stage. To realize this strategy, complexity in the deposition technology is inevitably introduced.

When we exploit the growth temperature to manipulate heteroepitaxial growth, much attention needs to be paid to interdiffusion, i.e., the diffusion of adatoms into the bulk of the substrate or, vice versa, the diffusion of substrate atoms into the film. These processes are highly unwanted if chemically sharp interfaces are to be grown. Therefore interdiffusion sets an upper limit on the manipulation of temperature. The interdiffusion has been estimated by Flynn using the concept of the penetration depth [116]. The penetration depth (L_b) is defined as the length over which the concentration of substrate atoms in the film decreases by $1/e$. This depth corresponds to $L_b = \sqrt{D_b t}$, where D_b is the bulk diffusion coefficient of the substrate atoms in the film and t is the typical time of a growth experiment. If L_b is smaller than the typical nearest-neighbor spacing of 2.5 Å, interdiffusion is negligible. On the other hand, pronounced interdiffusion is found when L_b is considerably larger than the nearest-neighbor distance. Using a deposition time of 10^4 s, Flynn's model gives a temperature limit of $T = (3/8)T_M$, where T_M is the melting temperature of the film material, for the onset of interdiffusion. In this derivation, an empirical formula $D_b(T) = 10^{-1/2} \times 10^{-7T_M/T}$ $\mathrm{cm^2\,s^{-1}}$ is used to relate to T_M. Using another empirical formula, $D_s(T) = 10^{-3} \times 10^{-3T_M/2T}$ $\mathrm{cm^2\,s^{-1}}$ for the surface diffusion, Flynn could also obtain lower temperature limits for step-flow growth and nucleation–growth-type layer-by-layer growth of $(3/8)\,T_M$ and $(1/8)\,T_M$, respectively. Here, reentrant layer-by-layer growth is not included in the considerations. The main results of Flynn's estimation are shown in Fig. 2.23, which can be used as a rough guide for choosing a growth temperature.

2.5.2 Application of Surfactants

As a result of the continuous endeavor to search for novel methods to produce smooth films, surfactants have been successfully applied in the epitaxial growth of ultrathin semiconductor and metal films. A typical example is that the use of surfactants, mostly Group V and IV elements (As, Sb, Bi, Pb, etc.),

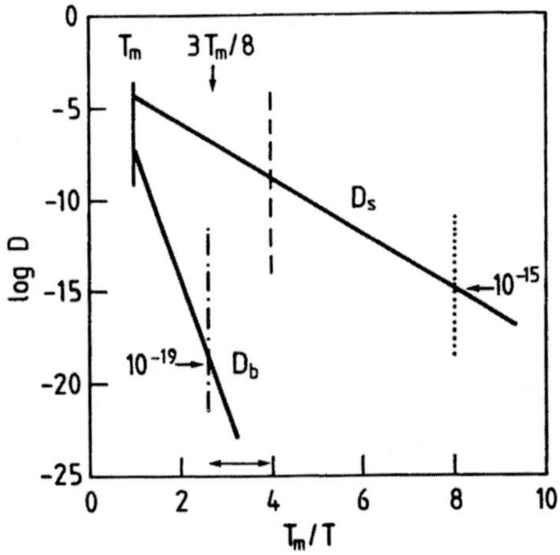

Fig. 2.23. Temperature dependence of the bulk diffusion coefficient D_b and surface diffusion coefficient D_s. The temperature dependence has been calculated using an empirical formula which relates the diffusion coefficients for bulk and surface diffusion to the melting temperature of the film material. The *dashed–dotted line* denotes the temperature limit above which interdiffusion between substrate and adsorbate atoms is observed ($T > (3/8)\ T_M$), while the *dotted* and *dashed lines* identify the low-temperature limits for nucleation–growth-type layer-by-layer growth ($T > (1/8)\ T_M$) and step flow growth ($T > (1/4)\ T_M$), respectively. The temperature interval denoted by the double-headed arrow therefore denotes the temperature regime where the surface mobility is high enough to enable the growth of smooth films but the interdiffusion is small enough to produce sharp interfaces. From [116]

leads to layer-by-layer growth of a Ge overlayer on a Si substrate even beyond the critical thickness [117]. A surfactant is a coating that floats on top of the growing film without being incorporated into it but whose presence changes the growth mechanism and promotes smooth film growth. Consequently, for use as surfactants, materials that have a low surface energy and a lack of chemical affinity to the growing material are required. A surfactant could be a layer of metal deposited on the substrate surface before the deposition of the film. Alternatively, it could be an adsorbed gas. For the growth of ultrathin magnetic films, gaseous surfactants are used more frequently than metals.

A lot of work has been devoted to understanding the function of the surfactant and many different models have been proposed [118, 119]. The proposed mechanisms include the following. (a) The surfactant modifies the thermodynamic balance of the surface and interface free energies, in such a way that the equilibrium state in the presence of the surfactant layer is a smooth, flat film; within this model, the adatoms exchange places locally with the surfactant atoms, and hence minimum surface diffusion results, thus preventing nucleation and growth of 3D islands. (b) The surfactant promotes interlayer diffusion. (c) The surfactant introduces a high density of nuclei and thus suppresses second-layer nucleation. (d) The effect of the surfactant comes from a small diffusion barrier for the adatoms on top of the surfactant and a large exchange barrier with the surfactant atoms.

Surfactant effects have been investigated in a number of studies on ultrathin magnetic films. For example, O, N, CO, C_2H_2 and a mixed gas of C_2H_2 and O_2 have been applied to grow ultrathin fcc Fe films on Cu(100) [120, 121, 122, 123]. Other examples include Pb for the growth of Fe on Cu(111) [119], oxygen for Fe on Cu_3Au(100) [124] and Ni on Cu(100) [125], and Pb for Co on Cu(111) [126, 127]. Noble metals (Au and Ag) and Cu, which show low surface energies, are also found to demonstrate surfactant effects in the growth of ultrathin ferromagnetic films such as Fe and Ni [128, 129, 130]. The application of surfactants not only modifies the growth and thus the film morphology (as shown in Fig. 2.24), but also exerts strong

Fig. 2.24. These STM images (500×500 Å^2) show that one ML of precovered Pb results in a high island density and a small film roughness for Fe grown on Cu(111) (*right panel*), compared with results on a clean Cu(111) substrate (*left panel*). From [119]

effects on the film structure. As a typical manifestation, the application of a surfactant can extend the film thickness range where a metastable phase is stabilized. This issue will be discussed again in Chap. 5.

References

1. L. Royer: Bull. Soc. Fr. Min. **51**, 7 (1928)
2. F. C. Frank and J. H. van der Merwe: Proc. Roy. Soc. A **198**, 205 (1949)
3. M. Volmer and A. Weber: Z. Phys. Chem. **119**, 6274(1981)
4. J. N. Stranski and L. Krastanov: Ber. Akad. Wiss. Wien **146**, 797(1938)
5. W. K. Burton, N. Cabrera and F. C. Frank: Trans. Roy. Soc. A **243**, 299(1951)
6. E. Bauer: Zeitschr. f. Kristallographie **110**, 372(1958)
7. M. Dupré: *Théorie mécanique de la chaleur*, (Paris, 1969) **71**, 657(1978)
8. W. R. Tyson: Surf. Sci. **62**, 267(1977)
9. J. J. Gilman: J. Appl. Phys. **31**, 2208(1960)
10. L. Z. Mezey, D. Marton and J. Giber: Period. Polytech. Mech. Eng. **21**, 189(1977)
11. S. H. Overbury, P. A. Bertrand and G. A. Somorjai: Chem. Rev. **75**, 547(1975).
12. A. R. Miedema: Z. Metallkd. **69**, 287(1978)
13. L. Z. Mezey and J. Giber: Jpn. J. Appl. Phys. **21**, 1569(1982)
14. H. H. Brongersma, M. J. Sparnaay and M. T. Buck: Surf. Sci. **71**, 657(1978)
15. N. D. Lang and W. Kohn: Phys. Rev. B **1**, 4555(1970)
16. N. D. Lang and W. Kohn: Phys. Rev. B **3**, 1215(1971)
17. R. Monnier and J. P. Perdew: Phys. Rev. B **17**, 2595(1978)
18. Z. Y. Zhang, D. C. Langreth and J. P. Perdew: Phys. Rev. B **41**, 5674(1990)
19. J. P. Perdew, H. Q. Tran and E. D. Smith **42**, 11627(1990)
20. M. S. Daw and M. I. Baskes: Phys. Rev. Lett. **50**, 1285(1983).
21. M. I. Baskes: Phys. Rev. Lett. **59**,2666(1987)
22. M. I. Baskes: Phys. Rev. B **46**, 2727(1992)
23. Byeong-Joo Lee, M.I. Baskes, Hanchul Kimand and Yang Koo Cho: Phys. Rev. B **64**, 184102(2001)
24. F. R. de Boer, R. Room, M. C. M. Mattens, A. R. Miedema and A. K. Niessen: *Cohesion in Metals*(North-Holland, Amsterdam 1988)
25. H. L. Skriver and N. M. Rosengaard: Phys. Rev. B **46**, 7157(1992)
26. M. Aldén, S. Mirbt, H. L. Skriver, N. M. Rosengaard and B. Johansson: Phys. Rev. B **46**, 6303(1992)
27. R. J. Needs and M. Mansfield: J. Phys.: Condens. Mater: **1**, 7555(1989)
28. M. Methfessel, D. Hennig and M. Scheffler: Phys. Rev. B **46**, 4816(1992)
29. M. J. Mehl and D. A. Papaconstantopoulos: Phys. Rev. B **54**, 4519(1996).
30. Wangyu Hu, Bangwei Zhang, Baiyun Huang, Fei Gao and David J Bacon:J. Phys.: Condens. Matter **13**, 1193(2001)
31. M. Christensen, S. Dudiy and G. Wahnström: Phys. Rev. B **65**, 045408(2002).
32. R. Stadler and R. Podloucky: Phys. Rev. B **62**, 2209(2000)
33. Siqing Wei and M. Y. Chou: Phys. Rev. B **50**, 4859(1994)
34. J. Kirschner: Nucl. Instr. and Meth. B **7/8**, 742(1985)
35. J. C. Heyraud, J. J. Métois: Surf. Sci. **177**, 213(1986)
36. G. Wulff, Zeitschr. f. Kristallographie **34**, 449(1901)

37. J. M. Bermond, J. J. Métois J. C. Heyroud and F. Floret: Surf. Sci. **416**, 430(1998)
38. R. Kaischew, Bull. Acad. Sc. Bulg. Ser. Phys. **2**, 191(1951)
39. S. V. Khare, S. Kodambaka, D. D. Johnson, I. Petrov, J.E. Greene: Surf. Sci. **522**, 75(2003)
40. C. Rottman and M. Wortis: Phys. Rev. B **24**, 299(1951)
41. M. Wortis, R. Vanselow and R. Howe (Eds): *Chemistry and Physics of Solid Surface*, Vol. 7 (Springer, New York 1987) pp 367
42. G. S. Icking-Konert, M. Giesen and H. Ibach: Phys. Rev. Lett. **83**, 3880(1999)
43. D. C. Schlößer, L. K. Verheij, G. Rosenfeld and G. Comsa: Phys. Rev. Lett. **82**, 3843(1999)
44. H. P. Bonzel and A. Emundts: Phys. Rev. Lett. **84**, 5804(2000)
45. R. Kern, G. Le Lay and J. J. Metois: Basic mechanisms in the early stages of epitaxy. In: *Current Topics in Materials Science*, vol 3, ed by E. Kaldis(North-Holland, Amsterdam New York Oxford 1979) pp 131–476
46. H. T. Johnson and L. B. Freund: J. Appl. Phys. **81**, 6081(1997)
47. D. E. Gray: *American Institute of Physics Handbook*, 2nd edn (McGraw-Hill Book Company, Inc, New York, Toronto, London 1996)
48. F. J. Himpsel, J. E. Ortega, G. J. Mankey and R. F. Wills: Advance in Physics **47**, 511(1998)
49. M. H. Grabow and G. H. Gilmer: Surf. Sci. **194**, 333(1988)
50. I. Daruka and A. Barabási: Phys. Rev. Lett. **79**, 3708(1997).
51. V. A. Shchukin, N. N. Ledntsov, P. S. Kop'ev and D. Bimberg: Phys. Rev. Lett. **75**, 2968(1995)
52. I. Daruka and A. Barabási: Appl. Phys. Lett. **72**, 2102(1998).
53. B. Lewis and J. C. Anderson: *Nucleation and Growth of Thin Film* (Academ. pr. New York 1978) pp 49–51
54. A. Pimpinelli and J. Villain: *Physics of Crystal Growth* (Cambridge University Press 1997) pp 111–125
55. S. C. Wang and G. Ehrlich: Phys. Rev. Lett. **70**,41(1993)
56. M. Karimi, T. Tomkowski, G. Vidali and O. Biham: Phys. Rev. B **52**, 5364(1995)
57. G. Ehrlich and F. Hudda: J. Chem. Phys. **44**, 1039(1966)
58. R. L. Schwoebe: J. Appl. Phys. **40**, 614(1969)
59. R. Stumpf and M. Scheffler, Phys. Rev. Lett. **72**, 254(1994)
60. Yinggang Li and A. E. DePristo: Surf. Sci. **319**, 141(1994)
61. E. Lundgren, B. Stanka, G. Leonardelli, M. Schmid and P. Varga: Phys. Rev. Lett. **82**, 5068(1999)
62. J. Jacobsen, K. W. Jacobsen, P. Stoltze and J. K. Nϕrskov: Phys. Rev. Lett. **74**, 2295(1995)
63. R. Kunkel, B. Poelsema, L. K. Verheij and G. Comsa: Phys. Rev. Lett. **65**, 733(1990)
64. M. Bott, Th. Michely and G. Comsa: Surf. Sci. **272**, 161(1992)
65. M. Giesen, G. Schulz Icking-Konert and H. Ibach: Phys. Rev. Lett. **80**, 552(1998)
66. M. Giesen, G. Schulze Icking-Konert and H. Ibach: Phys. Rev. Lett. **82**, 3101(1999)
67. T. Flores, S. Junghans and M. Wuttig: Surface Science **371**, 1(1997)
68. J. B. Hannon, C. Klünker, M. Giesen and H. Ibach: Phys. Rev. Lett. **79**, 2506(1997)

69. C. Klünker, J. B. Hannon, M. Giesen and H. Ibach: Phys. Rev. B **58**, R7556(1998)
70. G. Boisvert and L. J. Lewis: Phys. Rev. B **56**, 7643(1997)
71. H. Ibach, M. Giesen, T. Flores, M. Wuttig and G. Treglia: Surface Science **364**, 453(1996)
72. T. Flores, M. Hansen and M. Wuttig: Surface Science: **279**, 251(1992)
73. T. Flores, S. Junghans and M. Wuttig: Surface Science **371**, 14(1997)
74. G. Tréglia, B. Legrand, A. Saul, T. Flores and M. Wuttig: Surface Science **352-354**, 552(1996)
75. R. van Gastel, E. Somfai, W. van Saarloos and J. W. M. Frenken: Nature (London) **408**, 665 (2000)
76. R. van Gastel, E. Somfai, S. B. van Albada, W. van Saarloos and J. W. M. Frenken: Phys. Rev. Lett. **86**, 1562(2001)
77. T. Michely, M. Hohage, M. Bott, and G. Comsa: Phys. Rev. Lett. **70**, 3943 (1993); M. Kalff, G. Comsa, and T. Michely: Phys. Rev. Lett. **81**, 1255 (1998)
78. M. V. R. Murty and B. H. Cooper: Phys. Rev. Lett. **83**, 352(1999)
79. J. A. Venables: Phil. Mag. **27**, 693(1973)
80. J. A. Venables, G. D. T. Spiller and M. Hanbüchen: Rep. Prog. Phys. **47**, 399(1984)
81. G. Zinsmeister: Vacuum **16**, 529(1966); Thin Solid Films **2**, 497(1968); Ibid. **4**, 363(1969); Kristal. Tech. **5**, 207(1970); Thin Solid Films **7**, 51(1971)
82. R. M. Logan: Thin Solid Films **3**, 59(1969)
83. D. R. Frankl and J. A. Venables: Adv. Phys. **19**, 409(1970)
84. D. Walton: J. Chem. Phys. **37**, 2182(1962)
85. I. Hernández-Calderón and H. Höchst: Phys. Rev. B, **27**, 4961(1983)
86. J. H. Neave and B. A. Joyce: Appl. Phys. A **31**, 1(1983)
87. J. M. Van Hove, C. S. Lent, P. R. Pukite and P. I. Cohen: J. Vac. Sci. Technol. B **1**, 741(1983)
88. P. Bencok, et al.: Surf. Sci. **402-404**, 327(1998)
89. U. Korte, P. A. Maksym: Phys. Rev. Lett. **78**, 2381(1997)
90. W. Braun, L. Däweritz, K. H. Ploog, Phys. Rev. Lett. **80**, 4935(1998).
91. P. R. Pukite, C. S. Lent and P. I. Cohen: Surf. Sci. **161**, 39(1985)
92. T. Shitara, D. D. Vvedensky, M. R. Milby, J. Zhang, J. H. Neave and B. A. Joyce: Phys. Rev. B **46**, 6815(1992); ibid. **46**, 6825(1992).
93. W. Braun: *Applied RHEED – Reflection High-Energy Electron Diffraction During Crystal Growth*, (Springer, Berlin Heidelberg New York 1999)
94. G. Binnig, H. Rohrer, Ch. Gerber and E. Weibel: Phys. Rev. Lett. **49**, 57(1982)
95. F. Basenbacher: Rep. Prog. Phys **59**, 1737(1996)
96. H. -J. Güntherodt and R. Wiesendanger (Eds): *Scanning Tunnelling Microscopy I, II, III* , (Springer, Berlin Heidelberg New York 1994)
97. L. E. C. van de Leemput and H van Kempen: Rep.Prog. Phys. **55**, 1165(1992)
98. D. Drakova: Rep. Prog. Phys. **64**, 205(2001)
99. J. G. Simmons: J. Appl. Phys. **34**, 1793(1963)
100. E. Bauer: Ultramicroscopy **17**, 51(1985)
101. J. Tersoff, A. W. Denier van der Gon and R. M. Tromp, Phys. Rev. Lett. **70**, 1143(1993)
102. J. Tersoff and R. M. Tromp: Phys. Rev. Lett. **70**, 2782(1993)
103. M. Prutton: *Introduction to Surface Physics*, (Clarendon Press, Oxford 1994) page 24

104. M. P. Seah and C. P. Hunt: Rev. Sci. Instrum. **59**,217 (1988)
105. H. Landskron, G. Schmidt, K. Heinz, K. Müller, C. Stuhlmann, U. Beckers, M. Wuttig and H. Ibach: Surface Sci. **256**, 115(1991)
106. C. Argile and G. E. Rhead: Surface Science Reports **10**,277(1989)
107. W. Kirstein, B. Krüger and F. Thieme: Surface Sci. **176**, 505(1986)
108. J. A. Stroscio, D. T. Pierce and R. A. Dragoset: Phys. Rev. Lett. **70**, 3615(1993)
109. J. Hauschild, U. Gradmann and H. Elmers: Appl. Phys. Lett. **72**, 3211(1998)
110. S. Murphy, D. Mac Mathúna, G. Mariotto and I. V. Shvets: Phys. Rev. B **66**, 195417(2002)
111. J. Tersoff, A. W. Denier van der Gon and R. M. Tromp: Phys. Rev. Lett. **72**, 266(1994)
112. K. Bromann, H. Brune, H. Röder and K. Kern: phys. Rev. Lett. **75**, 677(1995).
113. J. Rottler and P. Maass: phys. Rev. Lett. **83**, 3490(1999)
114. T. Bernhard, R. Pfandzelter and H. Winter: Ncl. Instr. and Meth. in Phys. Res. B **203**, 111(2003)
115. G. Rosenfeld, R. Poelsema and G. Comsa: J. Cryst. Growth **151**, 230(1995)
116. C. P. Flynn: J. Phys F **18**, L195(1988)
117. M. Copel, M. C. Reuter, E. Kaxiras and R. M. Tromp: Phys. Rev. Lett. **63**, 632(1989); H. J. Osten, J. Klatt, G. Lippert, B. Dietrich and E. Bugiel: ibid., **69**, 450(1992).
118. D. Kandel and E. Kaxiras: Solid State Phys. **54**, 219 (2000)
119. M. C. G. Passeggi, Jr. J. E. Prieto and R. Miranda: Phys. Rev. B **65**, 35409(2001)
120. D. A. Steigerwald, I. Jacob and Jr. W. F. Egelhoff: Surf. Sci. **202**, 472(1988)
121. M. Wuttig, B. Feldmann, J. Thomassen, F. May, H. Zillgen, A. Brodde, H. Hannemann and H. Neddermeyer: Surf. Sci. **291**, 14(1993)
122. J. Thomassen, F. May, B. Feldmann, M. Wuttig and H. Ibach: Phys. Rev. Lett. **69**, 3831(1992)
123. A. Kirilyuk, J. Giergiel, J. Shen and J. Kirschner: Phys. Rev. B **52**, R11672(1995)
124. R. Da Prato, F. Bisio, P. Cantini, S. Terreni, M. Canepa and L. Mattera: Surf. Sci. **507-510**, 318(2002)
125. R. Nünthel et al.: Surf. Sci. **531**, 53(2003)
126. J. Camarero, T. Graf, J. J. de Miguel, R. Miranda, W. Kuch, M. Zharnikov, A. Dittschar, C. M. Schneider and J. Kirschner: Phys. Rev. Lett. **76**, 4428(1996)
127. W. F. Egelhoff, Jr. et al.: J. Appl. Phys. **80**, 5183(1996)
128. S. D. Bader and E. R. Moog: J. Appl. Phys. **61**, 3729(1987)
129. V. Blum et al.: Phys. Rev. B **59**, 15966(1999)
130. J. Roussel, A. Saúl, G. Tréglia and B. Legrand: Phys. Rev. B **55**, 10931(1997)

3 Structure of Ultrathin Films

In this chapter, various aspects of the actual film structure will be explored. These topics include the different possible structures of heteroepitaxial films and the precise determination of atomic positions. Furthermore, the characterization of film morphology will be treated. This subject has already been covered to some extent in the last chapter. In Sect. 3.2, it will be demonstrated that a single technique, namely low-energy electron diffraction, can give a wealth of information about the structure of ultrathin films.

3.1 Structural Properties of Epitaxial Films

In this section, the nomenclature of heteroepitaxial interfaces will be introduced. Afterwards, a classification of heterostructures based on the misfit will be presented and the response of the films to epitaxial strain will be reviewed. This information is applied latter to discuss the possibilities of producing metastable structures.

In order to describe the structure of a heteroepitaxial system, the relationship between the crystallographic directions of the film and the substrate must be specified. For this purpose, the conventional three-dimensional Miller indices are usually employed [1]. One first has to specify the substrate plane (HKL) and the overlayer plane (hkl) at the common interface. Then parallel directions in the substrate and overlayer planes must be identified. These directions are denoted by $[UVW]$ and $[uvw]$, respectively. These four sets of indices are given in the following sequence: $(HKL)\|(hkl)$; $[UVW]\|[uvw]$. For the growth of Ag(111) on Cu(100), for example [2], the experimentally observed geometry is denoted by Ag(111)$\|$Cu(100); Ag[110]$\|$Cu[110].

In the following, only cases will be considered where both the contact planes and the directions coincide, such as for the growth of fcc Fe(100) on Cu(100). In such a situation, the misfit is sufficient to characterize the epitaxial interface. As stated in (2.7), the misfit is defined as the difference between the *unstrained* lattice constants of the film and the substrate:

$$f = \frac{a_e(f) - a_e(s)}{a_e(f)} \ . \tag{3.1}$$

Here the index e is used to denote the equilibrium values of the unstrained film and substrate lattice constants, $a_e(f)$ and $a_e(s)$, respectively. For negative values of f, the initial layers of the film will be stretched, while for positive f the film is compressed.

Depending upon the size of the misfit, three different cases can be distinguished. These are depicted in Fig. 3.1. When the lattice constants of the film and the substrate are practically identical (matched), then the interfacial bonds are not strained. A matched structure is trivially obtained when the deposit and the substrate are identical, i.e., in the case of homoepitaxy. The case of lattice-matched heteroepitaxy, especially in semiconductor films, is more important. The largest success in growing lattice-matched heterostructures has been obtained with the epitaxial growth of $Ga_xAl_{1-x}As$ on GaAs. The ternary semiconductor has a band gap between 1.4 and 2.2 eV, depending upon the Al concentration, while the lattice constant hardly changes with the Al content [3]. Thus, this system is ideally suited for "band gap engineering" by simply controlling the composition in the epitaxial film.

Usually the lattice parameters of the film and substrate differ more substantially. If the deposit and the substrate have the same crystal structure, it is often possible to grow a limited number of coherent but strained layers. This strain is necessary to accommodate the difference in the lattice constants. Since the lattice constant of the film parallel to the surface is fixed at

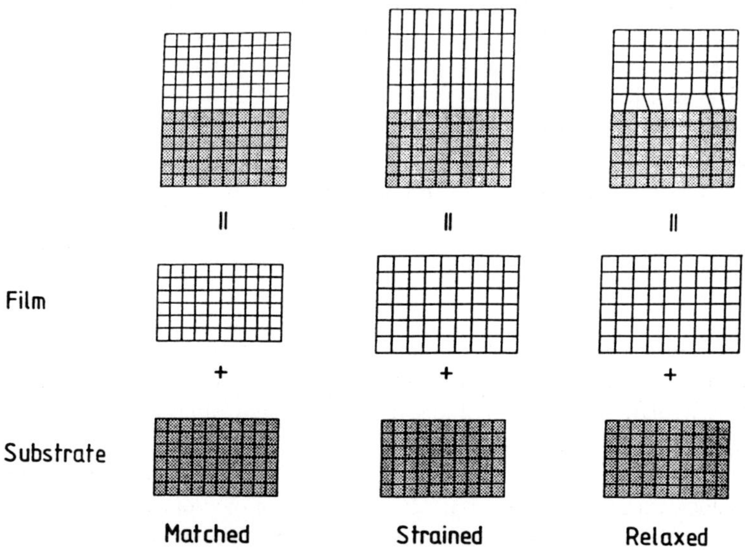

Fig. 3.1. Lattice-matched, strained and relaxed heteroepitaxial structures. For marginal mismatch between film and substrate, matched growth is expected. For films with larger misfit, strained epitaxy is encountered, which transforms into relaxed epitaxy with increasing thickness [1]

the in-plane lattice constant of the substrate, the film material will expand or contract perpendicular to the contact plane to keep the atomic volume almost constant.

Finally, when the mismatch is too large, the strain will be relaxed by introduction of misfit dislocations. In this case, the actual lattice constants of the film and the substrate will differ (incoherent growth). This type of epitaxy generally prevails for larger film thicknesses irrespective of the lattice mismatch. In these cases, the film will closely approach its equilibrium structure irrespective of the substrate.

For a determination of the structure of an epitaxial film, two questions are often important: (1) What is the *critical thickness* above which misfit dislocations are introduced? (2) How does the film respond to the epitaxial strain?

The critical thickness can be determined by minimizing the sum of the strain energy E_c and the energy of the dislocation network E_d with respect to the film strain ε_f. Matthews [4, 5] has derived approximate expressions for E_c and E_d, assuming that the shear moduli G of the film and the substrate are identical:

$$E_d = \frac{Gb(f - \varepsilon_f)}{2\pi(1 - \nu)} \ln\left(\frac{R_0}{b} + 1\right) , \tag{3.2}$$

$$E_c = \frac{2G(1 + \nu)}{1 - \nu} \varepsilon_f^2 d , \tag{3.3}$$

where $\boldsymbol{b}$ is the Burgers vector of the dislocation, R_0 is the radius from the dislocation where the strain field vanishes, ν is Poisson's ratio and d is the film thickness. While the dislocation energy has only a weak dependence upon film thickness, arising from R_0, the strain energy increases linearly with film thickness. Therefore, dislocation formation is energetically favorable above a critical thickness d_c. At this point, the film strain ε_f is approximately f and $R_0 \approx d_c$. Minimizing $E_c + E_d$ with respect to the film strain ε_f and rearranging to solve for the critical thickness d_c gives [5]

$$d_c = \frac{b}{8\pi(1 + \nu)f} \ln\left(\frac{d_c}{b} + 1\right) . \tag{3.4}$$

A graphical solution for the growth of fcc Fe on Cu(100) is shown in Fig. 3.2, assuming the formation of dislocations with a Burgers vector $\boldsymbol{b} = (1/2)\langle 110\rangle$. For partial dislocations such as Shockley and Frank dislocations with Burgers vectors of $(1/6)\langle 112\rangle$ and $(1/2)\langle 111\rangle$, respectively, stacking faults are created. If the stacking fault has a lower energy than an edge dislocation, then the critical thickness for the formation of these partial dislocations is even smaller. For films where the strain in the film is zero (complete relaxation), the density of dislocations n_δ in the interface is

$$n_\delta = \frac{f}{b_\parallel} , \tag{3.5}$$

where $b_\parallel$ is the component of the Burgers vector parallel to the interface.

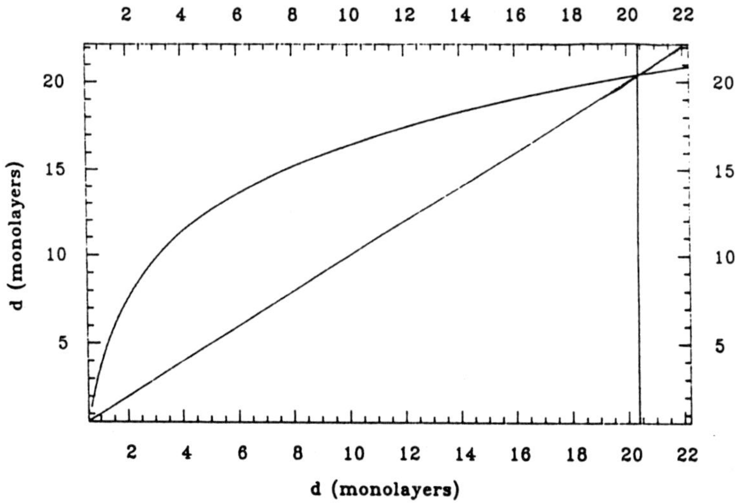

Fig. 3.2. Critical thickness for dislocation formation in fcc Cu on Cu(100). This figure shows the graphical solution of (3.4) for edge dislocations with Burgers vector $\boldsymbol{b} = 1/2\langle 110\rangle$. From the point at which the two curves cross, a critical thickness of approximately 20 ML is obtained. The lattice mismatch between Fe and Cu is about −0.7% at room temperature, and Poisson's ratio for fcc Fe is 0.45 [6]

In semiconductor epitaxy, considerably thicker dislocation-free films than estimated from (3.4) can be grown. This has been shown for Ge_xSi_{1-x} films (see Fig. 3.3). The reason is the large activation energy necessary for the formation of the extended dislocation array with well-defined spacings that was assumed by Matthews [5]. Instead, dislocations characterized by a certain width ω nucleate individually. For these isolated (screw) dislocations, the energy per unit area is [7]

$$E_d = \frac{Gb^2}{4\pi\omega} \ln \frac{d}{b} , \tag{3.6}$$

giving a critical thickness

$$d_c \simeq \frac{(1-\nu)b^2}{(1+\nu)8\pi\omega f^2} \ln \frac{d_c}{b} . \tag{3.7}$$

The solid line in Fig. 3.3 is the result of (3.7) for $\omega = 19.6$ Å. This equation gives an excellent fit to the experimental data [8]. For values of $d < d_c$, the films are defect-free (coherent), but strained, while above d_c they are incoherent. Owing to the detrimental effect of defects on the electronic properties of semiconductor heterostructures, it is extremely important in these films to establish coherent growth.

So far, we have considered heteroepitaxial systems where the crystallographic orientations of the crystalline film and the substrate coincide and

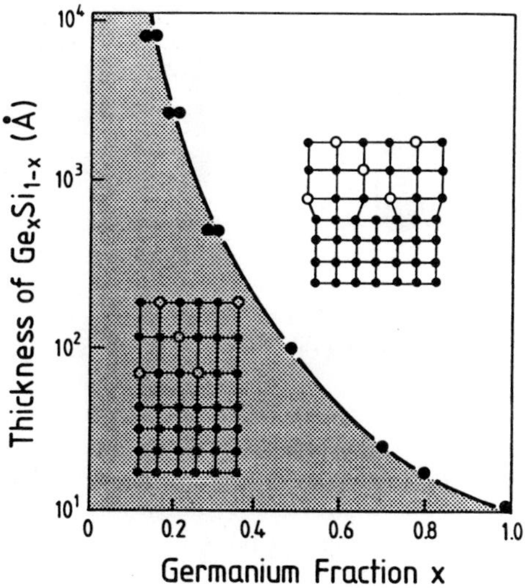

Fig. 3.3. Critical thickness of Ge_xSi_{1-x} on Si(100). The *shaded area* represents the region of defect-free strained-layer epitaxy of Ge_xSi_{1-x} on Si. The calculated and experimentally determined limits of defect-free growth are denoted by the *solid line* and the *dots*, respectively. Modified from [1]

the interfacial structure is characterized by the misfit. If the misfit is too large ($f > 15\%$) or the crystallographic orientations of the film and substrate differ considerably, other concepts need to be used to describe the interface. Coincidence models have been developed for these cases. They are based on the idea that the larger the number of atomic positions that coincide at the interface of the two crystals, the stronger the resulting bonding. As a starting point, the rigid lattices are considered, and relaxations are introduced only in a second step [9]. Meanwhile, computer modeling allows a more precise determination of the interface structure.

Three principal structures of heteroepitaxial films have been introduced above. Here, the focus will be on one of these cases, namely that of the coherently strained films. The question addressed is the response of the film to the strain arising from the misfit at the interface. If the misfit is not too large ($f < 15\%$), the deposit assumes the lattice constant of the substrate parallel to the contact plane. This usually leads to a strain which is equal to the misfit, at least below the critical thickness d_c. This strain is called epitaxial or coherency strain, and the growth type is sometimes called *pseudomorphism*.

When the deposit is strained to match the lattice constant of the substrate, it is subject to a *stress*. This stress is imposed by the substrate and acts in the contact plane, and thus it is a biaxial stress. The related expansion or contraction within the plane ($\varepsilon_{\|}$) will cause a contraction or expansion, respectively, of the spacing perpendicular to the contact plane ($\varepsilon_{\perp}$) to keep the atomic volume approximately constant. This is depicted schematically in

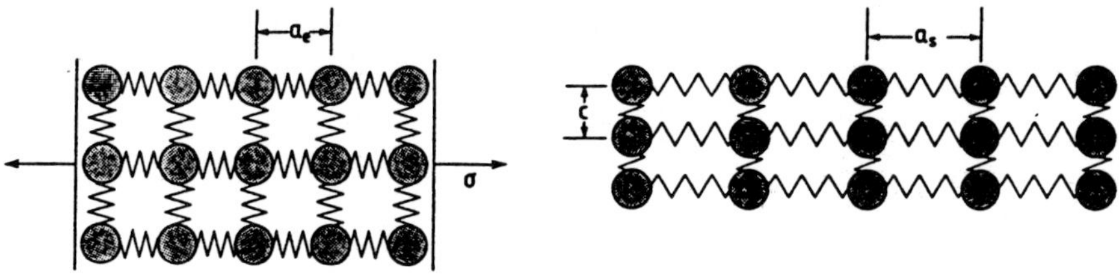

Fig. 3.4. Equilibrium and strained structure of thin films. The strained structure is realized here by epitaxial growth on a substrate with a larger lattice constant. a_e and a_s denote the lattice parameters of the film and the substrate in the unstrained equilibrium state

Fig. 3.4. For a film that is several layers thick, this expansion can be calculated from the macroscopic properties of the film material. For an fcc(100)||fcc(100) heterostructure, the following relation holds:

$$\frac{\varepsilon_\perp}{\varepsilon_\parallel} = -\frac{2\nu}{1-\nu}, \tag{3.8}$$

where $\varepsilon_\perp$ is the perpendicular strain, $\varepsilon_\parallel$ is the epitaxial strain in the contact plane and ν is Poisson's ratio [6]. By definition,

$$\frac{\varepsilon_\perp}{\varepsilon_\parallel} = \frac{\delta c/c}{\delta a/a} = \frac{(c - c_e)/c_e}{(a - a_e/a_e}. \tag{3.9}$$

The subscript e denotes the (unstrained) equilibrium distances, a is the lattice constant of the film and c is the lattice constant perpendicular to the film plane; δ means a change in the respective quantity. Since coherent growth is considered here, $a \equiv a_s$, i.e., the film adopts the in-plane lattice constant of the substrate. If the lattice constant perpendicular to the film is now measured, the equilibrium constant of the unstrained film can be determined by combining (3.8) and (3.9):

$$a_e = \frac{2\nu}{1+\nu}a + \frac{1-\nu}{1+\nu}c. \tag{3.10}$$

The importance of (3.10) stems from the fact that it allows the determination of the equilibrium lattice constant of a metastable phase. These metastable structures are phases that usually otherwise exist only as high-temperature or high-pressure modifications of the bulk material. An example is the growth of fcc iron on Cu(100). For bulk iron, this modification is stable only above 1185 K. However, even metastable phases which have no bulk analogue can be stabilized. This has been demonstrated by the growth of Co on GaAs, where a ferromagnetic bcc structure was obtained, even though Co has no naturally occurring bcc phase [10]. Another interesting example is the growth of α-Sn (gray tin) on the (100) surfaces of InSb and CdTe [11].

This tin modification is stable in bulk material only below 286 K, while the metallic β-phase (white tin) is stable above this temperature. Owing to the small misfit of α-Sn at the epitaxial interface, it can be stabilized on InSb and CdTe above 286 K. Since this tin modification is a semiconductor, novel heterostructures can be produced.

For the growth of such metastable phases, three essential requirements have been identified. These are a small lattice mismatch between the metastable phase and the substrate, a high supersaturation, and a large misfit between the metastable and the normal phase. This misfit introduces a strain energy barrier which kinetically hinders the transformation from the metastable to the stable phase [11]. The above examples illustrate that metastable structures expand the number and variety of materials, thus making new materials from "old" elements [12].

3.2 Thin-Film Characterization by Low-Energy Electron Diffraction

In the preceding section, structural aspects of heteroepitaxial systems, including films with epitaxial and perpendicular strain and misfit dislocations, have been presented. This discussion has shown the importance of an experimental characterization of the film structure, thus setting the scene for this section, where the possible ways to determine the structure of thin films by low-energy electron diffraction will be discussed. Figure 3.5 depicts schematically the potential of this technique.

In a low-energy electron diffraction experiment, a monoenergetic beam of electrons impinges on a single-crystal surface. The diffraction pattern, which is an image of the reciprocal unit cell, is usually displayed on a fluorescent screen. From the spot positions, the reciprocal unit cell can be directly determined, and hence the size and shape of the unit cell in real space. The positions of the diffraction spots, however, contain no information about the atomic positions within the unit cell, i.e., the surface structure. This information can only be obtained from the energy dependence of the intensities of the diffraction spots, the so-called I/V or $I(E)$ curves. A trial-and-error comparison of the calculated and measured I/V curves allows the identification of the "correct" structure. Traditionally, this approach has been applied only to well-ordered surfaces. In the last few years, it has been demonstrated that with the same theoretical approach, the local structure of disordered surfaces can also be addressed [14, 15]. While atomic positions can be derived from the energy dependence of the spot intensities, long-range correlations can be detected from the shape of the diffraction spots. Such spot profiles can be employed to investigate the structural perfection of the surface providing valuable information about steps, dislocations, faceting, mosaic width and adsorbate–adsorbate correlations [16].

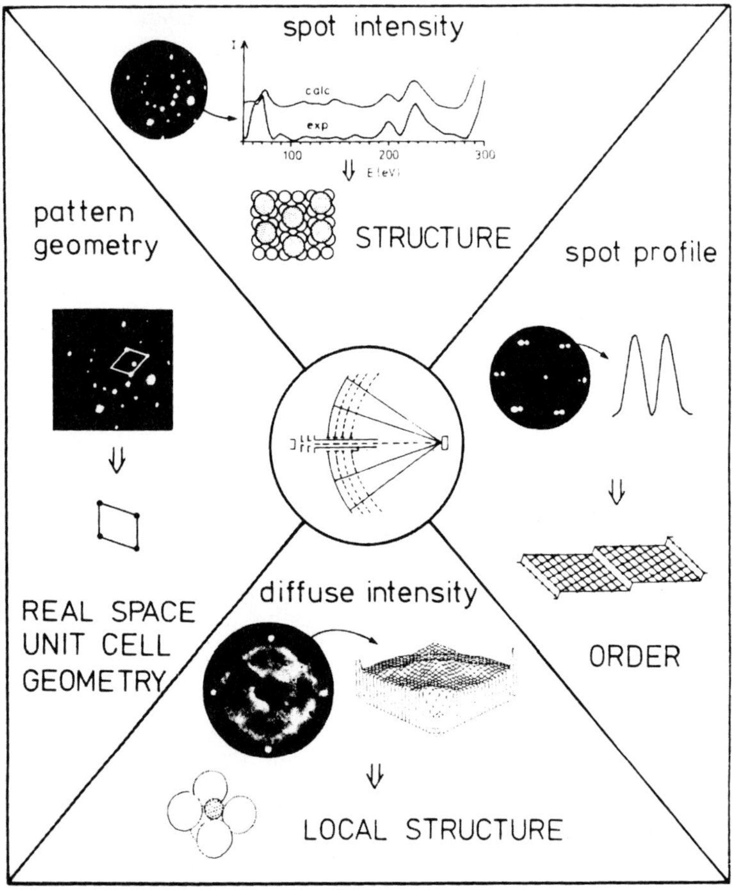

Fig. 3.5. Potential of LEED. The *center* displays the experimental set up. From the geometry of the diffraction pattern, the shape and size of the real-space unit cell are obtained (*left*). From the energy dependence of the spot intensities, atomic positions within the unit cell can be determined (*top*). Spot profiles reflect the surface perfection (*right*) and diffuse intensities are governed by the local structure (*bottom*). From [13]

To obtain deeper insight, a quantitative description of electron scattering is necessary. For an ensemble of atoms, the scattered intensity can be written as

$$I = |\Psi|^2 = \left| \sum_i \Psi_i \right|^2 , \tag{3.11}$$

where Ψ_i is the scattered amplitude from atom i and, in the equation below, Ψ_0 is the scattered amplitude from atom 0:

$$\Psi_i = \Psi_0 e^{i(\boldsymbol{k}_i - \boldsymbol{k}_f)\cdot \boldsymbol{R}_i} = \Psi_0 e^{i\triangle \boldsymbol{k}\cdot \boldsymbol{R}_i} . \tag{3.12}$$

Here $\boldsymbol{k}_i$ and $\boldsymbol{k}_f$ are the incident and outgoing wave vectors, respectively. $\triangle \boldsymbol{k} = \boldsymbol{k}_i - \boldsymbol{k}_f$ is the scattering vector and $\boldsymbol{R}_i$ is the vector from atom 0 to atom i. If the atoms occupy lattice sites of a three-dimensional crystal, then (3.11) can be rearranged to read:

$$I = \left| \sum_i \Psi_i \right|^2 = \left| \sum_{j=1}^{J} {}_{Basis}\Psi_{0j} \right|^2 \times \left| \sum_{i=1}^{N} {}_{Lattice} e^{i(\triangle \boldsymbol{k} \cdot \boldsymbol{R}_i)} \right|^2 = |F|^2 \times |G|^2 \, . \quad (3.13)$$

The first summation is over all atoms of the basis, while the second summation is over all unit cells of the lattice. The term $|F|^2$ is called the structure factor and $|G|^2$ is called the lattice factor. The lattice factor determines the positions of the LEED spots and their intensity profile. The structure factor term contains all multiple-scattering effects and the composition and arrangement of atoms within the unit cell.

So far, the description has been independent of the detailed nature of the interaction between the incoming particles and the atoms. For the scattering of low-energy electrons in the energy range considered here, the penetration depth in the solid is of the order of a few angstroms [17]. Therefore, as a first approximation, the summation in the lattice factor contains only two indices for the x and y directions which are assumed to be in the surface plane. Owing to the periodic arrangement of scatterers, the vector $\boldsymbol{R}_i$ can be written as a linear combination of unit vectors $\boldsymbol{a}_1$ and $\boldsymbol{a}_2$ that span the primitive two-dimensional unit cell of the surface, $\boldsymbol{R}_i = n_i\boldsymbol{a}_1 + m_i\boldsymbol{a}_2$. The lattice factor can then be written as

$$|G|^2 = \left| \sum_{n_i=0}^{\infty} \sum_{m_i=0}^{\infty} e^{i[\triangle \boldsymbol{k} \cdot (n_i\boldsymbol{a}_1 + m_i\boldsymbol{a}_2)]} \right|^2 . \quad (3.14)$$

G is significant only for $\triangle \boldsymbol{k} \cdot \boldsymbol{a}_1 = 2\pi$ and $\triangle \boldsymbol{k} \cdot \boldsymbol{a}_2 = 2\pi$. $\triangle \boldsymbol{k}$ can be written as a linear combination of the vectors $\boldsymbol{a}_1^*$ and $\boldsymbol{a}_2^*$ such that

$$\boldsymbol{a}_i^* \times \boldsymbol{a}_j = \delta_{ij}, \quad i, j \in \{1, 2\} \, , \quad (3.15)$$

where δ_{ij} is the Kronecker symbol. This is the equation that defines the reciprocal lattice in two dimensions. Thus electron diffraction from a single-crystal surface displays the two-dimensional reciprocal lattice of the surface on the LEED screen. This is depicted in Fig. 3.6. In contrast to the three-dimensional diffraction of neutrons and x-rays from single crystals, no condition holds for the third dimension. Thus the observed LEED spots lie on “Laue rods” (see Fig. 3.6). In reality, however, the electron beam “sees” the periodic arrangement within the first few layers, which gives rise to intensity modulations along the rods.

For an analysis of beam profiles, usually only the lattice factor is calculated. Multiple-scattering effects which are contained in the structure factor

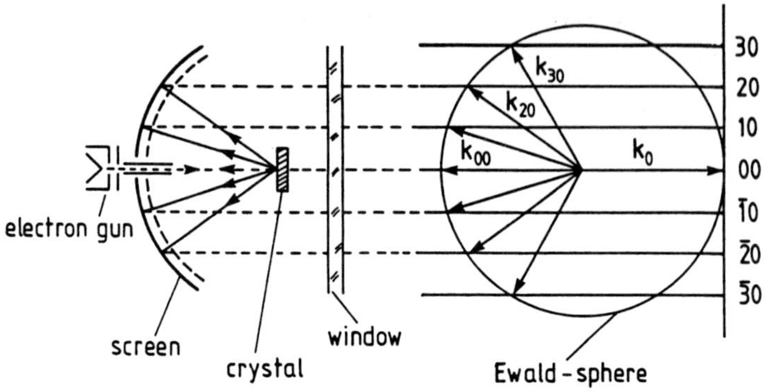

Fig. 3.6. Formation of a LEED pattern. The relationship between the LEED pattern displayed on the phosphor screen and reciprocal space is depicted. After [18]

are commonly neglected. The justification for this procedure comes from the fact that the structure factor varies much more slowly with the scattering vector than does the lattice factor. Since the lattice factor $|G|^2$ does not contain multiple-scattering effects, a kinematic analysis is sufficient.

Figure 3.7 shows some possible defect structures. By calculating the lattice factor for different defects, one can determine the corresponding modifications of the reciprocal-lattice rods and the related spot profile. As an example, we shall consider a stepped surface with a given step height d. A vector $\boldsymbol{g}$ can be defined which connects adjacent atoms across a step. This $\boldsymbol{g}$-vector will have the form $\boldsymbol{g} = x\boldsymbol{a}_1 + y\boldsymbol{a}_2 + d\boldsymbol{e}_3$ where $\boldsymbol{e}_3$ is the unit vector in the direction of the surface normal. As a result, there is a phase shift between two adjacent terraces of $\varphi = \Delta\boldsymbol{k} \cdot \boldsymbol{g}$. If $\varphi = 2\pi n$, the adjacent terraces interfere constructively, while destructive interference is observed when $\varphi = 2\pi(n + 1/2)$. For a given (h, k) spot, $\Delta\boldsymbol{k}$ depends only upon the incident energy. Thus this beam will oscillate between an out-of-phase and an in-phase condition with energy. For the $(0, 0)$ beam, an especially simple formula can be derived for the energies where constructive and destructive interference is observed [19]:

$$E_{00} = \frac{150.4(\mathrm{eV\AA}^2)}{4d^2\cos^2\theta} s^2 \,. \tag{3.16}$$

The energy at which constructive interference occurs is obtained by using integer values of s in the above equation, while half-integer values of s give the energies for destructive interference. For these destructive-interference conditions, the beam splits if a regular step sequence exists (Fig. 3.7a, b) or shows a broadening for a random step sequence (Fig. 3.7c, d). Thus both the step distribution and the step height, which is determined from the energies where the beams split or broaden, can be analyzed. Another example is the observation of different superstructure domains. In this case, the $\boldsymbol{g}$-vector be-

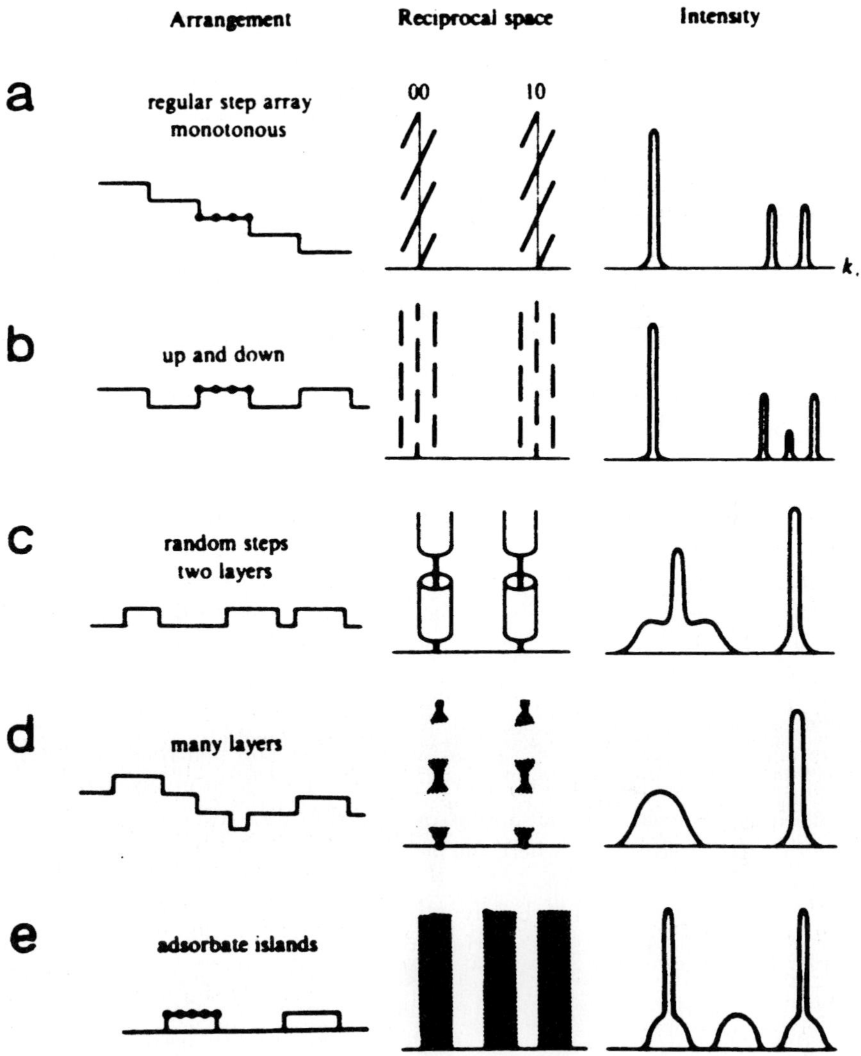

Fig. 3.7. Possible surface defect structures. Common defect structures, including regular and irregular step arrays and adsorbate islands, are shown on the *left-hand side*. The effect of these defects on the reciprocal-lattice rods is depicted in the *middle*, while the observed spot profile characteristic of the particular defect is shown on the *far right*. After [20]

tween adjacent domains lies in the $x-y$ plane. The absence of a z component in the $\boldsymbol{g}$-vector then leads to an *energy-independent* spot profile.

So far, the influence of the limited coherence length of the LEED experiment has been neglected. This leads to a finite transfer width, which is of the order of 50–200 Å in commercial LEED systems. In recent years, LEED systems have been designed that offer a considerably improved transfer width of 3000 Å [21]. By using grazing incidence, this value can be improved further. Since the observation of atomic correlations on the surface over distances larger than the transfer width is suppressed in a diffraction experiment, these improved systems offer a considerable advantage if very long-range correlations are to be investigated.

In Fig. 3.5, it was indicated that a precise determination of atomic positions is possible once the energy dependence of the LEED beams has been measured. Since a diffraction experiment measures scattered intensities rather than the scattered amplitudes, it is not possible to obtain the atomic positions from a Fourier transform of the measured I/V curves. This holds for every diffraction experiment, including x-ray diffraction and neutron scattering. Despite these difficulties, both of the latter techniques have been in part responsible for solving the structures of complex molecules, including DNA [22]. What facilitates structure determination in the latter two cases is the weak interaction between the incoming wave field and the atoms, which justifies the use of the kinematical approximation. Applying this approximation to the problem of surface structure determination by LEED leads to a third Bragg condition for the direction perpendicular to the surface. As a consequence, pronounced intensity maxima are expected for the energies which fulfill this condition. This is due to the neglect of both the inner potential and multiple scattering. Including the contribution from the former effect is straightforward. The difficulties arise from the contribution from the latter effect, because unlike the situation for x-rays and neutrons, the interaction between the incoming electron wave field and the atoms is strong. As a consequence, a tedious trial-and-error procedure has to be used to determine the surface structure, where the measured intensities are compared with calculated intensities.

In the following, it will be briefly outlined how these intensities are calculated. Figure 3.8 depicts the scattering potential in the vicinity of the surface. Additionally, some quantities that are often encountered in a description of the theory of full dynamical LEED calculations are shown. As in many band structure calculations, a spherical muffin-tin potential is used. The problem of calculating LEED intensities can be decomposed into three steps (see Fig. 3.9). First, the scattering of an incoming electron is calculated. Owing to the assumption of spherical symmetry, only the radial Schrödinger equation has to be solved.[1] The scattering strength is then described by l-dependent phase shifts, where l is the angular-momentum component. Usually a small number of l values is sufficient to describe the scattering of an electron by the

[1] The only difficult part in this problem is the contribution of exchange and correlation to the Coulomb interaction. This is usually introduced via a local $X\alpha$ potential. It is an advantage of low energy positron diffraction (LEPD) that this contribution is missing [23]. Otherwise, positron and electron diffraction can be treated on the same footing.

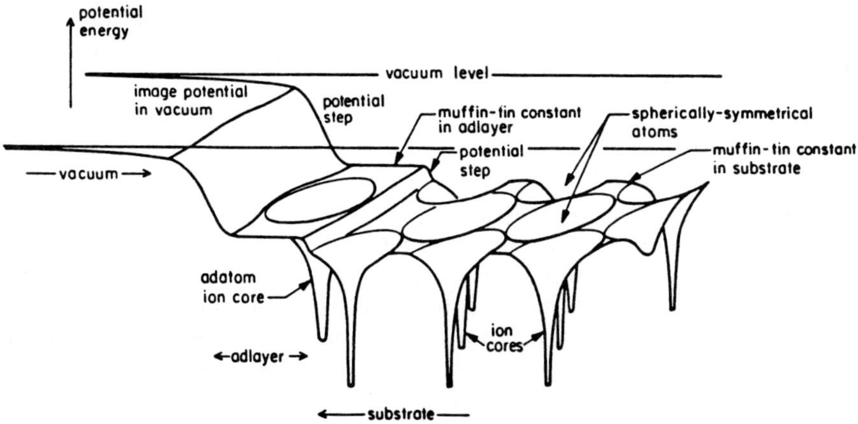

Fig. 3.8. Scattering potential for low-energy electrons: schematic illustration of the muffin-tin potential at a surface with an adlayer. After [23]

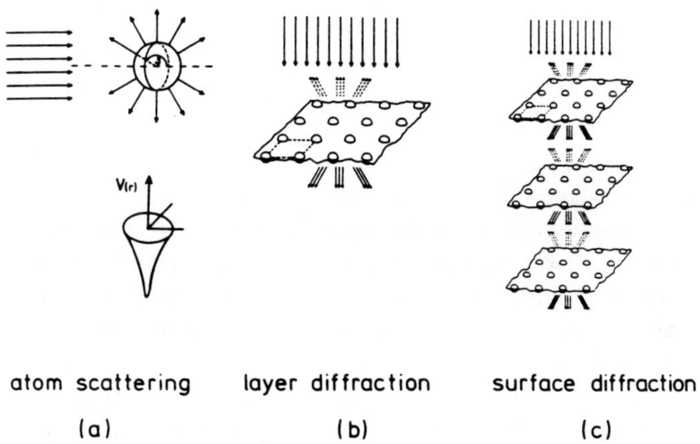

Fig. 3.9. Calculation of LEED intensities: the LEED intensities are calculated in three steps. First, the atomic scattering, represented by atomic phase shifts, is computed (**a**). Then, the layer diffraction matrices are calculated (**b**). Finally, the total surface diffraction is produced by a stacking of layers (**c**). From [13]

atomic potential. A rough estimation of the necessary number of phase shifts can be obtained from the condition $l_{max} \approx \pi d/\lambda$, where d is the muffin-tin diameter (atomic diameter) and λ is the wavelength of the electron. With $\lambda \approx 1$ Å ($E = 150$ eV) and $d = 3$ Å, $l_{max} \approx 9$. This value can be reduced in practice, since the outer part of the atom scatters relatively weakly.

In the next step, the diffraction amplitude of a single layer is calculated. This layer has the translation symmetry of the corresponding layer unit cell, so that diffracted beams can appear only in those directions determined by the reciprocal-lattice vectors. The number of possible wave vectors is further limited by the condition $k_{\|} < 2mE/h^2$, which is a consequence of energy conservation. Nevertheless, the number of beams that has to be included in the calculations can become very large if the real-space unit cell is large or high energies are included in the theoretical analysis. Diffraction by this layer can be described by four matrices $M^{\pm\pm}$, whose elements $M^{\pm\pm}(\boldsymbol{k}_{g'}, \boldsymbol{k}_g)$ give the amplitude for the diffraction process $\boldsymbol{k}_{g'} \to \boldsymbol{k}_{g}$. The pair of signs $++$ and $--$ describe forward scattering from inside and outside the crystal, respectively, while $+-$ and $-+$ denote backward scattering, again from inside and outside the crystal, respectively [13].

Finally, the diffraction matrices for several layers are combined to give the diffraction from the whole stack that has to be considered owing to the finite penetration depth of the low-energy electrons. Two different techniques are mainly used to form the stack of layers. These are layer doubling and renormalized forward scattering. The task is then to calculate the diffracted intensities for a model structure and to vary the structure until the calculated intensity curves closely reproduce the experimental data. To obtain good agreement, it is also necessary to vary at least three nonstructural parameters. One is the inner potential V_0, which causes a refraction when the electron wave enters or leaves the crystal. Furthermore, the limited mean free path of electrons leads to an attenuation of the LEED beams. This effect is included in the imaginary part of the inner potential. Finally, thermal vibrations are included by introducing temperature-dependent phase shifts which simulate the Debye–Waller factor.

To judge the agreement between theory and experiment, various reliability factors (R-factors) have been introduced [23]. These R-factors either compare intensities, such as R_1 and R_2 or they compare how well the positions of the maxima and minima in the I/V curves are reproduced. An example of the latter case is the Pendry R-factor, where the intensities of different maxima are even normalized, because of the understanding that the structural information in each maximum is independent of intensity [24]. For this R-factor, a statistical analysis has been performed which shows that structural models which differ by more than var R from the minimum R-factor can be excluded [23]. For the Pendry R-factor, var R is defined as $R_{min}\sqrt{8V_{0\mathrm{i}}/\triangle E}$, where $V_{0\mathrm{i}}$ denotes the imaginary part of the inner potential and $\triangle E$ describes the energy overlap between theory and experiment for all beams. Thus a large energy overlap should give a high accuracy of structure determination. Usually a structure determination is considered solved when the R-factor is fairly low, for example well below 0.4 for the Pendry R-factor. The visual comparison between theory and experiment should be good, and the R-factor should show a deep minimum upon parameter variation around the best-fit structure.

Initially, low-energy electron diffraction was mainly used to determine the structure of "simple" surfaces such as the clean, unreconstructed low-Miller-index surfaces of metals. This was due to the fact that with an increasingly large number of inequivalent atoms within the unit cell, the calculations become increasingly more time-consuming and it becomes almost impossible to scan the relevant parameter space. Recently, new approaches such as Tensor LEED have been developed [25], which largely overcome this difficulty by employing more efficient approximation schemes. This allows the investigation of considerably more complex surfaces.

A second important development has further enlarged the scope of low-energy electron diffraction. It has been realized recently [14] that the diffuse intensity for systems with partial disorder also contains structural information. More specifically, the measured diffuse intensity allows the determination of the adsorption site for adsorbates which are statistically distributed over identical lattice sites on a single-crystal surface. Since the measurement of these intensities can be performed similarly to the measurement of "ordinary" I/V curves for well-ordered surfaces [26], a direct comparison of the influence of order on the local adsorption geometry and the coverage dependence of adsorption sites is feasible.

These development should guarantee that LEED will remain one of the most versatile and wide-spread techniques for surface structure characterization for a very long time.

References

1. M. Ohring: *The Material Science of Thin Films*, (Academic Press, New York, 1992)
2. P. W. Palmberg and T. N. Rhodin: J. of Chem. Phys. **49**, 147(1968)
3. S. M. Sze: *Semiconductor Devices–Physics and Technology*, (McGraw–Hill, 1985)
4. J. W. Matthews: in *Epitaxial Growth*, ed. J. W. Matthews, (Academic Press, New york, 1985)
5. J. W. Matthews: J. Vac. Sci. Technol. **12**, 126(1975)
6. F. Jona and P. M. Marcus: In: *Surface Physics and Related Topics*, ed by Fu-Jia Yang et al., (World Scientific, Singapore, 1991) p213
7. J. C. Bean, In: *Silicon Molecular Beam Epitaxy*, ed by E. Kasper and J. C. Bean, (CRC press, Boca Raton, FL, 1988)
8. J. C. Bean: Physics Today **39**, No. 10, 2(1986)
9. W. Bollmann: *Crystal Defects and Crystalline Interfaces*, (Springer, Berlin 1970)
10. G. A. Prinz: Phys. Rev. Lett. **54**,1051(1985)
11. R. F. C. Farrow: J. Vac. Sci. Technol. B **1**, 222(1983)
12. L. M. Falicov et al.: J. Mater. Res. **5**, 1299(1990)
13. K. Heinz: Progr. in Surf. Sci. **27**, 239(1988)
14. P. J. Rous, J. B. Pendry, D. K. Saldin, K. Heinz, K. Müler and N. Bickel: Phys. Rev. Lett. **57**, 2951(1986)
15. M. A. Van Hove: In *Chemistry and Physics of Solid Surfaces VII*, eds. R. Vanselow and R. F. Howe (Berlin 1988) p513
16. M. Henzler: Applications of Surface Science **11-12**, 450(1982)
17. M. P. Seah and W. A. Dench: Surf. Interface Anal. **1**, 2(1979)

18. M. Henzler, W. Göpel: *Oberflächenphysik des Festkörpers*, (Teubner Studienbücher, Stuttgart, 1991)
19. M. Henzler: Surf. Sci. **22**, 12(1970)
20. A. Zangwill: *Physics at Surfaces*, (Cambridge University Press, Cambridge 1988)
21. U. Scheithauer, C. Meyer and M. Henzler: Surf. Sci. **178**, 441(1987)
22. J. D. Watson: *Molecular Biology of the Gene*, 3d ed. (Benjamin-Cummings, Menlo Park, CA 1976)
23. M. A. Van Hove, W. H. Weinberg and C. M. Chan: textitLow-Energy Electron Diffraction (Springer Series in Surface Science 6, Berlin 1986)
24. J. B. Pendry: J. Phys. C **13**, 937(1980)
25. P. J. Rous: Progr. in Surf Sci. **39**, 3(1992)
26. M. A. Van Hove: In *Chemistry and Physics of Solid Surfaces VII*, eds. R. Vanselow and R. F. Howe (Springer Berlin 1988)

4 Magnetism of Ultrathin Metal Films

The last two chapters have described the growth and structure of epitaxial films. Here, the basic aspects of ultrathin-film magnetism and their relationship to structure and growth will be the focus. First, theoretical models to describe the ferromagnetism and antiferromagnetism of 3d metals are introduced in Sect. 4.1. In addition, the relationship between structure and magnetism for bulk phases is discussed. In Sect. 4.2, magnetic properties of surfaces and thin films receive attention. In particular, the correlation between structure and magnetism will be of key interest. The magnetic anisotropy, which is important for technological applications, is also described in Sect. 4.2. Finally, our attention is turned to the most commonly used experimental techniques for the determination of magnetic properties of ultrathin films. These techniques cover the surface magnetooptic Kerr effect, ferromagnetic resonance measurements, and x-ray magnetic dichroism in both the edge absorption mode and the photoemission mode.

4.1 Itinerant-Electron Magnetism

The characteristic feature of a ferromagnet is the existence of a spontaneous magnetization below the Curie temperature T_c. The energy involved in the transition from the ferromagnetic to the paramagnetic state is of the order of $k_B T_c$ /atom$\simeq$ 0.1 eV/atom. This shows that ferromagnetism is not the consequence of a magnetic dipole interaction, which would give only an orientation-dependent energy difference of the order of 10^{-4} eV/atom. The mechanism responsible is the exchange interaction, which is a consequence of the Pauli principle and the Coulomb interaction between electrons. The Pauli principle forbids two electrons with parallel spins to occupy the same orbital state. Thus the effective Coulomb repulsion between a pair of electrons with parallel spins is different from that between a pair of electrons with antiparallel spins. In some cases the former is weaker than the latter. This favors a parallel alignment of spins and is the origin of ferromagnetism.

Depending upon the degree of localization of the electrons which carry the magnetic moments, two different models are applied to explain the ferromagnetism of solids. The Heisenberg model is well suited to describing the magnetism of mostly localized electrons found in materials such as 3d metal oxides, 4f metals and their compounds. Upon increasing delocalization of the electrons which carry the magnetic moments, the Heisenberg model becomes less applicable. Thus, the itinerant (delocalized) magnetism of the 3d metals (Fe, Co, Ni, Cr and Mn) is explained better by the band model, which was introduced by Stoner [1]. We shall outline the Stoner model as a theoretical basis in this section. Since the Stoner model is based on the electronic band structure, we shall first introduce a density functional theory, the local-density approximation (LDA), which is extensively used to calculate the electronic and magnetic properties of ultrathin metal films.

4.1.1 Local-Spin-Density Approximation

Solving the Schrödinger equation for an interacting many-particle system is impractical in most cases, so that approximate treatments need to be adopted to reduce the problem to the solution of single-particle equations. The LDA provides such a formalism [2]. In 1964, Hohenberg and Kohn showed that the ground-state properties, particularly the total energy $E[n_0(\boldsymbol{r})]$, of a system of interacting electrons are unique functionals of the electron density of the ground state $n_0(\boldsymbol{r})$. This forms the theoretical basis of the LDA. Since $n_0(\boldsymbol{r})$ is a single-particle density, the problem of determining the ground-state energy of an N-particle system is reduced to the problem of a single particle in an effective potential. The total energy can be decomposed into several different contributions [3]:

$$\begin{aligned} E[n(\boldsymbol{r})] &= T_0[n(\boldsymbol{r})] + \int \mathrm{d}\boldsymbol{r}\, V(\boldsymbol{r}) n(\boldsymbol{r}) \\ &+ \frac{1}{2} \int \mathrm{d}\boldsymbol{r}\, \mathrm{d}\boldsymbol{r}' \frac{1}{4\pi\varepsilon_0} \frac{n(r)n(r')e^2}{|\boldsymbol{r}-\boldsymbol{r}'|} + E_{xc}[n(\boldsymbol{r})] \,, \end{aligned} \tag{4.1}$$

where $V(\boldsymbol{r})$ describes the interaction of the electrons with the ion cores. $T_0[n(\boldsymbol{r})]$ is the functional for the kinetic energy of noninteracting electrons, and the double integral describes the electron–electron interaction within the Hartree approximation, which neglects the contribution of exchange and correlation. These two terms describe contributions to the Coulomb interaction beyond the Hartree approximation which arise as a consequence of the Pauli principle. This can cause electrons of the same spin to stay spatially apart, while electrons of opposite spin come more closely together. Therefore the *average* repulsion energy between two electrons of the same spin orientation is reduced by the intrinsic tendency of these electrons to avoid each other. The corresponding reduction of energy is called the *exchange energy*. This average quantity still neglects electron correlations, which are included in the *correlation energy*. The contribution of exchange and correlation to the electron–electron interaction and the contribution of correlation to the kinetic energy are contained in $E_{xc}[n(\boldsymbol{r})]$. Once (4.1) has been established, the ground-state electron density and energy of an interacting many-electron system can be determined by a self-consistent procedure using the following formulae:

$$\left(-\frac{\hbar^2}{2m}\nabla^2 + v_{\text{eff}}(\boldsymbol{r})\right)\psi_j(\boldsymbol{r}) = \epsilon_j\psi_j(\boldsymbol{r}) \,, \tag{4.2}$$

$$n(\boldsymbol{r}) = \sum_{j=1}^{N} |\psi_j(\boldsymbol{r})|^2 \,, \tag{4.3}$$

$$v_{\text{eff}} = V(\boldsymbol{r}) + \frac{e}{4\pi\varepsilon_0}\int \frac{n(\boldsymbol{r}')}{|\boldsymbol{r}-\boldsymbol{r}'|}\text{d}\boldsymbol{r}' + v_{xc}(r) \,, \tag{4.4}$$

where $v_{xc}(r)$ is the local exchange–correlation potential, defined as

$$v_{xc}(\boldsymbol{r}) = \frac{\delta E_{xc}[n(\boldsymbol{r})]}{\delta n(\boldsymbol{r})} \,. \tag{4.5}$$

The expression for v_{eff} is obtained from (4.1) by applying a variational principle. The essential difficulty in solving the above equations arises from the fact that E_{xc} is unknown. Thus the main challenge of reducing the N-particle problem to a one-particle problem is a suitable approximation for the exchange–correlation functional. Usually the local-density approximation is chosen. In this approximation,

$$E_{xc} = \int \text{d}\boldsymbol{r}\, n(\boldsymbol{r})\epsilon_{xc}(n(\boldsymbol{r})) \,, \tag{4.6}$$

where $\epsilon_{xc}(n(\boldsymbol{r}))$ is the exchange–correlation energy per electron of a homogeneous electron gas with a local density $n(\boldsymbol{r})$. Thus the effective potential of (4.4) becomes local. The simplest approximation for $\epsilon_{xc}(n(\boldsymbol{r}))$ is the $X\alpha$ approximation, where $\epsilon_{xc}(n(\boldsymbol{r}))$ is obtained by perturbation theory for a free-electron gas with density n:

$$\epsilon_{xc}(n) = -\alpha\frac{3}{4}\left(\frac{3}{\pi}\right)^{1/3} e^2 n^{1/3} \,, \tag{4.7}$$

where values between 2/3 and 1 are assumed for α.

This approximation includes the exchange energy but neglects the correlation energy. Therefore, more reliable forms of the exchange–correlation energy are necessary for a successful description of the properties of solids [4]. With such improved expressions for ϵ_{xc}, the crystal structure, lattice constants, cohesive energy and bulk moduli can be fairly well described. Lattice constants calculated by the LDA are underestimated by 3–4% for most metals. This is due to the fact that the exchange–correlation contribution to the electron–electron interaction favors overbinding.

To apply the LDA to magnetic materials, one has to consider spin densities instead of electron densities [5]. The total electron density can be decomposed into two parts, n^+ and n^-, the electron densities for up and down spins, respectively:

$$n(\boldsymbol{r}) = n^+(\boldsymbol{r}) + n^-(\boldsymbol{r}) \ . \tag{4.8}$$

If $m(\boldsymbol{r})$ is defined as

$$m(\boldsymbol{r}) = n^+(\boldsymbol{r}) - n^-(\boldsymbol{r}) \ . \tag{4.9}$$

then the local magnetization density is $\mu_B m(\boldsymbol{r})$. Since the magnetic anisotropy is irrelevant for the following discussion, the representation of the magnetization by a scalar quantity is justified. One can reformulate (4.2)–(4.4) for the spin densities n^+ and n^-. The important new feature of a description using spin densities rather than charge densities is a spin-dependent contribution from the exchange–correlation potential for spin-up and spin-down electrons. When only the effects of first order are considered, $v_{xc}(\boldsymbol{r})$ can be expanded with respect to $m(\boldsymbol{r})$ as [6]

$$v_{xc}^{\pm}(\boldsymbol{r}) = v_{xc}^{0}(\boldsymbol{r}) \mp \widetilde{v}[n(\boldsymbol{r})]m(\boldsymbol{r}) \ , \quad \widetilde{v}[n(\boldsymbol{r})] > 0 \ , \tag{4.10}$$

where v_{xc}^0 is the exchange–correlation contribution for the nonmagnetic case. As a consequence, spin-up electrons experience a more attractive potential than do spin-down electrons. Once it is feasible to calculate $v_{xc}^{\pm}(\boldsymbol{r})$ from a given spin-density distribution, the spin-specific band structure can be obtained self-consistently and a number of magnetic properties can be subsequently derived.

4.1.2 Stoner Model

In the Stoner model [1, 8], the potential shift in (4.10) is represented by a *constant*:

$$v_{xc}^{\pm}(\boldsymbol{r}) = v_{xc}^{0}(\boldsymbol{r}) \mp \frac{1}{2}IM \ , \quad M = \int_{V_{atom}} m(\boldsymbol{r})\,\mathrm{d}\boldsymbol{r} \ , \tag{4.11}$$

where $\mu_B M$ is the local atomic moment and I is the Stoner parameter, assumed to be independent of the wave vector $\boldsymbol{k}$. I can be considered essentially as the exchange energy. As a consequence of the constant potential difference, (4.2) can be expressed as

$$\left(-\frac{\hbar^2}{2m}\nabla^2 + v_{\mathrm{eff}}^{0}(\boldsymbol{r})\right)\psi_j(\boldsymbol{r}) = \left(\epsilon_j \pm \frac{1}{2}IM\right)\psi_j(\boldsymbol{r}) \ , \tag{4.12}$$

where the eigenvectors that solve (4.2) in the nonmagnetic case $\psi_{\boldsymbol{k}\nu}^{0}(\boldsymbol{r})$ also solve the magnetic case:

$$\psi_{\boldsymbol{k}\nu}^{\pm}(\boldsymbol{r}) = \psi_{\boldsymbol{k}\nu}^{0}(\boldsymbol{r}) \ , \tag{4.13}$$

where $\psi_{\boldsymbol{k}\nu}^{\pm}(\boldsymbol{r})$ is the solution for the magnetic case, and the subscripts $\boldsymbol{k}$ and ν denote the wave vector and the band index, respectively. The eigenvalues $\epsilon_{\boldsymbol{k}\nu}^{\pm}$, however, show a symmetric energy splitting:

$$\epsilon_{\boldsymbol{k}\nu}^{\pm} = \epsilon_{\boldsymbol{k}\nu}^{0} \mp \frac{1}{2}IM \ . \tag{4.14}$$

This results in *a spin-split band structure, where the densities of states for up and down electrons are only shifted by IM, and the functional form remains unchanged.*

If, starting from the nonmagnetic band structure (no spin splitting), a transfer of electrons from the spin-down band to the spin-up band is energetically favorable, spin splitting will occur, resulting in a nonzero $m(\boldsymbol{r})$. A sufficient condition for a metal to show ferromagnetism at $T = 0$ K is [8]

$$IN^0(\epsilon_F) > 1 \,. \tag{4.15}$$

This Stoner criterion states that metals for which the product of the Stoner parameter I and the nonmagnetic density of states at the Fermi level $N^0(\epsilon_F)$ is larger than one should show ferromagnetism.

4.1.3 Discussion

The Stoner model provides a simple way to calculate magnetic properties. The parameter I can be derived from the LDA formula with the correlation energy taken into account [7]. Hence, an explicit expression is obtained for I, which depends upon the electron energy ϵ. Thus $I(\epsilon)$, as well as $N^0(\epsilon_F)$, can be calculated by the LDA. Table 4.1 lists calculated results for the Stoner parameter, the nonmagnetic density of states and their product for a number of metals. This table shows that the Stoner criterion is fulfilled only for Fe, Co and Ni, precisely those metals that show itinerant ferromagnetism.

As mentioned before, the LDA is also able to reproduce the magnetic moments of these "band magnets" quite well. Figure 4.1 shows the densities of states of spin-up and spin-down electrons for Fe and Ni. It can be seen that the densities of states of spin-up and spin-down electrons are exchange-split but otherwise their form remains very similar. This shows that the

Table 4.1. Density of states at the Fermi level $N^0(\epsilon_F)$, the Stoner parameter at the Fermi level $I(\epsilon_F)$ and the product of these two quantities for various metals [7, 9]. Only Fe, Co and Ni fulfill the Stoner criterion, i.e., $I(\epsilon_F)N^0(\epsilon_F) > 1$

Element	$N^0(\epsilon_F)$ (eV^{-1})	$I(\epsilon_F)$ (eV)	$N^0(\epsilon_F)I(\epsilon_F)$
Na	0.23	1.82	0.41
Al	0.21	1.22	0.25
Cr	0.35	0.76	0.27
Mn	0.77	0.82	0.63
Fe (bcc)	1.54	0.93	1.43
Co	1.72	0.99	1.70
Ni	2.02	1.01	2.04
Cu	0.14	0.73	0.11
Pd	1.14	0.68	0.78
Pt	0.79	0.63	0.50

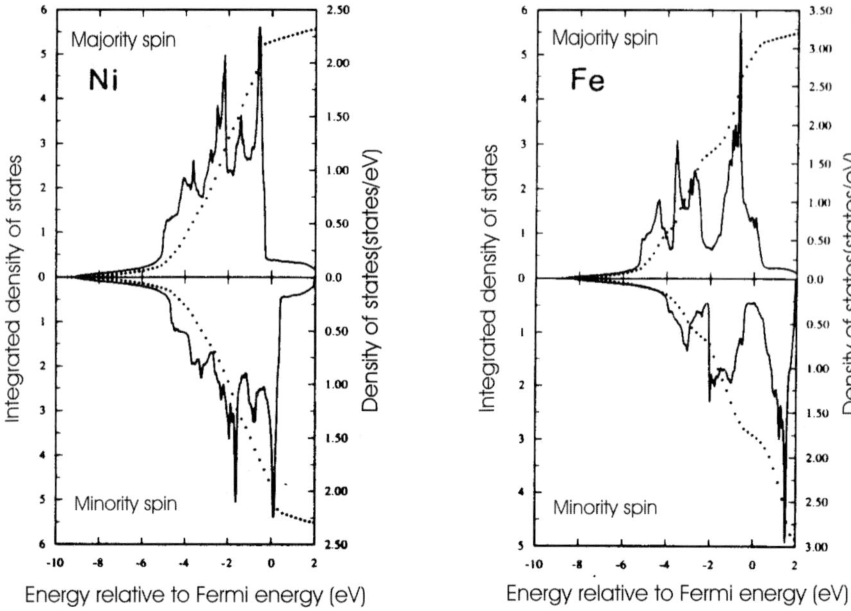

Fig. 4.1. Densities of states of spin-up and spin-down electrons for Fe and Ni. The densities of states for the spin-up and spin-down electrons are exchange-split but otherwise their forms are very similar. This shows the validity of the approximation made in the Stoner model. The *solid lines* describe the density of states and the *dotted lines* the integrated density of states, i.e., the number of electrons for a given spin direction. From [10], used with permission

approximation made in the Stoner model holds quite well. The difference between iron and nickel is that the spin-up band of Ni is filled and only 0.6 electrons are missing in the spin-down band, while for iron neither the spin-up nor the spin-down band is completely filled. This property makes iron a *weak* ferromagnet, while Ni, as well as Co, which also has a filled spin-up (majority) band, is called a *strong* ferromagnet.

The Stoner criterion also allows a straightforward discussion of the conditions under which conditions magnetism is favored. Certainly, a large Stoner parameter favors ferromagnetism. This parameter is element-specific and, in a first approximation, an intraatomic quantity. Thus it is independent of the local atomic environment. For the 3d, 4d and 5d metals, the following global trend for the Stoner parameter is found [7, 9]:

$$I_{3d} > I_{4d} > I_{5d} \,. \tag{4.16}$$

The most important quantity in most cases is the nonmagnetic density of states. For transition metals, the density of states close to the Fermi energy is dominated by the contribution of the d band. This density of states has pronounced structures, as shown for example by photoemission measurements.

Nevertheless, in a first approximation, the density of states of the d band is inversely proportional to the band width:

$$n_d^0 \sim \frac{1}{W_d} , \tag{4.17}$$

where W_d is the band width of the d electrons [14]. This width can be described within the nearest-neighbor tight-binding approximation by:

$$W_d = 2\sqrt{N_n} h_d . \tag{4.18}$$

Thus W_d depends on both the hopping matrix element to the nearest neighbors h_d, i.e., the hopping rate, and the number of nearest neighbors. The hopping rate decreases with increasing localization of the d electrons. For transition metals, the following trend is found:

$$h_{3d} < h_{4d} < h_{5d} . \tag{4.19}$$

Correspondingly, one has

$$W_{3d} < W_{4d} < W_{5d} , \tag{4.20}$$

leading to smaller band widths for 3d than for 4d and 5d metals. This is schematically depicted in Fig. 4.2, where the band widths of 4f and 5f electrons are also shown for comparison. From Fig. 4.2, one can see that the tendency towards magnetism is more pronounced in the later 3d metals, and in the 5f and 4f metals. The 3d and early 5f metals show a trend towards band magnetism (itinerant-electron magnetism), while the other rare earths are examples of localized magnets that can be described best by the Heisenberg model.

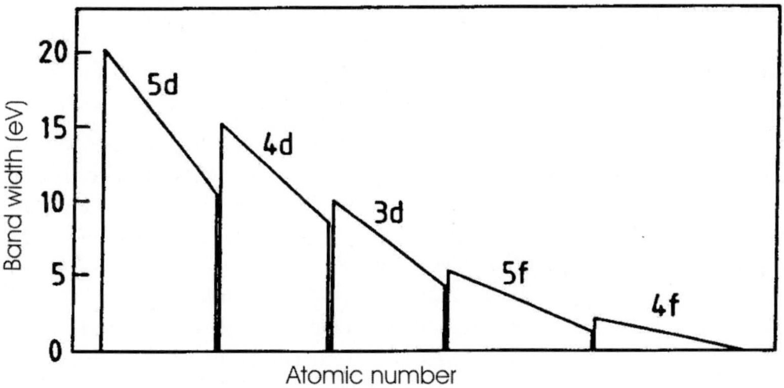

Fig. 4.2. Schematic representation of the band width W of transition metals, rare earths and actinides. After [6], used with permission

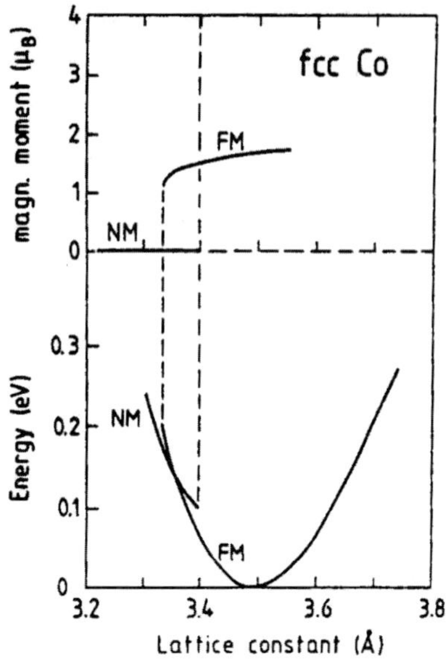

Fig. 4.3. Magnetic moment and ground-state energy as a function of lattice spacing: ground-state properties of fcc Co vs. cubic lattice constant in Å. Energy changes in eV per atom and the magnetic moment in Bohr magnetons are depicted. "NM" and "FM" denote the nonmagnetic and ferromagnetic solutions, respectively. The stability limits of these two phases are shown by *dashed lines*. From [11], used with permission

With this picture in mind, one can also understand the influence of increasing atomic spacing. When the lattice spacing in increased, the band width W decreases, and thus ferromagnetism becomes more favorable and the atomic moment per atom increases. This can be seen from Fig. 4.3, where calculations for the ground state of fcc Co are presented as a function of the lattice spacing. Above a lattice spacing of 3.35 Å, fcc Co is found to be ferromagnetic in this calculation. With increasing lattice spacing of the solid, the magnetic moment increases towards the atomic limit of $3\mu_B$.

Unfortunately, it is experimentally impossible to expand a crystal lattice which would correspond to applying a negative hydrostatic pressure. Even the small range of lattice compressions that are accessible by applying high pressures are insufficient to really probe the dependence of magnetism on interatomic spacings. Nevertheless, there is other indirect experimental evidence which points towards a pronounced correlation between magnetism and structure. In Fig. 4.4, the crystal structures of the 3d, 4d and 5d metals are shown together with predictions based on the tight-binding method.

First of all, it is pleasant to note that the tight-binding method correctly predicts the structure of most of the solids, but it also cannot be overlooked that it fails for elements such as Fe and Mn, which show magnetic behav-

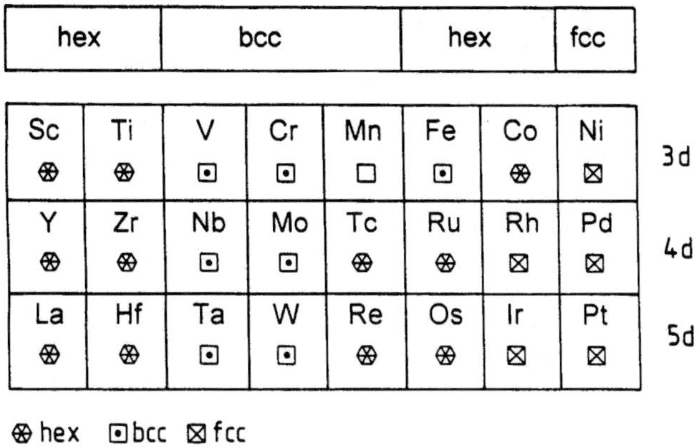

Fig. 4.4. Crystal structure of transition metals as a function of d-band filling. In the *upper portion* of the figure, the predicted structures of transition metals are depicted [12]. This prediction depends only on the d-band filling. The actual crystal structures observed at room temperature are shown in the *lower portion.* The *open square* for Mn denotes the complex cubic structure of this element. In general, the theory reproduces the observed structures quite well. It should be noted that magnetic 3d elements (Mn, Fe and Co) show deviations from the structure of their 4d and 5d counterparts. This indicates the strong influence of the magnetic ground state on the crystal structure. After [10], used with permission

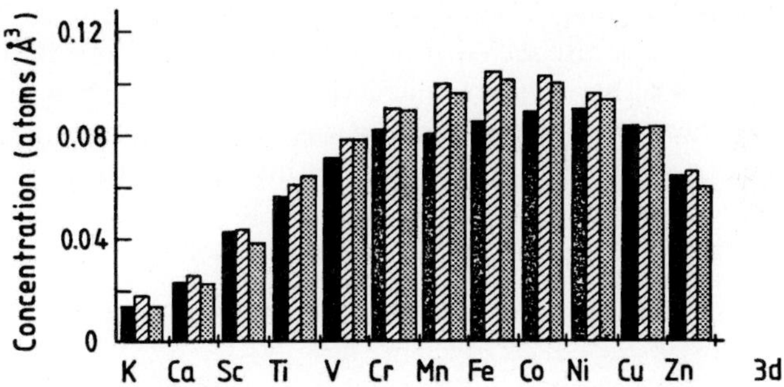

Fig. 4.5. Atomic densities of various transition metals.The atomic concentration of the element is shown for 3d, 4d and 5d metals. The magnetic elements Cr, Mn, Fe, Co and Ni have a smaller atomic density than have their nonmagnetic 4d and 5d counterparts. These latter elements are denoted by shaded bars. To make the presentation easier, the densities of 4d and 5d metals have been normalized by setting the densities of Cu, Ag and Au equal. After [13], used with permission

ior. Clearly, Fig. 4.4 gives only indirect evidence for a correlation between structure and magnetism. However, support for this claim comes from the fact that, typically, the structural phase diagrams of magnetic metals are richer than their nonmagnetic counterparts. In Fig. 4.5, the atomic densities of 3d, 4d and 5d metals are depicted. It can be seen that the atomic densities for those 3d metals that show magnetism (ferromagnetism or antiferromagnetism) are considerably smaller than the atomic densities of their nonmagnetic 4d and 5d counterparts.

4.1.4 Itinerant Antiferromagnetism – Spin Density Waves in Bcc Cr

Among the 3d transition metals, Fe, Co and Ni show ferromagnetism, while three other metals, namely bcc Cr, fcc Fe and α-Mn, with a complex crystal structure, show antiferromagnetism. In contrast to the few investigations of Mn and Fe, the antiferromagnetism of bcc Cr has been studied extensively. The antiferromagnetism of Cr is characterized by a static incommensurate *spin density wave* (SDW) as shown in Fig. 4.6. The spin density $m(\boldsymbol{r})$ shows a sinusoidal variation with position. The amplitude of the spin density oscillation was found to be 0.62 per atom at 4.2 K [15]. The wave vector can be in one of the six equivalent $\langle 100 \rangle$ directions, resulting in the formation of a domain structure. Cr shows a Néel temperature of 311 K, which is the highest among the three antiferromagnetic metals. At 123 K, there exists a spin-flip transition, at which the spin polarization direction changes from perpendicular to the wave vector at high temperatures to parallel to it at low temperatures. This change in magnetic ordering has been observed by polarized neutron scattering.

The SDW antiferromagnetism can be explained in the framework of band theory, and hence it is also called itinerant antiferromagnetism. The Stoner criterion tells us whether the ferromagnetic state or the paramagnetic state is energetically favorable, but it does not take the SDW state into account. Overhauser pointed out that there exists an instability of the paramagnetic state with respect to the formation of an SDW for an electron gas [16]. This instability occurs for an SDW having a wave vector $q \simeq 2k_F$, where k_F is

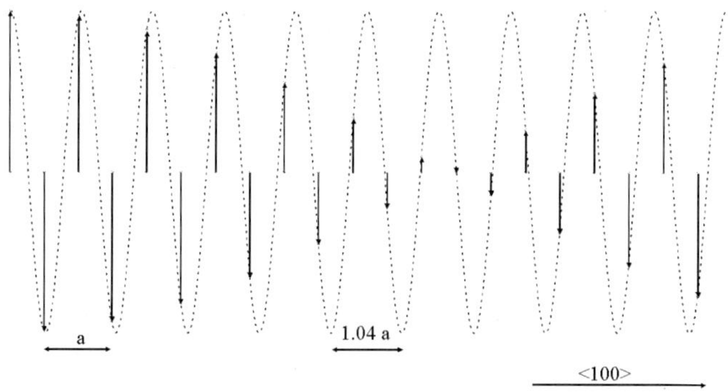

Fig. 4.6. Schematic incommensurate spin density wave in bcc Cr. The wave vector in the $\langle 100 \rangle$ direction has a length of $2\pi/1.04a$. The *arrows* indicate the amplitude of the spin density wave at the lattice sites

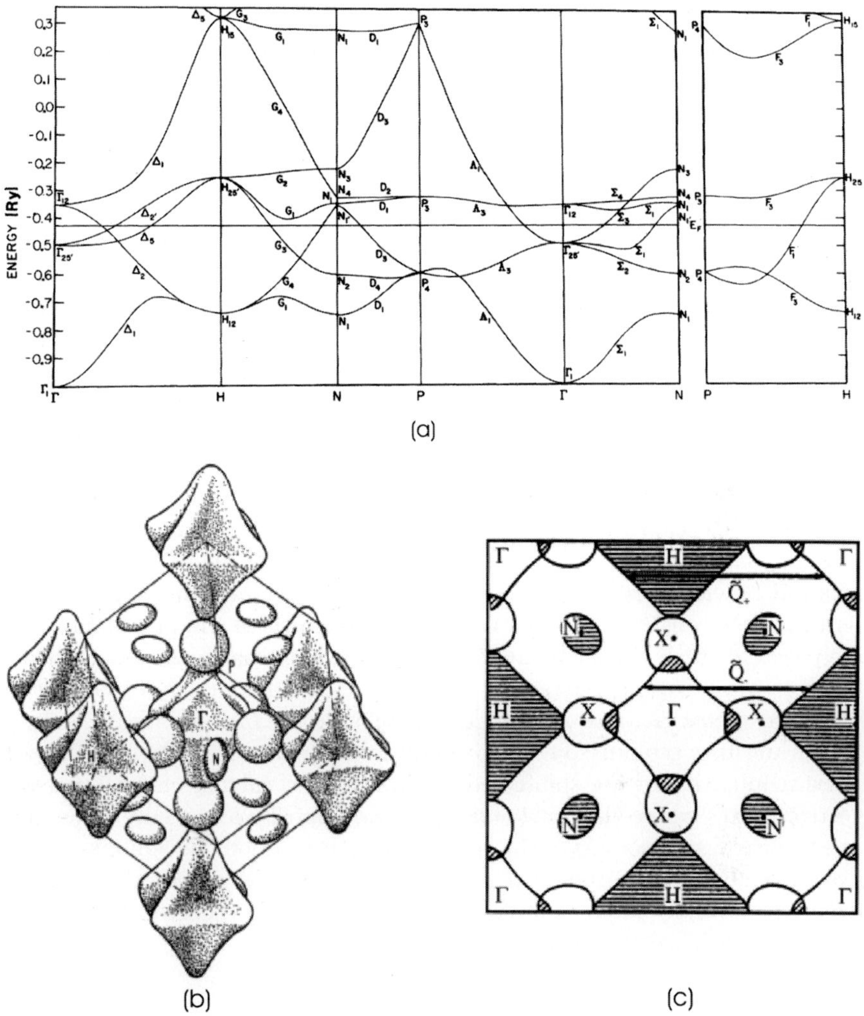

Fig. 4.7. Electronic structure of bcc Cr. (**a**) Energy bands along lines of high symmetry. (**b**) Three-dimensional sketch of the Fermi surface. An electron Fermi "pocket" is centered at the origin γ of the Brillouin zone, while four hole "jacks" are centered around the H points. See text for more details. (**c**) (001) cross section of the Fermi surface through the origin of the Brillouin zone. From [17, 18], used with permission

the diameter of the Fermi sphere. Let us illustrate this instability for a one-dimensional solid. Once an SDW with a wave vector of $2k_F$ is introduced, a periodic potential $V_{xc}^{\pm}$ with period π/k_F is established for spin-up and spin-down electrons. This potential produces an energy gap at the Fermi level. The energy of the occupied states near the Fermi level is lowered and that of the unoccupied states near the Fermi level is raised, so that the overall energy of the electron system decreases. The situation is quite analogous to the energy gaps that open at $k = \pm n\pi/a$ for a one-dimensional free-electron gas when the lattice-induced periodic potential is considered. On the other hand, such a spin-specific periodic potential will result in a periodically varying spin density, as we can see from the formulae of local spin density functional theory. This is an instability, which leads to an SDW ground state.

For bcc Cr, the situation is much more complicated than the one-dimensional case discussed above. Generally, the antiferromagnetism in bcc Cr arises from the particular electronic structure of its paramagnetic state, called a *nesting structure*. A nesting structure is encountered when two different sections of the Fermi surface can be brought into almost perfect coincidence by a rigid translation in $\boldsymbol{k}$-space. The electronic structure of bcc Cr is shown in Fig. 4.7. Cr has six valence electrons per atom ($3d^5 4s^1$). The first two bands are completely filled. Band three has a hole Fermi surface around H and N. We call it a hole Fermi surface because the states inside the "jack" centered at H are unoccupied, while those outside it are occupied. Band four presents an electron Fermi "pocket" centered at the origin of the Brillouin zone. In contrast to the hole Fermi "jack", the states inside the "pocket" are occupied by electrons, and those outside the "pocket" are unoccupied. The Fermi surface consists mainly of the states of band three and band four. The two pieces of the Fermi surface consisting of the hole "jack" and the electron "pocket" are separated by vectors $\boldsymbol{Q}$, which are approximately equal to the reciprocal vector $\boldsymbol{G}_{100}$, or its two equivalent vectors $\boldsymbol{G}_{010}$ and $\boldsymbol{G}_{001}$.

This nesting structure favors an SDW ground state [19]. Basically, a periodic modulation of the spin density will lead to a mixing of the two one-electron states whose spin wave functions are opposite and whose $\boldsymbol{k}$ vectors differ by the wave vector $\boldsymbol{q}$ of the spin density modulation. This mixing modifies the electron wave functions and their one-electron energy parameters, and it is especially important when $E_{\boldsymbol{k}} = E_{\boldsymbol{k}+\boldsymbol{q}}$ since then the interaction produces a first-order energy change

$$E_{\boldsymbol{k}} = E_{\boldsymbol{k}+\boldsymbol{q}} \;\rightarrow\; E_{\boldsymbol{k}} \pm V_{\boldsymbol{k}+\boldsymbol{q}} \,. \tag{4.21}$$

In addition, if $E_{\boldsymbol{k}}$ is close to the Fermi energy, the sum of the energies of the occupied states is lowered, and the total energy of the system is reduced. That is to say, the nesting structure for bcc Cr favors an SDW state with an wave vector corresponding to the nesting vector $\boldsymbol{Q}$.

4.2 Magnetism of Ultrathin Films

After discussing the general concepts of magnetism for 3d transition metals in the last section, we now turn to the characteristic features shown by ultrathin films. These new features discussed here include dimensionality effects, the modification of magnetic moments at surfaces and interfaces, and the contribution to the magnetic anisotropy arising from surfaces and interfaces.

4.2.1 Dimensionality Effects

When the thickness of a magnetic film decreases to the monolayer range, a dimensionality crossover from three-dimensional (3D) to two-dimensional (2D) behavior is expected. Statistical mechanics predicts different critical exponents for systems of different dimensionality. The critical exponent β, for example, describes how the magnetization M vanishes near T_c

$$M \propto \left(1 - \frac{T}{T_c}\right)^{\beta}, \quad T \rightarrow T_c\,. \tag{4.22}$$

Different β values are predicted when different models are used, but they are independent of the specific details of the chosen system. In the *Heisenberg model*, the orientation of the spins is not restricted. In this model, long-range magnetic ordering is expected only for three dimensional systems. The predicted β value is about 0.365. If the spin orientation is confined in a plane, one obtains the *xy model*. In this case, $\beta \approx 0.34$ for 3D system and $\beta \approx 0.23$ for finite 2D systems. [1] In the *Ising model*, a preferred spin orientation exists so that there are only two choices for the spins, parallel or opposite to the given direction. In this case, the critical exponents are 0.325 for a 3D system and 0.125 for a 2D system.

The critical exponents have been measured for a number of ultrathin films. These data, compiled by Himpsel et al. [20], are listed in Table 4.2. The measured exponents fall into two groups, which correspond to the 2D Ising model and the finite xy model.

Table 4.2. Experimentally measured critical exponents β for ultrathin metal films. Values close to 0.125 are characteristic of Ising-like systems, while values around 0.23 are indicative of 2D xy models

System	β	Reference
Fe/Pd(100)	0.127	[21]
Fe(110)/Ag(111)	0.137	[22]
Fe/W(110)	0.123	[23]
Fe/Au(100)	0.22	[24]
Ni/Cu(111)	0.24	[25]
Ni/Cu(111)	0.28	[26]
Ni/Cu(100)	0.24	[26]
Fe/W(100)	0.22	[27]

[1] At finite temperature and zero applied field, an infinite isotropic 2D xy system cannot sustain long-range order but exhibits a Kosterlitz–Thouless phase transition to a state with infinite correlation length at low temperature.

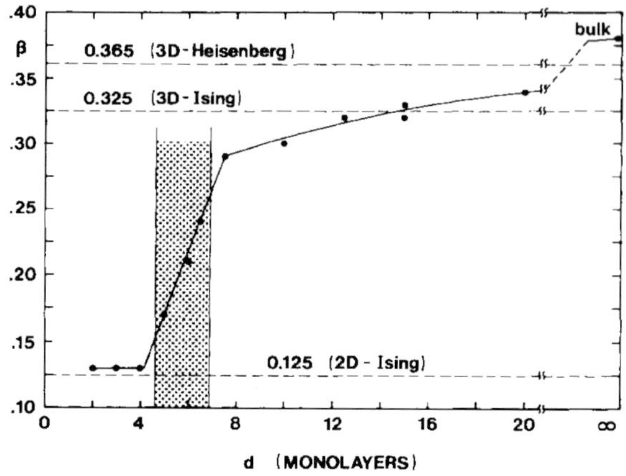

Fig. 4.8. Critical exponent β as a function of film thickness for Ni(111)/W(110). The *dashed lines* show the theoretical values for a 3D Heisenberg, a 3D Ising and a 2D Ising system. The *shaded regime* marks the crossover from 3D to 2D. From [28], used with permission

By measuring the critical exponent, Li and Baberschke [28] have demonstrated that a dimensionality crossover from 3D to 2D takes place with decreasing film thickness for Ni(111) films on a W(110) substrate. The critical exponent β as a function of thickness is shown in Fig. 4.8. The gradual decrease of β from 0.38 at 20 ML to 0.29 at 7.5 ML is interpreted as a crossover from 3D Heisenberg to 3D Ising behavior, due to an increasing anisotropy

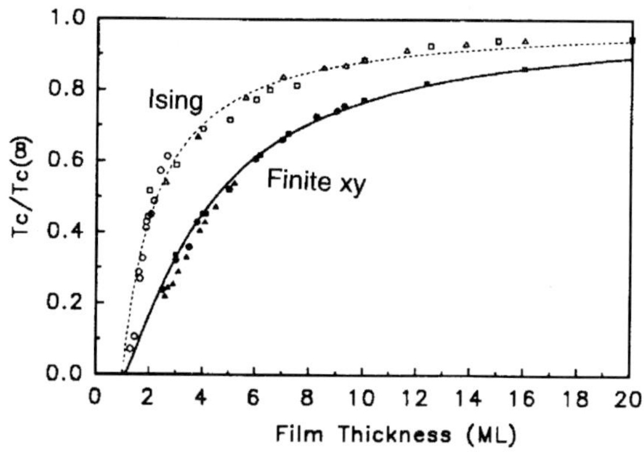

Fig. 4.9. Scaling behavior of the Curie temperature as a function of film thickness. The Curie temperatures have been normalized to their respective bulk values $T_c(\infty)$. The data have been collected for a number of systems: *open squares* for Ni(111)/W(110), *open circles* for Fe(110)/Ag(111), *open triangles* for Ni(111)/Re(0001), *solid squares* for Ni/Cu(100), *solid circles* for Ni_9Co_1/Cu(100), and *solid triangles* for Ni_3Co_1/Cu(100). From [26], used with permission

energy. A dimensionality transition from a 3D Ising system to a 2D Ising system is responsible for the dramatic reduction of β between 7 and 5 ML. Accompanying the dimensionality crossover is a reduction of the Curie temperature T_c with decreasing thickness. As shown in Fig. 4.9, the reduction of T_c displays finite-size scaling. When the Curie temperatures are normalized with respect to their bulk values, the thickness dependence for a variety of films can be scaled to two curves, one for the 2D Ising system with strong uniaxial magnetic anisotropy and the other for 2D xy films. As mentioned before, these two groups of films show critical exponents of 0.125 and 0.23, respectively. The Ising films show higher T_c values than do the xy films. For xy films, T_c is reduced to half of its bulk value at a thickness of about 5 ML and it decreases to zero at about 1 ML for both systems.

4.2.2 Magnetic Moments at Surfaces and Interfaces

For 3d transition metals, the itinerant magnetism leads to a magnetic moment that is determined by the electronic band structure. A positive exchange interaction splits the valence band into a majority band and a minority band. The difference in the number of occupied states between these two bands gives rise to the average magnetic moment per atom. At a metal surface, the broken symmetry and the reduced coordination number lead to a narrowing of the d band and localized surface states or surface resonance states, which affect the magnetic properties in the near-surface region. Several ab initio methods (most importantly the full-potential linearized augmented-plane-wave method) based on the local-spin-density functional theory have been developed to calculate the electronic and magnetic structure of materials (see [29] and references therein). One of the great successes of these theoretical studies is the prediction of a magnetic-moment enhancement at the surface of 3d transition metals. It is well recognized that the magnetic-moment enhancement comes from the reduced number of nearest neighbors and hence weaker interatomic hybridization, as we have discussed in the last section. The calculated results for the magnetic moments of 3d transition metal atoms at the surface and, for comparison, the respective bulk values are summarized in Table 4.3. In each case, the magnetic moment at the surface is enhanced. In accordance with the tight-binding model, the enhancement for a given bulk material is larger the smaller the number of nearest neighbors is. Thus, for fcc Ni, the more open (110) surface has a larger magnetic moment than has the (100) surface, which has one more nearest neighbor at the surface. For bcc crystals, the reverse situation is found. In this case the (100) surface is more open than the (110) surface and thus shows the larger increase in the magnetic moment. What is more difficult to explain within the tight-binding model is that the magnetic moment at the Cr and Fe surfaces has a much more pronounced enhancement than at the Ni and Co surfaces. In principle, this is due to the fact that for Ni and Co, the majority d band is completely filled while for Cr and Fe, both the minority and the

Table 4.3. Magnetic moment (in μ_B) at the surface and in the center layer, and the corresponding enhancement (in percent) for the surface atoms (from [29])

System	Surface	Center	Enhancement (%)
bcc Fe(001)	2.96	2.27	30
bcc Fe(110)	2.65	2.22	19
bcc Fe(111)	2.70	2.3	17
hcp Co(0001)	1.76	1.64	7
fcc Ni(001)	0.68	0.56	23
fcc Ni(110)	0.63	0.56	13
fcc Ni(111)	0.63	0.58	9
fcc Fe(001)	2.85	1.99	43
bcc Co(001)	1.95	1.76	11
bcc Co(110)	1.82	1.76	3
fcc Co(001)	1.86	1.65	13
bcc Cr(001)	2.49	0.59	322

majority band contain unoccupied states. In the latter situation, a d band narrowing can change the magnetic moment more easily. For Co and Ni, it is more difficult to change the magnetic moment since these metals are strong ferromagnets [14]. Spin-polarized low-energy electron diffraction (SPLEED) provides an experimental method to test the theoretical predictions. Qualitative agreement has been obtained for Fe(110), Ni(100) and Ni(111) surfaces [30, 31].

In the calculation of the magnetic moments, the distances between atoms at the surface are usually assumed to be identical to the corresponding bulk distance. This is very often not the case. Relaxations, a change of interlayer spacing in the vicinity of the surface, or even reconstructions, a change of the structural arrangement of the topmost layer(s), are the rule rather than the exception. All of these effects would modify the results for the magnetic moments of surface atoms. On clean metal surfaces, commonly an inward relaxation of the first interlayer spacing is found (see [32] and references therein), which has been attributed to a smoothing of the electron density at the surface [33]. Such a relaxation will influence the magnetic moment, as is shown in Fig. 4.10 for the Fe(100) surface [29]. An inward relaxation leads to a decrease of magnetic moment. Theoretically, an inward relaxation of 4% is determined for Fe(100) [34], which leads to a 6% decrease in the magnetic moment. Nevertheless, the resulting magnetic moment is still considerably larger than the bulk value of about $2.2\mu_B$ [29]. This example demonstrates that for an understanding of magnetic properties of surfaces, a detailed knowledge of the structure is desirable.

At the interface of two materials, the magnetization is influenced by the mutual interaction of these two materials. The mutual interaction may give rise to an enhanced or decreased interface magnetism and even to an os-

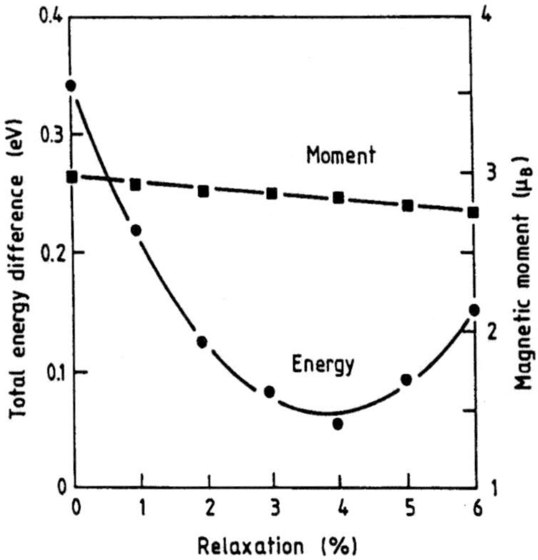

Fig. 4.10. Dependence of energy and surface magnetic moment on the first interlayer spacing. An inward relaxation of approximately 4% is found for the ground state of bcc Fe(001), leading to a small reduction of the surface magnetic moment compared with the unrelaxed surface. From [29], used with permission

cillatory magnetization profile in the interface region. Niklasson et al. have calculated the magnetic spin moments of magnetic 3d bilayer interfaces for a variety of systems using the local-spin-density approximation [35]. It was found that the magnetic moment of an atom at the interface is quite different from the bulk value. How the interface magnetic moment deviates from the bulk value depends on the adjacent material. To illustrate this effect, their results are given in Fig. 4.11.

Interface magnetism has also been studied experimentally during the last decade. A number of techniques have been applied to detect the interface magnetic moment. These are mainly spin-polarized electron spectroscopy (SPES), polarized neutron reflection, ferromagnetic resonance (FMR), conversion-electron Mössbauer spectroscopy (CEMS) and x-ray magnetic circular dichroism (XMCD). The combination of XMCD and conventional magnetometry such as superconducting quantum interference device (SQUID) measurements and vibrating-sample magnetometry (VSM) has proven to be a powerful tool to study interface magnetism. The latter measures the absolute value of the magnetization and the former separates the contributions of the two materials at the interface. Therefore the atomic magnetic moment of the interface element can be determined. Pd/Fe and Cr/Fe are the most frequently studied systems. The Fe moment has been found to be enhanced at the Pd/Fe interface, and a ferromagnetically coupled Cr bilayer with a huge moment has been reported, in agreement with the theoretical predictions. Induced magnetic ordering has also been found in "nonmagnetic" transition

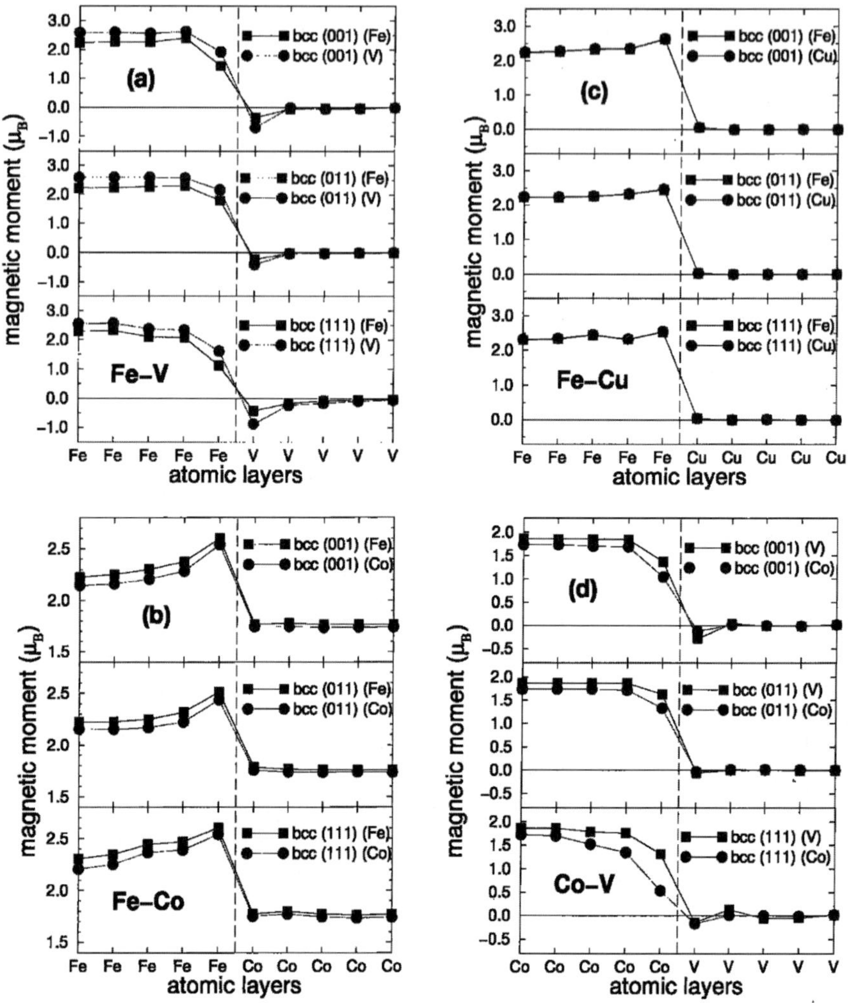

Fig. 4.11. The profiles of magnetic moments in the vicinity of the interface for various magnetic bilayers. Two different lattice constants, corresponding to the bulk lattice constants of the two materials forming the bilayer, have been assumed in the calculations and are given in parentheses in the figures. From [35], used with permission

metals such as V and Pd when they are next to a ferromagnetic layer at the interface. We summarize the experimental results on the interface magnetic moments in Table 4.4.

The properties of ultrathin epitaxial films can be discussed in a manner very similar to that for surface moments. For a monolayer on a substrate, an equation similar to (4.18) can be used. However, one has to distinguish

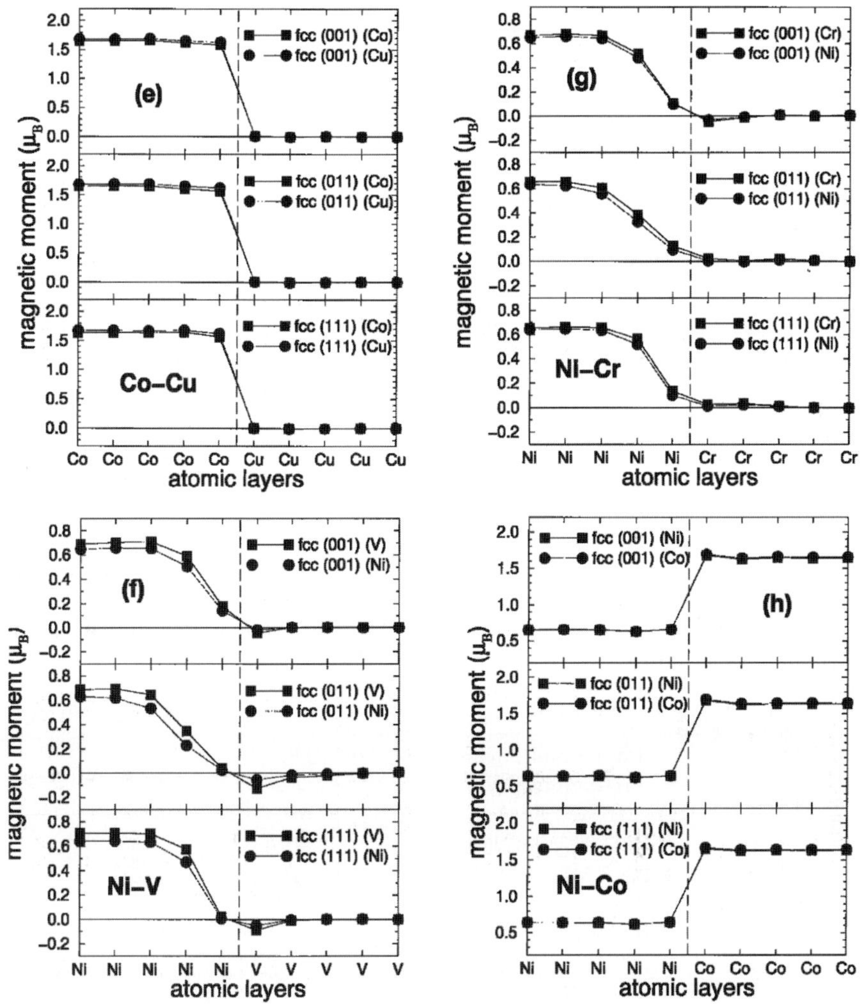

Fig. 4.11. Continued

between the hopping of d electrons within the monolayer $h_{\|}$ and the hopping element to the substrate $h_{\perp}$. Furthermore, the coordination numbers in the plane $N_{\|}$ and to the substrate $N_{\perp}$ have to be considered [14]. While the coordination numbers depend only on the structure of the surface, the hopping matrix elements will be specific to the chosen film–substrate pair. Hopping and therefore the film magnetism can thus be influenced by the choice of substrate. In particular, the use of noble-metal substrates with their low-lying d band should result in a small d–d hybridization between the film and substrate and therefore lead to a small $h_{\perp}$. In that case monolayer films show almost two-dimensional behavior. This is favorable for inducing or enhancing

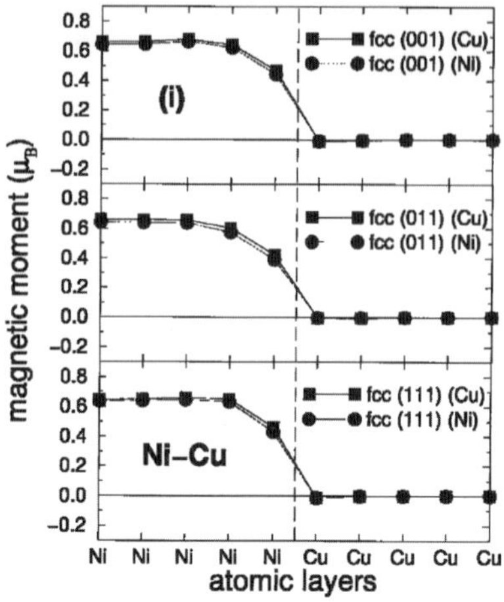

Fig. 4.11. Continued

Table 4.4. The measured magnetic moments (in Bohr magnetons) of interface atoms are obviously different from the bulk values, as demonstrated in this table. The bulk values are $2.3\mu_B$ for Fe, $0.6\mu_B$ for Ni and $0.59\mu_B$ for Cr. V and Pt are nonmagnetic metals

Element	System	Interface moment	Reference
Fe	Pd/Fe multilayers	2.8	[36]
Fe	Pd/Fe multilayers	2.3–3.2^a	[37]
Fe	Ag/5.5 ML Fe/Ag(001)	2.58^b	[38]
Fe	Cu/5.7 ML Fe/Ag(001)	2.50^b	[38]
Fe	Pd/5.7 ML Fe/Ag(001)	2.6^b	[38]
Fe	Fe/V superlattice	1.34–2.12^a	[39]
V	Fe/V superlattice	−0.27 to −1.06^{ac}	[39]
V	V/Fe(100)	−0.3^c	[40]
Cr	Cr/Fe(100)	4	[41]
Ni	Ni/Cu(100)	0.30–0.35^a	[42]
Ni	Ni/Fe(001)	0.69	[43]
Ni	Ni/Pt multilayers	0.24–0.54^a	[44]
Pt	Ni/Pt multilayers	0.09–0.21^a	[44]
Pt	Pt/Co multilayers	0.21	[45]

^{a} Average total moment, which varies from sample to sample, depending on the thicknesses of the component layers.
^{b} Average total moment.
^{c} The minus sign denotes that the V moment is antiparallel to the moment of the Fe substrate.

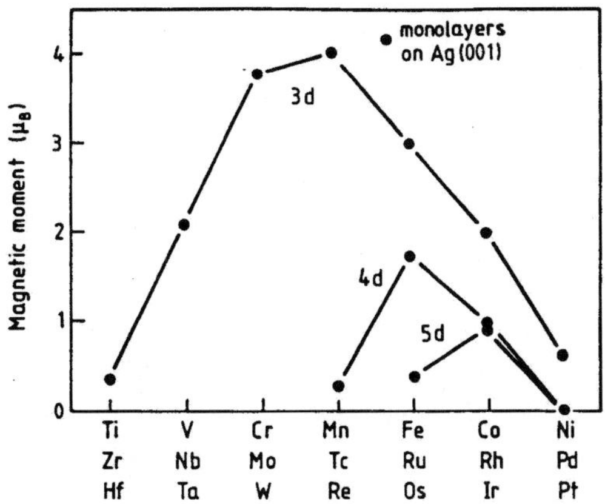

Fig. 4.12. Predicted magnetic moments of 3d, 4d and 5d metals on Ag(100). These calculations show that epitaxial monolayers of 4d and 5d elements can develop a magnetic moment on a noble-metal substrate. From [46], used with permission

magnetism. Indeed, it has been demonstrated by total-energy calculations within the local-spin-density approximation that even monolayers of 4d and 5d metals on Ag(100) show ferromagnetism [46] (see Fig. 4.12). In contrast, if a nonmagnetic transition metal is chosen as a substrate, the strong interaction between the film and substrate diminishes the ferromagnetism of the monolayer film [29].

To close this subsection, we would like to briefly discuss the orbital magnetic moment. It is well known that orbital moments are strongly suppressed in the bulk by the crystal splitting and by hybridization with the neighboring atoms, resulting in an almost complete quenching of the angular momentum. Calculations yield very small orbital moments for 3d ferromagnetic bulk metals: $0.082\mu_B$ for bcc Fe, $0.123\mu_B$ for hcp Co and $0.058\mu_B$ for fcc Ni [47]. Therefore the total moments come mainly from the spin contribution. However, the quenching of orbital moments can be partly lifted on the surface, resulting in enhanced surface orbital moments. The orbital moment of the fct Fe surface layer on Cu(100) has been measured to be three times larger than the bulk value by x-ray magnetic circular dichroism, which allows one to separate the spin and orbital contributions to the total moment [48]. Similarly, an orbital moment two times larger is found in ultrathin Co films grown on Cu(100) [49].

4.2.3 Magnetic Anisotropy in Ultrathin Films

In the preceding discussion, the direction of the magnetization with respect to the crystal axes was not considered. In reality, however, the total energy of a film is dependent on the spin orientation. The change in free energy for a crystal or film upon rotation of the magnetization is called the anisotropy energy. Though the magnitude of the anisotropy energy is of the order of only 10^{-6} to 10^{-3} eV/atom, the total anisotropy energy of a crystal or a film consisting of a large number of atoms is much larger than k_BT. Therefore the preferred magnetization direction is usually dictated by the anisotropy energy.

Two major contributions dominate the anisotropy energy – the spin–orbit coupling and the dipole–dipole interaction. Owing to its long-range character, the dipole–dipole interaction leads to a contribution to the anisotropy which depends upon the specimen shape, and is also called the shape anisotropy. The spin–orbit coupling relates the magnetization to the crystal lattice and results in a magnetocrystalline anisotropy. If the sample is stressed, the spin–orbit coupling is modified by the strain. This modification to the magnetocrystalline anisotropy is called magnetoelastic anisotropy, though it has the same microscopic origin as the magnetocrystalline anisotropy, i.e., the spin–orbit interaction. In the following, these three anisotropies will be discussed in more detail.

Shape Anisotropy. The shape anisotropy comes from the long-range magnetic dipolar interaction. The shape effect of the dipolar interaction in ellipsoidal ferromagnetic samples can be described via an anisotropic demagnetizing field, H_d, given by $\boldsymbol{H}_d = -\mathsf{N}\boldsymbol{M}$. Here $\boldsymbol{M}$ is the magnetization vector and N is a shape-dependent demagnetizing tensor. For a thin film, all tensor elements are zero except for the diagonal element corresponding to the direction perpendicular to the film plane, which is equal to unity. Since the magnetostatic energy can be expressed

$$E_d = -\frac{\mu_0}{2V}\int \boldsymbol{M}\cdot\boldsymbol{H}_d\,dv\,, \tag{4.23}$$

the shape anisotropy energy per unit volume of a film reads

$$E_d = \frac{1}{2}\mu_0 M^2\cos^2\theta\,. \tag{4.24}$$

Here θ is the angle of the magnetization $\boldsymbol{M}$ with respect to the film normal. According to this expression, the shape anisotropy favors an in-plane orientation for the magnetization. Since the film thickness does not enter into the expression, the shape anisotropy is a bulk quantity and is proportional to the number of atoms.

Magnetocrystalline Anisotropy. As early as 1954, Néel proposed a phenomenological description of the magnetic anisotropy [50]. In Néel's model, the spin–orbit interaction introduces an angle-dependent magnetic interaction. The interaction between two atomic magnetic moments separated by a vector $\boldsymbol{r}$ can be written as

$$E(r) \sim l(r)\cos^2\phi + q(r)\cos^4\phi\,, \tag{4.25}$$

where $l(r)$ and $q(r)$ are expansion coefficients and ϕ is the angle between the interatomic distance r and the parallel magnetic moments of the two atoms considered.

It should be pointed out that Néel's model is based on the localized-moment approximation and therefore is not well suited to itinerant magnets. Within the model of delocalized electrons, the magnetocrystalline anisotropy is explained as a modification to the band structure induced by the spin–orbit coupling. Spin–orbit coupling can be interpreted as the coupling between the spin of the electron and the magnetic field created by its own orbital motion around the nucleus. The orbital motion is coupled to the lattice via the electron potential of the ions. For ferromagnetic 3d metals (Fe, Co and Ni), the electronic states are degenerate for the 3d band. Spin–orbit coupling lifts the degeneracy. The splitting of the degenerate states at the Fermi level decreases the density of states (DOS) at the Fermi surface and increases the DOS below the Fermi level. Hence the total energy of the electrons decreases. The amplitude of the band splitting depends on the spin direction. The direction in which the spin–orbit coupling produces the maximum splitting and thus the minimum system energy gives rise to the easy axis.

The crystalline anisotropy energy can be expanded in successive powers of α_x, α_y and α_z, where α_i are the direction cosines of the orientation of the magnetization with respect to the crystal axes. Owing to the time-reversal symmetry, only even orders of α_i appear in this expansion. Furthermore, the expression can be simplified by the crystal symmetry. For a cubic crystal, the second-order term $(\alpha_x^2+\alpha_y^2+\alpha_z^2)$ is constant and does not give an anisotropy. In this case, the bulk crystalline anisotropy energy can be expressed as

$$\begin{aligned} E_K &= K_0 + K_1(\alpha_x^2\alpha_y^2 + \alpha_y^2\alpha_z^2 + \alpha_z^2\alpha_x^2) + K_2\alpha_x^2\alpha_y^2\alpha_z^2 \\ &\quad + K_3(\alpha_x^2\alpha_y^2 + \alpha_y^2\alpha_z^2 + \alpha_z^2\alpha_x^2)^2 + \cdots . \end{aligned} \tag{4.26}$$

The coefficients are temperature-dependent and vary from material to material. For Fe at 4.2 K, $K_1 = 4.02 \times 10^{-6}$ eV/atom, $K_2 = 1.44 \times 10^{-8}$ eV/atom and $K_3 = 6.6 \times 10^{-9}$ eV/atom. For a hexagonal crystal, the anisotropy energy has the form

$$E_K = K_0 + K_{u1}\sin^2\theta + K_{u2}\sin^4\theta + K_{u3}\sin^6\theta + K'_{u3}\sin^6\theta\cos^6\phi\,, \tag{4.27}$$

where θ is the angle between the magnetization vector and the hexagonal axis, and ϕ is the azimuthal angle of the magnetization. For a tetragonal lattice,

$$E_K = K_0 + K_{u1}\sin^2\theta + K_{u2}\sin^4\theta + K_{u3}\cos^2\alpha\cos^2\beta\,, \tag{4.28}$$

where the magnetization vector makes an angle of θ with respect to the tetragonal axis and forms angles of α and β with the other two axes.

At a surface, the translational symmetry is broken in the direction of the film normal, resulting in a uniaxial anisotropy [51]. This is termed surface anisotropy. Using the pair potential of (4.25), Néel derived this surface anisotropy energy for fcc (111) and fcc (100) surfaces as

$$E = K_s \cos^2 \theta \ , \tag{4.29}$$

with K_s differing for (111) and (100) surfaces. Here θ is the angle of the magnetization away from the surface normal. In general, the surface anisotropy energy can be written as

$$E = K_s \cos^2 \theta + K_{s,p} \sin^2 \theta \cos^2 \phi \ , \tag{4.30}$$

ϕ being the azimuthal angle [52, 53]. For $K_s < 0$, the surface normal is an easy axis. For $K_s > 0$, the film plane is an easy plane of the surface anisotropy. $K_{s,p}$ is the constant for the additional in-plane anisotropy.[2] $K_{s,p}$ disappears if the surface normal is an n-fold rotation axis with $n > 2$. If $K_{s,p}$ does not disappear, the x axis is the easy axis when $K_{s,p} < 0$ and the hard axis when $K_{s,p} > 0$.

The effect of the symmetry reduction at a surface on the anisotropy energy has also been examined within band theory. It is concluded that a large surface contribution can be expected, owing to the narrowed d band(see [54, 55] for example).

Magnetoelastic Anisotropy. The magnetoelastic anisotropy is nothing else than a strain-induced modification of the magnetocrystalline anisotropy. Strain in ultrathin films is mainly induced by the lattice mismatch between the film and the substrate, which is equal to $\eta = (a_f - a_s)/a_f$, where a_f and a_s are the lattice parameters of the film and the substrate, respectively. The strain state is film-thickness-dependent. This has been discussed in the previous chapter.

How the strain modifies the magnetocrystalline anisotropy can be analyzed by expanding the magnetocrystalline anisotropy energy in a Taylor series with respect to the strains:

$$E_K = (E_K)_0 + \sum (\partial E_K / \partial e_{ij})_0 e_{ij} + \cdots \ . \tag{4.31}$$

As an example, for a cubic crystal [56],

$$(E_K)_0 = K(\alpha_1^2 \alpha_2^2 + \alpha_1^2 \alpha_3^2 + \alpha_2^2 \alpha_3^2) \ , \tag{4.32}$$

and the first-order term in the Taylor expansion can be written as

$$E_{ME} = B_1(\alpha_1^2 e_{xx} + \alpha_2^2 e_{yy} + \alpha_3^2 e_{zz}) + B_2(\alpha_1 \alpha_2 e_{xy} + \alpha_2 \alpha_3 e_{xz} + \alpha_1 \alpha_3 e_{xz}) \ . \tag{4.33}$$

[2] The surface anisotropy energy has also been expressed in the literature as $E = K_s \sin^2 \theta + K_{s,p} \cos^2 \theta \cos^2 \phi$, so that it has a form consistent with that of the bulk anisotropy energy with uniaxial symmetry. In that case, the coefficient K_s has the opposite sign for a given anisotropy and the anisotropy energy differs from (4.30) by a constant.

Here K is the first-order cubic magnetocrystalline anisotropy constant. α_i are the direction cosines of the magnetization vector and e_{ij} denotes the strains. B_1 and B_2 are the magnetoelastic coupling constants. They are related to the first-order partial derivative through

$$\partial E_K/\partial e_{ii} = B_1\alpha_i^2 \ , \quad \partial E_K/\partial e_{ij} = B_2\alpha_i\alpha_j \ . \tag{4.34}$$

When the strains are known, the strain-induced modification E_{ME} is easy to calculate according to (4.33). As has been pointed out above, atoms at the surface and interface have a different atomic environment from the bulk atoms. Hence additional magneto–elastic coupling coefficients might be necessary to take this surface and interface effect into account. Therefore the effective magnetoelastic coupling constant should be written as $B_{\text{eff}} = B_{\text{bulk}} + B_s/t$, where t is the film thickness [57, 58].

From the above discussion we know that the dipole–dipole interaction contains a bulk contribution to the free energy, i.e., the induced anisotropy energy is shared by all the atoms. However, a distinction should be made for the magnetocrystalline anisotropy between the surface/interface atoms and the interior atoms, owing to the different atomic environments. For a cubic crystal, a square term appears in the expression for the anisotropy energy owing to the broken symmetry for the atoms at the surface and interface, while for bulk atoms, the lowest term is the quartic term. Therefore the surface/interface magnetocrystalline anisotropy could be much larger than the bulk contribution. To reflect this difference between interfacial atoms and bulk atoms, the magnetic anisotropy energy K can be phenomenologically separated into a volume contribution K_v and an interface and a surface contribution K_{in} and K_s, respectively. The relation between these contributions can be written as

$$K = K_{\text{eff}} = K_v + \frac{K_s + K_{in}}{t} \ , \tag{4.35}$$

where t is the film thickness. This relation represents the average value of the magnetic anisotropy energy of the interface atoms, surface atoms and inner atoms. In this way, K becomes an explicit function of the film thickness. It is possible that the lowest-energy state for the magnetization may change with film thickness, for example from in-plane to out-of-plane. This is to say, a spin reorientation transition could occur upon variation of the film thickness. On the other hand, the anisotropy coefficients are a function of a number of parameters, such as temperature, chemical composition and stress state. Spin reorientation transitions could also be induced by changes in those quantities. Spin reorientation transitions and a perpendicular magnetic anisotropy have been observed for a number of ultrathin films and have been intensively investigated.

4.3 Techniques for Magnetic Measurements on Ultrathin Films

4.3.1 Surface Magnetooptic Kerr Effect

The first application of the surface magnetooptic Kerr effect (SMOKE) to study surface magnetism took place only recently [59]. This is rather surprising, since our knowledge of magnetooptic effects is much older. In 1845, Michael Faraday discovered the influence of magnetized media on the transmission of linearly polarized light which leads to an elliptical polarization [60]. A similar effect upon reflection of linearly polarized light was discovered by Reverend John Kerr in 1876 [61]. The present interest in the magnetooptic Kerr effect stems from its use in reading the information on magnetooptic media. Additionally, it is now frequently used to investigate surface and thin-film magnetism. Bader [59] created the acronym SMOKE for this application, but it should kept in mind that the word "surface" better describes the present sensitivity rather than the information depth, which is of the order of 100–200 Å.

At present, SMOKE is used as a simple tool to search for new magnetic materials, to investigate surface magnetic anisotropy, critical exponents for magnetism in two dimensions and the thickness dependence of the Curie temperature, and for domain imaging [62].

The principle of the magnetooptic Kerr effect (MOKE) is depicted in Fig. 4.13. Linearly polarized incident light obtains a Kerr rotation and a Kerr ellipticity upon reflection from a magnetized medium. The microscopic origin of this effect is the spin–orbit coupling. Macroscopically, the effect can

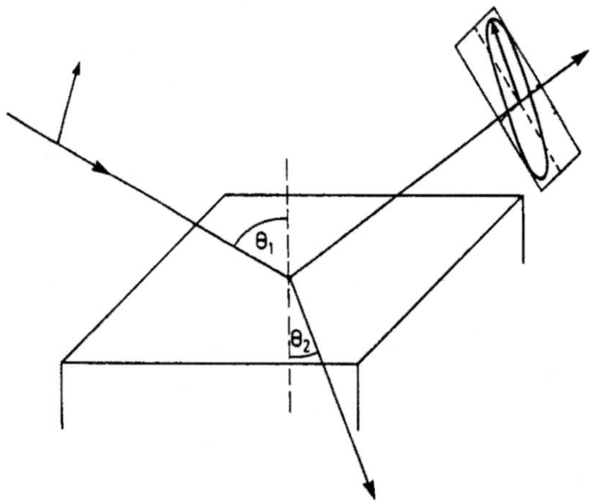

Fig. 4.13. Magneto-optic Kerr effect. Upon reflection of linearly polarized light from a magnetized sample, an elliptical polarization and rotation of the polarization plane occur

be described by the dielectric tensor ε which contains off-diagonal terms. In the most general case, where the angle of the incident light with respect to the magnetization has an arbitrary direction, the dielectric tensor has the following form:

$$\varepsilon = n^2 \begin{pmatrix} 1 & iQ_z & -iQ_y \\ -iQ_z & 1 & iQ_x \\ iQ_y & -iQ_x & 1 \end{pmatrix} , \tag{4.36}$$

where $\boldsymbol{Q} = (Q_x, Q_y, Q_z) = \boldsymbol{Q}(\alpha, \beta, \gamma)$ is called the Voigt vector [63], and α, β, γ are the direction cosines of the magnetization vector $\boldsymbol{M}$. This form of ε can be derived from symmetry arguments, which state that for a homogeneous magnetization in a given direction, the dielectric tensor must be cylindrically symmetric around this direction if the unmagnetized medium is optically isotropic. Argyres investigated the magnetooptic effect on the basis of band theory [64] and concluded that (i) the magnetooptic constants are proportional to the net magnetization M for a given specimen, and (ii) the proportionality constant depends on the frequency of the electromagnetic wave but is temperature-independent. It will be seen later that these two conclusions lay the foundation for the validity of SMOKE as a probe for studying magnetic phase transitions and critical behavior in ultrathin films. It should be pointed out, however, that this proportionality constant is determined by the electronic structure and thus varies from material to material.

Two normal modes can be derived from Maxwell's equations for an electromagnetic wave in a medium with a dielectric tensor of the form (4.36) [65]. These modes are left-handed circularly polarized light with a refractive index $n_L = n(1-(1/2)\boldsymbol{Q}\cdot\widehat{\boldsymbol{k}})$, and right-handed circularly polarized light with a refractive index $n_R = n(1+(1/2)\boldsymbol{Q}\cdot\widehat{\boldsymbol{k}})$. Therefore the magnetooptic effect can be equivalently described by two different complex indices of refraction for light of right- and left-handed circular polarization. The two circular modes travel in the medium with different velocities and are attenuated differently. As we know, linearly polarized light can be decomposed into left-handed and right-handed circularly polarized light with equal amplitudes. In the case of normal incidence of linearly polarized light, these two modes have different complex reflection coefficients

$$r^j = \frac{E_r^j}{E_i^j} = -\frac{n^j - 1}{n^j + 1} = |r^j| e^{i\Phi_j} , \quad j \in (\mathrm{L}, \mathrm{R}) , \tag{4.37}$$

where E_i and E_r denote the incoming and reflected wave fields, respectively, and n is the refractive index. Thus, upon emerging from the magnetized medium after reflection, the two altered modes recombine to yield a rotated axis of polarization and an ellipticity (see Fig. 4.13). The rotation and ellipticity are called the Kerr rotation α_{K} and the Kerr ellipticity ε_{K}, respectively; the Kerr ellipticity is given by the normalized difference between the amplitudes,

$$\varepsilon_K = -\frac{|r^L| - |r^R|}{|r^L| + |r^R|} , \tag{4.38}$$

and the Kerr rotation is given by

$$\alpha_K = -\frac{1}{2}(\Phi^L - \Phi^R) . \tag{4.39}$$

Upon magnetization reversal, both the Kerr ellipticity and the Kerr rotation change sign. This is the property that SMOKE exploits to probe magnetic films. However, in most practical cases, the linearly polarized light is reflected at a certain incidence angle. There are three extensively used Kerr configurations. These are the polar, longitudinal and transverse configurations, which describe three high-symmetry geometries as shown in Fig. 4.14. In the polar geometry, the magnetization direction is perpendicular to the plane of the film. In the longitudinal Kerr effect, the magnetization is in the film plane and also in the plane of incidence. In the transverse Kerr effect, the magnetization is in the film plane but perpendicular to the plane of incidence. The algebra describing the reflection is simplified in the above three geometries.

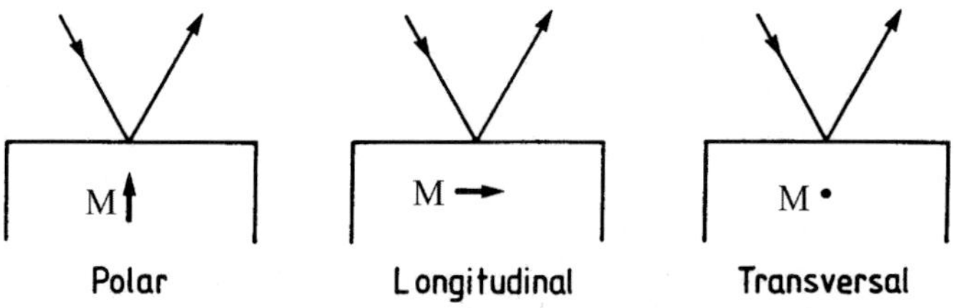

Fig. 4.14. Orientation of the magnetization and incidence plane for the three main Kerr geometries

In the case where the light is reflected at a nonzero incidence angle, it is convenient to decompose the incident and reflected light into p- and s-polarized plane waves, for which the polarization vector is in the plane of incidence and perpendicular to it, respectively (see Fig. 4.15). The reflection event is described by the following complex matrix relation:

$$\begin{pmatrix} E_s^{\mathrm{r}} \\ E_p^{\mathrm{r}} \end{pmatrix} = \begin{pmatrix} r_{ss} & r_{sp} \\ r_{ps} & r_{pp} \end{pmatrix} \begin{pmatrix} E_s^{\mathrm{i}} \\ E_p^{\mathrm{i}} \end{pmatrix} , \tag{4.40}$$

where E_s^{i} and E_p^{i} are the s and p components, respectively, of the electric vector of the incident light, and E_s^{r} and E_p^{r} are the corresponding quantities for the reflected light. The situation is simplified if the incident light contains only an s- or p-polarized component. In this case the Kerr rotation and ellipticity can be calculated from the following formulae [66]:

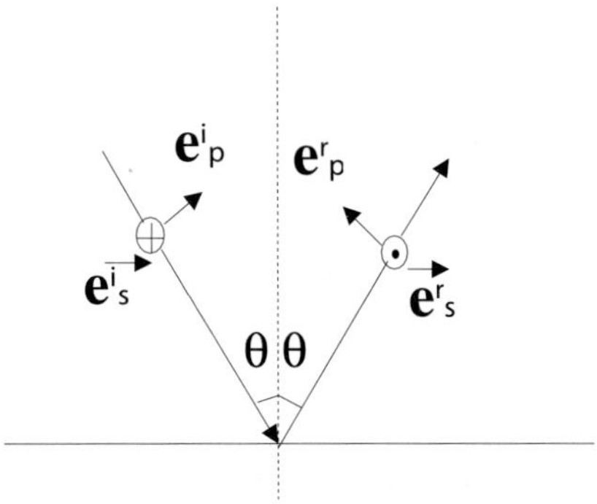

Fig. 4.15. Definition of the s and p directions for the incident and reflected waves at a boundary between two media

$$\alpha_{K_s} + i\varepsilon_{K_s} = \frac{r_{ps}}{r_{ss}} \tag{4.41}$$

for s-polarized incident light and

$$-\alpha_{K_p} + i\varepsilon_{K_p} = \frac{r_{sp}}{r_{pp}} \tag{4.42}$$

for p-polarized incident light.

In general, once the dielectric tensors and the geometries of the film system are given, the matrix r in (4.40) can be calculated from the electrodynamic theory. Otherwise, if the Kerr rotation and ellipticity are measured experimentally, the magnetooptic constants can be obtained by simulation. The formulae for the Kerr effect for a semi-infinite surface have been derived by Metzger et al. [66] and rewritten in a commonly used notation in [65]:

$$\alpha_{K_s} + i\varepsilon_{K_s} = \frac{in^2Q}{\epsilon_0 - 1} \times \frac{c(\epsilon_0 - \sin^2\theta)^{1/2} - b\sin\theta}{\epsilon_0 - \sin^2\theta + (\epsilon_0 - \sin^2\theta)^{1/2}\sin\theta\tan\theta} \tag{4.43}$$

and

$$-\alpha_{K_p} + i\varepsilon_{K_p} = \frac{-in^2Q}{\epsilon_0 - 1} \times \frac{c(\epsilon_0 - \sin^2\theta)^{1/2} + b\sin\theta}{\epsilon_0 - \sin^2\theta - (\epsilon_0 - \sin^2\theta)^{1/2}\sin\theta\, tan\,\theta}\,, \tag{4.44}$$

where the general direction of the magnetization is given by the cosines b and c with respect to the y and z axes respectively, and θ is the angle of incidence.

To apply the Kerr and Faraday effects to ultrathin magnetic films, the interface between the film and the substrate has to be incorporated (Fig. 4.16). The Kerr signal of the magnetic film is thus composed of the Kerr reflection at the surface and the Faraday effect that the light encounters upon transmission through the film, subsequent reflection at the film/substrate interface and another transmission through the film. The magnitudes of these two contributions are dependent on the film thickness. Qiu and Bader have

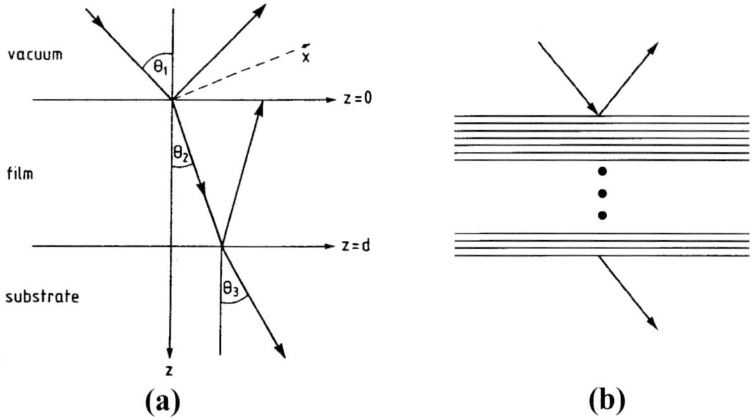

Fig. 4.16. (**a**) Magnetooptic Kerr effect for a thin magnetic film: both the Kerr reflection at the film surface and the Faraday effect of light that is transmitted through the film, subsequently reflected at the film–substrate interface and finally transmitted into the vacuum contribute to the Kerr signal. (**b**) Schematic illustration of a multilayer structure

given general formulae for elements of the matrix r for a multilayer structure in the ultrathin limit ($\sum_i n_i d_i \ll \lambda$) [62]. These formulae are quoted here:

$$r_{ss} = \frac{n_i \cos\theta_i - n_f \cos\theta_f}{n_i \cos\theta_i + n_f \cos\theta_f} , \tag{4.45}$$

$$r_{pp} = \frac{n_f \cos\theta_i - n_i \cos\theta_f}{n_f \cos\theta_i + n_i \cos\theta_f} , \tag{4.46}$$

$$r_{ps} = -\frac{4\pi}{\lambda} \frac{n_i \cos\theta_i}{(n_i \cos\theta_i + n_f \cos\theta_f)(n_f \cos\theta_i + n_i \cos\theta_f)} \times \left(\cos\theta_f \sum_m d_m n_m^2 Q_z^{(m)} - n_f n_i \sin\theta_i \sum_m d_m Q_y^{(m)} \right) , \tag{4.47}$$

$$r_{sp} = -\frac{4\pi}{\lambda} \frac{n_i \cos\theta_i}{(n_i \cos\theta_i + n_f \cos\theta_f)(n_f \cos\theta_i + n_i \cos\theta_f)} \times \left(\cos\theta_f \sum_m d_m n_m^2 Q_z^{(m)} + n_f n_i \sin\theta_i \sum_m d_m Q_y^{(m)} \right) . \tag{4.48}$$

Here n_i, θ_i, and n_f, θ_f are the complex refractive indices of and the incidence angles on the initial and final media, respectively. z is the surface normal direction (see Fig. 4.16), and d_m, n_m and $Q^{(m)}$ are the thicknesses, refractive indices and magnetooptic constants of the layers of the composite. For the special case of a magnetic single-layer film, the matrix elements of r are obtained by removing the summation symbol. In this case, several conclusions can be drawn from the above formulae. (i) Since the film thickness d_m is a real number, for a single ultrathin magnetic film both the Kerr rotation and the ellipticity depend linearly on the film thickness in the ultrathin limit. (ii) Q is in general a complex number and $\alpha_{K_s} + i\varepsilon_{K_s} = r_{ps}/r_{ss}$ can be considered as a product of Q and another complex number. Hence α_{K_s} and ε_{K_s} depend in a complex way upon Q [67]. However, if the real part of Q is much larger than the imaginary part so that the latter can be neglected, proportionality is a good approximation. At a wavelength of $\lambda = 6328$ Å, the magnetooptical constants Q are found to be $0.0376 + 0.0036i$ for Fe and $0.043 + 0.007i$ for Co [62]. (iii) Since the dielectric tensor is a function of frequency, the Kerr rotation and ellipticity are also functions of the frequency of the light. The frequency-dependent magnetooptic Kerr effect for Fe, Co and Ni has been calculated in [68] and measured in [67]. (iv) The polar magnetooptic effect is usually stronger than the longitudinal effect, since the n_m^2 term associated with Q_z is usually larger than $n_f n_i$ since $n_i = 1$. To obtain a strong Kerr effect, the angle of incidence should be small for the polar geometry and large for the longitudinal geometry.

To conclude, we shall describe one type of SMOKE setup and explain how it works. As shown in Fig. 4.17, a He–Ne laser ("laser" in the figure) operating in a single mode is used as a light source. The light becomes p-polarized

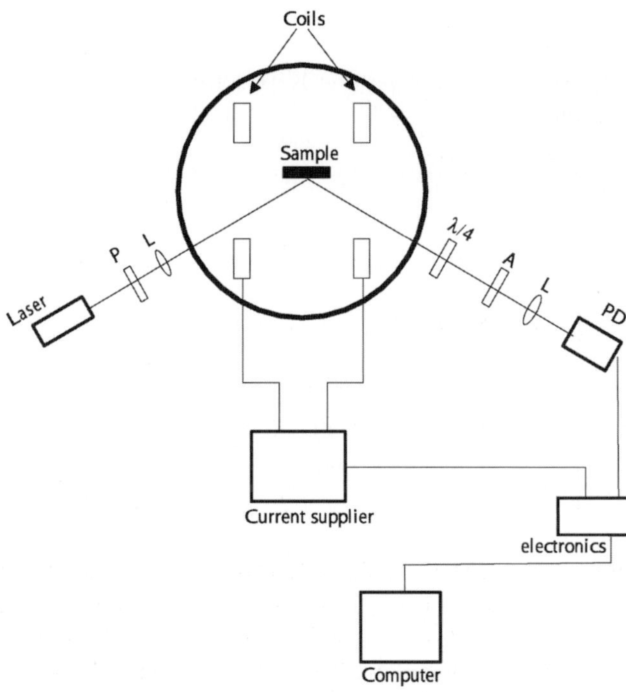

Fig. 4.17. Schematic illustration of a SMOKE setup. See the text for the explanation of the components

after passing through the first polarizer ("P"). Subsequently it is reflected from the sample surface ("sample"). Its intensity is detected by the photodiode ("PD"). The second polarizer ("A") is put between the sample and the photodiode and its polarization plane is rotated by an angle δ away from the extinction position. Two Helmholtz coils ("coils") are used to produce the magnetic field. The current which drives the coils is controlled by a computer ("computer"), and simultaneously the data for the light intensity detected by the photodiode are collected by the computer. In addition, two lenses ("L") are used to focus the light on the sample and on the photodiode. A $\lambda/4$ wave plate ("$\lambda/4$") is inserted into the light path to cancel the birefringence effect produced by the UHV windows through which the light passes.

Consider p-polarized light reflected from a sample surface. If the sample is nonmagnetic, the reflected light is purely p-polarized. If the sample is ferromagnetic then the reflected beam should consist of an s component (E_s) in addition to the dominant p component, with $|E_s|/|E_p|$ being the Kerr rotation. Since it is difficult to completely distinguish the s and p components and precisely measure them experimentally, the Kerr rotation is in practice measured by setting the polarizer (the analyzer) at a small angle (δ) from the extinction axis. In this case the intensity measured by the photodetector after the polarizer is

$$I = |E_p \sin\delta + E_s \cos\delta|^2 \approx |E_p\delta + E_s|^2 \,. \tag{4.49}$$

Since $E_s/E_p = -\alpha_{K_p} + i\varepsilon_{K_p}$ gives the Kerr rotation $-\alpha_{K_p}$ and ellipticity ε_{K_p}, (4.49) becomes

$$I = |E_p|^2|\delta - \alpha_{K_p} + i\varepsilon_{K_p}|^2 \approx |E_p|^2(\delta^2 - 2\delta\alpha_{K_p}) = I_0\left(1 - \frac{2\alpha_{K_p}}{\delta}\right) \,, \tag{4.50}$$

where

$$I_0 = |E_p|^2\delta^2 \tag{4.51}$$

represents the intensity at zero Kerr rotation. Since both α_{K_p} and ε_{K_p} scale linearly with the magnetization, the measured intensity as a function of H yields the magnetic hysteresis loop. The saturation Kerr rotation α_{K_p} can be determined by the relative change of the Kerr intensity $\triangle I$ obtained upon magnetization reversal. From (4.50) and (4.51), it is easy to obtain

$$\alpha_{K_p} = \frac{\delta}{4}\frac{\triangle I}{I_0} \,. \tag{4.52}$$

4.3.2 X-Ray Magnetic Dichroism

The successful production of tunable, intense and polarized ultraviolet synchrotron radiation has enabled a new technique for studying magnetism – x-ray magnetic dichroism [69]. Magnetic dichroism comprises those phenomena in which the measured quantities show a difference upon changing the polarization state of the impinging light or reversing the magnetization of the sample. To some extent, it is the x-ray counterpart of the magnetooptic Kerr effect. Magnetic dichroism can manifest itself in various forms in x-ray absorption and electron spectroscopy. For example, in x-ray absorption spectroscopy (XAS), the absorption coefficients, which are a function of photon energy, vary with a change of the helicity of a circularly polarized incident x-ray beam. By use of the magnetooptic sum rule, magnetic dichroism measured in the absorption mode allows the separation of the contributions of orbital and spin moments to the total magnetic moment [70]. In photoemission (PE), a difference in the electron emission intensity results from the reversal of either the helicity or the magnetization. Magnetic dichroism in core-level photoelectron spectroscopy can be exploited to probe the local magnetic properties on an atomic scale [71]. Magnetic dichroism in valence-band photoemission provides an efficient way to probe the electronic structure of magnetic materials, which is affected by spin–orbit coupling and exchange splitting. The attractiveness of the technique results from the general characteristics of electron spectroscopic techniques, which make magnetic dichroism element- and shell-specific. Hence magnetic dichroism is a powerful tool for obtaining a detailed understanding of the properties of ultrathin magnetic films, especially for the investigation of the magnetic properties of interfaces.

Basically, magnetic dichroism arises from the spin splitting and spin-orbit coupling. In this respect, it is similar to the magnetooptic Kerr effect. The microscopic mechanism for magnetic dichroism has been elucidated at different degrees of complexity [70, 72, 73, 74, 75]. We shall focus here on its applications in the study of ultrathin-film magnetism. We shall start with a simple review of the spin–orbit interaction in isolated atoms and in crystals because it plays a central role in the discussion of dichroism related to magnetic materials. Then we shall turn to core-level photoemission spectroscopy. Magnetic dichroism in x-ray absorption and the sum rule will be considered next. Finally, we shall consider magnetic dichroism in valence band photoemission.

Let us first consider the spin–orbit coupling in an isolated atom. For simplicity, a perturbation approach rather than a full relativistic treatment is adopted here. For an N-electron atom, the Hamiltonian can be approximated by

$$\mathbf{H} = \sum_{i=1}^{N} \left(-\frac{\hbar^2}{2m}\triangle_i - \frac{Ze^2}{4\pi\varepsilon_0}\frac{1}{r_i} + \xi(r_i)\mathbf{l}_i \cdot \mathbf{s}_i \right) + \sum_{i(i<j)=1}^{N} \frac{e^2}{4\pi\varepsilon_0}\frac{1}{r_{ij}} , \qquad (4.53)$$

where the $\xi(r_i)\mathbf{l} \cdot \mathbf{s}$ term describes the spin–orbit coupling. Analytic solutions are not available for this $3N$-dimensional Schrödinger equation. The self-consistent density functional approximation provides a numerical method for this many-electron problem, as we have discussed in the early part of this chapter. However, the basic features of atomic structure can be resolved in a much simpler way, in which we separate out a simple spherically symmetric term $\mathbf{H}_0$ and treat the rest as perturbations to $\mathbf{H}_0$. Thus the Hamiltonian of the system is rewritten as

$$\mathbf{H} = \mathbf{H}_0 + \mathbf{H}_1 + \mathbf{H}_2 \ , \tag{4.54}$$

where

$$\mathbf{H}_0 = \sum_{i=1}^{N} \left(-\frac{\hbar^2}{2m} \Delta_i + v(r_i) \right) \ , \tag{4.55}$$

$$\mathbf{H}_1 = \sum_{i(i<j)=1}^{N} \frac{e^2}{4\pi\varepsilon_0} \frac{1}{r_{ij}} - \sum_{i=1}^{N} (\frac{Ze^2}{4\pi\varepsilon_0 r_i} + v(r_i)) \ , \tag{4.56}$$

$$\mathbf{H}_2 = \sum_{i=1}^{N} \xi(r_i) \mathbf{l}_i \cdot \mathbf{s}_i \ . \tag{4.57}$$

$\mathbf{H}_0$ describes the unperturbed Hamiltonian with central symmetry. The physical meaning of $\mathbf{H}_0$ is that we consider each electron as moving in a central potential $v(r_i)$, created by the electric field of the nucleus and partially screened by the other electrons. $\mathbf{H}_1$ describes the noncentral part of the interelectronic repulsion. The spin–orbit coupling term $\mathbf{H}_2$ comes from the interaction between the electron spin and the magnetic field experienced by the electron when it moves around the atomic nucleus. $\mathbf{H}_0$ shows full rotational symmetry. Its eigenfunctions are products of a radial part and spherical harmonics,

$$\psi_{n,l,m} = R_{n,l}(r) Y_l^m(\theta, \phi) \ . \tag{4.58}$$

Each eigenfunction is labelled with three quantum numbers n, l, l_z, with $n = 1, 2, \cdots$, $l = 0, 1, 2, \cdots, n-1$ and $l_z = -l, -(l-1), \cdots, l-1, l$. Electrons having the same quantum number n are said to belong to the same shell, which is often also specified by K, L, M, N, $\cdots$ for $n = 1, 2, 3, 4, \cdots$. For a given n, each possible l defines a subshell. This subshell consists of a $(2l+1)$-dimensional irreducible subspace of the full rotation group of $\mathbf{H}_0$. Therefore the energy eigenvalue of $\mathbf{H}_0$ is $(2l+1)$-fold degenerate. The subshells with $l = 0, 1, 2, 3, \cdots$ are also denoted by s, p, d, f, $\cdots$. Besides the spatial degree of freedom, the electron possesses another intrinsic degree of freedom – the electron spin. The electron spin can be introduced by multiplying the space wave function ψ_{n,l,l_z} by the spin eigenfunctions α and β, which belong to the spin quantum numbers $s_z = (1/2)$ and $s_z = -(1/2)$ respectively. For the Hamiltonian $\mathbf{H}_0$, the spatial part and the spin part are completely independent. An electron state can be described by a complete set of quantum numbers n, l, l_z and s_z with a total $2(2l+1)$-fold degeneracy.

In the case where the spin–orbit interaction is much larger than the noncentral part of the interelectron repulsion, one has to add $\mathbf{H}_2$ and only $\mathbf{H}_2$ to the total Hamiltonian. This holds, for example, for the core shells of heavy atoms. The inclusion of the perturbation $\mathbf{H}_2$ partly lifts the degeneracy with respect to l_z and s. $\mathbf{H}_2$ may have matrix elements different from zero between eigenstates of $\mathbf{H}_0$ with equal l, s and $l_z + s_z$. Now neither l_z nor s_z is a good quantum number anymore. Instead, $\mathbf{j}^2 = (\mathbf{l} + \mathbf{s})^2$ and $\mathbf{j}_z = \mathbf{l}_z + \mathbf{s}_z$ commute with $\mathbf{H}_2$, and j and j_z become good quantum numbers. The energy level which is determined by n ($n \neq 0$) and l under $\mathbf{H}_0$, then splits into two levels

$$j = l + \frac{1}{2}, \quad E_2 = \frac{\zeta_{nl}}{2} l, \tag{4.59}$$

$$j = l - \frac{1}{2}, \quad E_2 = -\frac{\zeta_{nl}}{2}(l+1), \tag{4.60}$$

where E_2 denotes the spin–orbit energy, with $\zeta_{nl} = \hbar^2 \int R_{nl}^2(r)\xi(r)\,\mathrm{d}r$. The allowed values for j_z are $j, j-1, \cdots, -(j-1), -j$, so the degrees of degeneracy are $2l+2$ and $2l$, respectively, for these two split energy levels. This is called the j–j coupling scheme. Figure 4.18 schematically shows the energy levels of copper resulting from spin–orbit interaction. Since the spin–orbit interaction is proportional to the fourth power of the effective nuclear charge (Z^4) but the noncentral part of the interelectron repulsion is proportional to Z only, the spin-orbit interaction is the dominant perturbation for the states of core levels for 3d transition metals, while it is smaller for the valence electrons, since the effective nuclear charge is high for core levels but much lower for

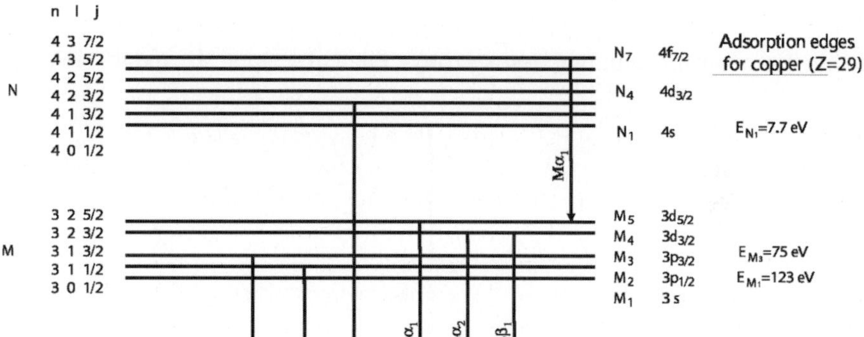

Fig. 4.18. The schematic energy levels (not drawn to scale) resulting from j–j coupling for copper. The standard notations used in the literature for the levels are given together with the allowed x-ray emission transitions. After [81], used with permission

valence electrons. Therefore the j–j coupling scheme is a good approximation for core-level states.

The degeneracy of the energy levels with respect to j_z in j–j coupling can be lifted by interaction with the exchange-split valence band when a ferromagnetic metal is considered [71, 76]. Thus the spin–orbit-split level is further split into sublevels, labeled by j_z. In an atomic model, this splitting is caused by the electrostatic interaction of the core level with the magnetically polarized valence electrons. This splitting can be detected experimentally by photoemission dichroism of the core level and thus becomes a diagnostic of atomic magnetic order in ultrathin films, and even at surfaces and interfaces [71, 77, 78, 79, 80].

The core levels can act as the initial states of photoelectron transitions in core-level magnetic-dichroism experiments. The final states could be free-electron states when the photon energy is well above the binding energy of the core electrons, as in x-ray core-level photoemission spectroscopy (see Fig. 4.19a). Alternatively, they could be the empty valence states just above the Fermi level, as in x-ray edge absorption spectroscopy (see Fig. 4.19b). The x-ray photoelectron transition probability can be dealt with in the dipole approximation [81, 82]. The transition must obey the selection rules

$$\triangle J = 0, \pm 1 \,, \quad \triangle J_z = \sigma \widehat{\boldsymbol{q}} \cdot \widehat{\boldsymbol{m}} \,, \quad J_i + J_f \geq 1 \,, \tag{4.61}$$

where J and J_z are the quantum number of total angular momentum and the magnetic quantum number, respectively. The subscripts "i" and "f" denote

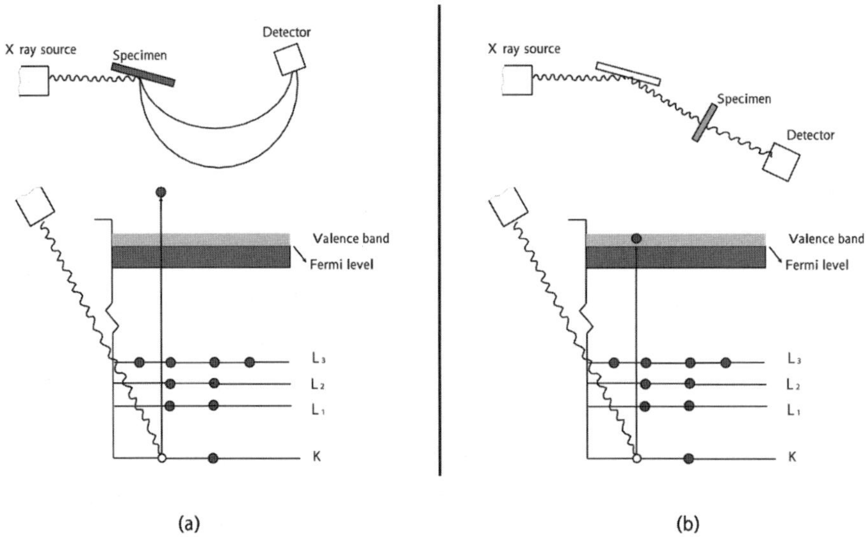

Fig. 4.19. Schematic experimental setup and energy-level diagram for (**a**) x-ray photoemission spectroscopy (XPS) and (**b**) x-ray absorption spectroscopy. After [81], used with permission

the initial state and the final state involved in the transition. $\widehat{\boldsymbol{q}}$ is the unit vector in the propagation direction of the light, $\widehat{\boldsymbol{m}}$ is the unit vector of the magnetic axis, and σ represents the helicity of the light. σ equals 1 for right-handed circularly polarized light and -1 for left-handed circularly polarized light. The case $\sigma = 0$ corresponds to linearly polarized light. When only a single electron is involved in the transition, the selection rules can be specified further:

$$\triangle l = \pm 1 \,, \quad \triangle s = 0 \,, \quad \triangle J = 0, \pm 1 \,, \quad \triangle J_z = \sigma \widehat{\boldsymbol{q}} \cdot \widehat{\boldsymbol{m}} \,. \tag{4.62}$$

This specification is a consequence of the fact that the spin does not change in the electric-dipole approximation, since the electric-dipole operator acts only on the spatial part of the electron wave function. If circularly polarized light is used and the axis of the magnetic field is parallel or antiparallel to the propagation direction of the light, we simply have

$$\triangle J_z = J_{z\mathrm{f}} - J_{z\mathrm{i}} = \pm 1 \,. \tag{4.63}$$

In Fig. 4.20, a hypothetical system is used to illustrate the origin of magnetic dichroism in x-ray absorption, where the spin degree of freedom is dropped and the degeneracy with respect to the magnetic quantum number m is lifted. It can be clearly seen that the energy of the photons resonantly absorbed by the system changes when the helicity of the circularly polarized radiation varies. Hence, magnetic circular dichroism is observed.

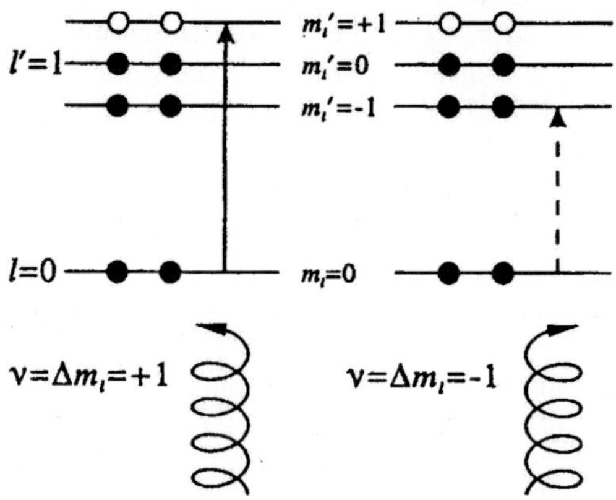

Fig. 4.20. Energy-level scheme of a hypothetical system exhibiting magnetic circular dichroism. Electrons at the core level $l = 0$ can be excited to the empty sublevel $l' = 1$, $m' = +1$ by right-handed circularly polarized light, but the transition to the sublevel $l' = 1$, $m' = -1$ by left-handed circularly polarized light is forbidden, since the sublevel is fully occupied, indicating a difference in absorption when the helicity of the light changes. From [84], used with permission

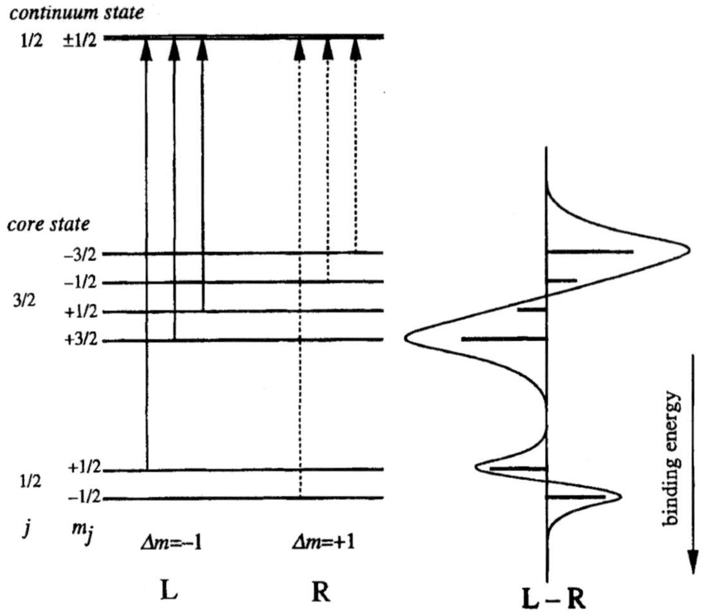

Fig. 4.21. Dipole transitions from a core level to a continuum free-electron state with left-handed and right-handed circularly polarized light, and the resulting circular dichroism in the photoemission. From [83], used with permission

For core-level photoemission, the final free-electron state is characterized by $J = 1/2$ and $J_z = \pm 1/2$. The dipole transitions from an exchange-split core p level to a continuum free-electron state with left-handed and right-handed circularly polarized light are shown in Fig. 4.21. Transitions are allowed from the sublevels $J_z = -3/2$ and $-1/2$ of the core level with right-handed circularly polarized light ($\triangle J_z = +1$) and from the sublevels $J_z = +1/2$ and $+3/2$ with left-handed circularly polarized light ($\triangle J_z = -1$). Consequently, the difference spectrum $(L - R)$, which is defined as the difference between the intensities for the left- and right-handed circularly polarized light used, shows a $+ - -+$ structure. This feature has been exploited to detect magnetic order at a surface, at an interface and in thin films [71]. This method is element-specific and probes the local magnetic properties on an atomic scale. On the other hand, it is surface-sensitive. The depth of information can be tuned by changing the photon energy.

Figure 4.22 shows magnetic dichroism measured by Baumgarten et al. for the Fe(110) surface using circularly polarized x-rays [71]. A grazing incidence of photons was used in this measurement, so that the photon spin and the magnetization were nearly parallel. The electron takeoff angle was 55°. The difference spectrum demonstrates obviously the $+ - -+$ feature, which is explained as coming from the interference of the spin–orbit interaction in

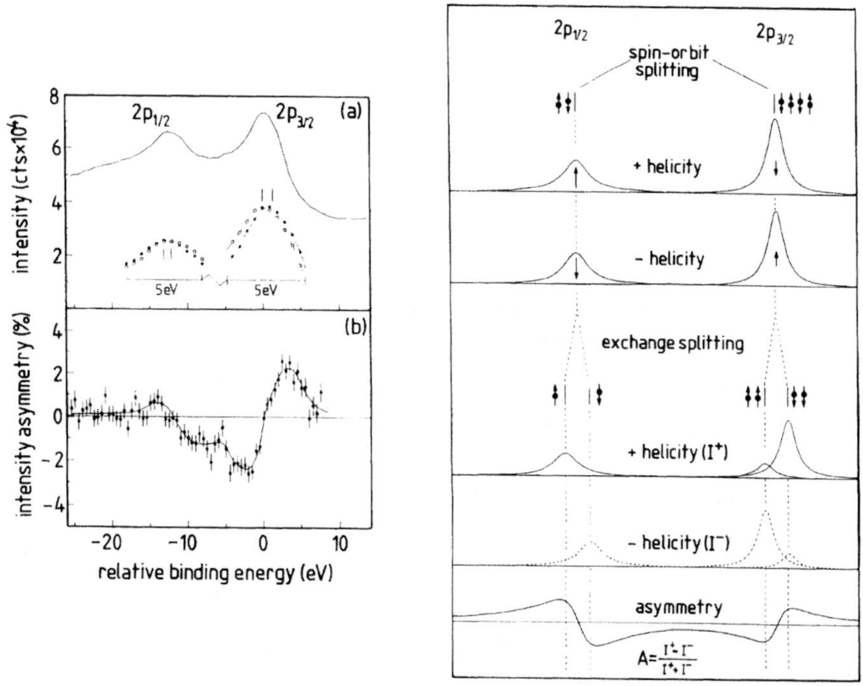

Fig. 4.22. (**a**) and (**b**): the magnetic x-ray dichroism in photoemission observed for an Fe(110) single crystal. (**a**) The *upper curve* shows the total intensity spectrum in the region of the spin-orbit-split Fe 2p level. The *lower curves* show, on an enlarged scale, the partial intensities in the peak regions for the photon spin parallel (*solid lines*) and antiparallel (*dashed lines*) to the sample magnetization. (**b**) Asymmetry A of the measured intensity for the photon spin parallel to the sample magnetization (I^+) and antiparallel to it (I^-). $A = (I^+ - I^-)/(I^+ + I^-)$. The asymmetry curve was obtained from two separate runs with opposite photon helicity. (**c**) Schematic explanation of the origin of the observed magnetic dichroism in photoemission. From *top* to *bottom* the figure shows the expected relative intensities of Lorenz-type photoelectron lines when excited by circularly polarized light in the presence of spin–orbit splitting only (*top*), and with additional exchange splitting (*bottom*). From [71], used with permission

the core levels and the exchange interaction with the 3d valence electrons, together with the dipole selection rules for circularly polarized light.

Besides the circularly polarized magnetic dichroism exemplified in Fig. 4.22, magnetic dichroism has also been observed in angle-resolved photoemission experiments with linearly polarized light (LMDAD) [77, 85, 86]. Since the LMDAD effect averages out if a full angular integration of the photoemission current is performed, angular selection of the photoelectrons is required. On the other hand, a particular experimental geometry is also necessary to define a chirality where the magnetization $\boldsymbol{M}$ is parallel to $\boldsymbol{E} \times \boldsymbol{P}$,

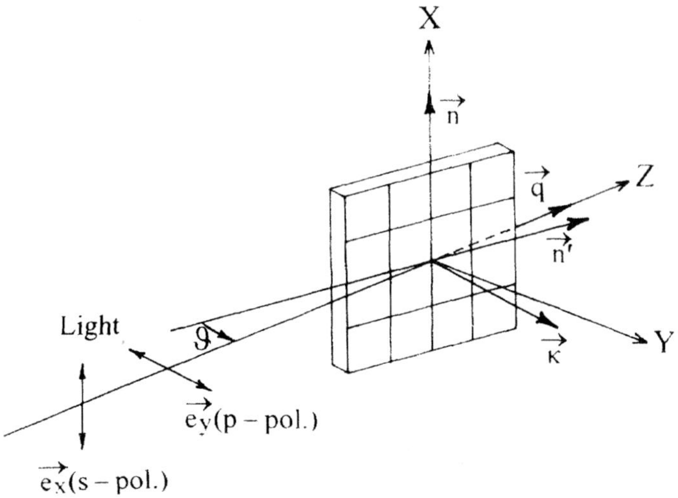

Fig. 4.23. An experimental geometry for linear magnetic dichroism in angularly resolved core-level photoemission. From [87], used with permission

with $\boldsymbol{E}$ and $\boldsymbol{P}$ being the polarization vector of the photon and the photoelectron momentum, respectively (see Fig. 4.23). The interpretation of the LMDAD effect is based on atomic photoionization theory. Generally, the dependence of the spectra on the direction of the magnetization is also a result of the interplay between the spin–orbit interaction of the core-level electrons and their exchange interaction with the valence electrons. A quantitative relationship has been derived via density matrix theory for the experimental geometry shown in Fig. 4.23 [87]. Three unit vectors $\boldsymbol{n}$, $\boldsymbol{k}$ and $\boldsymbol{q}$ are used in the formula to define the directions of the sample magnetization, the electron emission and the photon beam, respectively. The LMDAD for a one-electron sublevel (or a core hole) of an $n\mathrm{p}_{1/2}$ or $n\mathrm{p}_{3/2}$ core level reads

$$\begin{aligned} I_j^{LMDAD} &= I_j(\boldsymbol{k},\boldsymbol{n}) - I_j(\boldsymbol{k},-\boldsymbol{n}) \\ &= \frac{\sigma_{nlj}(\omega)}{2\pi}\frac{3i}{2}C_{221}^{j}(2j+1)^{1/2}\rho_{10}^{n} \\ &\times (\boldsymbol{k}\cdot\boldsymbol{q})(\boldsymbol{q}\cdot[\boldsymbol{k}\times\boldsymbol{n}]) \pm [(\boldsymbol{k}\cdot\boldsymbol{q})(k_x n_y + k_y n_x) - 2(\boldsymbol{n}\cdot\boldsymbol{q})k_x k_y] \,. \end{aligned} \tag{4.64}$$

Here the upper and lower signs refer to s- and p-polarized light, respectively, $\sigma_{nlj}(\omega)$ is the partial photoionization cross section, and $C_{221}^{j}(2j+1)$ is a dimensionless parameter. ρ_{N0}^{n} is the state multipole describing the polarization properties of a sublevel with a given projection J_z (see Fig. 4.21) [88]. The values of ρ_{10}^{n} for $n\mathrm{p}_{1/2}$ and $n\mathrm{p}_{3/2}$ states are listed in Table 4.5. For s-polarized light, (4.64) gives a zero result for both directions of magnetization $\boldsymbol{n}$ and $\boldsymbol{n}'$ in Fig. 4.23, in accordance with experiment [85]. For p-polarized

Table 4.5. State multipoles ρ_{10}^n for magnetic sublevels $n\mathrm{p}_{1/2}$ and $n\mathrm{p}_{3/2}$

	$J=1/2$		$J=3/2$			
J_z	$1/2$	$-1/2$	$3/2$	$1/2$	$-1/2$	$-3/2$
ρ_{10}^n	$1/\sqrt{2}$	$-1/\sqrt{5}$	$3/(2\sqrt{5})$	$1/(2\sqrt{5})$	$-1/(2\sqrt{2})$	$-3/(2\sqrt{5})$

light and $\boldsymbol{n} \perp \boldsymbol{q}$, (4.64) becomes

$$I_j^{LMDAD} = \frac{\sigma_{nlj}}{2\pi} 3iC_{221}^j (2j+1)^{1/2} \rho_{10}^n \sin\theta \cos\theta \,, \tag{4.65}$$

θ being the angle of grazing incidence of the light.

LMDAD measurements have been performed mostly on shallow 3p core levels for ferromagnetic 3d transition metals, with the motivation of being used of a probe for surface magnetism. The validity of LMDAD as a surface magnetometer has been justified experimentally [86, 89]. The magnitude of the LMDAD asymmetry of a given sublevel, which is defined as $A^{LMDAD} = (I(\boldsymbol{n}) - I(-\boldsymbol{n}))/(I(\boldsymbol{n}) + I(-\boldsymbol{n}))$, is related directly to the magnetic order parameter $\langle \boldsymbol{M} \rangle$, whilst the splitting between the sublevels is related to the atomic magnetic moment. Figure 4.24 shows the Fe 3p LMDAD spectra from an Fe(100) surface. The measurements were performed at a temperature of 150 K and a linearly polarized monochromatic x-ray beam

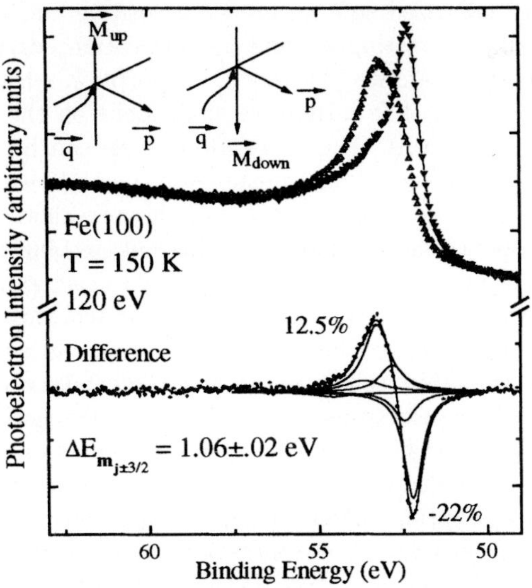

Fig. 4.24. LMDAD dichroism of Fe $3p$ from the Fe(100) surface measured at 150 K with monochromatic synchrotron radiation of an energy of 120 eV. The *top curves* show the LMDAD spectra for the two opposite magnetization directions. The *lower curves* show the difference. From [86], used with permission

with an energy of 120 eV was used. The experimental chiral geometries are presented in the inset. For the 3p core levels, the spin–orbit splitting and the exchange splitting have comparable magnitudes, so that the $3p_{1/2}$ and $3p_{3/2}$ LMDAD spectra overlap with each other. To obtain the LMDAD dichroism signal for a specific sublevel, a line shape analysis is always required. The difference spectrum in Fig. 4.24 has been fitted by a sextuplet of sublevels, which are weighted on the basis of (4.64) [85]. The fitting shows that the positive and negative extrema of the dichroism spectrum correspond to the energetic positions of the $J_z = \pm 3/2$ sublevels. The width of the dichroism spectrum is conveniently defined by the energy difference between these two peaks. The proportionality between the magnetic asymmetry and the order parameter of magnetization has been studied by measuring the temperature dependence of the asymmetry [86]. The LMDAD magnetic asymmetry shows the same temperature dependence as spin-polarized secondary-electron spectroscopy (SPSEP), but is quite different from the MOKE result. This has also been taken as evidence that the LMDAD spectrum is surface-sensitive, like SPSEP, while MOKE measures the average (bulk) properties.

Up to now, we have discussed the magnetic dichroism in the core-level photoemission spectrum, where the inner electrons are excited to free-electron states by x-ray photons. The inner electrons can also be excited to empty states in the valence band (see Fig. 4.19). These transitions are measured by the absorption spectrum. The x-ray magnetic circular dichroism in the absorption mode is the difference between the total absorption cross sections of circularly polarized soft x-rays with positive and negative helicity at the magnetically interesting L_2 and L_3 absorption edges in the case of 3d transition metals or at the M_4 and M_5 absorption edges in the case of the rare earths (see Fig. 4.18). Here we shall focus on the $L_{2,3}$ absorption edge of magnetic 3d metals. The absorption edges can be measured using a variety of techniques, including transmission, partial electron yield, total electron yield, partial fluorescence yield, sample current and reflection. The different measurement schemes represent different probing depths, allowing depth-dependent studies, and are complementary to each other. A discussion of the relative merits of the various measurement schemes is given in [90]. The importance of measuring the magnetic dichroism in edge absorption is based on the fact that this technique can give the orbital and spin magnetic moments of the valence electrons separately, by applying the magnetooptic sum rule. Therefore we shall first give a simple introduction to this sum rule.

The sum rule in magnetooptics refers to an expression for the difference in the integrated absorption coefficient when the magnetization direction or photon polarization is reversed. Recently, a new sum rule for x-ray circular magnetic dichroism has been derived by Thole, Carra and their coworkers [91, 92], based on the local atomic picture and the electric-dipole approximation. Rigorously, the valence electrons in 3d metals are itinerant rather than localized and thus the band structure should be considered for the final states in core-level edge absorption. However, the validity of the sum rules even for the 3d magnetic metals (Fe, Co and Ni) has been justified both by experiments [93] and by first-principles band structure calculations [94, 95]. The sum rule for the orbital moment is given by [91]

$$\rho = \frac{\int_{\text{edge}} \mathrm{d}\omega(\mu^+ - \mu^-)}{\int_{\text{edge}} \mathrm{d}\omega(\mu^+ + \mu^- + \mu^0)} = \frac{1}{2}\frac{c(c+1) - l(l+1) - 2}{l(l+1)(4l+2-n)}\langle L_z \rangle \ . \tag{4.66}$$

In this formula, μ^+, μ^- and μ^0 are the x-ray absorption coefficients, when the direction of magnetization is parallel (μ^+), antiparallel (μ^-) and perpendicular (μ^0) to the photon angular-momentum vector. In an equivalent experimental geometry, μ^+, μ^- and μ^0 can also be measured as the absorption coefficients for right-handed circularly polarized light, left-handed circularly polarized light and linearly polarized light with the polarization vector parallel to the magnetization axis. In this case the magnetization direction is fixed. μ^0 will be the average of μ^+ and μ^-. The integral should be taken over a complete core-level edge. In the case where the edge is spin–orbit split, the integration must be over the two split components, i.e., the $l+1/2$ and $l-1/2$ components. In the case of 2p–3d absorption, these two components are the L_2 and L_3 edges. Here l is the orbital quantum number of the valence state and c that of the core state. n is the number of electrons in the valence shell, for which a value obtained from band structure calculations is usually employed. Using (4.66), the expectation value of the orbital angular momentum and thus the orbital magnetic moment can be determined by measuring the magnetic dichroism in x-ray core-level edge absorption. For $L_{2,3}$, $l = 2$ (3d) and $c = 1$ (2p), leading to

$$\langle L_z \rangle = 2 \int_{L_{2,3}\ \text{edge}} \mathrm{d}\omega(\mu^+ - \mu^-) \times \frac{10 - n}{\int_{L_{2,3}\ \text{edge}} \mathrm{d}\omega(\mu^+ + \mu^- + \mu^0)} \ . \tag{4.67}$$

The sum rule related to the spin moment has also been derived in a similar way [92]. Its general expression reads

$$\langle S_z \rangle = A \left[\frac{\int_{J^+}(\mu^+ - \mu^-)\mathrm{d}\omega - [(c+1)/c]\int_{J^-}(\mu^+ - \mu^-)\mathrm{d}\omega}{\int_{J^+ + J^-}(\mu^+ + \mu^- + \mu_0)\mathrm{d}\omega} \right] - B\langle T_z \rangle \ , \tag{4.68}$$

where

$$A = \frac{3c(4l+2-n)}{l(l+1) - 2 - c(c+1)}$$

and

$$B = \frac{l(l+1)[l(l+1) + 2c(c+1) + 4] - 3c(c-1)^2(c+2)^2}{6lc(l+1)(4l+2-n)} \ .$$

For the $L_{2,3}$ edges, one obtains

$$\langle S_z \rangle = \frac{3}{2}\left[\int_{\mathrm{L}_3}(\mu^+ - \mu^-)\mathrm{d}\omega - 2\int_{\mathrm{L}_2}(\mu^+ - \mu^-)\mathrm{d}\omega\right]$$
$$\times \frac{(10-n)}{\int_{\mathrm{L}_2+\mathrm{L}_3}}(\mu^+ + \mu^- + \mu^0)\mathrm{d}\omega) - 3.5\langle T_z \rangle\ . \qquad (4.69)$$

In contrast to the sum rule for the orbital moment, complications are involved in (4.68) in two ways. First, for the spin sum rule, it is necessary to separate the L_3 and L_2 edges owing to the different coefficients appearing in front of the integrals of $\int_{J^+}(\mu^+ - \mu^-)\mathrm{d}\omega$ and $\int_{J^-}(\mu^+ - \mu^-)\mathrm{d}\omega$ in (4.69). For the magnetic 3d transition metals in which we are interested, these two edges partially overlap. Therefore certain assumptions or models are required to separate the two edges. This will inevitably introduce uncertainties into the result. Secondly, in addition to the expectation value of the spin quantum number, another quantity, $\langle T_z \rangle$, the expectation value of the magnetic-dipole operator $\mathbf{T}$ projected onto the quantum axis ($\mathbf{T} = \sum_i(\mathbf{s}_i - 3\mathbf{r}_i(\mathbf{r}_i \cdot \mathbf{s}_i)/r_i^2)$), is present in the sum rule. A nonzero $\langle T_z \rangle$ could be induced by an anisotropic charge distribution around the atom or by spin–orbit interaction [96]. The contribution from the spin–orbit interaction is small for 3d metals. The magnetic dipole induced by the anisotropic charge distribution is zero in bulk systems with cubic symmetry, but can be strongly enhanced at surfaces and interfaces owing to the symmetry breaking. A theoretical calculation shows that an error of up to 50% will be induced for the Ni(100) surface if the spin sum rule is applied neglecting the $\langle T_z \rangle$ term [95]. Therefore, to determine the surface spin magnetic moment using (4.68), a knowledge of $\langle T_z \rangle$ is required. Fortunately as shown by Stöhr and König, the value of $\langle T \rangle$ along three orthogonal high-symmetry directions, $T_x + T_y + T_z$, is zero even for a system with a crystal symmetry lower than cubic but higher than D_{2h} [97]. This allows a direct determination of $\langle S \rangle$ by averaging the measurement along the x, y and z axes or by magnetizing the sample along the "magic" direction with a large enough external magnetic field [97, 98]. The "magic" direction for a cubic crystal is the direction of the body diagonal. This diagonal makes the same angle with the three cubic axes and thus the projection along this direction is $(T_x + T_y + T_z)\cos\theta = 0$.

The x-ray absorption spectra for the $\mathrm{L}_{2,3}$ edge and the magnetic-circular-dichroism spectra measured by Chen et al.[93] are shown in Fig. 4.25. The thin Fe film was evaporated on a 1 μm thick semitransparent parylene substrate so that the spectra could be measured in the transmission mode to eliminate the experimental artifacts which could be introduced in the reflection mode. In the measurement, photons with a circular-polarization degree of 76% hit the sample at an incidence angle of 45°. The applied magnetic field was parallel to the intersection of the incidence plane and the sample surface. The transmission x-ray absorption spectra, normalized with respect to the incident photon flux, were taken with the projection of the helicity of the incident photons parallel (I_+ in Fig. 4.25a) and antiparallel (I_- in Fig. 4.25a) to the magnetization direction. The magnetization-independent spectra of

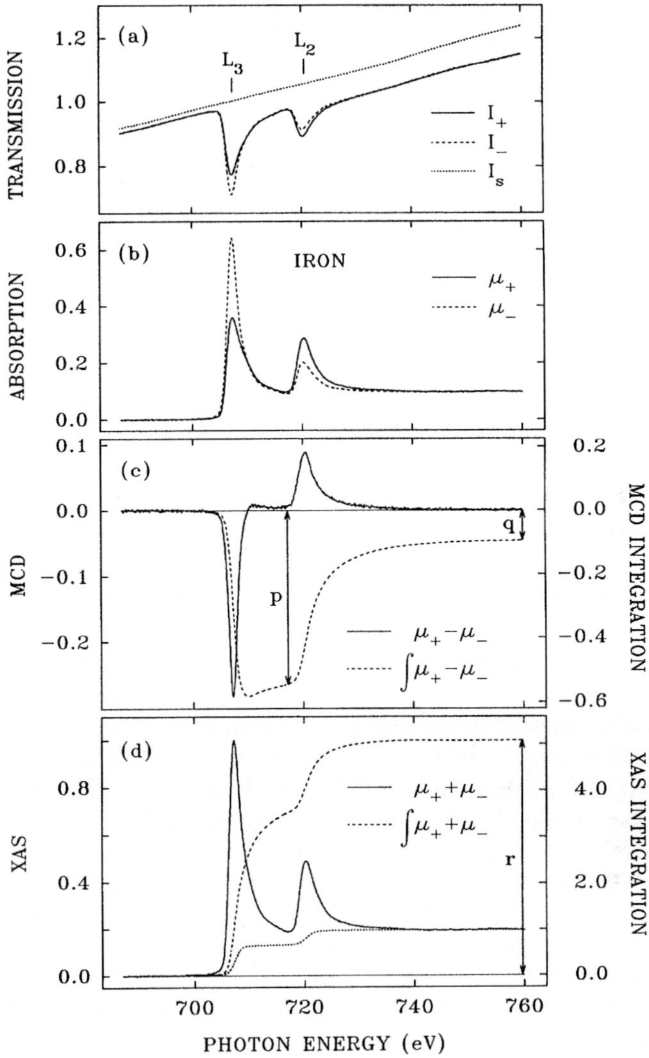

Fig. 4.25. $L_{2,3}$-edge x-ray absorption spectrum and magnetic-circular-dichroism (MCD) spectrum of Fe/parylene film. (**a**) Transmission spectra of Fe/parylene and of the parylene substrate alone, taken at two opposite saturation magnetizations; (**b**) the absorption spectra calculated from the transmission data shown in (**a**); (**c**) and (**d**) are the MCD and summed XAS spectra calculated from the spectra shown in (**b**). The *dotted line* shown in (**d**) is the two-step-like function used for edge-jump removal before the integration. The p and q shown in (**c**) and r shown in (**d**) are the three integrals needed in the sum-rule analysis. From [93], used with permission

the substrate (I_s in Fig. 4.25a) were also measured in order to calculate the relative absorption cross section using the relation $\mu^{\pm} \propto -\ln(I_{\pm}/I_s)$. The difference spectra $\mu^+ - \mu^-$ have been multiplied by $(1/\cos 45°)/0.76$ after taking into account the photon incidence angle and the circular-polarization degree, while the sum spectrum $\mu^+ + \mu^-$ has been left untreated. Three integrals are needed in the sum-rule analysis. They are $p = \int_{\mathrm{L}_3}(\mu^+ - \mu^-)\mathrm{d}\omega$, $q = \int_{\mathrm{L}_2+\mathrm{L}_3}(\mu^+ - \mu^-)\mathrm{d}\omega$ and $r = \int_{\mathrm{L}_2+\mathrm{L}_3}(\mu^+ + \mu^-)\mathrm{d}\omega$ (see Fig. 4.25c, d). Once p, q and r are known, the individual orbital and spin magnetic moments can be calculated as $m_{orbit} = -4q(10-n)\mu_B/3r$ and $m_{spin} = -(6p-4q)(10-n)\mu_B/r$, respectively by neglecting $\langle T_z \rangle$ and taking $\mu^0 = (1/2)(\mu^+ + \mu^-)$, where n is the number of 3d electrons and μ_B is the Bohr magneton. The integral q for the whole range, $\mathrm{L}_3 + \mathrm{L}_2$, can be precisely determined from the integrated spectrum. Since the L_3 edge may overlap with the L_2 edge, there is a minor uncertainty in determining the integral p for the L_3 edge alone, which is usually calculated by choosing a cutoff. The quantity r is obtained by integrating the $\mu^+ + \mu^-$ spectrum after a double step function has been subtracted from it. The double step function is used to describe the s-state contribution to the absorption edge. In Fig. 4.25d, the thresholds for the two-step function have been set to the peak positions of the L_3 and L_2 white lines ($\mu^+ + \mu^-$). The heights of the L_3 and L_2 steps have been set to 2/3 and 1/3, respectively, of the average intensity of the last 15 eV of the spectra, in accordance with their quantum degeneracy $2J+1$. The two-step-like function has then been convoluted with a Voigt function to simulate the intrinsic linewidth and the experimental resolution. In this way, the individual orbital and spin magnetic moments can be calculated. The results determined by this procedure show good agreement with those obtained by gyromagnetic-ratio measurements.

Though quantitative measurement of spin and/or orbital moments is possible by applying the sum rule to the x-ray absorption spectra, this method is only applicable for ferromagnetic materials. For antiferromagnetic and paramagnetic materials, the branching ratio, which is defined by the integral intensity ratio $I(\mathrm{L}_3)/[I(\mathrm{L}_2 + I(\mathrm{L}_3)]$ for 3d elements, is more interesting. Thole and van der Laan have studied the relationship between the branching ratio and the atomic magnetic moment by a full atomic approach [99]. It was concluded for the $\mathrm{L}_{2,3}$ edge of 3d elements that the branching ratio deviates from the statistical value (2/3) when an electrostatic interaction between the core hole and the valence electrons takes place. If spin–orbit splitting in the valence band can be neglected, high-spin states have on average a larger branching ratio than low-spin states. The branching ratio reaches a maximum for the Hund's-rule state of a free atom. The relationship between the branching ratio and the atomic magnetic moment has been studied by experiments for Fe compounds [100]. Therefore, qualitative information about the local magnetic moment could be obtained by measuring the branching ratio and comparing it with a material whose atomic moment is known.

Magnetic dichroism has also been studied in valence band photoemission. In this case, the local atomic picture becomes invalid owing to the itinerant properties of the valence electrons, and the electron spin loses its property of being a good quantum number owing to strong spin–orbit interactions. To explain the measured spectra, a fully relativistic calculation of the band structure is usually required. A review of this subject, designed especially for experimentalists, has been given by Kuch and Schneider [72].

4.3.3 Ferromagnetic Resonance (FMR)

Since the first experimental observation of ferromagnetic resonance by Griffiths [101] in 1946, FMR has been a standard technique for studying the ground-state properties of magnetic materials, especially for the investigation of magnetic anisotropy. Owing to its submonolayer sensitivity, in situ ultrahigh-vacuum FMR has also been developed for the investigation of ultrathin films and superlattices [102, 103, 104]. The magnetic anisotropy, interlayer exchange coupling and relaxation of magnetization of ultrathin films and superlattices have been successfully studied by FMR. In this subsection, we shall first give a compact introduction to the precessional motion of the magnetization, which provides the theoretical basis for data analysis in FMR measurements. This is followed by a description of the measuring technique. The important applications of FMR in the study of ultrathin magnetic films are then illustrated.

Equation of Motion of Magnetization. The magnetic moment ($\boldsymbol{m}$) of electrons, atoms or ions is always associated with an angular momentum ($\boldsymbol{P}$), which could be an electron spin moment, an orbital moment, or a mixture of both. The magnetic moment is related to the angular momentum by

$$\mu_0 \boldsymbol{m} = -\gamma \boldsymbol{P} \,, \tag{4.70}$$

where γ is the gyromagnetic ratio; $\gamma = (\mu_0|e|/2m_e)g = 1.015 \times 10^5 g$. g is called the Landau splitting factor, which is around 2 for the spin moment and 1 for the orbital moment. In the case where both the spin moment and the orbital moment contribute to the total magnetic moment, the orbital angular momentum $\boldsymbol{P}_L$, spin angular momentum $\boldsymbol{P}_S$, orbital magnetic moment $\boldsymbol{\mu}_L$ and spin magnetic moment $\boldsymbol{\mu}_S$ precess around the total angular momentum $\boldsymbol{P}_J$ (see Fig. 4.26). Therefore the total magnetic moment $\boldsymbol{\mu}_J$ is the sum of the components of $\boldsymbol{\mu}_L$ and $\boldsymbol{\mu}_S$ along the direction of $\boldsymbol{P}_J$. In this case, the g-factor is given by

$$g = 1 + \frac{J(J+1) + S(S+1) - L(L+1)}{2J(J+1)} \,, \tag{4.71}$$

where S, L and J are, respectively, the quantum numbers of the spin, orbital and total angular momentum, giving the spin, orbital and total angular momentum as

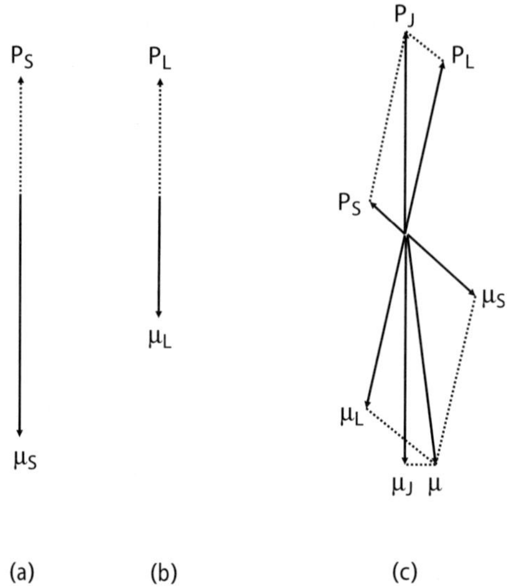

Fig. 4.26. The magnetic moment $\boldsymbol{m}$ of an electron is related to the angular momentum $\boldsymbol{P}$ by $\mu_0 \boldsymbol{m} = -\gamma \boldsymbol{P}$, where the gyromagnetic ratio $\gamma = (\mu_0|e|/2m_e)g = 1.015 \times 10^5 g$. (**a**) If only the electron spin is involved, $g \approx 2$; (**b**) if only the orbital moment is involved, $g = 1$; (**c**) for a mixed situation, g is given by (4.71)

$$
\begin{aligned}
P_S &= \sqrt{S(S+1)}\hbar \, , \\
P_L &= \sqrt{L(L+1)}\hbar \, , \\
P_J &= \sqrt{J(J+1)}\hbar \, , \quad J = |L-S|, |L-S+1|, \cdots, |L+S| \, .
\end{aligned}
\tag{4.72}
$$

In a magnetic field $\boldsymbol{H}$, the magnetic moment experiences a torque $\boldsymbol{L}$, which is given by

$$\boldsymbol{L} = \mu_0 \boldsymbol{m} \times \boldsymbol{H} \, . \tag{4.73}$$

Therefore the angular momentum, which is related to the magnetic moment by (4.70), will change with time t:

$$\frac{\mathrm{d}\boldsymbol{P}}{\mathrm{d}t} = \mu_0 \boldsymbol{m} \times \boldsymbol{H} \, . \tag{4.74}$$

Using (4.70), one obtains

$$\frac{\mathrm{d}\boldsymbol{m}}{\mathrm{d}t} = -\gamma \boldsymbol{m} \times \boldsymbol{H} \, . \tag{4.75}$$

This is the equation of motion of a single magnetic moment. Equation (4.75) can also be derived from quantum mechanics.

In ultrathin magnetic films, the film thickness is small compared with the exchange length,[3] and thus the strong exchange interaction keeps all magnetic moments parallel across the film so that the film can be regarded as a giant moment [104]. In this case, the equation of motion of the magnetization $\boldsymbol{M}$ has the same form as (4.75),

$$\frac{\mathrm{d}\boldsymbol{M}}{\mathrm{d}t} = -\gamma \boldsymbol{M} \times (\boldsymbol{H}_{\mathrm{eff}}) \,. \tag{4.76}$$

The only difference is that the applied static field in (4.75) is replaced by an effective magnetic field $\boldsymbol{H}$ in (4.76). The effective magnetic field includes the effective anisotropy field, in addition to the external field $\boldsymbol{H}$. Owing to the existence of magnetic anisotropy, the magnetization prefers to align along the easy direction to minimize the anisotropy energy, just as in the case where an external field exists in this direction. The effective field could also include a term arising from an inhomogeneous exchange interaction, but this is not the case for ultrathin films, and this term will be excluded from our discussion. In general, the effective field can be derived from the expression for the Helmholtz free energy F. For magnetic media, the derivative of the free energy is written as $\mathrm{d}F = -S\mathrm{d}T - \mu_0 \boldsymbol{H} \cdot \mathrm{d}\boldsymbol{M}$, and thus

$$\boldsymbol{H}_{\mathrm{eff}} = -\frac{1}{\mu_0}\frac{\partial F}{\partial \boldsymbol{M}} = -\frac{1}{\mu_0}\left(\frac{\partial F}{\partial \boldsymbol{M}_x}\boldsymbol{i} + \frac{\partial F}{\partial \boldsymbol{M}_y}\boldsymbol{j} + \frac{\partial F}{\partial \boldsymbol{M}_z}\boldsymbol{k}\right) \,. \tag{4.77}$$

The free energy includes contributions from the Zeeman energy $F_Z = -\mu_0 \boldsymbol{M} \cdot \boldsymbol{H}$, the anisotropy energy (see Sect. 4.2.3) and the exchange interaction $F_{exch} = -(A/M_s^2)\boldsymbol{M}_s \cdot \nabla^2 \boldsymbol{M}_s$, where A is the exchange stiffness coefficient. It should be pointed out that in (4.76), $\boldsymbol{H}_{\mathrm{eff}}$ becomes a function of $\boldsymbol{M}$ and thus (4.77) is now not valid anymore. In the following, the general expression for the free precessional frequency will be derived from the free energy in spherical coordinates (see Fig. 4.27) [105, 106, 107]. The Cartesian components of the effective field can be expressed using the polar coordinations θ and ϕ as

$$\begin{aligned}
H_{\mathrm{eff}\ x} &= -\frac{1}{\mu_0}\frac{\partial F}{\partial M_x} = -\frac{1}{\mu_0}\left(\frac{\partial F}{\partial \theta}\frac{\partial \theta}{\partial M_x} + \frac{\partial F}{\partial \phi}\frac{\partial \phi}{\partial M_x}\right) = \frac{\sin\phi}{\mu_0 M_s \sin\theta}\frac{\partial F}{\partial \phi} \,, \\
H_{\mathrm{eff}\ y} &= -\frac{1}{\mu_0}\frac{\partial F}{\partial M_y} = -\frac{\cos\phi}{\mu_0 M_s \sin\theta}\frac{\partial F}{\partial \phi} \,, \\
H_{\mathrm{eff}\ z} &= -\frac{1}{\mu_0}\frac{\partial F}{\partial M_z} = \frac{1}{\mu_0 M_s \sin\theta}\frac{\partial F}{\partial \theta} \,.
\end{aligned} \tag{4.78}$$

By substituting $\boldsymbol{H}_{\mathrm{eff}}$ with (4.78), the vectorial equation (4.76) can be transformed into three scalar equations. Of these equations, only two are independent; the third equation can be derived from the other two. From any two of

[3] The exchange length in the Heisenberg model equals the lattice spacing times the square root of the exchange energy divided by a typical demagnetizing energy, i.e., $l_{ex} = a\sqrt{2A/\mu_0 M^2}$.

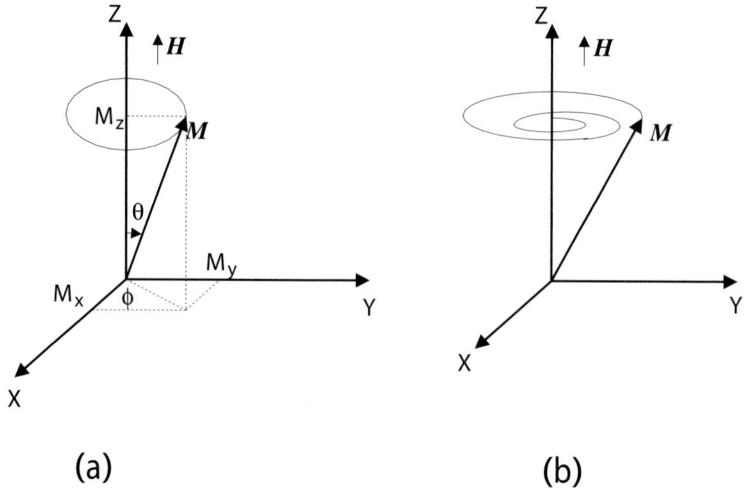

Fig. 4.27. The motion of the magnetization in a static magnetic field. (**a**) In the absence of damping, the magnetization precesses around the static magnetic field and its component along the magnetic field remains constant. (**b**) In the presence of damping, the track of the tip of the magnetization vector becomes a spiral rather than a circle as in (**a**), and the damping torque forces $\boldsymbol{M}$ to approach the direction of $\boldsymbol{H}$ gradually

these three scalar equations, we obtain

$$\frac{\partial\theta}{\partial t} = \gamma H_\phi \, ,$$
$$\frac{\partial\phi}{\partial t} = -\gamma H_\theta \, , \tag{4.79}$$

where

$$H_\theta = -\frac{1}{\mu_0 M_s \sin\theta}\frac{\partial F}{\partial\theta} \, ,$$
$$H_\phi = -\frac{1}{\mu_0 M_s \sin\theta}\frac{\partial F}{\partial\phi} \, . \tag{4.80}$$

Assuming that, for some reason, the magnetization $\boldsymbol{M}$ deviates from a stable direction (θ_0, ϕ_0) by a small angle and points in the direction ($\theta_0 + \triangle\theta$, $\phi_0 + \triangle\phi$), (4.79) becomes

$$\frac{\partial(\triangle\theta)}{\partial t} = \frac{\partial\theta}{\partial t} = -\gamma\frac{1}{\mu_0 M_s \sin(\theta_0+\triangle\theta)}\frac{\partial F}{\partial\phi}$$
$$= -\gamma\frac{1}{\mu_0 M_s \sin(\theta_0+\triangle\theta)}\left[\left(\frac{\partial F}{\partial\phi}\right)_{(\theta_0,\phi_0)}\right.$$

$$
\begin{aligned}
&+ \left(\Delta\theta \frac{\partial}{\partial\theta} + \Delta\phi \frac{\partial}{\partial\phi}\right) \left(\frac{\partial F}{\partial\phi}\right)_{(\theta_0,\phi_0)} + \cdots \Bigg] \\
&\approx -\gamma \frac{1}{\mu_0 M_s \sin\theta_0} \left[\Delta\theta \left(\frac{\partial^2 F}{\partial\theta\,\partial\phi}\right)_{(\theta_0\phi_0)} + \Delta\phi \left(\frac{\partial^2 F}{\partial\phi^2}\right)_{(\theta_0,\phi_0)} \right], \qquad (4.81)
\end{aligned}
$$

In the derivation of (4.81), the conditions for the stable magnetization direction (θ_0, ϕ_0)

$$
\left(\frac{\partial F}{\partial\theta}\right)_{(\theta_0\ ,\phi_0)} = 0, \quad \left(\frac{\partial F}{\partial\phi}\right)_{(\theta_0,\phi_0)} = 0 \qquad (4.82)
$$

have been exploited. For the case where the magnetization $\boldsymbol{M}$ of a crystal precesses around a stable direction, let us write the deviation of $\boldsymbol{M}$ from the stable direction (θ_0, ϕ_0) as $\Delta\theta = \widehat{\Delta\theta}\, e^{i\omega t}$ and $\Delta\phi = \widehat{\Delta\phi}\, e^{i\omega t}$, (4.81) then becomes

$$
\begin{aligned}
&\left(i\omega + \frac{\gamma}{\mu_0 M_s \sin\theta_0} \left(\frac{\partial^2 F}{\partial\theta\,\partial\phi}\right)_{(\theta_0,\phi_0)} \right) \widehat{\Delta\theta} \\
&\quad + \left(\frac{\gamma}{\mu_0 M_s \sin\theta_0} \left(\frac{\partial^2 F}{\partial\phi^2}\right)_{(\theta_0,\phi_0)} \right) \widehat{\Delta\phi} = 0\,, \\
&\left(-\frac{\gamma}{\mu_0 M_s \sin\theta_0} \left(\frac{\partial^2 F}{\partial\theta^2}\right)_{(\theta_0,\phi_0)} \right) \widehat{\Delta\theta} \\
&\quad + \left(i\omega - \frac{\gamma}{\mu_0 M_s \sin\theta_0} \left(\frac{\partial^2 F}{\partial\theta\,\partial\phi}\right)_{(\theta_0,\phi_0)} \right) \widehat{\Delta\phi} = 0\,. \qquad (4.83)
\end{aligned}
$$

Similarly to the derivation for (4.77), the solvability condition for (4.83) gives

$$
\left(\frac{\omega}{\gamma}\right)^2 = \frac{1}{(\mu_0 M_s \sin\theta_0)^2} \left[\frac{\partial^2 F}{\partial\theta^2} \frac{\partial^2 F}{\partial\phi^2} - \left(\frac{\partial^2 F}{\partial\phi\,\partial\theta}\right)^2 \right]_{(\theta_0,\phi_0)} . \qquad (4.84)
$$

With (4.84), we have reached an important formula for the free precessional frequency of the magnetization of a crystal, which is determined by the second derivatives of the free energy for the stable direction. This frequency is also called the Larmor frequency. It is called the free precessional frequency because no damping mechanism has been taken into account. In this case, the component of the magnetization along the stable direction does not change with time. However, any real magnetic system contains damping mechanisms, so that the magnetization will finally relax to its equilibrium direction (see Fig. 4.27b). The damping of the motion of the magnetization could be caused by an energy transfer from the magnetic system to the lattice motion via magnetoelastic interactions, and possibly to conduction electrons in the metal via spin–orbit coupling. This is the direct damping mechanism (Fig. 4.28). It is also possible that the precessional motion of the magnetization decays, and

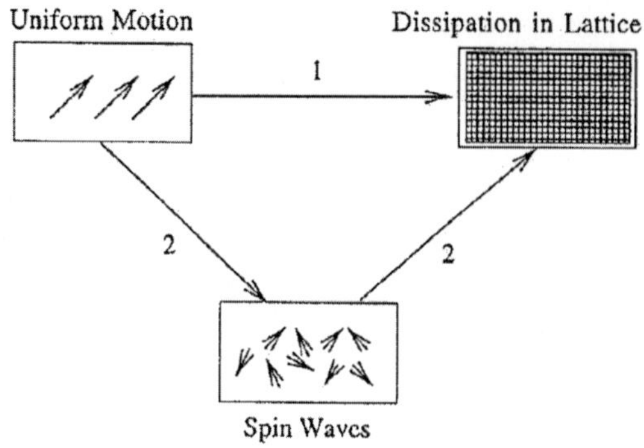

Fig. 4.28. Two damping mechanisms. (1) Direct relaxation to the lattice. (2) Decay first into magnons, which finally decay to the lattice. From [108], used with permission

that magnons (spin waves) are excited by the decay energy as a result. The excited magnons finally annihilate, and the energy is transferred to the environment. This is the indirect mechanism. Usually, the damping effect is treated phenomenologically by introducing a dispersion term into (4.76):

$$\frac{\mathrm{d}\boldsymbol{M}}{\mathrm{d}t} = -\gamma \boldsymbol{M} \times (\boldsymbol{H}_{\mathrm{eff}}) + \frac{\alpha}{M_s} \boldsymbol{M} \times \frac{\mathrm{d}\boldsymbol{M}}{\mathrm{d}t} \,. \tag{4.85}$$

Equation (4.85) is called the Laudau–Lifshitz–Gilbert (LLG) equation; α is the Gilbert damping constant. In the literature, another expression is frequently used for the LLG equation:

$$\frac{1}{\gamma} \frac{\mathrm{d}\boldsymbol{M}}{\mathrm{d}t} = -\boldsymbol{M} \times (\boldsymbol{H}_{\mathrm{eff}}) + \frac{G}{\gamma^2 M_s^2} \boldsymbol{M} \times \frac{\mathrm{d}\boldsymbol{M}}{\mathrm{d}t} \,, \tag{4.86}$$

where G is called the Gilbert damping coefficient. The damping term is small and is usually neglected for simplicity, but it becomes essential when spin dynamics are concerned. We shall return to this topic later in this subsection.

Ferromagnetic Resonance. Resonant absorption occurs when a ferromagnetic system is placed in an electromagnetic field whose frequency equals the precessional frequency of the ferromagnetic system. This is ferromagnetic resonance. The resonance frequency is in the microwave range. In a typical ferromagnetic-resonance experiment, the sample is inserted into a microwave cavity or installed as a component of the cavity. The microwave frequency is fixed at an eigenresonance of the cavity. A homogeneous DC static magnetic field is applied perpendicular to the wave vector of the microwave field. Usually, the static field can rotate within a certain angular range. A high-sensitivity microwave spectrometer is used to measure the microwave losses of the cavity when the static magnetic field is scanned. Microwave absorption peaks are observed at certain values of the static field, where the ferromagnetic-resonance condition is satisfied.

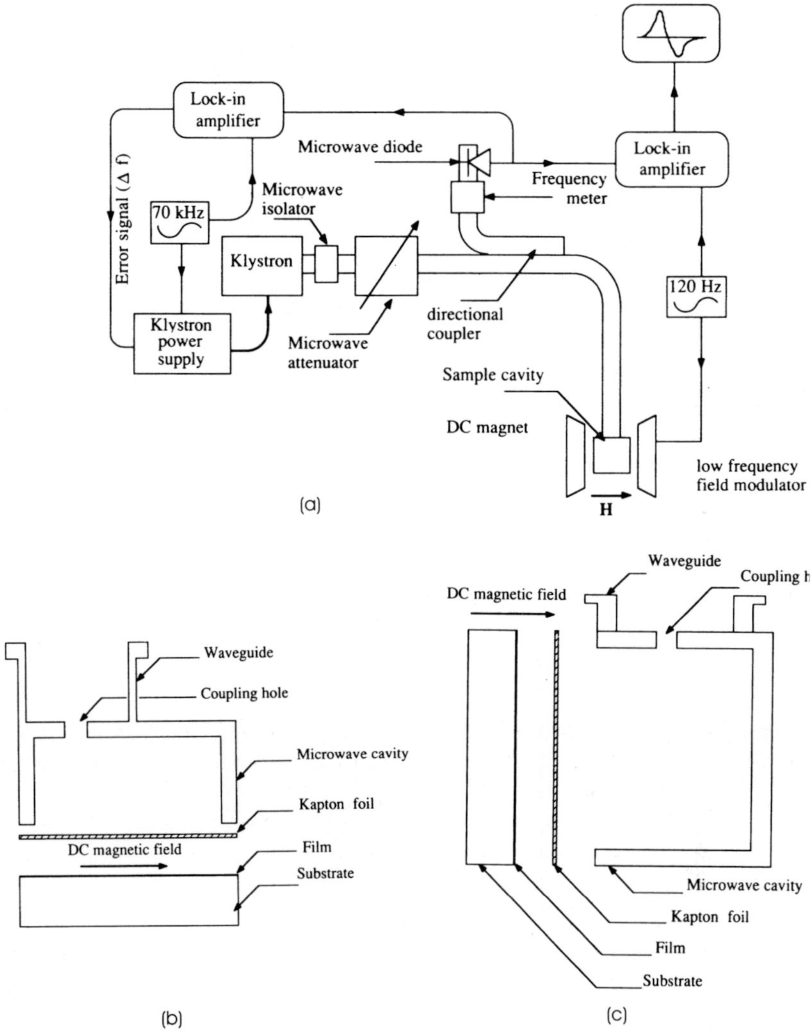

Fig. 4.29. (**a**) Block diagram of a microwave spectrometer used in FMR measurements. See text for explanation. (**b**) An expanded view of a sample resonance cavity for in-plane FMR measurements. The sample forms the end wall of a TE_{01n} cylindrical cavity. The microwave power is coupled through the coupling hole. The Kapton foil and the sample are held against the cavity body by a spring-loaded mechanism. (**c**) An expanded view of the sample resonance cavity for a perpendicular FMR measurement. From [109], used with permission

Figure 4.29 shows a block diagram of a microwave spectrometer used in FMR measurements. The reflected microwave power from a sample resonance cavity is directed by a microwave directional coupler to a diode, which is used to detect the FMR signal. A low-frequency modulation (100–200 Hz) is used to monitor the field derivative of the out-of-phase microwave susceptibility, $d\chi''/dH$. The klystron repeller is modulated by a 70 kHz voltage, allowing one to lock the microwave frequency to the eigenresonance of the sample cavity. Lock-in amplifiers are employed for monitoring the FMR signal and for providing the dc error voltage which locks the klystron microwave frequency to the sample resonance cavity. In the ideal case, the cavity stores all microwave energy at the eigenresonance and no microwave power is reflected. When the FMR condition is satisfied by tuning the applied magnetic field, the ferromagnetic-resonance absorption in the spin system changes the dielectric function and the permeability in the cavity and thus the resonance characteristics, and will be "resonantly" enhanced. Figure 4.30b illustrates a typical FMR signal in the form of the derivative of the absorption. The most important parameter for an FMR measurement is the resonance field H_{res}. The magnetic anisotropy is usually investigated by measuring how the resonance field changes with variation of the direction of the static field. The FMR linewidth $\triangle H_{res}$ may also contain useful information. It has been found that the FMR linewidth provides a measure of the magnetic homogeneity of a magnetic system [103]. The narrowest linewidth is usually observed for the structurally and magnetically most homogeneous sample. Since the mi-

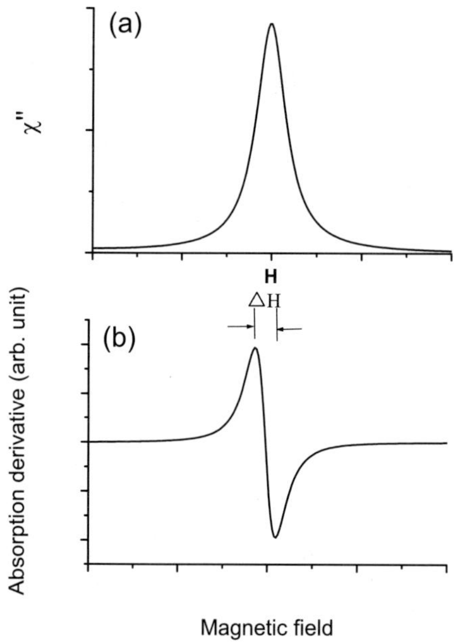

Fig. 4.30. (**a**) The imaginary part of the microwave susceptibility shows a peak around the resonant magnetic field, indicating a resonant absorption of the microwave power by a ferromagnetic system. (**b**) The corresponding derivative ($\propto d\chi''/dH$) which is detected in a FMR measurement describes the absorption

croscopic damping mechanism has not been well established, a quantitative analysis of the FMR linewidth is difficult.

Examples of Applications. FMR is a well-established method for the quantitative determination of magnetic anisotropy. To extend this technique to the investigation of ultrathin metal films, setups suitable for in situ FMR measurements in ultrahigh vacuum have been developed [109, 111]. In the following, we shall cite a publication by Gerhardter et al. [116] to show the details of how the magnetic anisotropy of a thin Fe/W(110) film have been studied by FMR. The experimental geometry is shown in Fig. 4.31. Considering the uniaxial surface contribution and the cubic bulk contribution to the magnetocrystalline energy, the free-energy density of the system is given by

$$\begin{aligned} E = & -\mu_0 HM(\sin\vartheta_H \cos\varphi \sin\vartheta + \cos\vartheta_H \cos\vartheta) + \frac{1}{2}\mu_0 M^2 \cos^2\vartheta \\ & + K_s \cos^2\vartheta + K_{s,p} \sin^2\vartheta \sin^2\varphi + K_1 \left[\frac{1}{4}\sin^4\vartheta \sin^4\varphi + \frac{1}{4}\cos^4\vartheta \right. \\ & \left. + \sin^4\vartheta \sin^2\varphi \cos^2\varphi + \sin^2\vartheta \cos^2\vartheta \cos^2\varphi - \frac{1}{2}\sin^2\vartheta \cos^2\vartheta \sin^2\varphi \right], \quad (4.87) \end{aligned}$$

where the first term is the Zeeman energy, and the second the demagnetization energy. The terms containing the coefficients K_s and $K_{s,p}$ represent the out-of-plane and in-plane uniaxial anisotropy energies, respectively. The remaining term is the cubic magnetocrystalline anisotropy energy. When the

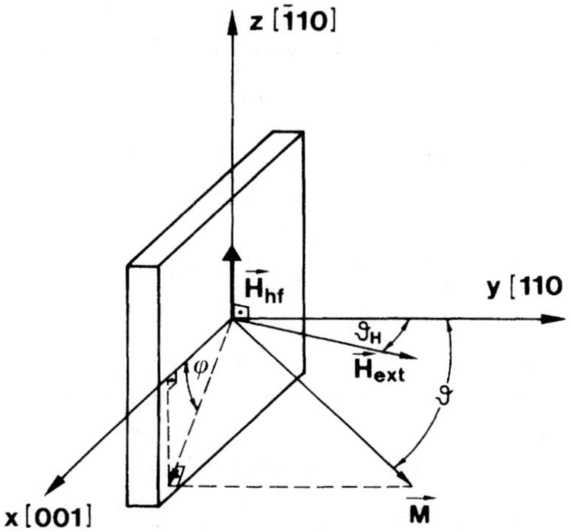

Fig. 4.31. Coordinate system used to describe the ferromagnetic resonance field. $\boldsymbol{H}_{hf}$ is parallel to the z axis. The external magnetic field can be rotated in the x–y plane. The y axis is the surface normal. From [116], used with permission

resonance field at a given microwave frequency has been experimentally measured, the equilibrium position (ϑ_0, φ_0) of the magnetization can be found using (4.82). Since the resonance field is large enough to overcome the in-plane anisotropy field and to turn the magnetization into the x–y plane, the equilibrium angle φ_{eq} is equal to zero. Another equilibrium angle ϑ_{eq} is obtained from

$$\frac{\partial E}{\partial \vartheta} = -\mu_0 HM \sin(\vartheta_H - \vartheta_{eq}) - \left(\frac{1}{2}\mu_0 M^2 + 2K_s\right) \sin\vartheta_{eq} \cos\vartheta_{eq} + K_1(\cos^3\vartheta_{eq} \sin\vartheta_{eq} - 2\sin^3\vartheta_{eq} \cos\vartheta_{eq}) = 0 \,. \quad (4.88)$$

The resonance field can be derived from (4.84) and (4.87), yielding

$$\left(\frac{\omega}{\gamma}\right)^2 = \left\{-H\cos(\vartheta_H - \vartheta_{eq}) + M_{\text{eff}}(2\sin^2\vartheta_{eq} - 1) + \frac{K_1}{\mu_0 M}(12\sin^4\vartheta_{eq} - 15\sin^2\vartheta_{eq} + 3)\right\} \times \left[H\frac{\sin\vartheta_H}{\sin\vartheta_{eq}} + \frac{2K_{\text{eff}}}{\mu_0 M} + \frac{K_1}{\mu_0 M}(\sin^2\vartheta_{eq} - 1)\right] , \quad (4.89)$$

where

$$M_{\text{eff}} = M + \frac{2(K_s + K_1)}{\mu_0 M} , \quad K_{\text{eff}} = K_{s,p} + K_1 \,. \quad (4.90)$$

By performing measurements of the angular dependence of the resonance field H and fitting it with (4.88) and (4.89), the magnetic anisotropies can be quantitatively determined. Figure 4.32 shows angularly dependent FMR spectra from a 130 Å Fe/W(110) film measured at 500 K. It can be seen that the resonance field is strongly dependent on the angle ϑ_H.

Besides the magnetic anisotropies, the gyromagnetic ratio γ, and thus the spectroscopic splitting factor g,[4] can also be obtained by fitting the frequency- and angle-dependent FMR results [118, 119]. When the g-factor has been measured, the ratio of the orbital to the spin magnetic moment is immediately known via [117]

$$\frac{\mu_L}{\mu_S} = \frac{g-2}{2} \,. \quad (4.91)$$

As has been discussed in the preceding subsection, x-ray magnetic-circular-dichroism measurements can also separate the spin contribution and orbital contribution from the total magnetic moment. Compared with XMCD, the advantage of the FMR method is that the setup is much simpler and the measurement can be performed in the laboratory. Figure 4.33 shows the fre-

[4] The Landau factor is called the spectroscopic splitting factor (g) if it is measured by FMR or the magnetomechanical factor (g') if it is measured by gyromagnetic experiments. These two factors are related by $1/g + 1/g' = 1$. For details, see [117].

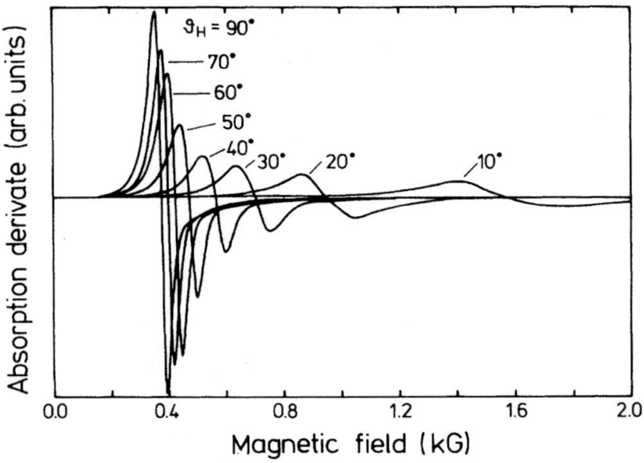

Fig. 4.32. FMR spectra for a 130 Å Fe film as a function of the orientation of the external magnetic field (ϑ_H) with respect to the film plane. The microwave frequency was 9 GHz. From [116], used with permission

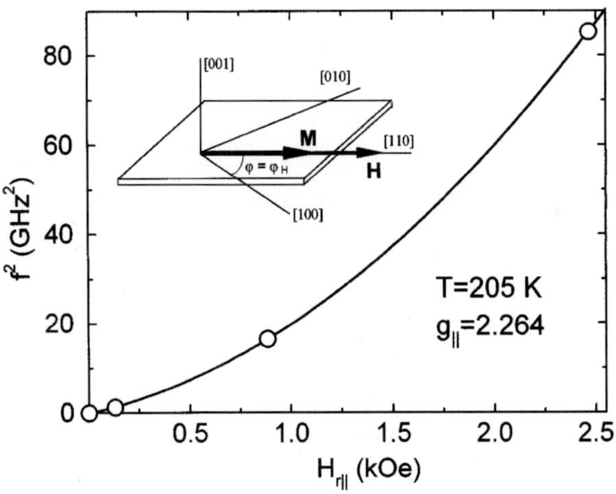

Fig. 4.33. The g-factor is obtained by fitting the microwave frequency as a function of the resonance field. From [119], used with permission

quency dependence of the resonance field measured for a bcc (001) Fe/V bilayer [119]. For a parallel experimental configuration as shown in the inset of Fig. 4.33, the resonance condition is

$$\frac{\omega^2}{\gamma_{\|}^2} = H_{r\|}^2 + AH_{r\|} + B \ , \tag{4.92}$$

where A and B are determined by the anisotropy energy and are constants at a given temperature. $\gamma_{\|}$ denotes the gyromagnetic ratio measured with an in-plane magnetization. Using (4.92), the spectroscopic splitting factor $g_{\|}$ can be obtained by fitting $f^2 = (\omega/2\pi)^2$ as a function of $H_{r\|}$ (Fig. 4.33). For a perpendicular configuration, in which the static field and the magnetization are set to be out-of-plane, the resonance condition becomes

$$\frac{\omega}{\gamma_{\perp}} = H_{r\perp} + C\ , \tag{4.93}$$

where C is a constant. $g_{\perp}$ can be measured in the same way as $g_{\|}$ is measured. g is in general anisotropic for noncubic structures [53].

With the magnetization reversal processes employed in spintronic devices and magnetic recording media becoming faster and faster, it is becoming more and more important to understand the spin dynamics and magnetic relaxation processes of ultrathin metal films, which make up the core components in the above applications. The spin dynamics in the classical limit is described by the LLG equation, and FMR is therefore a convenient tool for studies of magnetization reversal and magnetic relaxation by measuring the Gilbert damping coefficient. The FMR linewidth provides useful information about magnetic relaxation processes. The dependence of the FMR linewidth on the microwave frequency is found experimentally to follow a linear dependence [109, 110],

$$\triangle H = \triangle H_0 + 1.16\frac{\omega}{\gamma}\frac{G}{\gamma M_s}\ , \tag{4.94}$$

where the second term describes the viscous damping, which can be directly derived from the LLG equation and is therefore called the intrinsic linewidth. The first term describes an additional frequency-independent linewidth which arises from magnetic inhomogeneities, and hence is called the extrinsic linewidth. The extrinsic damping has been extensively discussed in terms of the two-magnon scattering model [112, 113, 114]. In the the two-magnon scattering model, a uniform precession of $\boldsymbol{M}$ (FMR mode) is first excited by the resonant microwave field, and, then sample inhomogeneities such as defects at the surface and interface scatter the FMR modes ($k = 0$) into magnons with finite wave vectors (see Fig. 4.34) [120]. The spin wave mode, which has a finite wave vector and is degenerate with the FMR mode, does not always exist. When the wave vector of the spin wave is parallel to the magnetization vector, the energy of a magnon with a finite wave vector is always greater than that of the FMR mode [121]. Therefore, the two-magnon scattering mechanism does not work in this case. In ultrathin films, the wave vectors of spin waves are constrained to be in the film plane. Hence, if the magnetization of an ultrathin film is perpendicular to the film, the extrinsic linewidth will vanish since the two-magnon scattering mechanism does not work in this case.

Figure 4.35 shows experimental results obtained by Urban et al., where FMR linewidths were measured with three microwave frequencies and in four

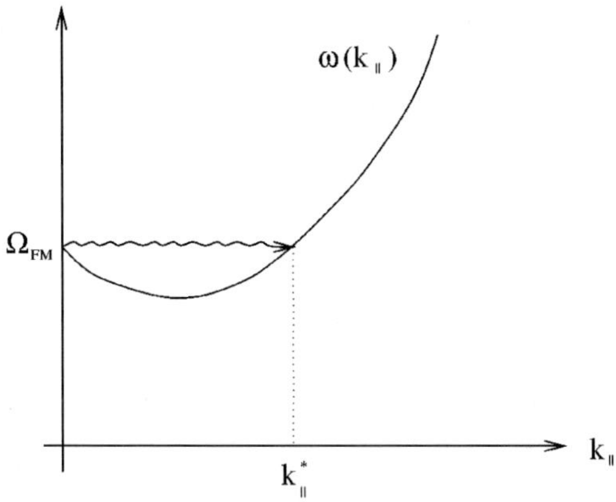

Fig. 4.34. Graphical representation of the scattering process from a uniform mode, excited by a microwave field, to a spin wave mode degenerate with it. The uniform mode can be considered as a spin wave mode with zero wave vector. The magnetic dipole–dipole interaction (demagnetization energy) has to be taken into consideration in the dispersion relation of the spin waves to obtain a spin wave mode degenerate with the uniform mode. From [113], used with permission

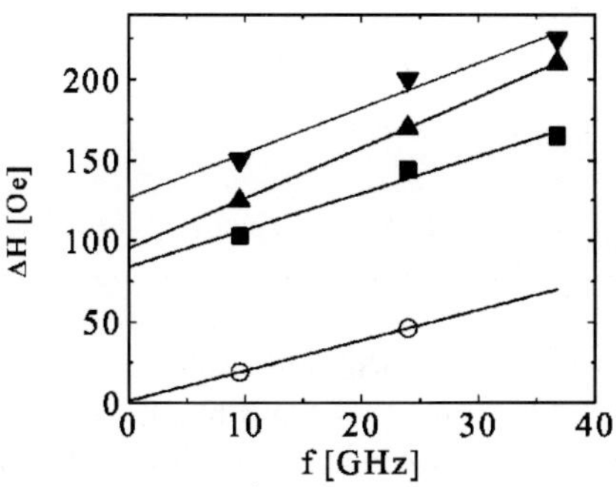

Fig. 4.35. The FMR linewidth for a 20 ML Cr/15 ML Fe/GaAs(001) film as a function of microwave frequency. The measurements were performed with the magnetization along the cubic axis [100] (*solid downward-pointing triangles*), the hard uniaxial axis [1$\bar{1}$0] (*solid upward-pointing triangles*), the in-plane easy axis [110] (*solid squares*) and the film normal [001] (*open circles*). From [115], used with permission

different configurations [115]. The data were linearly fitted using (4.94). The slopes of the fitting lines gives the effective Gilbert damping coefficient G, and the intercepts at the ordinate yield the so-called zero-frequency linewidth $\triangle H_0$. The results are listed in Table 4.6. One can see that the effective Gilbert damping coefficient varies with the magnetization orientation. This could be explained by theoretical prediction of Arias and Mills [113] that the extrinsic linewidth actually depends on the resonance frequency, so that the linear

Table 4.6. The values of $\triangle H_0$ and the effective Gilbert damping coefficient G_{eff} for the three principal in-plane and perpendicular orientations of the 20 ML Cr/15 ML Fe/GaAs(100) film

Direction	$\triangle H_0$ (Oe)	$G_{\text{eff}}(10^8\text{s}^{-1})$
$[1\bar{1}0]$	95 ± 1	2.46 ± 0.01
[110]	84 ± 9	1.81 ± 0.26
[100]	127 ± 11	2.26 ± 0.34
[001]	0	1.51 ± 0.02

fitting is really an oversimplified treatment. The intrinsic Gilbert damping coefficient is measured to be $(1.51 \pm 0.02) \times 10^8\ \text{s}^{-1}$ using a perpendicular configuration, where the extrinsic linewidth is zero.

We have discussed three techniques, i.e., the magnetooptic Kerr effect, magnetic dichroism and ferromagnetic resonance, in this section. In fact, numerous techniques, for example various electron scattering techniques and electron spectroscopies, polarized neutron reflection, Mössbauer spectroscopy, Brillouin light scattering, and second-harmonic generation, have also been successfully applied in studies of film magnetism. To keep this book as compact as possible, we shall not present these methods here, but shall include a brief introduction to each technique in subsequent sections when results obtained by that technique are cited.

References

1. E. C. Stoner: Proc. R. Soc. A **154**, 656(1936)
2. R. O. Jones and O. Gunnarsson: Rev. Mod. Phys. **61**, 689(1989)
3. W. Kohn and L. J. Sham, Phys. Rev. **140 A**, 115(1965)
4. E. K. U. Gross and R. M. Dreizler: *Density Functional Theory–NATO ASI Series B: physics Vol. 337*, (Plenum Press, New York and London 1995) pp 33–64
5. U. von Barth and L. Hedin: J. Phys. C **4**, 2064(1971)
6. R. Zeller: in *Magnetismus von Festkörpern und Grenzflächen*, (IFF/KFA Jülich 1993)
7. O. Gunnarsson: J. Phys. F: Metal Phys. **6**, 587(1976)
8. E. C. Stoner: Proc. R. Soc. A **169**,339(1939)
9. J. F. Janak: Phys. Rev. B **16**, 255(1977)
10. V. L. Moruzzi, J. F. Janak and A. R. Williams: *Calculated Electronic Properties of Metals* ,(Pergamon Press, New York 1978)
11. P. M. Marcus and V. L. Moruzzi: J. Appl. Phys. **63**, 4045(1988)
12. D. G. Pettifor: Metallurgical Chemistry, ed by O. Kubaschewski (Her Majesty's Stationery Office, London, 1972) pp191
13. B. Heinrich, A. S. Arrott, C. Liu and S. T. Purcell: J. Vac. Sci. Technol. A **5**, 1935(1987)

14. S. Blügel: in *Magnetismus von Festkörpern und Grenzflächen*, (IFF/KFA Jülich 1993)
15. E. Fawcett: Rev. Mod. Phys. **60**, 209(1988)
16. A. W. Overhauser: Phys. Rev. **128**, 1437(1962)
17. D. G. Laurent, J. Callaway, J. L. Fry and N. E. Brener: Phys. Rev. B **23**, 4977(1981)
18. H. Zabel: J. Phys.: Condens. Matter **11**, 9303(1999)
19. W. M. Lomer: Proc. Phys. Soc. **80**, 489(1962)
20. F. J. Himpsel, J. E. Ortega, G. J. Mankey and R. F. Wills: Advance in Physics **47**, 511(1998)
21. S. D. Bader: Proc. Inst. elect. electron. Engrs. **78**, 909(1990)
22. Z. Q. Qiu, J. Pearson and S. D. Bader: Phys. Rev. Lett. **67**, 1646(1991)
23. H. J. Elmers, J. Hauschild, H. Höche and U. Gradmann: Phys. Rev. Lett. **73**, 898(1994)
24. W. Dürr , M. Taborelli, O. Paul, R. Germar, W. Gudat, W. Pescia and M. Landolt: Phys. Rev. Lett. **62**, 206(1989)
25. C. A. Ballentine, R. L. Fink, J. Araya-Pochet and J. L. Erkskine: Phys. Rev. B, **41**, 2631(1990)
26. F. Huang, M. T. Kief, G. J. Mankey and R. F. Willis: Phys. Rev. B **49**, 3962(1994)
27. H. J. Elmers, J. Hauschild, G. H. Liu and U. Gradmann: J. Appl. Phys. **79**, 4985(1996)
28. Y. Li and K. Baberschke: Phys. Rev. Lett. **68**, 1208(1992)
29. A. J. Freeman and Ru-qian Wu: J. Mag. Mag. Mat. **100**, 497(1991) and references therein
30. E. Tamura, R. Feder, G. Waller and U. Gradmann: Phys.Stat.sol. (b) **157**, 627(1990); R. Feder, S. F. Alvarado, E. Tamura and E. Kisher:, Surf. Sci. **127**, 83(1983); G. A. Mulhollan, A. R. Köymen, D. M. Lind, F. B. Dunning, G. K. Walters, E. Tamura and R. feder: Surf. Sci. **204**, 503(1988)
31. U. Gradmann, K. Wagner, N. Weber and H. J. Elmers, J. Mag. Mag. Matt. **167**, 21(1997)
32. H. L. Davis, J. B. Hannon, K. B. Ray and E. W. Plummer: Phys. Rev. Lett. **68**, 2632(1992) and reference therein
33. M. W. Finnis and V. Heine: J. Phys. F **4**, L37(1974)
34. C. Li: Bull. Am. Phys. Soc. **36**, 677(1991)
35. A. M. N. Niklasson, B. Johansson and H. L. Skriver: Phys. Rev. B **59**, 6373(1999)
36. L. Cheng, Z. Altounian, D. H. Ryan, and J. O. Ström-Olsen: J. Appl. Phys. **91**, 7188(2002)
37. A. Simopoulos, E. Devlin, A. Kostikas, A. Jankowski, M. Croft and T. Tsakalakos: Phys. Rev. B **54**, 9931(1996)
38. J. A. C. Bland, C. Daboo, B. Heinrich, Z. Celinski, and R. D. Bateson: Phys. Rev. B **51**, 258(1995)
39. A. Scherz, et al.: Phys. Rev. B **64**, 180407(R)(2001); J. Geissler, et al.: Phys. Rev. B **65**, 020405(R)(2001)
40. P. Fuchs, K. Totland and M. Landolt: Phys. Rev. B **53**, 9123(1996).
41. C. Turtur and G. Bayreuther: Phys. Rev. Lett. **72**, 1557(1994).
42. S. S. Dhesi, H. A. Dürr, G. van der Laan, E. Dudzik and N. B. Brookes: Phys. Rev. B **60**, 12852(1999)

43. J. I. Lee, S. Hong, A. J. Freeman and C. L. Fu: Phys. Rew. B **47**, 810(1993)
44. F. Wilhelm, et al.: Phys. Rev. Lett. **85**, 413(2000)
45. J. Geissler, et al.: Phys. Rev. B **65**, 020405(R)(2001)
46. S. Blügel: Phys. Rev. Lett. **68**, 851(1992)
47. O. Hjortstam et al.: Phys. Rev. B **53**, 9204 (1996)
48. J. Hunter Dunn, D. Arvanitis, and N. Mårtensson: Phys. Rev. B **54**, R11157(1996)
49. M. Tischer, et al.: Phys. Rev. Lett. **75**, 1602(1995)
50. L. Néel: J. Phys. Rad. **15**, 376(1954)
51. Z. Q. Qiu and S. D. Bader: Surf.Sci. **438**, 319(1999)
52. U. Gradmann, J. Korecki and G. Waller: Appl. Phys. A **39**, 101(1986)
53. G. T. Rado: Phys. Rev. B **26**,295(1982)
54. For example, Š. Pick and H. Dreyssé: Phy. Rev. B **46**, 5802(1992)
55. G. H. O. Daalderop, P. J. Kelly and M. F. H. Schuurmans: Phys. Rev. B **50**, 9989(1994)
56. C. Kittel: Rev. Mod. Phys. **21**, 541(1949)
57. O. Song, C. A. Ballentine and R. C. O'Handley: Appl. Phys. Lett. **64**, 2593(1994)
58. D. Sander: Rep. Prog. Phys. **62**, 809(1999)
59. S. D. Bader: J. Magn. Magn. Materials **100**, 515(1991)
60. M. Faradays: Trans. Roy. Soc. (London) **5**, 592(1846)
61. J. Kerr: Phil. Mag. **3**, 339(1887); **5**, 161(1878)
62. Z. Q. Qiu and S. D. Bader: Rev. Sci. Instrum. **71**,1243(2000)
63. W. Voigt, Handbook der Electrizität und des Magnetismus (barth, Leipzig, 1915), Vol. IV. 2, p. 39
64. P. N. Argyres: Phys. Rev. **97**, 334(1955)
65. E. R. Moog, C. Liu, S. D. Bader and J. Zak: Phys. Rev. B **39**, 6949(1989)
66. G. Metzger, P. Pluvinage, and R. Tortueg: Ann. Phys. (Paris) **10**, 5(1965)
67. G. S. Krinchik and V. A. Artem'ev: Soviet Physics JETP **26**, 1080(1968)
68. T. Gasche, M. S. S. Brooks and B. Johansson: Phys. Rev. B **53**, 296(1996)
69. G. Schütz, W. Wagner, W. Wilhelm, P. Kienle, R. Zeller, R. Frahm and G. Materlik: Phys. Rev. Lett. **58**, 737(1987).
70. H. Ebert: Circular Magnetic X-ray Dicroism in Transition Metal Systems. In: **Spin-Orbit Influenced Spectroscopies of Magnetic Solids**, Lecture Notes in Physics, Vol. 466, edited by H. Ebert and G. Schütz (Springer, Berlin, 1996) pp 159–196
71. L. Baumgarten, C. M. Schneider, H. Petersen, F. Schäfers and J. Kirschner: Phys. Rev. Lett. **65**, 492(1990)
72. W. Kuch and C. M. Schneider: Rep. Prog. Phys. **64**, 147(2001)
73. J. Henk, T. Scheunemann, S. V. Halilov and R. Feder: J. Phys.: Condens. Matter **8**, 47(1996)
74. R. Feder and J. Henk: Magnetic dichroism and spin polarization in valence band photoemission. In: **Spin-Orbit Influenced Spectroscopies of Magnetic Solids**, Lecture Notes in Physics, Vol. 466, edited by H. Ebert and G. Schütz (Springer, Berlin, 1996) pp 85–104
75. T. Scheunemann, S. V. Halilov, J. Henk, and R. Feder: Solid State Commun. **91**, 487(1994)
76. B. T. Thole and G. van der Laan: Phys. Rev. B **49**, 9613(1994)

77. Ch. Roth, F. U. Hillebrecht, H. Rose, and E. Kisker: Phys. Rev. Lett. **70**, 3479(1993)
78. G. van der Laan, M. A. Hoyland, M. Surman, C. F. J. Flipse and B. T. Thole: Phys. Rev. Lett. **69**, 3827(1993)
79. F. Sirotti, and G. Rossi: Phys. Rev. B **49**, 15682(1994)
80. M. Getzlaff, Ch. Ostertag, G. H. Fecher, N. A. Cherepkov and G. Schönhense: Phys. Rev. Lett. **73**, 3030(1994)
81. Jean-Michel Mariot and Christian Brouder: Spectroscopy and Magnetism: An Introduction. In: **Magnetism and Synchrotron Radiation** , Lecture Notes in Physics, Vol. 565, edited by E. Beaurepaire, F. Scheurer, G. Krill and J.-P. Kappler (Springer, Berlin, 2001) pp 24–59
82. H. Ebert, V. Popescu and D. Ahlers: Phys. Rev. B **60**, 7156(1999)
83. G. van der Laan: Magnetic Ground State Properties from Angular Dependent Magnetic Dichroism in Core Level Photoemission. In: **Spin-Orbit Influenced Spectroscopies of Magnetic Solids**, Lecture Notes in Physics, Vol. 466, edited by H. Ebert and G. Schütz (Springer, Berlin, 1996) pp 125–143
84. S. W. Lovesey and S. P. Collins: *X-ray Scattering and Absorption by Magnetic Materials*, (Claredon, Oxford 1996)
85. G. Rossi, F. Sirotti, N. A. Cherepkov, F. C. Farnoux and G. Panaccione: Solid state Commun. **90**, 557(1994)
86. F. Sirotti, G. Panaccione and G. Rossi: Phys. Rev. B **52**, R17063(1995)
87. N. A. Cherepkov: Phys. Rev. B **50**, 13813(1994)
88. K. Blum: *Density Matrix Theory and Applications*, (Plenum Press, New York and London, 1981)
89. G. Panaccione, F. Sirotti, E. Narducci and G. Rossi: Phys. Rev. B **55**, 389(1997)
90. Y. U. Idzerda, C. T. Chen, H. -J. Lin, G. Meigs, G. H. Ho and C. -C. Kao: Nuclear Instruments and Methods in Physics Research Section A: Accelerators, Spectrometers, Detectors and Associated Equipment **347**, 134(1994)
91. B. T. Thole, P. Carra, F. Sette and G. van der Laan: Phys. Rev. Lett. **68**, 1943(1992).
92. P. Carra, B. T. Thole, M. Altarelli and Xindong Wang:; Phys. Rev. Lett. **70**, 694(1993)
93. C. T. Chen, Y. U. Idzerda, H. -J. Lin, N. V. Smith, G. Meigs, E. Chaban, G. H. Ho, E. Pellegrin and F. Sette: Phys. Rev. Lett. **75**, 152(1995)
94. Ruqian Wu, Dingsheng Wang and A. J. Freeman: Phys. Rev. Lett. **71**, 3581(1993)
95. Ruqian Wu and A. J. Freeman: Phys. Rev. Lett. **73**, 1994((1994)
96. G. van der Laany: J. Phys.: Condens. Matter **10** , 3239(1998)
97. J. Stöhr and H. König: Phys. Rev. Lett. **75**, 3748(1995)
98. M. Sacchi and J. Vogel: Dichroism in X-ray Absorption. In: **Magnetism and Synchrotron Radiation**, Lecture Notes in Physics, Vol. 565, edited by E. Beaurepaire, F. Scheurer, G. Krill and J.-P. Kappler (Springer, Berlin, 2001) pp 87–108
99. B. T. Thole and G. van der Laan: Phys. Rev. B **38**, 3158(1988)
100. G. van der Laan, M. P. Bruijn, J. B. Goedkoop and A. A. MacDowell: Proc. Soc. Photo-Opt. Instrum. Eng. **733**, 354(1987)
101. J. H. Griffiths: Nature **158**, 670(1946)
102. B. Heinrich and J. F. Cochran: Adv. Phys. **42**, 523(1993)

103. M. Farle: Rep. Prog. Phys. **61**, 755(1998)
104. J. Lindner and K. Baberschke: J. Phys.: Condens. Matter **15**, R193(2003)
105. H. Suhl: Phys. Rev **97**, 555(1955)
106. J. Smit and H. G. Beljers: Philips Res. Rep. **10**, 113(1955)
107. Shaobin Liao: *Ferromagnetism III*, (Scietific Press 1996) (In Chinese)
108. H. Suhl: IEEE Trans. Magn. **34**, 1834(1998)
109. B. Heinrich: Ferromagnetic Rresonance in Ultrathin Film Structures. In: **Ultrathin Magnetic structures**, Vol. II, edited by B. Heinrich and J. A. C. Bland (Springer-Verlag, 1993) pp 195–222
110. B. Heinrich, J. F. Cochran and R. Hasegawa: J. Appl. Phys. **57**, 3690(1985)
111. M. Farle and K. Baberschke: Phys. Rev. Lett. **58**, 511(1987)
112. R. D. McMichael, M. D. Stiles, P. J. Chen and W. F. Egelhoff, Jr: J. Appl. Phys. **83**, 7037(1998)
113. R. Arias and D. L. Mills: Phys. Rev. B **60**, 7395(1999)
114. R. Urban, G. Woltersdorf and B. Heinrich: Phys. Rev. Lett. **87**, 217204(2001)
115. R. Urban, B. Heinrich, G. Woltersdorf, K. Ajdari, K. Myrtle and J. F. Cocjran: Phys. Rev. B **65**, 020402(2001)
116. F. Gerhardter, Yi Li and K. Baberschke: Phys. Rev. B **47**, 11204(1993)
117. C. Kittel: Phys. Rev. **76**, 743(1949)
118. A. J. P. Meyer and G. Asch: J. Appl. Phys. **32**, 330S(1961)
119. A. N. Anisimov, M. Farle, P. Poulopoulos, W. Platow and K. Baberschke: Phys. Rev. Lett. **82**, 2390(1999)
120. R. D. McMichael, D. J. Twisselmann and A. Kunz: Phys. Rev. Lett. **90**, 227601(2003)
121. M. Sparks: *Ferromagnetic Relaxation Theory*, (McGraw-Hill, New York San Francisco Toronto London 1964) pp 14–18

5 Epitaxy-Stabilized Structures

Ultrathin films can be grown pseudomorphically on suitable substrates under certain growth conditions, even when the structure grown differs from the equilibrium structure of the bulk material. This provides a new method to stabilize nonequilibrium structures at room temperature and ambient pressure. The stabilized nonequilibrium structure can be a metastable structure if the substrate lattice matches the metastable structure. A metastable structure corresponds to a state of a local minimum of the free energy. Nevertheless, the energy of this structure is higher than that in the thermodynamic equilibrium state. Owing to this energy difference, there is a thickness limit for the metastable pseudomorphic film. With increasing thickness, the energy necessary to introduce defects or break interface bonds will become smaller than the energy difference between the metastable and the equilibrium phase. Hence the system will relax to the equilibrium state. In this chapter, we shall first present the theoretical predictions for various metastable phases of 3d metals (from V to Ni). This is followed by a summary of the experimental results for epitaxy-stabilized metastable[1] ultrathin films, including their growth, structure, and magnetic properties.

5.1 Theoretical Search for Metastable Structures

Magnetic transition metals are particularly rich in metastable structures. This feature is reflected by the existence of various high-temperature phases in their thermodynamic equilibrium phase diagrams. For example, Mn occurs in four allotropic modifications, i.e., the α, β, γ and δ phases, and Fe has three modifications, i.e., α-Fe, γ-Fe, and δ-Fe, in the phase diagram. As a general rule, each of these different crystallographic phases possesses different magnetic properties. This occurs because the difference in electronic

[1] In this book, we generally call a pseudomorphic ultrathin film with a Bravais lattice which differs from the bulk equilibrium phase an epitaxy-stabilized metastable film. However, one should realize that these epitaxy-stabilized metastable films often have a lower free energy than the bulk equilibrium structure, owing to the interface interaction. They are actually thermodynamic equilibrium states in the ultrathin-film limit. This distinguishes epitaxy-stabilized structures from quenched metastable bulk phases.

energy which accompanies a structural phase change is often of the same order of magnitude as the energy associated with a magnetic phase change [1]. This leads to a close correlation between structure and magnetism in magnetic 3d metals, which is of prime interest to both physicists and materials scientists. Recently, the Invar effect in iron–nickel alloys has been found to originate from this correlation [2]. Upon reduction of the volume of the lattice cell, the spin moments change continuously from a ferromagnetic state to a disordered noncollinear configuration. From the general correlation between structure and magnetism, materials scientists have seen the potential of exploiting materials with new magnetic properties by stabilizing metastable structures [3].

There are several possible approaches to stabilizing a metastable phase and studying its magnetic properties. The approach that has been studied most thoroughly is alloying. For example, by investigating fcc iron alloys with varying iron contents, one can extrapolate back to the properties of pure fcc iron. The main drawback of this method is the unwanted influence of the second alloy component. A second possibility is the production of precipitates in a host material. By choosing different host matrices, different structural modifications can be produced. Microcrystalline particles of fcc iron can precipitate from solid solution in copper and retain this structure at room temperature [4]. However, a precise structural and magnetic characterization of the precipitates is a difficult task. The third method does not suffer from this drawback. Here single crystals are used as substrates to stabilize metastable structures by epitaxial growth. Substrates with different lattice constants and elemental compositions can be employed to stabilize different metastable structures in thin films of the deposited material. As we shall see later in this chapter, a number of ultrathin films with metastable structures have been prepared at room temperature by this strategy. The main advantage of this method is that both the structure and the magnetic properties can be determined experimentally and thus the correlation between them can be investigated in detail.

For the selection of a suitable substrate for the epitaxial growth of a metastable phase, knowledge of the structure of the metastable phase is required. However, obtaining such knowledge from experiments is rather cumbersome. At room temperature, high-temperature equilibrium phases become metastable phases. Their structural parameters can be determined either by rapidly quenching the high-temperature structure to room temperature or by extrapolating the high-temperature crystallographic parameters to room temperature. However, not all metastable phases are present in the phase diagram as high-temperature phases. In principle, for any presumed crystallographic symmetry, if there exists a local minimum of the internal energy with respect to variation of the lattice parameters, then the structure associated with this minimum energy is called a metastable structure. In contrast, in an equilibrium phase diagram, only the structures with the lowest Gibbs free energy at a given temperature and pressure are present. Furthermore, there might be metastable phases that are uniquely associated with epitaxial growth. By epitaxial growth, one can expand the in-plane lattice of the film by forcing growth on a substrate with a larger nearest-neighbor distance than in a bulk equilibrium phase. This is somewhat equivalent to a negative pressure, which is inaccessible in the phase diagram. On the other hand, one should realize that the in-plane tensile stress will result in a biaxially strained structure. From this viewpoint, the epitaxy-induced distortion is actually different from that produced by a negative hydrostatic pressure. The biaxially distorted structure in an epitaxially grown ultrathin film may itself be a metastable structure, as has already been proven by both theoretical calculations and experiments. Since the number of substrates for epitaxial growth is limited and the measurement of stress in ultrathin films is still difficult, it is impossible to identify these distorted metastable phases experimentally. On the other hand, owing to the tremendous improvement in computer power and the significant developments in calculation methods, theoretical calculations have become a feasible and even an efficient method to find metastable structures. One has to note that additional difficulties arise for 3d magnetic materials in such calculations, where correlation and exchange interactions of the electrons need to be described adequately.

Extensive investigations of metastable structures have been carried out by first-principles electronic band structure calculations. Most of these calculations are based on the local-spin-density approximation within the density functional theory. In such calculations, the total energy E of a system is determined precisely as a function of the atomic volume and magnetic moment, yielding the binding surface. Each local minimum on the binding surface presents a locally stable structure (see Fig. 5.1). These structures are metastable structures except for the one with lowest Gibbs free energy, which is the thermodynamic equilibrium structure. From the binding surface, not only the lattice parameters but also the magnetic properties of the metastable phases can be derived. The numerous calculation results up to 1990 have been reviewed by Moruzzi and Marcus [6]. Most of these calculations were concerned with the three typical structures bcc, fcc and hcp. However, it has been recently realized that other stable or metastable structures can be induced by epitaxial growth as well. Among them, the most important epitaxy-induced stable structure appears to be the tetragonal structure [7, 8, 9].

Epitaxially grown ultrathin films are always more or less strained because the surface mesh of the substrate seldom perfectly matches that of the film material in its unstrained state, i.e., in one of its stable or metastable states. The in-plane strain imposed by the substrate usually leads to an expansion or contraction along the direction perpendicular to the plane, since the structure in this direction is allowed to relax freely until this stress component vanishes. For a film material with a stable cubic structure, this situation corresponds to a strained tetragonal structure. Let us consider a cubic lattice undergoing a

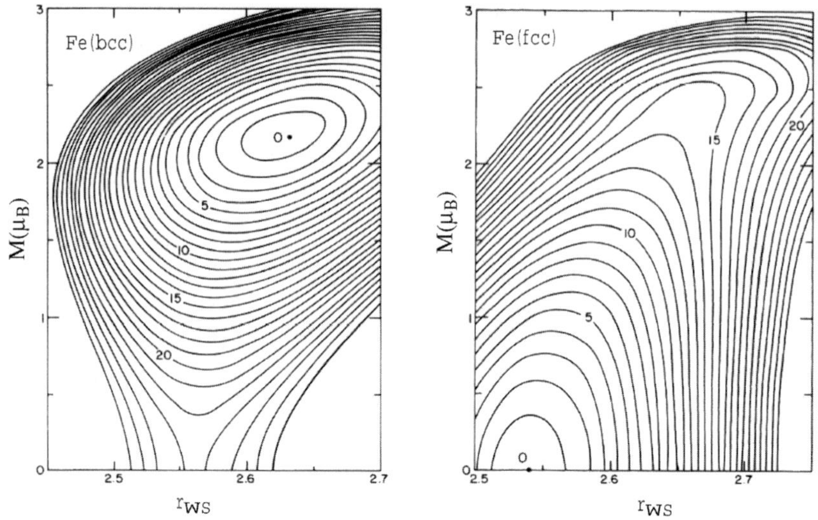

Fig. 5.1. Binding-energy surfaces for Fe in the bcc and fcc structures plotted as a function of Wigner–Seitz radius r_{WS} and atomic magnetic moment M. The contour lines are at 1 mRy (1 Rydberg = 13.6 eV) intervals and are labeled relative to the minimum energy. For the bcc structure, the energy contours at constant r_{WS} yield a minimum at a finite magnetic moment M throughout the range of r_{WS} studied, and hence bcc Fe is magnetic in this r_{WS} range. In contrast, the equilibrium point at $M = 0$ indicates that fcc Fe is normally nonmagnetic. Used with permission by [5]

tetragonal distortion by continuously expansion or contraction of the in-plane lattice constant. If at a certain point the in-plane stress during tetragonal deformation vanishes, then this distorted structure is called a tetragonal phase [10]. This state corresponds to a local minimum of the energy with respect to tetragonal deformation and will be stable or metastable, depending on whether it has the lowest energy compared with the other minima.

Tetragonal phases have been studied by first-principles calculations mainly along the epitaxial Bain path [10, 11, 12]. The best way to explain the epitaxial Bain path may be to describe how it is produced. To obtain this path, the in-plane lattice constant(s) is (are) varied within a range in which we are interested, while the axial parameter is determined by minimizing the total energy. In general, this procedure can be applied to various surfaces of various crystal structures, but it has so far been limited to {001} surfaces of tetragonal structures, such as in the case of pseudomorphic epitaxy of a cubic material on a substrate surface with a square unit mesh. In this case the epitaxial Bain path is a curve in the tetragonal plane, which has the in-plane lattice parameter a and the out-of-plane parameter c as coordinates (see Fig. 5.3a). This curve passes through all the possible tetragonal states produced by epitaxial strain (see Fig. 5.2). The total energy of the

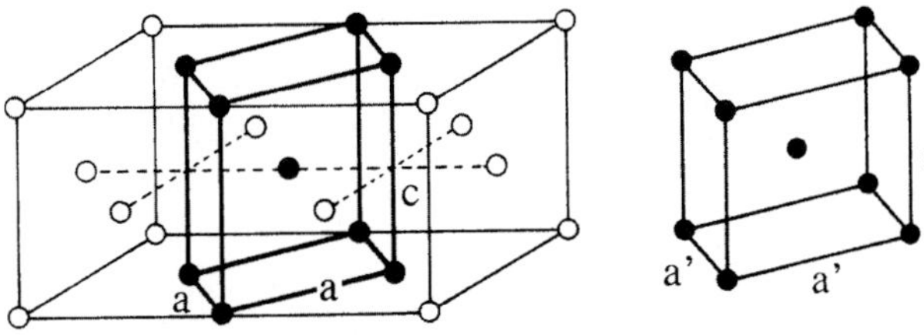

Fig. 5.2. The transition between the bcc and fcc structures can be realized via a continuous tetragonal distortion, called the Bain path. Both the fcc and the bcc lattice can be described as special cases of a body-centered tetragonal unit cell. The Bain transformation between the fcc and bcc lattices is thus described by the continuous change of a single parameter, namely the c/a ratio, so that $c/a = \sqrt{2}$ refers to the fcc and $c/a = 1$ to the bcc structure. From [9], used with permission

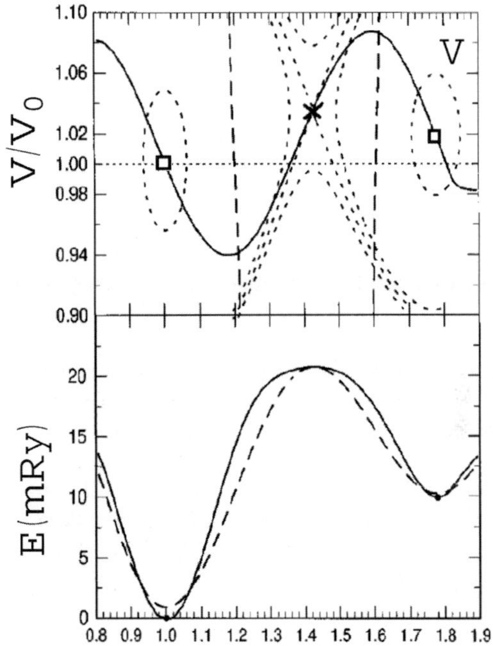

Fig. 5.3. (**a**) Epitaxial Bain path for vanadium (solid line) with a stable bcc state at $c/a = 1$, a metastable bct state at $c/a = 1.78$ and an unstable fcc state at $c/a = 1.414$ (saddle point, marked with a cross). Energy contours are drawn as *short-dashed lines* at $\triangle E = 1$ mRy above the stable and metastable points, through the saddle point, and at $\pm\triangle E$ from the saddle point. The *open square* on the left hand side is the experimental result for the V/Cu(001) system. (**b**) Total energy E of tetragonal V as a function of c/a along the epitaxial Bain path (*solid line*) and at constant volume $V = 1.04V_0$ (*dashed line*). From [13], used with permission and references therein

possible tetragonal states is usually plotted as a function of the c/a ratio (see Fig. 5.3b). Such plots seem to always show two total-energy minima: the lower minimum defines the stable tetragonal equilibrium state, and the other the metastable state. A more instructive way discuss the epitaxial Bain path is to plot total energy contours as a function of the normalized volume V/V_0 and the axial ratio c/a (see Fig. 5.3a). For simplicity, the calculations here have also been carried out assuming a constant atomic volume [14]. The reason for such a simplified treatment is that the c/a ratio at the minima of the total energy depends little on the atomic volume.

The main results concerning metastable structures of magnetic 3d metals predicted by theoretical calculations are summarized in the following and in Tables 5.1–5.3 and 5.5–5.7.

5.1.1 Vanadium

Elemental vanadium crystallizes in a bcc α-phase at normal pressure over the whole temperature range. The equilibrium cell parameters have been calculated for fcc and hcp structures as well. Along the epitaxial Bain path, besides the bcc ground state, a metastable tetragonal phase with an axial ratio of up to 1.8 has been predicted. Quite interestingly, the fcc structure ($c/a = 1.414$) is present as a total-energy maximum on the epitaxial Bain path (see Fig. 5.3). This means that the fcc structure is unstable against tetragonal deformation.

Table 5.1. Theoretical predictions for the structure of metastable V phases. The lattice parameters are given in atomic units (a.u.). The Bohr radius, equal to 0.5292 Å, is used as an atomic length unit. r_{WS} denotes the Wigner–Seitz radius, which is related to the average atomic volume by $V = (4\pi/3)r_{WS}^3$. The following abbreviations are used for the calculation method: LMTO, linear muffin-tin orbital method; FSM, fixed spin moment; ASA, atomic-sphere approximation; FLAPW, full-potential linearized augmented plane wave. "NM" denotes the nonmagnetic phase. The bcc structure, marked by an asterisk, exists as the ground-state structure

Structure	Lattice parameter(s)(a. u.)	Magnetic phase	Atomic magnetic moment (μ_B)	Calculation method	Reference
bcc*	$r_{WS} = 2.82$	NM			
hcp	$r_{WS} = 2.827$	NM		LMTO+FSM	[16]
fcc	$r_{WS} = 2.80$			LMTO+ASA	[17]
bct	$c/a = 1.78$ $V/V_0 = 1.02$			FLAPW	[11]
bct	$c/a = 1.8$			FLAPW	[9]

5.1.2 Chromium

Cr crystallizes in a bcc structure under ambient conditions. Bcc Cr is one of only a few itinerant antiferromagnetic metals. Early in the 1970s, it was suggested that thin Cr films grown on Au and Ir(111) by vapor evaporation had an fcc structure [18]. Furthermore, these thin Cr films were reported to show superconductivity with a transition temperature of up to 3.5 K [19, 20]. These observations or claims have motivated strong interest in theoretical studies of Cr with various structures. The electronic structure of fcc Cr was investigated by Xu et al. with a self-consistent linearized muffin-tin orbital method [21]. A Stoner factor of 0.82, estimated from the density of states, leads to the conclusion that magnetism is not expected for fcc Cr. For hcp Cr, a weak itinerant ferromagnetism with a magnetic moment of a few tenths of μ_B has been predicted [16], while the antiferromagnetic solution has been excluded for this structure [22]. Guo and Wang have recently reexamined the structural and electronic properties of fcc and hcp Cr crystals using density functional theory with generalized gradient corrections [23]. In that work, much attention was paid to the structural stability of the two metastable phases. The elastic constants were determined by a polynomial fitting of the calculated total energy as a function of the strains, showing that fcc Cr is unstable against any shear deformation and hcp Cr is unstable against only symmetry-breaking lattice distortions. A bct tetragonal metastable structure has been predicted on the epitaxial Bain path [24]. Besides the bcc, fcc and hcp structures, Paxton et al. have extended the calculations to simple cubic, simple hexagonal and diamond cubic structures for the first-row transition metals [25]. Stable solutions exist for all six structures, but the last three structures show a higher total energy than do the first three structures for chromium.

Table 5.2. Theoretical predictions for the structure and magnetic properties of metastable Cr phases. The following abbreviations are used to describe the calculation method: LAPW, linear augmented plane wave; DFT-GGA, density functional theory with generalized gradient approximation; LMTO, linear muffin-tin orbital method; FSM, fixed spin moment. The bcc structure, marked by an asterisk, exists as the ground-state structure. "SDW" denotes the spin-density-wave state and "FM" the ferromagnetic phase

Structure	Lattice parameter(s)(Å)	Magnetic phase	Atomic magnetic moment (μ_B)	Calculation method	Reference
bcc*	$a = 2.88$	SDW	0.62	Experiment	[26]
fcc	$a = 3.57$			LAPW	[27]
fcc	$a = 3.62$			DFT-GGA	[23]
hcp	$r_{WS} = 1.41$	FM	0.12	LMTO+FSM	[16]
hcp	$a = 2.49$ $c/a = 1.78$			DFT-GGA	[23]

5.1.3 Manganese

At atmospheric pressure, manganese occurs in four structures with increasing temperature, i.e., the α, β, γ and δ phases [28]. The first two phases, α and β, have complex structures with 29 and 20 atoms, respectively, per primitive unit cell. The transformation between the α and β phases takes place at 1073 K. The γ phase is a simple fcc phase. The quenched γ phase shows antiferromagnetism at room temperature, with a magnetic moment of $2.30\mu_B$. The Wigner–Seitz radius r_{WS} equals 2.752 atomic units (a.u.) at room temperature, and the Néel temperature is 540 K. Solid Mn exists in the bcc δ phase above 1407 K, its Wigner–Seitz radius is 2.867 a.u. at 1417 K. The magnetic structure of the δ phase has not been determined as yet.

Efforts have been made to clarify the magnetic ground-state structure of the high-temperature phases of Mn. By analyzing the linear response of the paramagnetic phase, Fry et al. found an instability towards the formation of a ferromagnetic state at $T = 0$ K for bcc Mn [29]. From a conventional spin-polarized band structure calculation, they also showed that ferromagnetic bcc Mn has a moment ranging from $0.76\mu_B$ to $4.93\mu_B$ in the range of lattice constants between 5.2 and 8.0 a.u. They further predicted an abrupt magnetic transition around $a = 6$ a.u. (3.18 Å) from a low-spin state for small unit cells to a high-spin state for large unit cells. For the fcc γ phase, the antiferromagnetic (AF) solution was found to be lowest in energy from the equilibrium volume ($r_{WS} \simeq 2.58$ a.u.) up to the free-atom limit [30].

Using the generalized gradient approximation, Asada and Terakura calculated the structural and magnetic properties of three typical crystallographic phases of Mn, i.e., fcc, bcc and hcp [31] (see Fig. 5.4). They concluded the following. (i) The most stable phase of the three structures investigated is the fcc antiferromagnetic phase. For the hcp structure, the total energy of the AF ground state is marginally lower than that of the nonmagnetic (NM) phase, in contrast to the results of Goniakowski et al., which predicted an NM ground state for hcp Mn [16, 22]. Ferromagnetic (FM) solutions are found for $r_{WS} > 2.85$ a.u. (ii) The fcc AF phase is the next lowest phase in total energy. For this structure, the total-energy difference between the AF ground state and the next lowest NM state is 3.4 mRy (46.2 meV). (iii) For the bcc structure, the ground state is FM, with $r_{WS} = 2.699$ a.u. and $\mu = 0.99\mu_B$. The NM minimum, with $r_{WS} = 2.658$ a.u., is located 1.4 mRy (19.0 meV) above the FM minimum, and the AF minimum lies 3.5 mRy (47.6 meV) further above the NM minimum, with $r_{WS} = 2.797$ a.u. and $\mu = 2.73\mu_B$. These theoretical predictions seem to be somewhat puzzling, since we know that the actual ground state of Mn is the complex cubic α phase and the hcp phase does not appear in the phase diagram at all. The discrepancy between the experimental results and the theoretical calculation was explained by an important contribution of the lattice vibrations to the free energy.

The tetragonal lattice distortion of Mn was first studied by Oguchi and Freeman [32]. Recently, Qiu and Marcus reexamined the tetragonal phases of

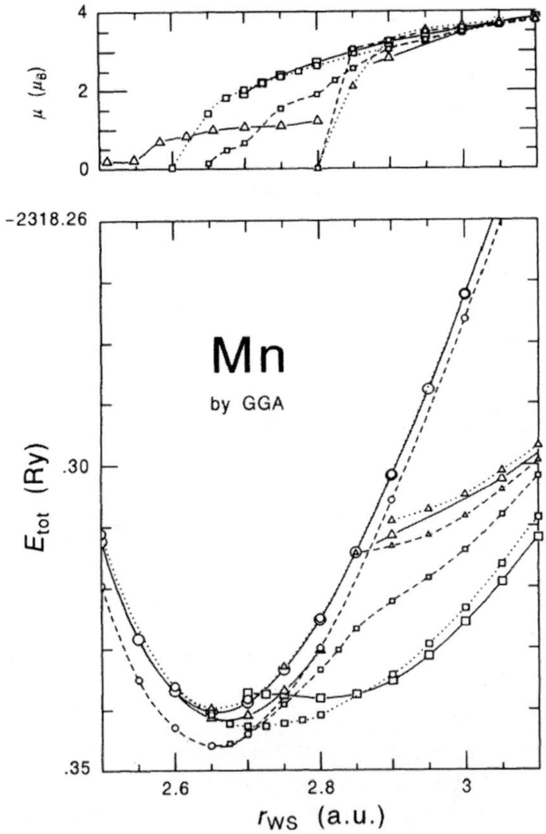

Fig. 5.4. Total energy E_{tot} and magnetic moment μ of Mn as a function of r_{WS}. The *solid curves* correspond to the bcc structure, the *dotted curves* to the fcc structure and the *dashed curves* to the hcp structure. The *circles* indicate the nonmagnetic phase, the *triangles* the ferromagnetic phase, and the *squares* the antiferromagnetic phase. From [31], used with permission

Mn using the full-potential linearized augmented-plane wave method within the local-spin-density approximation [14]. They first assumed a constant volume for the calculation and predicted an AF fct tetragonal equilibrium state at $c/a = 0.958$ and an fct metastable AF state at $c/a = 0.600$. They found that the AF bcc state was inherently unstable, since the energy has a local maximum there. In a subsequent work, the volume of the tetragonal phases on the epitaxial Bain path was also calculated from first principles and the lattice parameters a and c were determined independently for the stable and metastable tetragonal structures [33]. In the calculation by Qiu and Marcus [14, 33], only one AF configuration was considered, namely the AF1 configuration (see Fig. 5.5). Sliwko et al. have suggested that bcc Mn has a complex magnetic structure [35]. Their calculation shows that bcc Mn is FM

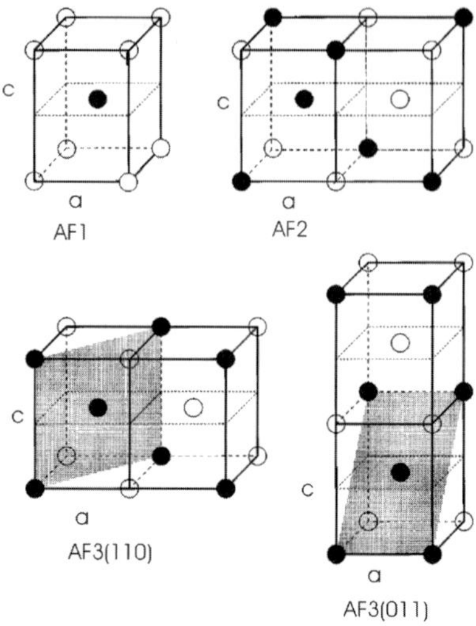

Fig. 5.5. The antiferromagnetic structure types AF1, AF2, AF3(011) and AF3(110) for a body-centered cubic lattice. In the bcc limit, i.e., at $c/a = 1$, this notation coincides with the standard notation. AF3(110) and AF3(011) are equivalent for the bcc structure, and they are called simply AF3 in this case. The *solid* and *open circles* indicate the lattice points of the up and down spin sublattices, respectively. From [34], used with permission

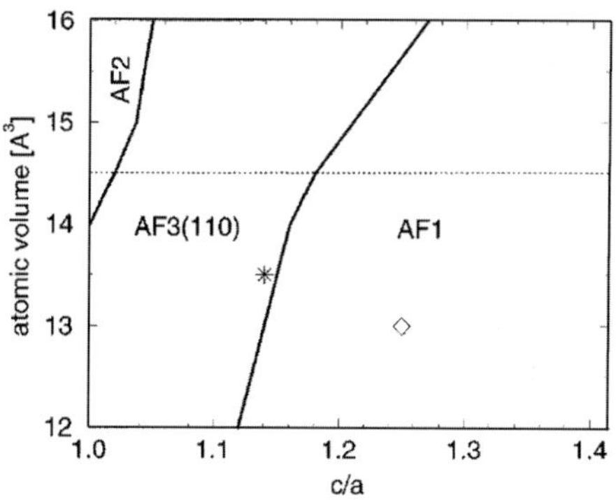

Fig. 5.6. The ground state of bct bulk Mn as a function of the atomic volume and the axial ratio c/a. The *asterisk* indicates the experimental data point for Mn on Fe(001), with $V = 13.5\,\text{Å}^3$ and $c/a = 1.14$, and the *diamond* indicates the data for Mn on Pd(001), with $V = 13.0\,\text{Å}^3$ and $c/a = 1.25$. From [34], used with permission

at the theoretical equilibrium atomic volume of 11 Å^3. The material becomes ferrimagnetic for expanded lattices up to an atomic volume of 12.5 Å^3, from where on the AF2 state is the most stable. Krüger et al. [34] investigated the magnetic structures of tetragonal Mn with multiple AF configurations considered and derived a magnetic phase diagram, as shown in Fig. 5.6.

Table 5.3. Theoretical predictions for the structure and magnetic properties of metastable Mn phases. The lattice parameters are given in two kinds of units, namely atomic units and Å. The Bohr radius, equal to 0.5292 Å, is used as an atomic unit of length. r_{WS} denotes the Wigner–Seitz radius, which is related to the average atomic volume by $V = (4\pi/3)r_{WS}^3$. The following abbreviations are used for the calculation method and magnetic phases: ASW, augmented spherical wave; FSM, fixed spin moment; GGA, generalized gradient approximation; LMTO, linear muffin-tin orbital method; LSDA, local-spin-density approximation; LS, low-spin state; FM, ferromagnetic; AF, antiferromagnetic; NM, nonmagnetic

Structure	Lattice parameter(s)	Magnetic phase	Atomic magnetic moment (μ_B)	Calculation method	Reference
bcc	$r_{WS} = 2.59$ a.u.	LS		ASW+FSM	[15]
bcc	$r_{WS} = 2.669$ a.u.	FM	0.99	GGA	[31]
fcc	$r_{WS} = 2.718$ a.u.	AF	2.13	GGA	[31]
hcp	$r_{WS} = 2.654$ a.u.	AF	0.20	GGA	[31]
hcp	$r_{WS} = 2.588$ a.u.	NM		LMTO+FSM	[16]
fct[a]	$c/a = 0.958$	AF		LSDA	[14]
fct	$a = 3.676$ Å c = 3.490 Å	AF		GGA	[33]
fct	$c/a = 0.600$	AF		LSDA	[14]
fct	a = 4.338 Å c=2.585 Å	AF		GGA	[33]

[a] The face-centered tetragonal (fct) phase is actually not a Bravais lattice. It can be classified as a body-centered tetragonal (bct) Bravais lattice. However, the term "fct lattice" appears frequently in the literature.

5.1.4 Iron

As schematically depicted in Fig. 5.7, bulk iron shows allotropic behavior, i.e., it has a rich variety of crystallographic and magnetic phases as a function of temperature and pressure [36]. At room temperature and atmospheric pressure, iron crystallizes in the bcc α phase. This phase is ferromagnetic with a Curie temperature of 1042 K. At 1185 K, there is a structural transition into the paramagnetic γ phase. Surprisingly enough, this fcc phase has a smaller atomic volume than the bcc modification. At higher temperatures (1667 K), yet another structural transition is observed. Between 1667 K and the melting point at 1808 K, the δ phase, with a bcc structure, is stable. It is only at high pressures that Fe adopts the hexagonal structure (ε phase) that its isoelectronic counterparts ruthenium and osmium assume under standard-state conditions (298 K, 1 atm). Table 5.4 lists the crystal structures and lattice spacings for the different iron modifications. The fcc γ phase has a lattice constant of 3.65 Å at 1185 K. By using the known thermal expansion coefficient at high temperature and extrapolating to room temperature, a lattice constant of 3.59 Å for fcc Fe at 300 K can be predicted.

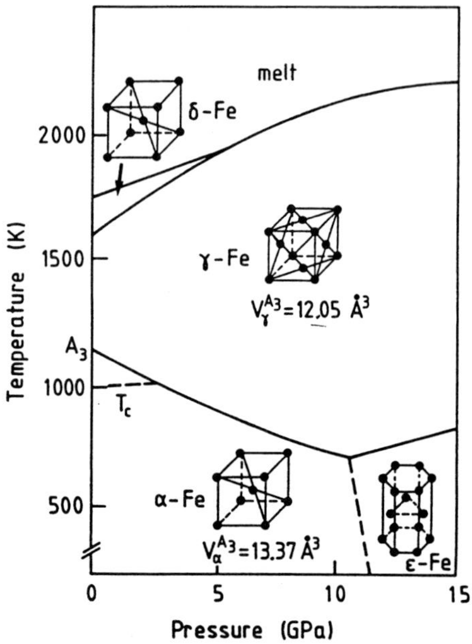

Fig. 5.7. Phase diagram of Fe. After [37], used with permission

Table 5.4. Crystal structure and magnetic ground state of bulk Fe at atmospheric pressure (10^5 Pa) and various temperatures. After [36]

Phase	Structure	Temperature range (K)	Lattice constant in Å at (temperature in K)	Magnetic ground state
α	bcc	< 1042	2.87(293)	Ferromagnetic
β	bcc	1042–1185	-	Paramagnetic
γ	fcc	1185–1667	3.646(1462)	Paramagnetic
δ	bcc	> 1667	2.932(1941)	Paramagnetic

Fe is one of the most intensively investigated systems. The continuous interest shown by theorists was originally connected to the special location of Fe in the periodic table of the elements, which is between the antiferromagnetic and ferromagnetic 3d elements. Furthermore, understanding the behavior of Fe appears to be crucial to understanding the Invar problem. Subsequently, the failure of local-spin-density approximations (LSDA) to predict the correct ground state for Fe [38, 39, 40, 41, 42] has provoked many attempts to improve the calculations [43, 44, 45]. Among these attempts, the most important may be the inclusion of the gradient correction to the exchange–correlation functional. The successful epitaxial growth of ultrathin Fe films on Cu(100) and the observed complex correlation between structure and magnetism [46] has further stimulated the research interest in this system and extended the investigation of bulk Fe to ultrathin films [47, 48, 49]. For this system, the calculation results are quite sensitive to the calculation method employed. Choosing a different density functional for the exchange–correlation potential can give a completely different sequence in total energy for the various crystal structures and magnetic phases considered. This situation reflects the fact that the actual energies of these structures and/or magnetic phases are very close to each other. Since so many calculations exist for bulk Fe and their results are not without controversy, it is difficult to summarize the theoretical predictions for the structure and magnetic properties of Fe in compact form. Therefore we just selectively list the main conclusions and the corresponding references in Table 5.5. The theoretical lattice parameters and magnetic properties of the bcc phase are also listed in the table, if they are available, so that one can assess the calculation methods by comparing the results with the experimental values. The theoretical predictions for the correlation between structure and magnetism in Fe will be discussed in the next chapter.

In contrast to the numerous studies of bulk Fe, only a few papers have been concerned with Fe films. Recently, Friák et al. [56] have performed calculations of the total energy as a function of c/a and the atomic volume. In this work, the total energy was first calculated at constant atomic volume with variation of c/a, and then the volume was varied and the calculation was repeated. Four magnetic structures are considered. These were the ferromagnetic, nonmagnetic, single-layer antiferromagnetic (AFM1) and double-layer antiferromagnetic (AFMD) phases. By comparing the calculated total energies of different magnetic structures at a given c/a and atomic volume, a magnetic phase diagram was obtained (see Fig. 5.8). This diagram is helpful for predicting the magnetic states of Fe overlayers on (001) substrates.

Since both the c/a ratio and the atomic volume are determined uniquely by a, the structures which could be produced by epitaxial growth lie on a line in the (V/V_{exp})–(c/a)) plane in Fig. 5.8, which coresponds to the epitaxial Bain path. The epitaxial Bain path for Fe has been found by Qiu et al. with four magnetic phases considered, i.e., the nonmagnetic, the ferromagnetic, the type-I single-layer antiferromagnetic (AF1) and the double-layer antiferromagnetic (SAF2) phase [14, 47]. The results are shown in Fig. 5.9. Note that a different definition of c/a has been used here. The fcc structure is considered here as a tetragonal phase with $c/a = 1$, while $c/a = \sqrt{2}/2$ corresponds to the bcc structure.

5.1.5 Cobalt

Co is found in the hcp structure at room temperature. It undergoes a martensitic transformation into an fcc structure above 700 K. Both the hcp and the fcc phase are ferromagnetic. Fcc Co shows a Curie temperature of 1388 K [57]. Both the ground-state structure and the magnetic behavior of Co are in contrast to those of the isoelectronic 4d and 5d transition elements Rh and Ir. Rh is a paramagnetic metal and Ir is a superconductor, and both have

Table 5.5. Theoretical predictions for the structure and magnetic properties of the metastable phases of bulk Fe. The first two rows give the experimental data for the bcc ground state and metastable fcc phase. The following abbreviations are used to denote the calculation method: LSDA, local-spin-density approximation; GGA, generalized gradient approximation; LMTO, linear muffin-tin orbital method; FSM, fixed-spin method; GC, gradient-corrected exchange–correlation functional; GEA, gradient-expansion approximation

Structure	Magnetic phase	Lattice parameter(s)	Atomic magnetic moment (μ_B)	Calculation method	Reference
bcc	FM	$a = 2.87$ Å	2.22	Experiment	[50]
fcc	AFM	$a = 3.59$ Å	0.7	Experiment	[4, 51]
bcc	FM	$a = 2.76$ Å	2.08	LSDA	[38]
fcc[a]	NM	$a = 3.34$ Å		LSDA	
	AFM	$a = 3.33$ Å	0.64	LSDA	
	FM	$a = 3.33$ Å		LSDA	
hcp	NM	$r_{WS} = 2.54$ a.u.		LMTO+FSM	[16]
bcc	FM	$r_{WS} = 2.63$ a.u.	2.15	LSDA	[5]
fcc	NM	$r_{WS} = 2.54$ a.u.		LSDA	
bcc	FM	$a = 2.835$ Å	2.24	GGA	[52]
fcc	NM	$a = 3.454$ Å		GGA	
bcc	FM	$a = 2.884$ Å	2.32	GGA	[40]
fcc	AFM	$a = 3.55$ Å	1.21	GGA	
hcp	AFM			GGA	
fcc	SSDW	a[b]		LMTO	[53]
fcc	SSDW			LMTO+GC	[54]
bcc	FM	$a = 2.766$ Å	2.12	LSDA	[41]
fcc[c]	NM	$a = 3.388$ Å		LSDA	
bcc	FM	$a = 2.96$ Å	2.35	GEA	
fcc	NM	$a = 3.509$ Å		GEA	
bcc	FM	$a = 2.90$ Å	2.15	DFT+Stoner	[43]
fcc	NM	$a = 3.593$ Å		DFT+Stoner	
bcc	FM	$a = 2.850$ Å	2.46	GGA	[42]
fcc	FM	$a = 3.652$ Å	2.61	GGA	
bcc	FM	$a = 2.84$ Å	2.17	GGA	[44]
fcc	AFM1[d]	$a = 3.50$ Å	1.30	GGA	
	AFM2	$a = 3.53$ Å	1.8	GGA	

[a] In this calculation, the NM, AFM and FM phases with the fcc structure are degenerate and are incorrectly calculated to be the lowest-energy phases.
[b] A spiral spin-density wave was found to be the ground state of fcc Fe at certain atomic volumes, for example when r_{WS} equals 2.72, 2.69 and 2.66 a.u.
[c] In this work, the LSDA calculation predicts a wrong ground state, i.e., the fcc NM phase, while the GEA method correctly produces the bcc FM phase as the ground state.
[d] In this work, two kinds of AFM spin configurations were considered, namely the usual type-I AFM with alternating layers of up and down spins and a double-layer AFM in which the spin orientation is alternated for every two layers. The most favorable fcc phase has AFM spin order. The total energies of the two AFM states are indistinguishable within the accuracy of the calculation.

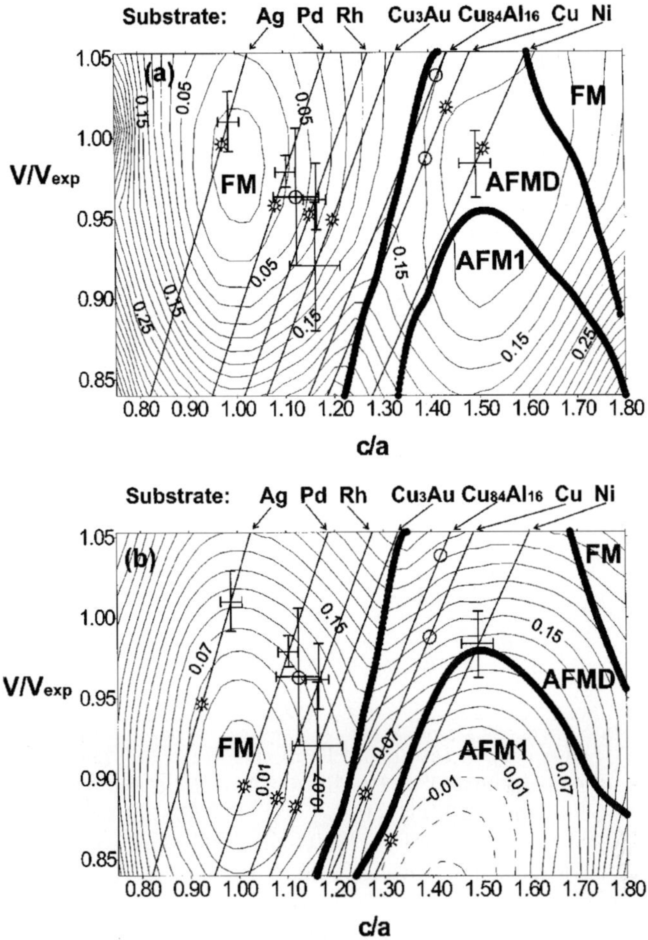

Fig. 5.8. Total energy of iron as a function of c/a and atomic volume relative to the FM bcc equilibrium state, calculated within the generalized gradient approximation (GGA) (**a**) and LSDA (**b**). The atomic volume is normalized to the experimental value of 11.72 Å^3 for bcc Fe. The contour interval for the energy is 20 meV. *Thick lines* separate the different magnetic phases. These are the ferromagnetic (FM), single-layer antiferromagnetic (AFM1), and double-layer antiferromagnetic (AFMD) phases. The *straight lines* correspond to the constant lattice parameters of various (001) substrates. The *crosses*, which are made up of the vertical and horizonal error bars and centered on those straight lines, represent the experimentally determined structures of Fe films on the corresponding substrates. One of these crosses, which does not fall on the corresponding line and which represents 3.3 ML Fe/Cu_3Au [55], is marked by an *open circle*. The two *open circles* in the AFMD region show the experimental structure where no error bars are given. The *smaller open circles combined with asterisks* represent the theoretical predictions by Friák et al. The contours with negative energy (relative to FM bcc Fe) are drawn as *dashed lines* in (**b**). From [56], used with permission

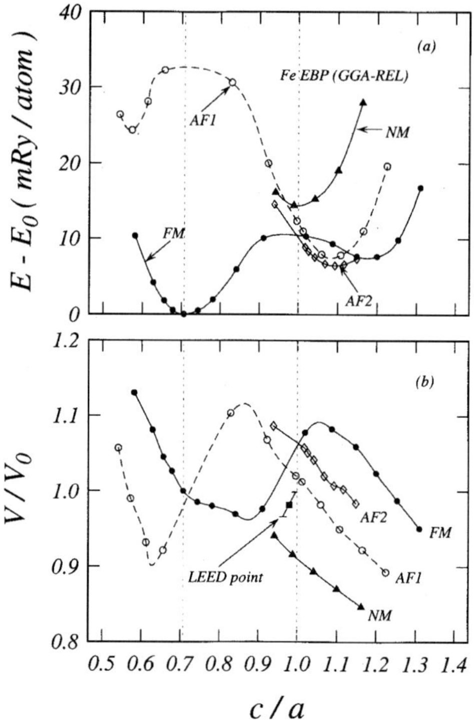

Fig. 5.9. (**a**) The energy of tetragonal Fe in the NM, FM, AF1 and AF2 phases as a function of c/a, calculated by the GGA, where E_0 is the energy per atom in the bcc FM ground state. (**b**) The $(V/V_0) - (c/a)$ curves of tetragonal Fe for the different magnetic phases, where V_0 is the theoretical atomic volume in the ground state, 11.57 Å^3. The data point with the error bar is the experimental value of V/V_0 plotted against c/a for Fe films on Cu(001), where $V_0 = 11.73$ Å^3 is the measured ground-state atomic volume. From [47], used with permission

a stable fcc structure. Self-consistent paramagnetic total-energy band structure calculations by Min et al. show that the fcc phase is lowest in energy in the paramagnetic state and is only slightly lower (1 mRy) in energy than the hcp phase ([58] and references therein). The bcc phase was the highest in energy. However, spin-polarized LSDA calculations in the same study produce different results. Now the hcp phase becomes more stable than the fcc phase by about 2 mRy and the total energy of bcc Co becomes close to that of the fcc phase. It is therefore suggested that the stability of the hcp and bcc phases results from their magnetic order. The three main structures of Co, i.e., hcp, fcc and bcc, are all theoretically found to be ferromagnetic. For bcc Co, the antiferromagnetic configuration is slightly lower in energy than the nonmagnetic configuration, but both are much higher in energy than the ferromagnetic configuration [59]. A similar sequence is predicted for bcc Fe

Table 5.6. Theoretical predictions for the structure and magnetic properties of metastable Co phases. The following abbreviations are used to denote the calculation method: LSDA, local-spin-density approximation; LMTO, linear muffin-tin orbital method; KKR, Korringa–Kohn–Rostoker method; LAPW, linear augmented plane wave; FLAPW, full-potential linearized augmented plane wave.

Structure	Magnetic phase	Lattice parameter	Atomic magnetic moment (μ_B)	Calculation method	Reference
hcp	FM	$r_{WS} = 2.62$ a.u.	1.58^a	Experiment	[58]
fcc	FM	$r_{WS} = 2.61$ a.u.	1.61	Experiment	[58]
bcc	FM	$r_{WS} = 2.63$ a.u.	1.41	Experiment	[61]
hcp	FM	$r_{WS} = 2.61$ a.u.	1.63	LSDA–LMTO	[58]
fcc	FM	$r_{WS} = 2.61$ a.u.	1.64	LSDA–LMTO	[58]
bcc	FM	$r_{WS} = 2.62$ a.u.	1.73	LSDA–LMTO	[58]
fcc	FM	$r_{WS} = 2.56$ a.u.	1.56	KKR	[6]
bcc	FM	$r_{WS} = 2.61$ a.u.	1.68		[62]
bcc	FM		1.72	LSDA–LMTO	[63]
hcp			1.59	LSDA–LMTO	[63]
bcc	FM	$a = 2.76$Å	1.73	LSDA–LAPW	[59]
hcp	FM	$a = 4.59$ a.u.	1.45	LSDA–FLAPW	[64]
fcc	FM	$a = 6.49$ a.u.	1.51	LSDA–FLAPW	[64]
bcc	FM	$a = 5.18$ a.u.	1.62	LSDA–FLAPW	[64]

[a] This is the electron spin moment. The total moment for hcp Co is $1.71\mu_B$.

[59]. Unlike the case of bcc Fe, the majority band of bcc Co is fully occupied, leading to an atomic magnetic moment that is relatively stable against variation of the atomic volume [59]. Bcc Co is found to be elastically unstable with respect to the Bain strain [60]. The total energy of tetragonal Co decreases monotonically as the c/a ratio is varied from 1, which corresponds to the bcc structure, to $\sqrt{2}$, which corresponds to the fcc structure.

5.1.6 Nickel

Studies of Ni are relatively scare, perhaps owing to the absence of allotropic modifications. The stable structure of Ni is fcc, which is ferromagnetic below 631 K. The atomic magnetic moment of Ni at 0 K is $0.616\mu_B$, much lower than the values for Fe ($2.216\mu_B$) and Co ($1.715\mu_B$).

Self-consistent energy-band calculations show an extremely large density of states at the Fermi level for metastable hcp Ni, which is therefore predicted to be ferromagnetic [16, 65]. For bcc Ni, Moruzzi et al. predicted that it is nonmagnetic at its equilibrium volume and becomes ferromagnetic when the unit cell expands above $r_{WS} = 2.60$ a.u. [5]. However, these results were

Table 5.7. Theoretical predictions for the structure and magnetic properties of metastable Ni phases. The following abbreviations are used to denote the calculation method: LMTO, self-consistent linear muffin-tin-orbital method; FSM, fixed-spin-moment scheme; APW, self-consistent augmented-plane-wave method

Structure	Magnetic phase	Lattice parameter(s)	Atomic magnetic moment (μ_B)	Calculation method	Reference
fcc	FM	$a = 3.524$ Å	0.616	Experiment	[67]
hcp	FM	$a = 2.50$ Å $c = 4.04$ Å	0.76	APW+Stoner	[65]
hcp	FM	$r_{WS} = 2.57$ a.u.	0.59	LMTO+FSM	[16]
bcc	FM	$r_{WS} = 2.60$ a.u.		APW+FSM	[66]

modified in a subsequent study by Moruzzi and Marcus [66], in which a ferromagnetic ground state was predicted.

5.2 Epitaxy-Stabilized Ultrathin Magnetic Films

Motivated by the prospect of finding new magnetic phases and a chance to understand the interplay between the lattice structure and the magnetic behavior, extensive experimental investigations have been carried out on epitaxy-stabilized ultrathin 3d metal films, which show different structures from the corresponding bulk material. The results are summarized in this section.

5.2.1 Ultrathin V Films

To verify the existence of the tetragonal metastable phase on the epitaxial Bain path predicted by first-principles calculations [11], Tian et al. have prepared ultrathin vanadium films on Ni(001) substrates [68]. The stable structure of vanadium is bcc with a lattice constant of 3.03 Å. A bct metastable phase is predicted on the epitaxial Bain path with $a = 2.43$ Å and $c = 4.33$ Å [12]. Therefore Ni(001) would be a suitable substrate to stabilize this bct structure, since fcc Ni, with a lattice constant of 3.52 Å, forms a primitive square mesh with an atomic spacing of 2.49 Å in the (001) plane. Experiments show that vanadium grows pseudomorphically on Ni(001) to a thickness estimated to be about eight atomic layers, albeit with a high density of defects and disorder. A QLEED (quantitative LEED) analysis reveals that the structure of the V film is bct with a = 2.49 Å(dictated by the Ni substrate) and **c** between 4.32 and 4.23 Å. The observed axial ratio is quite close to the predicted value of 1.8. The bct metastable phase has also been stabilized on a Cu(100) substrate ($a = 2.556$ Å) with a thickness of up to 13 Å [12]. The magnetic properties of these tetragonal phases have not been investigated.

5.2.2 Ultrathin Cr Films

There have been a number of attempts to synthesize metastable Cr structures in ultrathin epitaxial films [69, 70, 71]. However, the films, prepared on carefully chosen substrates, always turned out to show a bcc structure, the stable equilibrium phase. Only recently has ferromagnetic hcp Cr successfully been stabilized by an interface interaction in Cr/Ru multilayers [72]. The difficulty in stabilizing metastable Cr structures by pseudomorphic growth perhaps indicates that the metastable structures have a considerably higher energy than the bcc ground state.

In [72], hcp Cr was found to grow pseudomorphically on hcp Ru up to 4 Å. After 4 Å of Cr, a structural relaxation from hcp to bcc starts, following the Nishiyama–Wassermann epitaxy mode. That is to say, the lattice parameter along the $[11\bar{2}0]$ direction gradually changes from $\sqrt{3}a_{hcp}$ to $\sqrt{2}a_{bcc}$ and that along $[1\bar{1}00]$ changes from a_{hcp} to a_{bcc}. The relaxation process towards the bulk bcc (110) phase is completed above 10 Å of Cr. This structural evolution with thickness was first deduced from RHEED patterns and further confirmed by x-ray absorption near-edge structure (XANES) measurements. Correspondingly, the magnetic properties of the superlattice change with the thickness of the Cr layer (see Fig. 5.10). The superlattice with a 4 Å Cr layer is ferromagnetic, as is clearly demonstrated by the magnetic hysteresis (see the inset of Fig. 5.10). When the Cr layer thickness increases above 4 Å, the magnetic moment decreases. Element-specific x-ray magnetic circular dichroism measurements show that the magnetic moment comes from the Cr layers. It is quite noticeable that the ferromagnetic hcp phase of Cr has an anomalously large volume (+14%) compared with that of bcc Cr.

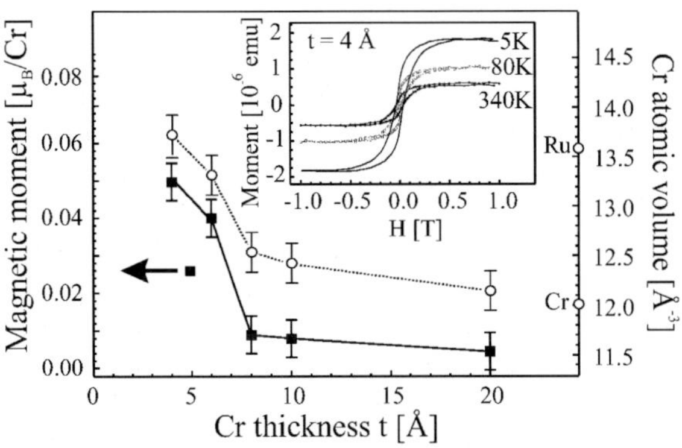

Fig. 5.10. Atomic volume (*open circles*) and atomic magnetic moment (*solid squares*) of Cr as a function of Cr layer thickness t in $[\mathrm{Cr}(t)\mathrm{Ru}(10\,\text{Å})]_{20}$ multilayers. The bulk values for bcc Cr and hcp Ru are also indicated on the *right-hand axis*. The *inset* shows temperature-dependent hysteresis loops for a Cr/Ru superlattice with $t = 4$ Å. From [72], used with permission

The reported evidence for fcc Cr is quite interesting. Fcc Cr was claimed to have been prepared on Ir(111) and Au(111) substrates in 1972, according to the LEED pattern and the calculated lattice parameters [18]. Subsequently, Brodsky et al. reported that superconductivity was found in Au/Cr/Au sandwiches with T_c up to 3 K, on the basis of the observed resistance anomalies [19, 20] (see Fig. 5.11). The structure of the sandwiches was examined by transmission electron microscopy, but the Au(100) lattice observed for the Cr layer does not allow one to tell whether the thin Cr layer has an fcc or a bcc structure (see Fig. 5.12). The structure of Cr was assumed to be fcc on the basis of a previous report by Thomas and Haas [18]. Supporting evidence for an fcc Cr phase was provided later by synchrotron x-ray diffraction experiments performed on Au/Cr/Au(100) sandwiches [73]. In these measurements, weak satellites were observed straddling the Au[200] reflection. These satellites were attributed to a thin fcc layer. However, extended x-ray absorption fine structures (EXAFS) measured in the same study showed features identical to those of bcc Cr, leading to the conclusion that the fcc domain in the Cr layer coexisted with microcrystalline bcc Cr. Nevertheless, the existence of fcc Cr in Au/Cr/Au sandwiches has been questioned by several authors. The lattice constant of fcc Au (a = 4.078 Å) is larger than that one of bcc Cr (a = 2.88 Å) by almost exactly a factor of $\sqrt{2}$ (see

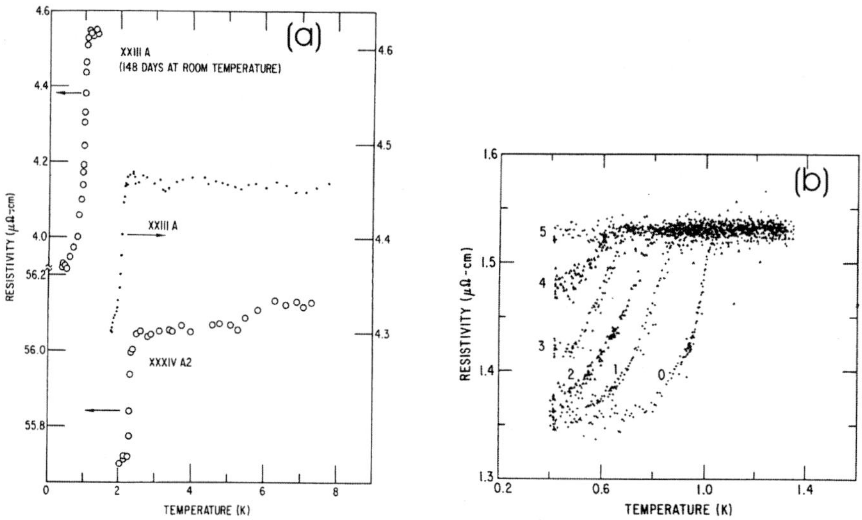

Fig. 5.11. (**a**) Low-temperature resistivity data for Au(73 nm)/Cr/Au(100) sandwich structure (XXIIIA) and [Au (3.9 nm)/Cr (1.9 nm)]$_{10}$ multilayers (XXXIVA2). (**b**) Magnetic-field dependence of resistivity as a function of temperature for sample XXIIIA after 215 days at room temperature. The field was applied perpendicular to the plane of the film. The numbers alongside the data points represent the field in kOe. From [19], used with permission

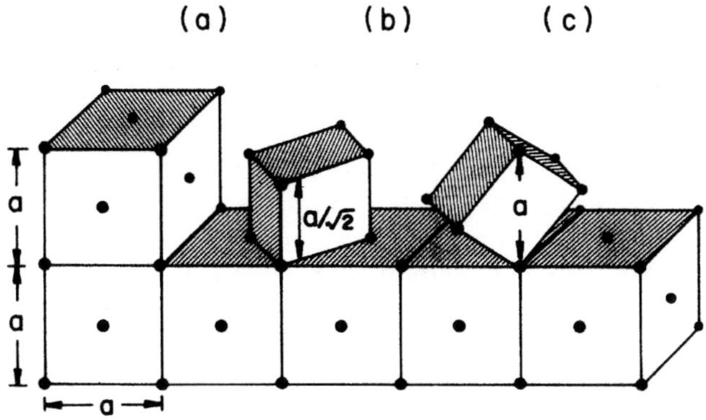

Fig. 5.12. Schematic representation of possible epitaxial arrangements of Cr on Au(100): (**a**) (100) fcc Cr on (100) fcc Au; (**b**) (100) bcc Cr on (100) fcc Au; (**c**) (110) bcc Cr on (100) fcc Au. From [73], used with permission

Fig. 5.12), while the theoretically predicted lattice of fcc Cr (a$\sim$ 3.6 Å) is smaller than that of fcc Au by 12%. From the viewpoint of lattice matching, bcc Cr therefore seems to be more favorable. Zajac et al. performed measurements by low-energy electron-diffraction, Auger electron spectroscopy, ultraviolet photoelectron spectroscopy and electron energy loss spectroscopy on Cr overlayers on Au(100) and found that the usual bcc Cr structure was adequate for explaining all the measured results [74]. On the basis of their measurements, these authors suggested further that the superconductivity observed in Au/Cr/Au sandwiches could come from the disordered Cr at the interface. The bcc structure was also supported by work by Hanf et al.[75]. Hanf et al. found further that interfacial interdiffusion and alloy formation occur in Cr/Au(100) films even at room temperature [75].

5.2.3 Ultrathin Mn Films

Mn is an element with interesting crystallographic and magnetic properties. As we have discussed in Sect. 5.1.3, the lower-temperature phases of Mn (α and β) have complex cubic structures, while the cubic phases, namely fcc γ-Mn and bcc δ-Mn, are stable only at higher temperatures. The γ and δ phases cannot be stabilized at room temperature by quenching pure Mn from high temperatures. Epitaxial growth of Mn films on suitable substrates is more suitable for stabilizing the metastable Mn phases at room temperature. Regarding potential magnetic properties, the free Mn atom has a half-filled 3d shell and hence a maximum moment of $5\mu_B$. Therefore it is quite attractive to attempt to obtain novel magnetic phases by epitaxial growth of ultrathin Mn films. The stabilized structures include bct Mn on V(001), Fe(001), Ag(001) and Pd(001); trigonal or hexagonal Mn on Ru(001),

Cu(111), Pd(111), Ir(111) and Au(111); and fcc Mn on GaAs(001). The structural and magnetic properties of these films will be discussed in this subsection. Mn is also known for the easy formation of surface alloys on a variety of substrates at low coverage [76, 77] and the interesting magnetic properties of its monolayer on ferromagnetic substrates [78, 79]. We shall discuss this aspect in detail in Chap. 7.

Mn/V(001) [80]. Ultrathin films of Mn have been grown on V(001) pseudomorphically up to 15 Å. The structure of the film has been determined to be body-centered tetragonal, with $a = 3.023$ Å, forced by the V(001) substrate, and $c = 2.92 \pm 0.06$ Å. A slight expansion ($\triangle d_{12} = 0.11 \pm 0.03$ Å) is found for the first interlayer spacing.

Mn/Fe(001). Mn grows on an Fe(001) substrate via a layer-by-layer mode [81, 82]. The film structure has been determined to be body-centered tetragonal by LEED and RHEED [82, 83]. The in-plane lattice parameter is determined by the substrate to be 2.87 Å. The axial parameter *(c)* was found to be 3.27 Å by fitting the energies at which the primary Bragg reflections possess maximum intensity as a function of n^2 (the square of the index labeling the maxima) [82]. A quantitative LEED analysis measured c to be 3.23 Å [83]. The values determined by these two different methods are quite close to each other. It has also been reported that Mn layers in Fe/Mn multilayers on Ge(001) show a bcc structure with only a slight tetragonal distortion (a = 2.86 Å, c = 2.79 Å) [81]. Quite interestingly, the LEED $I(V)$ curves indicate that the interlayer spacing for very thin Mn films, for example a 3.5 Å thick Mn film, is considerably smaller than the bulk value for thicker films, consistent with the theoretical prediction of a 1.3 Å interlayer spacing for a Mn bilayer on Fe(001) [84]. This is reflected by the shift of the major peaks in the $I(V)$ curve first towards higher energy and subsequently towards lower energy with increasing film thickness (see Fig. 5.13).

Mn/Ag(100). Tetragonal Mn has been stabilized in Mn/Ag superlattices on Ag(001) with Mn layer thicknesses up to 23 Å [85]. The Ag(001) substrate films were grown on 5 ML Fe(001) seed layers, which were grown on a GaAs(001) substrate with ZnSe(001) epilayers as buffer layers. X-ray photoelectron diffraction (XPD) results show that the Mn layer grows as a body-centered tetragonal phase with $a = 2.89$ Å and $c = 3.31$ Å. XPD measurements also show that the growth at room temperature is not ideal layer-by-layer, but with some population of the second Mn layer occurring before the first is completed. The magnetic behavior of the superlattices was examined by temperature-dependent ferromagnetic resonance, measurement of the magnetic susceptibility and the use of a superconducting quantum interference device (SQUID). None of the samples studied showed a detectable ferromagnetic signal at temperatures down to 5 K, even though one sample had a total Mn thickness of 1035 Å. Therefore it was concluded that the

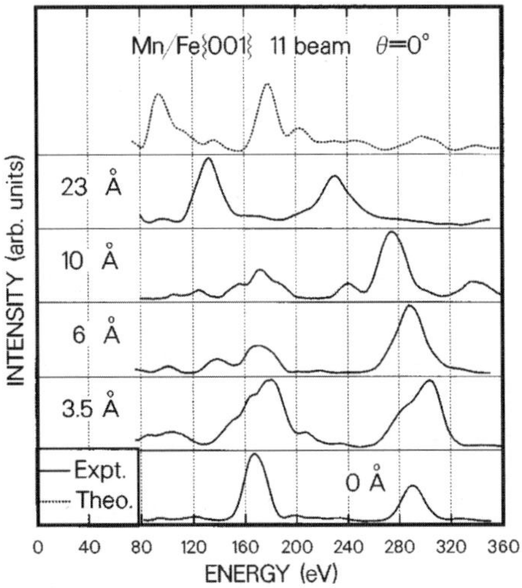

Fig. 5.13. Evolution of the LEED spectrum of the (11) beam with increasing thickness of Mn films on Fe(001). The film thickness is given in each panel, hence the *bottom curve* is the spectrum for clean Fe(001). Note the trend of the major peaks to shift towards higher energies for the thinnest film and towards lower energies for the thicker films. The *upper dotted curve* is a theoretical spectrum calculated for a bilayer of Mn with an interlayer spacing of 1.3 Å on Fe(001). From [83], used with permission

tetragonal Mn prepared was either nonmagnetic or antiferromagnetic at low temperature.

Mn/Pd(001) [86]. In the room-temperature growth of ultrathin Mn films on Pd(001), the LEED pattern changes with increasing coverage. Starting from the 1×1 pattern typical of clean Pd(001), the pattern shows a weak $c(2 \times 2)$ structure at a Mn coverage of about 1.5 ML. This pattern remains until a coverage of 6 ML is reached. Above 7 ML, the LEED pattern becomes 1×1 again, and remains so until it is obliterated by the background. A quantitative LEED analysis gives the following structural parameters: $a = 2.75$ Å, $d_{bulk} = 1.715 \pm 0.03$ Å, $d_{12} = 1.805 \pm 0.03$ Å and $d_{23} = 1.715 \pm 0.04$ Å, where d_{12}, d_{23} and d_{bulk} denote the first interlayer spacing, the second interlayer spacing and the bulk interlayer spacing, respectively [86].

Mn/Cu$_3$Au(100) [87]. The growth of Mn on $Cu_3Au(100)$ at various temperatures has been studied, mainly by medium-energy electron diffraction (MEED) (see Fig. 5.14). At 300 K, pronounced oscillations are observed, which are characteristic of layer-by-layer growth. Oscillations are observed up to 20 ML at a deposition temperature of 223 K. With a further decrease in growth temperature, these oscillations are less pronounced and are restricted to the first 10 ML. Below 160 K, a continuous decrease in intensity

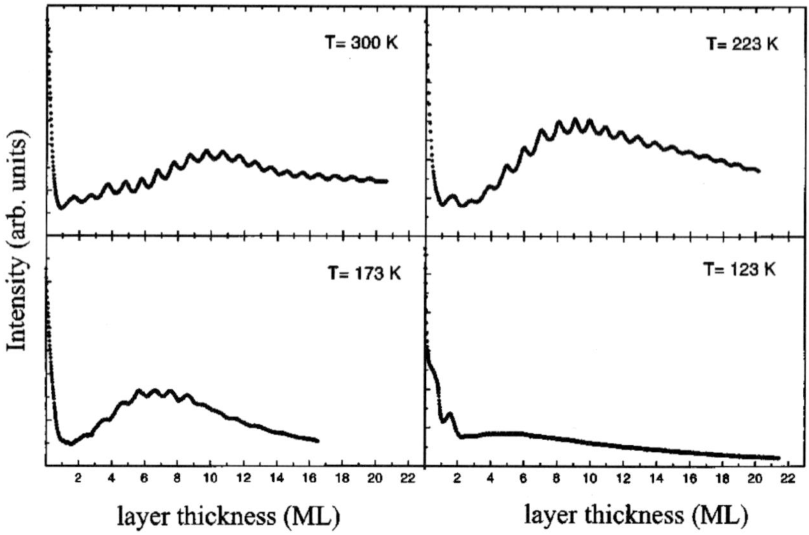

Fig. 5.14. MEED curves of the (0,0) spot intensity during deposition of Mn on $Cu_3Au(100)$ at various growth temperatures. The curves were recorded with an electron energy of 3 keV and an angle of incidence of 81.4° with respect to the surface normal. From [87], used with permission

is observed. The disappearance of intensity oscillations and the decreasing intensity result from three-dimensional growth caused by the decreasing mobility of deposited atoms.

Though LEED patterns show no superstructure, irrespective of film thickness, and the position of the LEED beams does not vary with thickness within the range investigated (<20 ML), a full dynamical LEED $I(V)$ analysis reveals that a structural transition, manifested mainly by an obvious change in interlayer spacing, occurs at a critical thickness between 5.2 and 8 ML. For a 5.2 ML film, the interlayer spacings have been determined to be $d_{34} = 1.92$ Å, $d_{23} = 1.91$ Å and $d_{12} = 1.925$ Å. The resulting c/a ratio closely resembles that of an almost cubic structure. In contrast, a 16 ML film shows an interlayer spacing of 1.772 Å for the bulk layer of Mn, indicating a tetragonally distorted fcc structure. The atomic volumes of these two structures observed for small thickness and large thickness differ by 7%. No ferromagnetic signal has been detected by MOKE for film thicknesses up to 25 ML and temperatures as low as 100 K.

Mn/Ru(001) [88]. The growth of Mn on the (001) surface of a hexagonal Ru substrate has been studied by RHEED. During deposition of the first two layers, the RHEED patterns only change slightly in intensity and sharpness, indicating that the first two epitaxial layers of Mn have the same close-packed structure as the Ru substrate, though the stacking of these two layers with

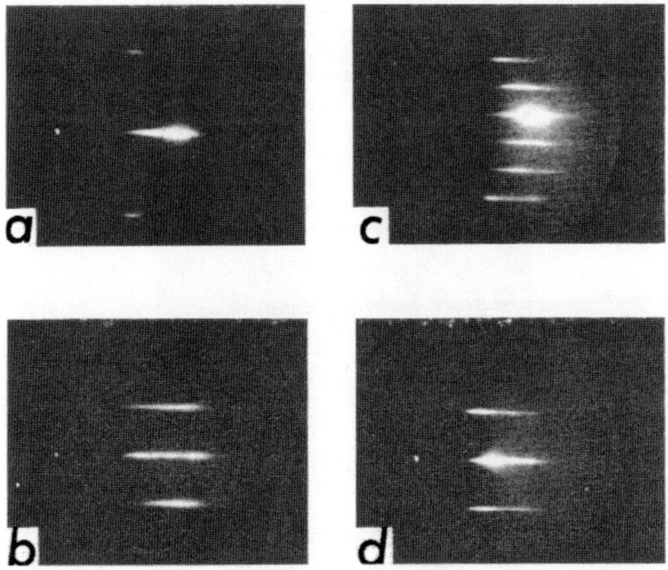

Fig. 5.15. RHEED patterns for the Ru(001) surface before Mn growth ((**a**) and (**b**)) and for Mn grown on the Ru(001) surface after the trimerization ((**c**) and (**d**)). The electron beam is parallel to Ru[120] in (**a**) and (**c**), and to Ru[110] in (**b**) and (**d**). During the growth of the first two monolayers, the pattern remains the same as before the growth. Subsequently, new patterns, as shown in (**c**) and (**d**), develop. A $(\sqrt{3} \times \sqrt{3})R30°$ Mn phase for thick films has been proposed to explain the evolution of the RHEED patterns. From [88], used with permission

respect to one another or with respect to the substrate cannot be determined (see Fig. 5.15). For growth beyond the second layer, the RHEED pattern shows a change in the unit cell to one of the same hexagonal symmetry. The new structure is characterized by $(\sqrt{3} \times \sqrt{3})R30°$, or equivalently by $(3 \times 1)R90°$, depending on how the unit cell is chosen. The new RHEED pattern becomes perceptible at a 2.2 ML coverage of Mn. It is fully developed by the completion of the fourth monolayer and persists up to 40 ML. The pattern is stable against annealing to 450 °C, showing evidence that the $(\sqrt{3}\times \sqrt{3})R30°$ pattern is likely to result from a metastable phase rather than surface reconstruction. We shall refer to this phase as the $(\sqrt{3} \times \sqrt{3})R30°$ phase hereafter. A structural model based on the Zn_2Mg-type Laves phase has been proposed to explain the RHEED pattern (see Fig. 5.16 for a detailed explanation).

The properties of the local magnetic moments of the first two layers of Mn and the new phase have been examined by x-ray photoemission spectra. Before we describe the results, however, we would like to insert a short description of outer-core photoemission spectroscopy. For a paramagnetic or magnetically ordered material, the multiple splitting of an outer-core level

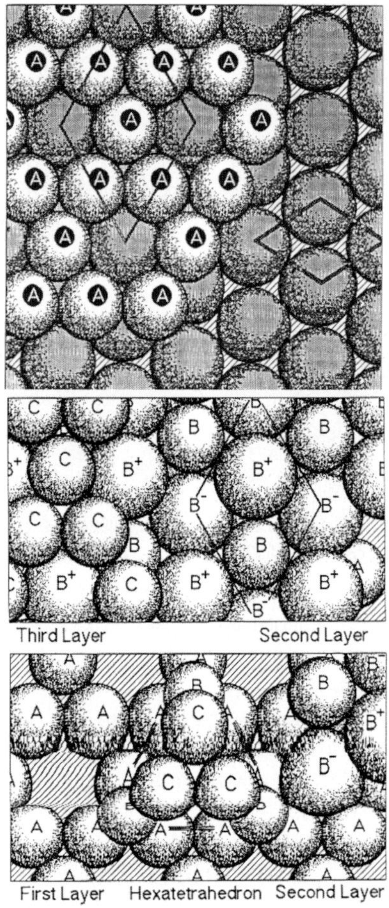

Fig. 5.16. *Top*: the development of the first layer of the $(\sqrt{3} \times \sqrt{3})R30^\circ$ phase of Mn on the Ru(001) surface. The nearest-neighbor distances are 2.70 Å in the lower substrate layer and $2.70 \times \sqrt{3}/2 = 2.34$ Å in the open hexagonal structure of the upper Mn layer. The unit cell in the lower layer is indicated by the *diamond* on the *lower right*. The unit cell in the upper layer, indicated by the *diamond* in the *upper left* corner of the figure, is three times larger and rotated by 30° (or 90°). There are atoms sitting in the holes, which are not shown. *Middle and bottom*: building up the Zn_2Mg-type Laves phase of Mn. The basic building block is the hexatetrahedron, shown in the middle of the *bottom panel*. The atoms labeled A, B and C are the smaller Mn atoms in the first, second and third layer, respectively. The atoms labeled B^- and B^+ are the larger Mn atoms. The start of the second layer is depicted at the *bottom right*. The B^- atoms sit in the hole of the open hexagonal lattice of the A atoms. The B^+ atoms, as well as the B atoms, sit on top of trios of A atoms. The B layer has the B^- atoms slightly below the layer, the B atoms in the layer, and the B^+ atoms slightly above the layer. The C layer is the same as the A layer, except for a shift in position. From [88], used with permission

such as a 3s and 3p level for a transition metal of the 3d series is related to the local magnetic moment of the atoms or ions. This effect was first interpreted in terms of a strong intraatomic L–S term splitting in the final state of a system with a core hole. For the simplest case of 3s emission from an atom with spin S, the splitting is due to the exchange interaction between the unpaired 3d electrons and the remaining 3s electron. This model predicts a doublet with a separation proportional to $(2S+1)\times$(3s–3d exchange integral), as a direct result of the Van Vleck theorem [89, 90]. However, the measured energy separation is usually a factor of two smaller than that predicted from a calculation of the exchange integral, and high-resolution x-ray photoemission spectroscopy also shows additional low-intensity satellite peaks at higher binding energy. It has now been found that, besides the exchange interaction, the intrashell electron correlation [91, 92] and final-state screening effects [93] also contribute to the multiple-splitting structure. Therefore one should take much care when applying the rule of proportionality between the local magnetic moment and the splitting. A systematic investigation of the photoemission spectra of insulating Mn and Fe compounds has revealed that the proportionality holds approximately for ionic crystals, while it breaks down as soon as the covalency becomes strong [93]. As a rule of thumb, the 3s splitting reflects the local magnetic moment only when the 2p core-level satellite is negligible. For metallic bcc Fe, the spin-resolved 3s spectra, show both minority- and majority-spin character, with the minority-spin peak located at lower binding energy [94]. The measured intensity ratio of the spin-resolved peaks is in agreement with the d-electron band structure.

After this discussion of the interpretation of the multiplet structure in 3s photoemission spectroscopy, let us now return to the measurement of Mn/Ru(001) ultrathin films. Figure 5.17 shows both the 3s and the $2p_{3/2}$-level x-ray photoemission spectra of Mn for three structures, namely, a two-layer film of Mn on Ru(001), a thicker film with a $(\sqrt{3}\times\sqrt{3})R30°$ RHEED pattern and an MnAg alloy. Since it is well established that Mn atoms in dilute Ag alloys have a high moment approaching Hund's limit of $5\mu_B$ [95], the spectra of the MnAg alloy serve as a reference for the determination of the local moments of the ultrathin Mn films on Ru(001). The 3s Mn spectra show nearly the same splitting for these three structures, though the satellite peaks have different intensities. These observations were explained by assuming that all materials contain Mn in a state with an atomic magnetic moment of $5\mu_B$, but the fraction of Mn in that state differs from material to material. The difference in the 2p spectra between the two-layer film and the $(\sqrt{3}\times\sqrt{3})R30°$ phase is also insignificant. However, the 2p spectrum of the MnAg alloy is quite different, perhaps indicating a change in the screening mechanism of the final state as the Mn atom becomes completely surrounded by Ag.

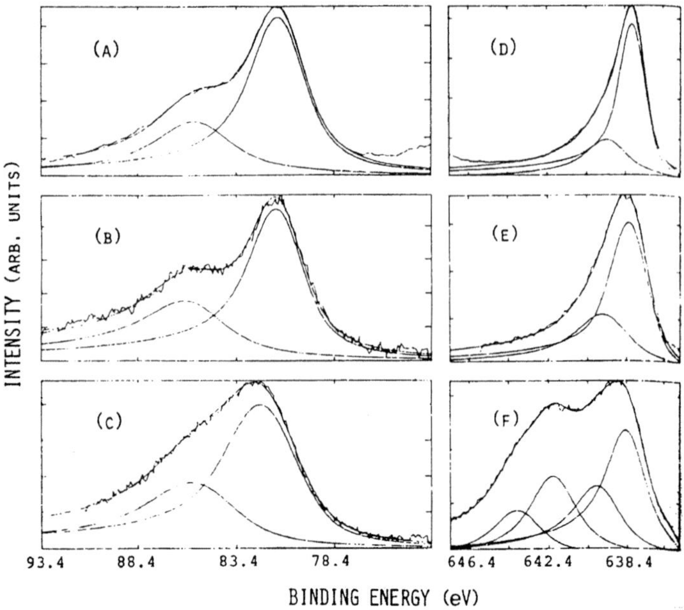

Fig. 5.17. Mn 3s ((A), (B) and (C)) and $2p_{3/2}$ ((D), (E) and (F)) photoemission spectra. Spectra are shown for the $(\sqrt{3}\times\sqrt{3})R30°$ phase ((A) and (D)); the corresponding spectra for two layers of Mn ((B) and (E)) on Ru(001) and an AgMn alloy (17 at % Mn) are shown for comparison ((C) and (F)). From [88], used with permission

Mn/Cu(111). Tian et al. have studied the growth and electronic properties of Mn thin films on Cu(111) [96]. After deposition of about 1 Å Mn on the unheated Cu(111) substrate, the LEED pattern shows evidence of the presence of an incommensurate $(\sqrt{3}\times\sqrt{3})R30°$ phase (see Fig. 5.18). With increasing Mn coverage, the beams produced by the substrate decrease in intensity and finally disappear when the thickness of the Mn film exceeds 26 Å, leaving only the LEED pattern of the $(\sqrt{3}\times\sqrt{3})R30^o$ Mn phase. The in-plane periodicity of the $(\sqrt{3}\times\sqrt{3})R30°$ phase along the $\langle 110\rangle$ direction of the Cu(111) net is $(6.3\pm1)\%$ larger than the in-plane lattice constant of Cu(111), i.e., $2.55\text{Å}\times1.063 = 2.71$ Å. The lack of perfect sixfold symmetry of the LEED pattern, which is characteristic of the $(\sqrt{3}\times\sqrt{3})R30°$ phase observed on other (111) substrates [97], was ascribed to the presence of small amounts of fcc γ-Mn, since an fcc (111) surface gives rise to a threefold LEED pattern. The incommensurate property of the Mn overlayer on the Cu(111) substrate was confirmed by normal-incidence x-ray standing-wave measurements [98]. Quantitative modeling of the data excluded the possibility of the Mn overlayer being α-Mn, γ-Mn (fcc) or δ-Mn (bcc). However, the data are consistent with a Laves phase structurally related to Zn_2Mg.

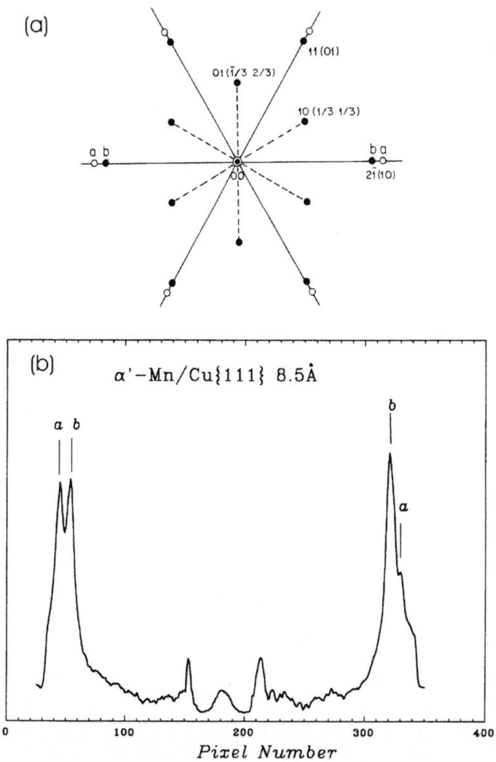

Fig. 5.18. (**a**) Schematic LEED pattern showing spots from the Cu(111) substrate (*open circles*) and spots from the Mn film phase (*filled circles*). Two ways can be chosen to index the LEED spots from the Mn film. One way, which is to index them with reference to the substrate, is given in parentheses. Alternative indices can also be chosen with respect to the $(\sqrt{3} \times \sqrt{3})R30°$ phase. (**b**) Profile of the intensity in the LEED pattern along a horizontal line through the (00) spot. The electron energy was 65.1 eV. The feature in the center was produced by the backsurface of the electron gun. The peaks marked "a" and "b" originate from the Cu(111) substrate and the Mn film. From [96], used with permission

Mn/Pd(111) [97]. The fcc Pd substrate has a lattice parameter of 3.89 Å, and shows a trigonal in-plane lattice with $a = 2.75$ Å. The lattice constant of fcc γ-Mn is 3.73 Å at room temperature when extrapolated from temperatures above 1095°C [86] and therefore has a misfit of 4.2% on Pd. The Mn films deposited at room temperature on Pd(111) were both pseudomorphic and epitaxial. The 1×1 LEED pattern with threefold symmetry can persist up to 16 ML. The actual structure of the films is not strictly fcc, because the in-plane lattice constant is determined by the substrate (2.75 Å), but the bulk interlayer spacing was measured to be 2.16 Å by fitting the LEED $I(V)$ spectra; this spacing; is smaller than the interlayer spacing of Pd(111) (2.25 Å) by

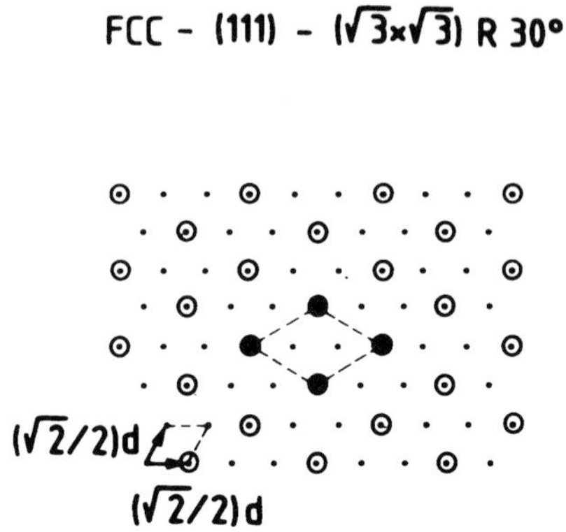

Fig. 5.19. Schematic diagram of a $(\sqrt{3} \times \sqrt{3})R30°$ surface structure in real space. The adsorbates are indicated by *circles* and the surface atoms by *points*. The unit cells of both lattices are indicated, and d is the fcc bulk lattice parameter. From [99], used with permission

3.8%. Quite interestingly, if the deposition temperature is raised, for example to 150°C, the film presents a different structure, with a $(\sqrt{3} \times \sqrt{3})R30°$ LEED pattern (see Fig. 5.19). The symmetry of the LEED pattern changes gradually from threefold to sixfold with increasing thickness. By the time the surface coverage reaches 7–8 ML, the symmetry of the LEED pattern becomes entirely sixfold. The origin of the $\sqrt{3} \times \sqrt{3})R30°$ LEED pattern remains to be clarified. The authors of [99] have proposed that it could be related to α-Mn.

Valence-band photoemission spectra from both the (1×1) phase and the $\sqrt{3} \times \sqrt{3})R30°$ phase were measured with s-polarized 90 eV synchrotron radiation (see Fig. 5.20). The spectrum from the (1×1) phase shows a pronounced peak at about -4 eV, whereas that from the $(\sqrt{3} \times \sqrt{3})R30°$ phase exhibits instead a broad peak around -3 eV. The latter shows features similar to those of α-Mn. This is why the $\sqrt{3} \times \sqrt{3})R30°$ phase was proposed to be structurally related to α-Mn. On the other hand, the 3s core-level spectra are almost the same for both phases. The splitting of the 3s level is 5.6 eV for both phases; this is larger than the value of 4.08 eV measured on the polycrystalline α-Mn film. This splitting corresponds to a spin magnetic moment of $3.9\mu_B$ per atom, larger than the value of $2.5\mu_B$ found in α-Mn.

Mn/Ir(111) [100]. Similarly to Mn overlayers on Ru(001), two phases of Mn were identified in ultrathin Mn films grown on Ir(111) by their different LEED patterns. For Mn coverages below 4 ML, the electron diffraction pattern is (1×1), indicating epitaxial growth within this thickness range. Above 4 ML, the pattern changes to $(\sqrt{3} \times \sqrt{3})$, and this pattern persists up to a coverage of 20 ML.

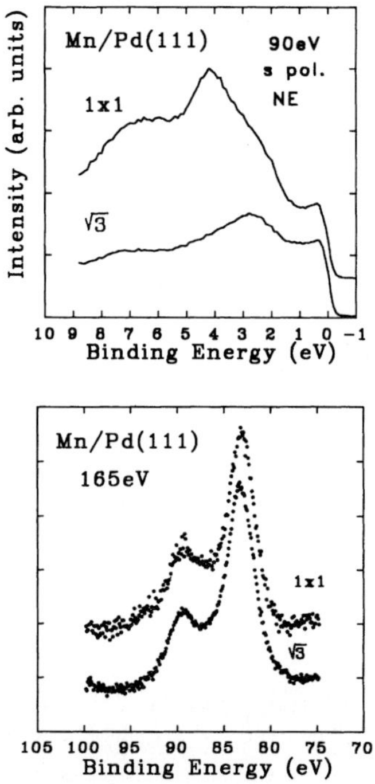

Fig. 5.20. *Top*: angle-resolved normal-emission valence photoemission spectra measured with s-polarized photons of energy 90 eV on a 20 ML (1 × 1) Mn/Pt(111) system and a 12 ML $(\sqrt{3} \times \sqrt{3})R30^\circ$ Mn/Pd(111) system. *Bottom*: 3s core-level spectra for these two films. From [97], used with permission

The splitting of the Mn 3s photoemission peak is 4.5 and 4.0 eV for the 1 and 10 ML films, respectively, implying that the magnetic moment of Mn in the 1 ML film is larger than that in the 10 ML film. This conclusion has been confirmed by Mn $L_{2,3}$ absorption spectroscopy (Fig. 5.21). The L_3 branching ratio in the $L_{2,3}$ absorption spectroscopy, which is defined as $I(L_3)/[I(L_3) + I(L_2)]$, provides another method to estimate the local magnetic moment of Mn. As a general rule, the L_3 branching ratio decreases with decreasing moment, as is shown by both atomic calculations [101] and experimental observations (e.g. [102]). The gradual decrease in the L_3 branching ratio was ascribed to an enhanced surface and/or interface magnetic moment, whose contribution is gradually attenuated with film thickness.

Mn/Au(111). The growth and structure of ultrathin Mn films on Au(111) has recently been investigated using STM and LEED in a thickness range up to 8 ML at room temperature [103]. In the initial stage, the Mn atoms preferentially decorate the elbows of the $(22 \times \sqrt{3})$ Au(111) reconstruction and form clusters aligned in rows along the $[11\overline{2}]$ direction (see Fig. 5.22a). The decoration is attributed to a site-selective exchange of impinging adatoms

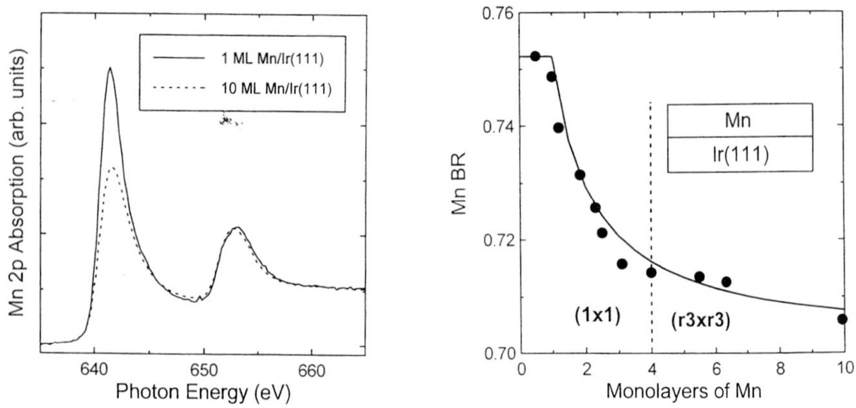

Fig. 5.21. *Left*: Mn $L_{2,3}$ absorption spectra for 1 ML and 10 ML Mn films on Ir(111). The spectra have been normalized to a constant step height above the L_2 edge. *Right*: the L_3 branching ratio (BR), defined as $I(L_3)/[I(L_3)+I(L_2)]$, plotted as a function of film thickness. Also shown are the regions where different LEED patterns are observed. From [100], used with permission

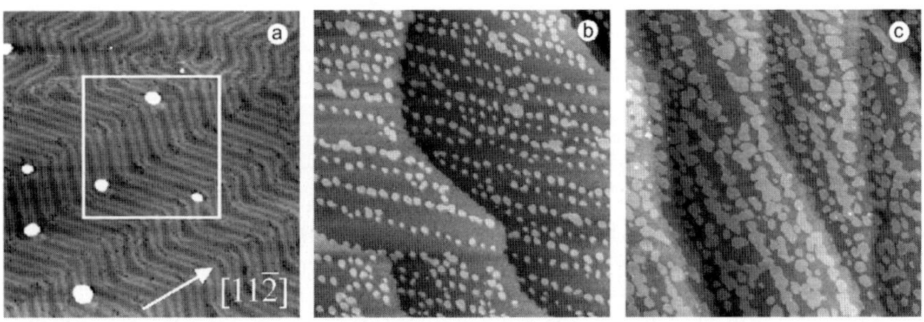

Fig. 5.22. STM images after deposition of Mn on Au(111) at room temperature (**a**) after 0.01 ML Mn (110×110 nm^2), (**b**) 0.2 ML Mn (200×200 nm^2) and (**c**) 0.5 ML Mn (200×200 nm^2). From [103], used with permission

with gold atoms at the elbow sites, which act as nuclei for further aggregation. This effect is typical of the deposition of 3d transition metals, such as Fe, Co and Ni, on the Au(111) substrate. The deposition of Mn in the submonolayer regime results in the formation of two-dimensional islands at the elbows of the herringbone structure. The height of the islands was determined to be 2.90 ± 0.10 Å. In agreement with the symmetry of the Au(111) surface, the islands show a hexagonal shape with edges along the close-packed $\langle 0\bar{1}0 \rangle$ directions of the substrate. With increasing Mn film thickness (1–3 ML), the growth mode changes from layer-by-layer to island growth. Between 4 and 6 ML Mn, a structural transition takes place, as shown by the corresponding change in the LEED pattern from (1×1) for 0–4 ML Mn to $(\sqrt{3}\times\sqrt{3})R30°$ for thicker Mn films (see Fig. 5.23).

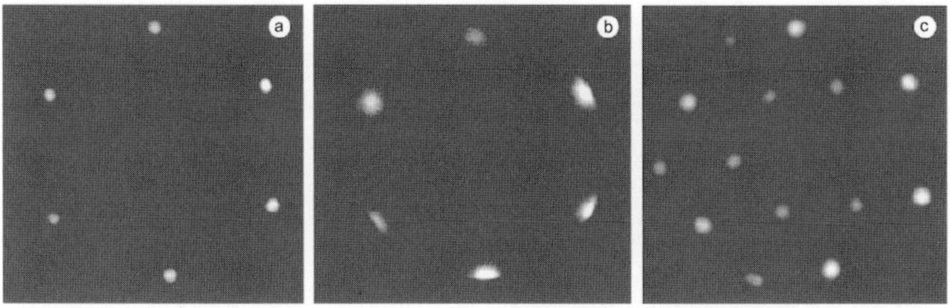

Fig. 5.23. LEED pattern for (**a**) pure Au(111) surface ($E = 53$ eV), (**b**) 1 ML Mn on Au(111) ($E = 46$ eV) and (**c**) 6 ML Mn on Au(111) ($E = 57$ eV). From [103], used with permission

We have already mentioned the $(\sqrt{3} \times \sqrt{3})R30°$ structure several times. Without exception, this structure is associated with Mn films grown on close-packed surfaces, namely hcp (001) and fcc (111) surfaces. This structure usually occurs in two situations. In one case, the Mn film grows pseudomorphically for the first several monolayers, and then the $(\sqrt{3} \times \sqrt{3})R30°$ LEED or RHEED pattern replaces the (1×1) structure with increasing thickness. This structural transition has been observed in Mn/Ru(001), Mn/Ir(111) and Mn/Au(111). In the other case, the $(\sqrt{3} \times \sqrt{3})R30°$ structure occurs during the deposition of Mn on heated substrates, while the Mn film prepared at room temperature shows a (1×1) structure. This is the case for Mn/Pd(111). Mn/Cu(111) is an exception. It shows an incommensurate $(\sqrt{3} \times \sqrt{3})R30°$ structure for the whole thickness range studied. There is reason to consider this $(\sqrt{3} \times \sqrt{3})R30°$ structure as a metastable phase, which has a lower energy than the high-temperature fcc phase or the hcp structure. This new structure appears to be related to the close-packed plane of the substrate and thus can be stabilized by it. A hexagonal Zn_2Mg-type Laves structure has been proposed for this new phase observed in Mn/Ru(001) films [88]. A recent normal-incidence x-ray standing-wave study on a Mn/Cu(111) film [98] has eliminated the possibility of the $(\sqrt{3} \times \sqrt{3})R30°$ phase being α-Mn, γ-Mn (fcc) or δ-Mn (bcc). On the other hand, the data support the Laves structure since it is indicated that there exist several different interlayer spacings [98]. However, the precise structure of the $(\sqrt{3} \times \sqrt{3})R30°$ phase is yet to be determined.

Mn/GaAs(001). Fcc Mn has been successfully prepared on a GaAs(001) substrate using molecular-beam epitaxy [104]. The fcc film can be stabilized for more than 120 ML. The fcc structure was verified by the RHEED pattern and by x-ray diffraction (XRD). The epitaxial Mn film obeys an orientation relationship of $(001)_{Mn} \parallel (001)_{GaAs}$ and $[100]_{Mn} \parallel [110]_{GaAs}$ with respect to the substrate. The lattice parameter was determined to be 3.68 Å by RHEED and 3.62 Å by XRD. The transition region is composed of a Mn–Ga–As al-

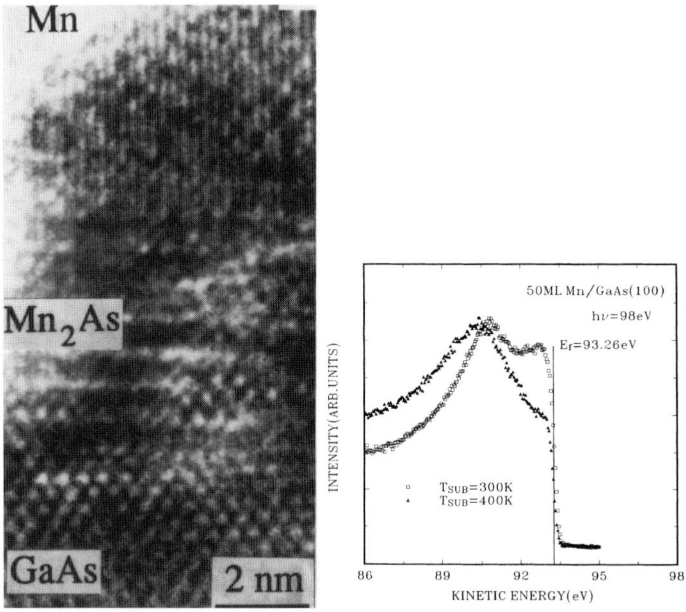

Fig. 5.24. *Left*: HRTEM image showing a transition region between the fcc Mn film and the substrate. *Right*: comparison of the valence band spectra reveals a downward shift of the electron density of the occupied 3d state for fcc Mn with respect to α-Mn. From [105, 106], used with permission

loy formed at the Mn/GaAs interface. This alloy layer is believed to play an important role in the formation of fcc Mn. The existence of a transition region has been confirmed by high-resolution transmission electron microscopy (HRTEM) (see Fig. 5.24a), and this region is believed to have a structure related to the Mn_2As-type structure, whose in-plane lattice parameter (3.8 Å) is close to that of the fcc Mn formed [105]. The substrate temperature was found to be very important for the preparation of fcc Mn. The favorable temperature is around 400 K. At a lower temperature (300 K), polycrystalline α-Mn is formed. At a higher temperature (500 K), the chemical reaction and interdiffusion in the system are so enhanced that no pure metallic Mn phase can be obtained.

The electronic structure of fcc Mn was studied using synchrotron radiation. Compared with α-Mn, fcc Mn has a lower electron density at the Fermi edge but a far higher density below the Fermi edge (see Fig. 5.24b). However, the Mn 3s core-level spectra for the fcc phase and α-phase show no obvious difference.

5.2.4 Ultrathin Fe Films

Since the energies of various magnetic phases of fcc Fe, namely the ferromagnetic, nonmagnetic and antiferromagnetic state's, are very close to each other, the magnetic properties of fcc Fe are very sensitive to slight changes in structure. Studying the modification of the magnetic properties of fcc Fe films upon a structural change provides a unique way to understand the correlation between structure and magnetism. As a consequence of this motivation, ultrathin fcc Fe films have been intensively investigated in the last two decades, with Fe/Cu(100) as a prototype system, in which complex correlations have been observed between the magnetic properties on one hand and the film thickness and growth conditions on the other hand. The published literature on fcc Fe films is so extensive that we shall discuss it in a separate chapter. In this subsection, we focus on another metastable structure of Fe, the hcp phase.

Hcp Fe occurs as the ε phase in the Fe phase diagram at a pressure larger than 135 kbar (see Fig. 5.7). Its atomic volume is reduced to 10.9 Å^3. Mössbauer effect measurements have shown that the high-pressure ε phase is in a paramagnetic state down to liquid-helium temperature [107]. Theoretically, equilibrium hcp Fe has been predicted to be nonmagnetic [16] or antiferromagnetic [40]. Experimentally, substrates with an hcp structure such as Ru and Re have been used to stabilize hcp Fe.

Growth and Structure of Fe/Ru(0001) Films

The growth and structure of Fe overlayers on a Ru(0001) surface were studied early with the motivation of exploiting new catalysts [108]. The growth of the first monolayer of Fe at 520 K follows a layer-by-layer mechanism, as is indicated by the clear breakpoint in the Auger uptake curves. During the deposition of Fe up to 1 ML, the LEED pattern shows no other long-range structure except the hexagonal (1×1) diffraction spots. This means that the first Fe monolayer has a close-packed structure commensurate with that of the Ru underlayer. When the Fe coverage exceeds one monolayer, however, new spots due to a $(6\sqrt{3}\times 6\sqrt{3})R30°$ superstructure appear around the main spots. This superstructure can be explained by the formation of a bcc (110) surface structure for the second Fe overlayer with an epitaxial relation of $[\bar{1}10]$ Fe $\parallel$ [120] Ru.

The room-temperature growth of Fe on Ru(0001) shows some differences [109, 110, 111]. The breakpoints in the Auger uptake curves [110] and the streaks in the RHEED pattern [109] still suggest a layer-by-layer mode for the initial growth (see Fig. 5.25), while the hexagonal (1×1) LEED or RHEED pattern persists to a much larger Fe thickness (14 Å in [109], 8 ML in [110] and 4 ML in [111]), indicating a significant increase in the thickness limit for pseudomorphic growth upon lowering the deposition temperature. Structural relaxation to bcc (110) also occurs for larger Fe thicknesses, as demonstrated

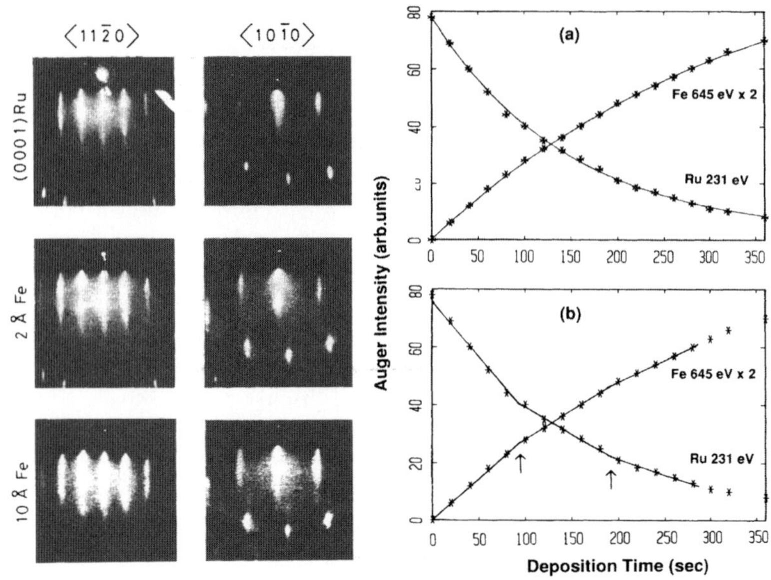

Fig. 5.25. *Left*: RHEED streaks observed at 30 keV electron energy for Fe layers grown on Ru(0001) at room temperature. The incidence azimuth and the Fe thickness are given in the figure. From [109]. *Right*: Measurements of Auger intensity vs. deposition time for Fe grown on Ru(0001). (**a**) shows an exponential least-squares fit of the data, and (**b**) shows two breaks in each curve for the same data as in (**a**). From [110], used with permission

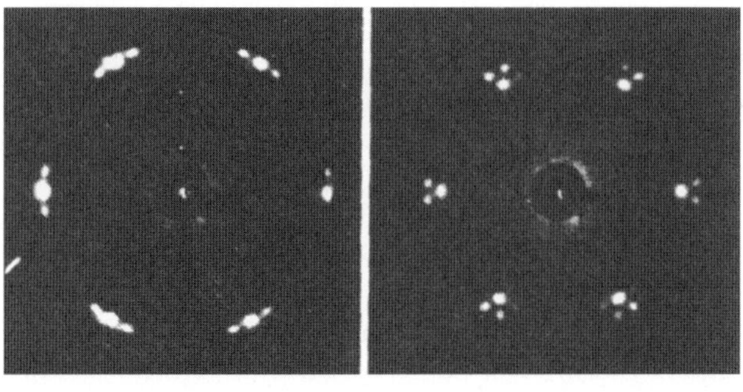

Fig. 5.26. LEED patterns for a 15 ML Fe film grown on Ru(0001) at room temperature. The electron energies are 67 eV for the *left-* and 89 eV for the *right-hand side*, respectively. From [111], used with permission

by the complex LEED patterns shown in Fig. 5.26 for a 15 ML Fe film on Ru(0001). These complex LEED patterns were interpreted as arising from multiple relaxation domains (see Fig. 5.27 for a detailed explanation). The existence of relaxation domains is also indicative of an island growth mode for thick Fe films on Ru(0001).

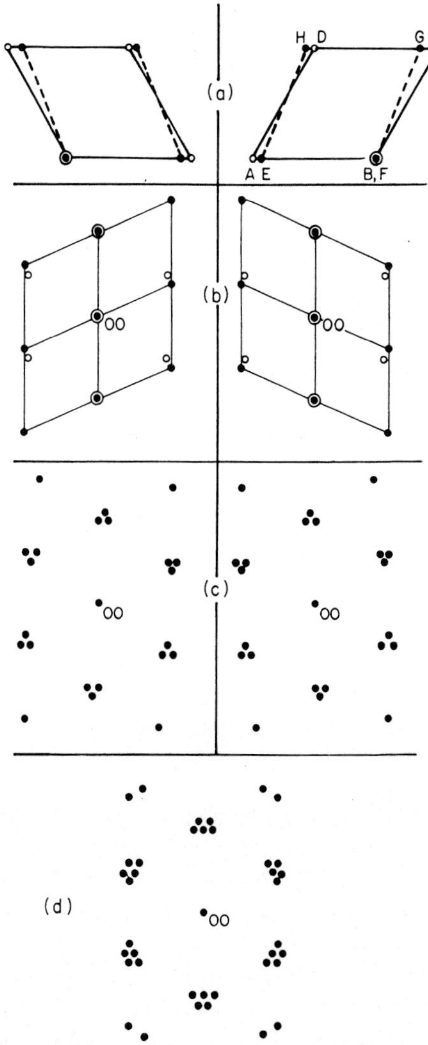

Fig. 5.27. (**a**) Two equivalent Kurdjumov–Sachs orientations of the bcc (110) lattice on Ru(0001). The *open circles* at A, B, C and D represent atoms in the hexagonal net of Ru(0001) and the *solid circles* at E, F, G and H represent the atoms in the bcc (110) net. (**b**) Schematic LEED patterns of relaxed single domains with the real-space orientations shown in (**a**). The *open circles* now represent the reciprocal lattice of the hexagonal net and the *solid circles* represent that of the bcc (110) net. (**c**) Schematic LEED patterns expected from superposition of rotationally equivalent domains based on the sixfold symmetric Ru(0001) substrate. The equivalent domains are obtained by rotating each orientation in (**a**) by 120° or 240°. Note that the two configurations in (**a**) are not rotationally equivalent but are related by a mirror plane. (**d**) Schematic LEED pattern expected from superposition of both sets of domains in (**c**). From [111], used with permission

Much attention has been paid to the precise determination of the structure of Fe/Ru(0001) films, which show a (1×1) diffraction pattern. Since it is impossible to tell just from the LEED or RHEED pattern whether the stacking sequence is hcp (ABABA $\cdots$) or fcc (ABCABC $\cdots$), other techniques must be used to reach this goal. One method is to perform high-resolution x-ray diffraction on Fe/Ru superlattices. In one work [109], the c parameter of Fe/Ru superlattices with varying thicknesses of the Fe (<12 Å) and Ru layers had been measured by θ–2θ scans. Extrapolating the c values to pure Fe films using $c(x,y) = (xc_{Fe} + yc_{Ru})/(x+y)$ gives $c_{Fe} = 4.15$ Å, where x and y are the thicknesses of the individual Fe and Ru layers, respectively. In another work, anomalous x-ray diffraction near the Fe absorption K edge was used to determine the interlayer spacing for Fe and Ru in Fe/Ru superlattices [112]. The c parameter was determined to be 4.11 Å for Fe and 4.33 Å for Ru. The former c value is consistent with that derived from LEED $I(V)$ curves (4.10 Å) [111], but slightly larger than that of the ε-Fe phase (4.06 Å) after the effect of the pressure is taken into account. If an fcc stacking sequence is assumed, c_{Fe} should be around 3.75 Å to maintain the atomic volume of fcc Fe of 11.3 Å^3. This result is indicative of an hcp structure for this Fe film. In one of the above studies [109], the hcp stacking was further confirmed by the observation of the $(11\bar{2}3)$ Bragg peak in powder x-ray diffraction patterns. The powder was obtained from films that were lifted off from the mica substrate on which the superlattices were prepared. Therefore it was concluded in those studies that Fe layers in Fe/Ru superlattices retain a hexagonal structure with an anomalously large volume of about 12.5 Å^3 when the Fe layer is thin enough (<12 Å).

The local atomic environment of Fe atoms in Fe/Ru(0001) superlattices has been further investigated by extended x-ray-absorption fine structure (EXAFS) and x-ray-absorption near-edge-structure (XANES) measurements [113]. First, we would like to present a short explanation of why these techniques can be used to determine the local atomic environment and thus the stacking sequence [114, 115]. X-ray absorption fine structure (XAFS) refers to the oscillatory structure in the x-ray absorption coefficient above an x-ray absorption edge (Fig. 5.28). EXAFS describes the fine structure in the absorption about 30 eV above the x-ray edge. This fine structure contains precise information about the local atomic structure around the absorbing atom, such as interatomic distances and coordination numbers. In contrast, the fine structure lying within the first 30 eV of the edge position is referred to as XANES. XANES is often dominated by strong scattering processes and local atomic resonances, and its quantitative interpretation is still a difficult task. Therefore we shall focus on EXAFS. If multiple-scattering effects are neglected, which is correct at least in the high-energy part of the spectrum, the EXAFS modulations can be described in terms of short-range interference between the outgoing and backscattered photoelectron wave functions. Within the so-called plane-wave approximation, the EXAFS spectrum for the

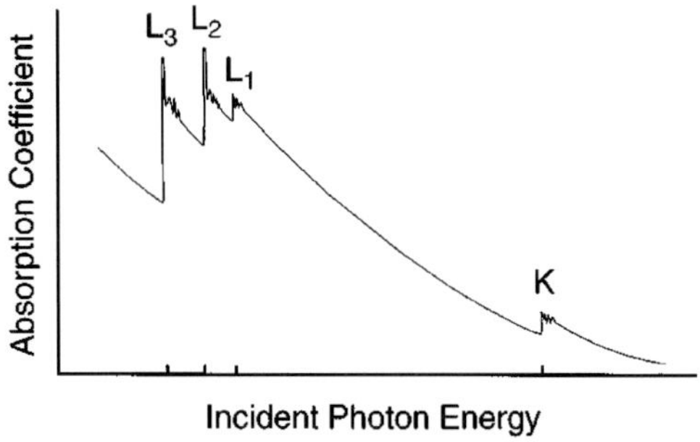

Fig. 5.28. Schematic view of the x-ray absorption coefficient as a function of the incident photo energy. Four x-xay edges are shown: K, L_1, L_2 and L_3. From [115], used with permission

excitation of an s-state electron at, for example, the K edge, is given by

$$\chi(k) = -\frac{1}{k}\sum_j \frac{3(\widehat{\epsilon}\cdot\widehat{R}_j)^2 f_j(\pi,k)}{R_j^2}\exp(-2\sigma_j^2k^2)\exp\left(\frac{-2R_j}{\lambda(k)}\right)\sin\Phi_j(k)\,, \tag{5.1}$$

where

$$\Phi_j(k) = 2kR_j + 2\delta(k) + \arg[f_j(\pi,k)]\,. \tag{5.2}$$

Here, k is the photoelectron momentum, R_j the distance of atom j from the absorbing atom, σ_j its Debye–Waller factor, and $\lambda(k)$ the mean free path of the photoelectron. $\widehat{\epsilon}$ and $\widehat{R}_j$ are the unit vectors of the electric field of the x-ray beam and of the displacement of the atom j from the excited atom, respectively. $f_j(\pi,k)$ is the complex backscattering amplitude of atom j, which allows the chemical identification of the backscattering atom. The phase shift $\Phi_j(k)$ involves three components: the first is the path to the neighboring atom j, and the other two terms come from the phase lag due to the travel through the potentials of the absorbing and backscattering atoms. $\chi(k)$ is defined experimentally as a normalized x-ray absorption coefficient $\Delta\mu/\mu_0$, where μ_0 is a structureless (atomic-like) background. As is shown by (5.1) and (5.2), once the phase shift and the backscattering-amplitude function are both known, the local atomic environment of the absorbing atom can be determined by fitting EXAFS data using the above equations. The phase shifts and the backscattering-amplitude functions have been obtained and tabulated for a large number of elements from both experiments on samples of known structure and theoretical calculations. The good agreement between experimental and theoretical values qualifies EXAFS as a well-established and element-specific method to study local atomic environments.

Figure 5.29a shows EXAFS data from two Fe/Ru superlattices measured with an incident beam polarized in-plane and out-of-plane. Two Fe–Fe distances can be deduced from these data: 2.73 Å for in-plane neighbors and 2.5 Å for out-of-plane neighbors. If an hcp stacking sequence is assumed, the

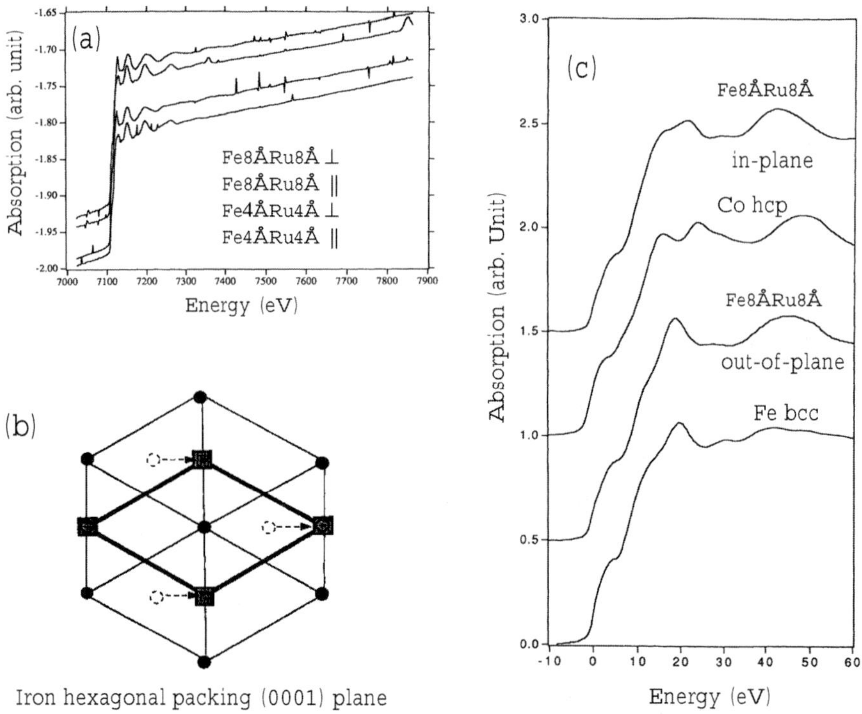

Fig. 5.29. (**a**) EXAFS raw data for two Fe/Ru superlattices measured with the electric field vector of the incident x-ray beam parallel and perpendicular to the film plane. (**b**) An AB'AB' hexagonal packing model proposed on the basis of the EXAFS results, in which the B' plane is shifted from the center of the triangle to the middle of one side of the triangle. (**c**) The XANES spectra of an Fe/Ru superlattice with the incident x-ray beam polarized in-plane and out-of-plane compared with corresponding spectra for hcp Co and bcc Fe. From [113], used with permission

c parameter is calculated to be 3.88 Å, much smaller than the values determined by x-ray diffraction and by LEED. To remove the discrepancy, a new hexagonal stacking, named AB'AB' packing, has been proposed. In this model the B' atoms sit in a bridging position just above the midpoint of a side of the triangle (see Fig. 5.29b). Such a stacking gives $c = 4.18$ Å. In this AB'AB' stacking, the relationship between two neighboring planes is similar to that for bcc (110) planes. Therefore the XANES spectra of the Fe/Ru superlattice (see Fig. 5.29c), which on the one hand show a similarity to hcp Co when the incident photons are polarized in-plane and on the other hand resemble the spectra of bcc Fe when the incident photons are polarized out-of-plane, have been taken as evidence for the AB'AB' model.

However, the AB'AB' stacking model derived from XANES spectra is not conclusive. A later x-ray diffraction study, which was performed in the transmission mode for in-plane Bragg reflections and with a grazing incidence angle for out-of-plane measurements [116], provided evidence for the regular hcp structure and against the AB'AB' stacking. Since the precise structure of the metastable Fe stabilized by hcp Ru is therefore still controversial, we shall also include an early hypothesis about the growth and structure of Fe/Ru(0001) films, which was proposed on the basis of an analysis of LEED $I(V)$ spectra [111]. In this model, only the first monolayer of Fe is pseudomorphic on the hcp Ru substrate. Its growth follows the layer-by-layer mode. Additional Fe grows into three-dimensional bcc Fe(110) domains, following the Kurdjumov–Sachs orientation with respect to the underlying pseudomorphic Fe monolayer. Since the bcc domains are too small to be detected by LEED, the (1×1) LEED pattern persists to a much larger thickness than the actual pseudomorphic thickness. A similar suggestion was made in another study, in which, however, the pseudomorphic growth thickness was 2 ML rather than 1 ML [117].

Magnetic Properties of Fe Films on Ru(0001). As we have seen above, a pseudomorphic ultrathin Fe film on a Ru(0001) substrate has an anomalously expanded atomic volume. A high-spin ferromagnetic phase is expected from the enlarged atomic volume according to theoretical calculations [5, 6]. Interestingly, in contrast to these theoretical predictions, it has been found that pseudomorphic Fe/Ru(0001) layers are magnetically dead when the thickness of the Fe layer is less than 2 ML for thin films [110] or 4 ML for superlattices [118]. Figure 5.30 shows thickness-dependent MOKE results for ultrathin Fe films on Ru(0001). For the 1 ML and 2 ML Fe/Ru(0001) systems, no MOKE hysteresis was detected down to 100 K. However, ferromagnetism was detected at room temperature for films thicker than 2 ML, and the easy axis of magnetization was found to be perpendicular for films of less than 4.5 ML. A change to in-plane was detected for thicker films. In this work, the (1×1) LEED pattern was observed for film thicknesses of up to 8 ML.

Two possible mechanisms could explain the magnetically dead layers. One connects the loss of ferromagnetism to a strong Fe–Ru hybridization, which may destroy the magnetism of the Fe layer at the interface. This mechanism is supported by the fact that an Fe–Ru solid solution is paramagnetic for an Fe concentration of up to 75%. For the same reason, interfacial interdiffusion may also destroy the ferromagnetism of the Fe overlayer. Another possible mechanism associates the magnetically dead layers with the hcp structure. This is supported by the fact that the high-pressure ϵ phase is nonmagnetic. A recent study of Fe/Ru multilayers strongly suggests that the magnetically dead layers are tightly correlated with an hcp or hcp-like structure [119]. The multilayers in that study were prepared on a 25 Å Fe(001) seed film on a MgO(001) substrate by dc magnetron sputtering. The thickness of the Fe layer was kept constant at 7.5 Å for all of the multilayers, while the thickness of the Ru layers (t_{Ru}) was varied from 0 to 20 Å. It was found by x-ray diffraction that the inter-planar spacing of the multilayers increases linearly from that of bulk Fe (1.43 Å) to 1.48 Å when t_{Ru} is increased to 3.5 Å.

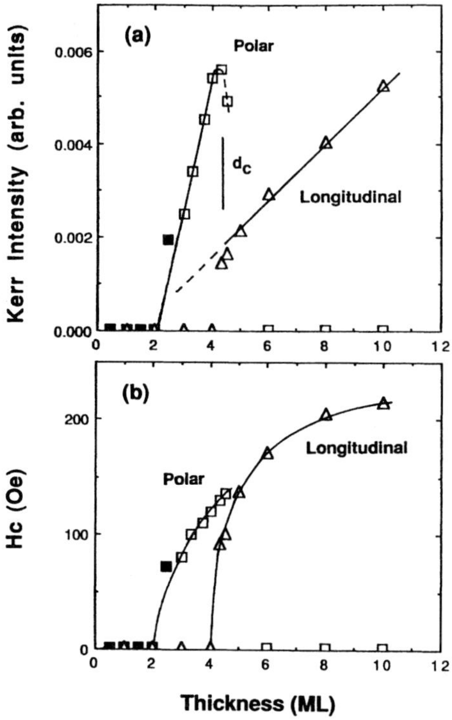

Fig. 5.30. Thickness dependence of the Kerr intensity (**a**) and coercivity (**b**) for Fe/Ru(0001) grown at room temperature (*open symbols*) and 100 K (*solid symbols*). *Squares* represent the polar Kerr signals and *triangles* the longitudinal signals. From [110], used with permission

This indicates that the Fe/Ru superlattices have a coherent bct structure in this thickness range. Around $t_{Ru} = 3.5\,\text{Å}$, a structural transition from bct (001) to hcp (102) was observed. Because no bcc-Fe-related features were observed for $t_{Ru} > 3.5\,\text{Å}$, it was concluded that the Fe/Ru multilayer again has a coherent crystal structure after the structural transition, with both the Ru and the Fe adopting an hcp structure, as indicated by the observed hcp (102) reflection peak. Changes in magnetic properties were observed, accompanying the evolution of the structure with t_{Ru}. The overall Kerr rotation first increases with increasing t_{Ru}, but then falls rapidly to zero beyond $t_{Ru} = 3.5\,\text{Å}$ (see Fig. 5.31). Consistently, the Fe atomic magnetic moment measured by element-specific x-ray magnetic circular dichroism also drops to zero after the structural transition (see the right panel of Fig. 5.31). The coercive field also exhibits a cusp at $t_{Ru} = 3.5\,\text{Å}$, which is characteristic of a structural transition.

When the Ru layers in the Fe/Ru superlattices were replaced by Re, a similar structural phase transition was observed around $t_{Re} = 9\,\text{Å}$ [120].

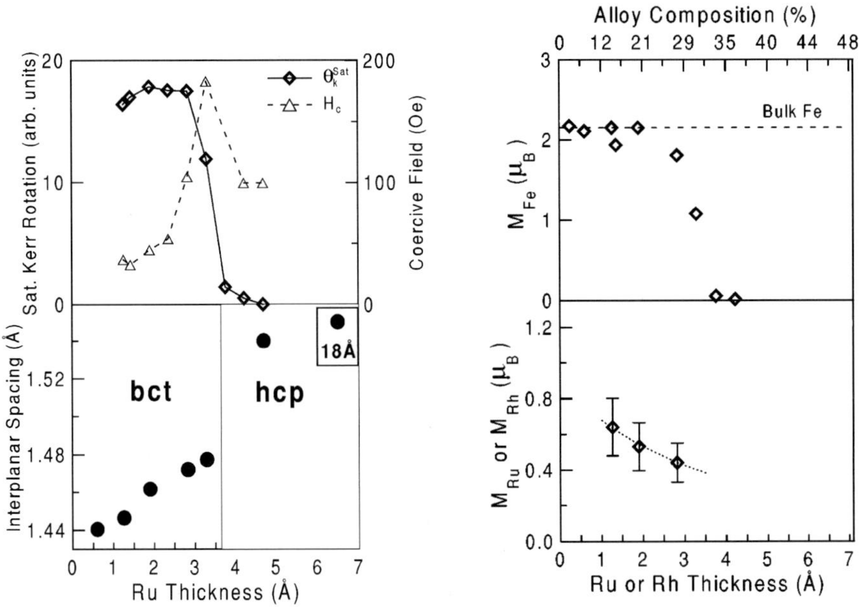

Fig. 5.31. *Left bottom*: the interplanar spacing of an Fe/Ru superlattice as a function of Ru thickness. The Fe layer thickness is maintained constant at 7.5 Å. When the Ru layer thickness is below 3.5 Å, the film has a bct (001) orientation with a lattice spacing that increases linearly with increasing Ru thickness. When the Ru thickness becomes larger than 3.5 Å, the multilayer shows an abrupt change in lattice constant, and a structural transition to hcp (102). *Left top*: saturation Kerr rotation θ_K^{sat} and coercivity H_c for the multilayers over the same Ru thickness range. From [119]. *Right*: summary of the element-specific magnetic moments in the multilayers as measured by XMCD. Modified from [119], used with permission

However, the magnetic properties of Fe/Re superlattices exhibit different features from Fe/Ru multilayers (see Fig. 5.32) [120]. The main difference is that the Fe/Re superlattice remains ferromagnetic after its coherent structure has changed into hcp. The overall Kerr rotation first decreases with increasing t_{Re}. It reaches a minimum in the phase transition region, with two-thirds of the signal amplitude being lost. After the completion of the structural phase transition from bct to hcp at $t_{Re} \approx 11$ Å, the Kerr rotation begins to increase slowly with t_{Re}. It is somewhat unexpected that the Fe atomic magnetic moment in Fe/Re superlattices measured by XMCD remains almost constant with t_{Re} despite the structural phase transition, which is in contrast to the conspicuous change in Kerr rotation.

The magnetic properties of hcp Fe may be deduced from the different magnetic behaviors shown by Fe/Ru and Fe/Re multilayers [120]. Bulk hcp Re (c = 4.46 Å and a = 2.76 Å) has a larger atomic volume (14.71 Å^3) than bulk Ru (12.30 Å^3). Correspondingly, a larger atomic volume of hcp stabilized

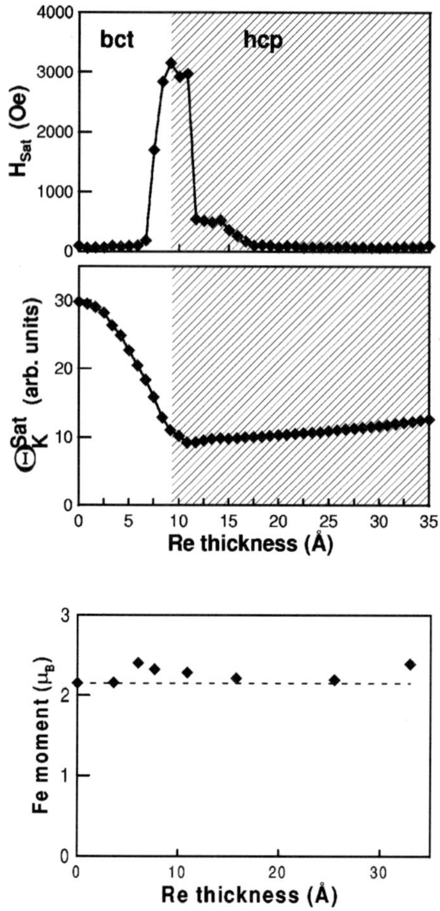

Fig. 5.32. *Top*: Dependence of saturation field H_{Sat} (*top*) and of Kerr rotation at saturation Θ_K^{Sat} (*middle*) on Re thickness for Fe/Re multilayers. The Fe layer thickness is maintained constant at 8 Å. *Bottom*: summary of the Fe magnetic moment per atom in Fe/Re multilayers as measured by XMCD. The *dashed line* indicates the bulk value of the magnetic moment of Fe. From [120], used with permission

Fe is expected in Fe/Re multilayers. Actually, for an Fe/Re superlattice with $t_{Re} = 33$ Å, the average volume per atom for the stabilized hcp Fe was measured to be 14.57 Å^3. This is much larger than that for fcc Fe, and also larger than that for bulk Ru, which can be considered as the upper limit for the hcp stabilized Fe in Fe/Ru multilayers. On the basis of the moment–volume instability, which we shall discuss in the next chapter, a large atomic volume favors a ferromagnetic state. Therefore it was deduced that hcp Fe in an Fe/Ru superlattice still does not have a sufficient atomic volume to support a ferromagnetic moment even though its atomic volume is larger than that for fcc Fe, while the hcp Fe in an Fe/Re superlattice does have a sufficient atomic volume.

These interesting experimental observations in Fe/Ru films and Fe/Ru and Fe/Re multilayers have provoked a number of theoretical investigations. One theoretical calculation using the VASP simulation package [121] predicted a nonmagnetic ground state with lattice parameters of $a = 2.44$ Å and $c/a = 1.69$ for regular bulk hcp Fe, in good agreement with another calculation by the tight-binding linear muffin-tin orbital method [123], and a ferromagnetic ground state with $a = 2.562$ Å and $c/a = 1.612$ for AB'AB' stacked hexagonal bulk Fe. The FM AB'AB' stacked Fe is less stable than the nonmagnetic hcp phase at the equilibrium volume but it has a lower energy in an expanded state. The calculation for $Fe_{5ML}Ru_{5ML}$ multilayers [121] showed that FM order is the most stable magnetic ordering for both regular hcp Fe and AB'AB'-stacked Fe when the in-plane lattice parameter is fixed at the calculated Ru bulk value of 2.70 Å. This is in contradiction with the experimental observation that the FM signal disappears after the bct–hcp structural phase transition in Fe/Ru multilayers [119]. The calculations for Fe_{nML}/Re_{7ML} (n = 1, 3, 5) multilayers indicated that, when an in-plane lattice spacing of bulk Re (2.75 Å) and a regular hcp structure were assumed, AFM order appeared at $c/a = 1.38$ and FM order at a slightly higher atomic volume ($c/a = 1.42$) [123]. Both AFM and FM order appear to be more stable than the NM state only when the Fe atoms begin to adopt a volume expanded by about 27%, which is accompanied by an axial distortion (c/a= 1.42). The FM and AFM order remain almost degenerate up to $c/a = 1.50$, and then FM order becomes more stable at larger atomic volumes. These predictions are consistent with the different magnetic behaviors observed when Fe/Ru and Fe/Re multilayers are compared. For a single overlayer of Fe, both on Ru(0001) [122] and on Re(0001) [123], the in-plane AFM alignment was predicted to be the ground state. The FM phase becomes more stable when the Fe thicknesses increased [123]. This may explain the magnetically dead layer observed experimentally. For the structure of an Fe/Ru(0001) film, it was predicted [121] that at the Ru/Fe interface the Fe plane continues to grow in the regular hcp stacking but that further Fe layers prefer to adopt the AB'AB' stacking sequence. It was also found theoretically that the magnetic state is insensitive to structural misalignment and intermixing at the interfaces [121, 123]. Therefore the magnetically dead layer is less likely to come from interdiffusion and defects at the interface.

5.2.5 Ultrathin Co Films

At room temperature, the stable phase of Co has an hcp structure (a = 2.507 Å, c = 4.070 Å). Above 470°C, the fcc phase becomes the stable phase instead of the hcp phase. Fcc Co (a = 3.544 Å) can be prepared at room temperature by rapid quenching of molten samples [124]. To the authors' knowledge, the less stable bcc Co phase has not yet been prepared by this quenching method. Nevertheless, both metastable structures have been stabilized at room temperature by pseudomorphic growth. Fcc Co has been successfully prepared, mainly on Cu(100), Cu(110) and Cu(111), and bcc Co on GaAs(001), Fe(100), FeAl(001), TiAl(010), W(100), W(110), W(111) etc.

Fcc Co Films on Cu(100), Cu(110) and Cu(111). Owing to the small lattice mismatch between fcc Cu and fcc Co (−1.9%) and the low miscibility of Co in bulk Cu, excellent layer-by-layer growth has been identified by a number of investigators for the Co/Cu(001) system, at least within a thickness range between 2 and 10 monolayers [125, 126, 127, 128]. This is the reason why Co/Cu(100) has frequently been chosen as a prototype to study two-dimensional magnetism.

In contrast to the wide consensus on the layer-by-layer mode for the growth above 2 ML, the initial growth behavior of Co/Cu(001) is still controversial. In the early studies [125, 129, 130], the growth of Co on Cu(100) below 2 ML was also considered to follow the layer-by-layer mode, mainly on the basis of Auger studies, and a sharp interface was assumed on the basis of the bulk immiscibility between these two elements. Later, a study based on angle-resolved X-ray photoemission scattering (ARXPS)[2] reported on the formation of double-layer islands during the initial growth [131]. The existence of bilayer islands was confirmed by subsequent STM investigations [127, 128, 132], though the extent to which the initial growth deviates from the ideal layer-by-layer mode varied between different reports. In Fig. 5.33, STM images of ultrathin Co films that were evaporated at room temperature are shown. At a Co coverage as low as 0.6 ML, bilayer islands (white tone) already represent a high percentage, about 20%, of the total coverage (see Fig. 5.33a). A crossover from a layer-by-layer to a bilayer growth mode has been reported upon increasing the deposition rate (see Fig. 5.34). The bilayer growth occurs at a higher deposition rate owing to the hindered diffu-

[2] This technique is based on the fact that the intensity of core-level photoelectrons and Auger electrons in x-ray photoelectron spectroscopy is enhanced along the nearest- and next-nearest-neighbor axes. This enhanced intensity can be explained by a forward-scattering enhancement by lattice atoms due to constructive interference of the outgoing electrons. By measuring the angular dependence of the photoelectron intensity, element-specific information about the local atomic environment can be obtained. For an fcc (001) thick film, intensity maxima are expected at polar angles of 0° and 45° when the azimuthal angle is fixed in the [100] direction. However, for one monolayer of layer-by-layer-grown Co on Cu(100), one should observe an isotropic Co $2p_{3/2}$ ARXPS pattern, since there is no other Co layer to produce forward-scattering enhancement. With the addition of a second layer, a peak at 45° along the [101] crystallographic direction will be seen, but no enhancement along the normal direction. Only at the onset of the growth of the third layer will the 0° peak be observed. These features can be used to characterize the growth mode of the film.

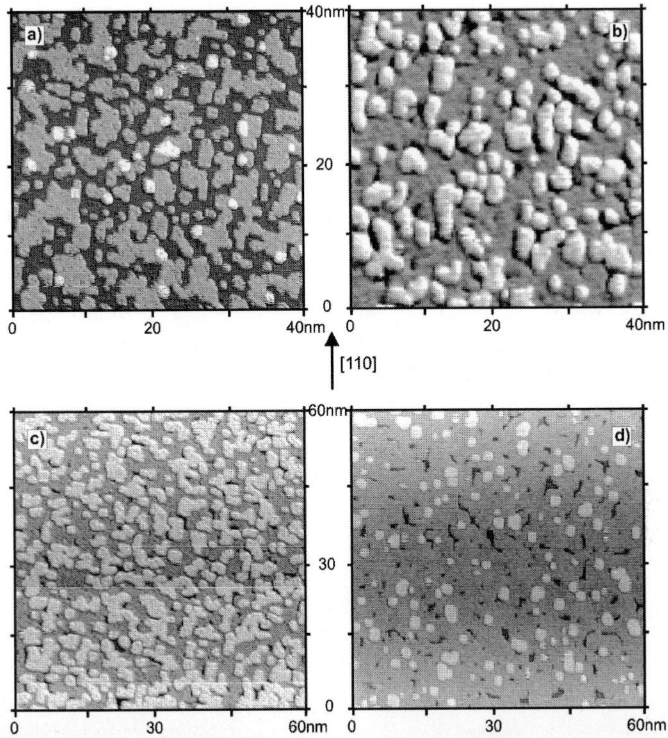

Fig. 5.33. STM images of room-temperature-evaporated Co/Cu(100) ultrathin films with a thickness of (**a**) 0.60 ML, (**b**) 1.25 ML, (**c**) 1.45 ML and (**d**) 2.03 ML. The deposition rate was around 0.2 ML per minute. From [128], used with permission

sion from the first layer to the substrate. As another important modification of the early conclusions about the growth of Co on Cu(100), intermixing between Co and Cu has been observed in the initial growth of Co on Cu(100) [126, 133, 134]. It has been proposed that, during the initial stages of growth, some of the deposited Co atoms exchange sites with atoms of the Cu substrate surface. Direct evidence for this exchange process has been obtained from STM observations [133, 134]. An STM image of 0.12 ML Co deposited on Cu(100) at room temperature shows small dark indentations besides the bright islands of the first layer (see Fig. 5.35) [134]. These small indentations were interpreted as substitutional Co atoms after site exchanges with Cu atoms. Consequently, there are three different adatom species on the surface: substitutional Co, on-surface Co and on-surface Cu. On the basis of the results of density functional theory (DFT) calculations [134] and the bias-dependent contrast in STM images of 0.12 ML Co on Cu(100) [133], it was concluded that the incorporated Co atoms act as pinning centers for diffusing surface adatoms and hence as nucleation centers for island formation. More-

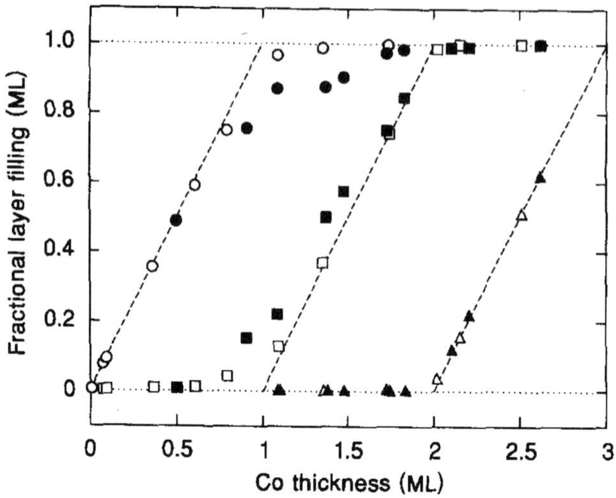

Fig. 5.34. Fractional layer filling, evaluated from STM observations, as a function of Co thickness for different deposition rates. *Empty* and *filled circles* denote slow deposition (0.003 ML s^{-1}) and fast deposition (0.3 ML s^{-1}), respectively. The *dashed lines* correspond to the fractional layer filling expected for perfect layer-by-lyer growth. From [133], used with permission

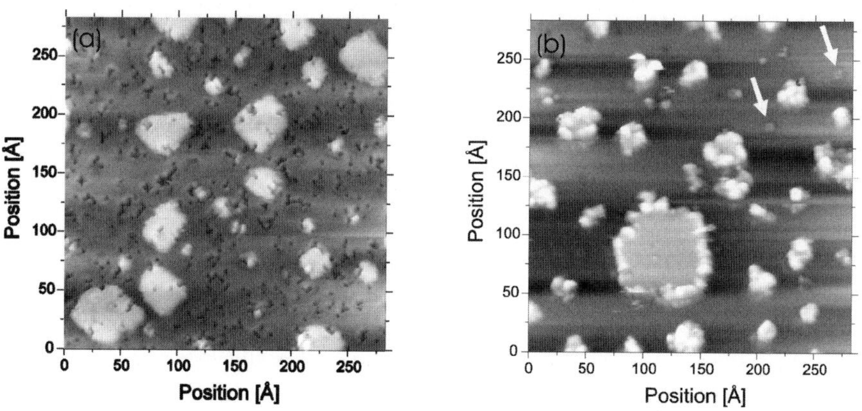

Fig. 5.35. (**a**) STM image (voltage U = 0.9 V, current I = 0.4 nA) of 0.12 ML Co deposited on Cu(001) at room temperature. The small indentations (*dark gray*) are interpreted as incorporated Co atoms. (**b**) STM image (U = 3.5 V, I = 0.4 nA) of 0.11 ML Co/Cu(100) deposited at 415 K and afterwards exposed to about 20 L CO. From [134], used with permission

over, the DFT calculations showed that a Co adatom is always bound more strongly on the surface than a Cu adatom is, owing to a stronger d–d and d–sp hybridization. Furthermore, the pinning effect of the incorporated Co atoms is stronger for Co adatoms than for Cu. The higher adsorption energy of Co results in a lower mobility of Co on Cu(100). These results may explain the observed bimodal growth of Co/Cu(100) in the initial stages of growth [134, 133]. The term "bimodal growth" refers to the observation of two kind of islands with different sizes (see Fig. 5.33a) and chemical composition. The bimodal growth is clearly demonstrated by CO titration experiments [134]. Figure 5.35b shows an STM image of a 0.11 ML Co/Cu(100) film after it has been exposed to 20 L CO (1 L (Langmuir) = 10^{-6} Torr s). The bright clouds are observed only after the exposure and are attributed to the adsorbed CO molecules. The small islands are completely covered with CO, while CO is adsorbed only at the edges of the large islands. This observation indicates that the small islands consist mainly of Co, and the large ones of Cu with Co decoration at the edges.

The structure of Co/Cu(100) ultrathin films has been investigated by multiple techniques. Only the (1×1) LEED pattern has been observed during the film growth. The **c**(2×2) pattern reported early for the growth of the first layer was later attributed to surface contamination [125, 130, 135]. Quantitative analysis of LEED I/V curves indicated a 6% contraction of the interlayer spacing for a 1 ML Co overlayer on Cu(100) compared with the Cu bulk value, and a 6% contraction in the top layer and 3% in the second layer for an 8 ML film [130]. An angle-resolved XPS investigation showed that high-quality fcc Co can be stabilized on Cu(100) up to at least 20 ML [131]. Extended x-ray absorption fine structure measurements revealed that the exact structure of the Co/Cu(100) film is a face-centered tetragonal structure [136]. A constant tetragonal distortion was observed for films with a thickness from 2 to 15 ML, with an in-plane lattice parameter $a_{\|}$ of 3.61 Å and a perpendicular lattice parameter $a_{\perp}$ of 3.46 Å. The critical thickness for pseudomorphic growth was reported to be located between 10 and 20 ML, depending on the substrate temperature [131, 137]. Above the critical thickness, strain relaxation starts by dislocation formation.

It is not surprising that an ultrathin fcc Co film is ferromagnetic at room temperature, considering that the high-temperature phase (>700 K) with the same structure is ferromagnetic up to 1388 K. The magnetic moment per atom for a Co film on Cu(100) is close to the bulk value of $1.7\mu_B$ [138]. The research interest in the Co/Cu(100) system has focused mainly on the magnetic behavior of the films below 2 ML. Such films are expected to act as a physical realization of a two-dimensional magnet [139]. Since a well-known result by Mermin and Wagner states that in monatomic thin films, long-range order is forbidden at any finite temperature for an isotropic magnetic interaction [140], the magnetic properties of monolayer films have received considerable interest. One monolayer of Co on Cu(100) was reported to be ferromagnetic [129, 141]. However, this result was questioned later by another study [142], which showed that a single monolayer of Co does not exhibit ferromagnetism down to T = 50 K, while films above 1.5 ML are ferromagnetic but with a much lower Curie temperature than that of bulk Co. The thickness for

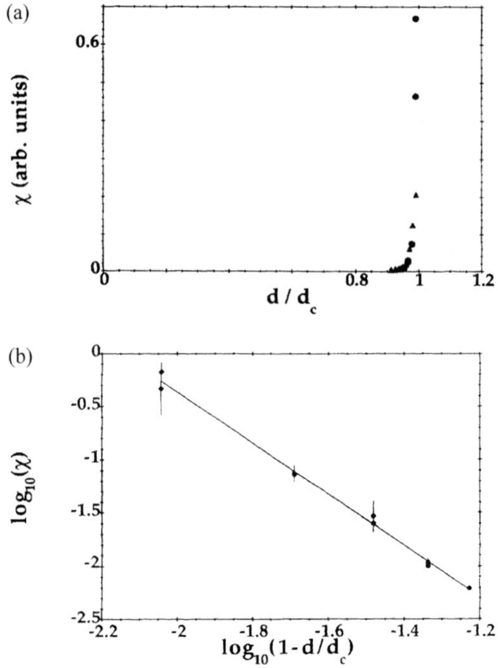

Fig. 5.36. (**a**) The magnetic susceptibility χ of Co/Cu(100) films, plotted as a function of the reduced film thickness d_c, which is approximately 1.3 ML. The *circles* and *triangles* represent the data from two separate experiments. (**b**) A log–log plot of the data shown as circles in (**a**). The slope gives the critical exponent $\gamma = 2.41$, which is defined by a power law $\chi \propto (1 - d/d_c)^{-\gamma}$. From [144], used with permission

the onset of room-temperature ferromagnetism has been reported to be 1.6 ML [143, 144] and 1.8 ML [145]. Now it seems clear that the onset of ferromagnetism is associated with a change in the film morphology. Schumann et al. have measured the paramagnetic susceptibility in a thickness range around the critical thickness at which the onset of ferromagnetism occurs (see Fig. 5.36a) [144]. They found that their data could be explained well by the onset of a geometrical percolation phase transition. The critical exponent $\gamma = 2.41 \pm 0.07$ derived from the experiments (see Fig. 5.36b) is very close to the theoretical value of 2.389 for the two-dimensional percolation phase transition. An even more direct and detailed connection between magnetism and morphology has been demonstrated by a recent combined study by MOKE and STM [128]. The critical thickness for ferromagnetic order has been identified as the thickness at which coalescence of the islands of the second layer sets in. Annealing the film at about 400 K for 10 minutes increases the Curie temperature and the ferromagnetic signal for films with a thickness around and below the critical value (see Fig. 5.37). This is due to the improvement

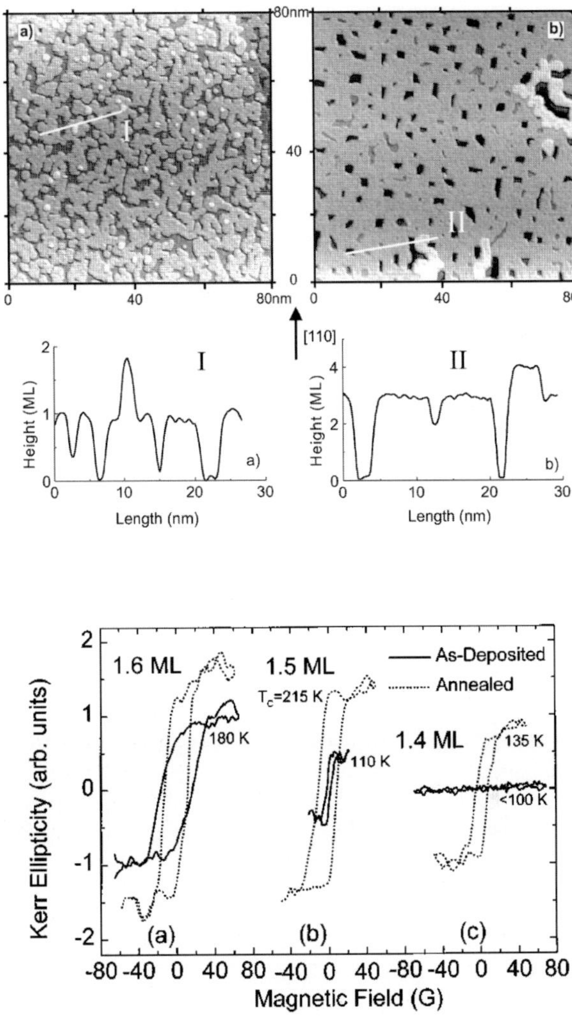

Fig. 5.37. *Top*: surface morphology of a 1.7 ML Co/Cu(100) film in the as-deposited (left) and annealed (right) states. Characteristic line scans along the paths indicated in the STM images are also shown. *Bottom*: MOKE hysteresis loops of Co/Cu(100) ultrathin films with thickness of (**a**) 1.6 ML, (**b**) 1.5 ML and (**c**) 1.4 ML. The *solid lines* correspond to films in the as-deposited state. The Curie temperatures of the films are also indicated. The loops were measured at about 100 K. From [128], used with permission

of the coalescence of the islands. On the basis of these observations, it was proposed that the first layer of the Co/Cu(100) film is actually an alloy and that the second layer consists mainly of pure Co. This argument is consistent with the growth studies that we have discussed above.

The magnetic anisotropy of ultrathin fcc Co films on Cu(100) has been measured by Krams et al. [143]. The anisotropy energy of the fcc Co(001) film can be expressed to the lowest order as

$$E = \frac{1}{4} K_{in-plane} \sin^2 2\phi \sin^4 \theta + \frac{2k_s}{d} \sin^2 \theta \ , \quad K_{in-plane} = K_p + \frac{2}{d} k_p \ , \ (5.3)$$

where $K_{in-plane}$ is the total in-plane anisotropy, including the volume contribution K_p and the surface contribution k_p; k_s is the out-of-plane surface anisotropy. The measurement results for $K_{in-plane}$ and k_s are shown in Fig. 5.38 for Co/Cu(100) films both with and without a Cu capping layer. Fitting the thickness-dependent data for $K_{in-plane}$ gives $K_p = (-2.3 \pm 0.15) \times 10^6$ erg/cm^3 and $k_p = 0.034 \pm 0.004$ erg/cm^2 for the uncovered films, and $K_p = (-2.2 \pm 0.15) \times 10^6$ erg/cm^3 and $k_p = 0.031 \pm 0.003$ erg/cm^2 for the films covered with 2 ML Cu. The average value for the out-of-plane anisotropy constant k_s is -0.46 ± 0.09 erg/cm^2 for uncovered films and changes to 0.15 ± 0.04 erg/cm^2 upon adding 2 ML of Cu. The obvious change in k_s upon covering the film with a Cu overlayer allows one to separate $k_s = -1.06 \pm 0.17$ erg/cm^2 for the Co/vacuum interface from $k_s = 0.15 \pm 0.04$ erg/cm^2 for the Co/Cu interface. These data indicate that the uncovered Co/Cu(100) film should have an in-plane anisotropy indepen-

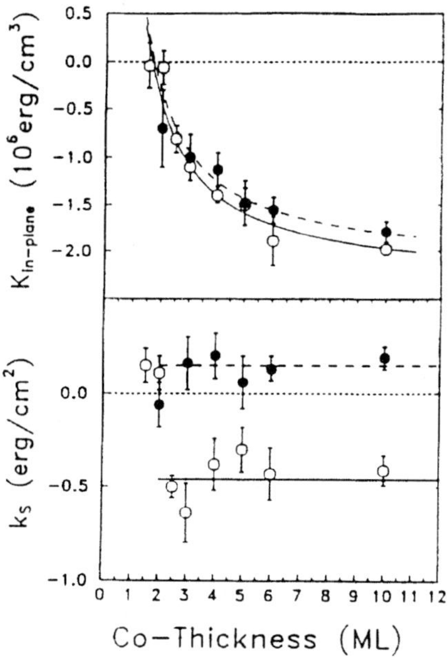

Fig. 5.38. In-plane (*upper panel*) and out-of-plane (*lower panel*) magnetic anisotropy constants as a function of the Co layer thickness for Co/Cu(100) films without (*open circles*) and with (*solid circles*) a 2 ML Cu capping layer. From [143], used with permission

dent of the film thickness, while the covered film may show a perpendicular anisotropy at low film thickness. However, MOKE measurements on 1.8 ML Co/Cu(100) films covered with 2 ML Cu detected only longitudinal hysteresis loops [146].

The growth of Co on the Cu(111) substrate is quite different from that on Cu(100). There exists much more pronounced interfacial intermixing on Cu(111) than on Cu(100) [126, 147, 148]. STM observations [149] have shown that the growth of Co on Cu(111) at room temperature starts with the formation of bilayer islands, which decorate the steps at both the lower and the upper step edges. The islands exhibit triangular shapes, with their edges aligned with the close-packed $\langle 1\bar{1}0 \rangle$ directions of the Cu substrate. Low-energy ion scattering measurements [148] have shown that these islands are not stable against interdiffusion even at room temperature. Cu substrate atoms tend to migrate to the surface to form a capping layer on top of the Co islands. In contrast to the excellent layer-by-layer growth of Co on Cu(100) above 2 ML, the initially formed 2 ML high islands on Cu(111) do not coalesce even at a Co coverage of up to 5 ML. The islands grow both in height and in the lateral dimensions, producing an epitaxial but granular film [149].

While Co simply adopts an fcc stacking on Cu(100) owing to the absence of two-dimensional square symmetry in the hcp structure, the (111) surface supports both fcc (ABC) and hcp (ABAB) stacking. Medium-energy ion scattering [150], extended x-ray absorption fine structure [152] and x-ray-photoelectron diffraction [153] provide evidence that the first layer of the film continues the fcc stacking sequence of the substrate. However, domains with an hcp structure develop subsequently, though it is not certain whether the hcp stacking starts at the second [150, 151] or the third [152, 153] deposited layer. The fraction of the atoms in the hcp structure in the top layer increases with coverage [147, 150]. It was observed that the hcp stacking had become dominant for a 5 ML Co film [150], but traces of fcc domains can still be detected for a 50 ML film [147]. Two or three additional Cu capping layers on 5 ML Co can stabilize the fcc domains and switch the hcp stacking back to fcc, forming twinned fcc domains [151].

The Co/Cu(111) system provides excellent conditions for studing the effect of surfactants on film growth and the resulting magnetic properties. Pre-deposition of one layer of Pb on Cu(111), which leads to a (4×4) LEED pattern, allows one to improve the later growth of Co to a considerable extent [154, 155]. STM investigations have shown that, though bilayer islands still exist in the initial stages, the growth of Co on (4×4)Pb/Cu(111) at room temperature proceeds in the layer-by-layer mode by two-dimensional nucleation from the second monolayer on (see Fig. 5.40, and compare Fig. 5.39). Consequently, the film roughness is reduced dramatically. The effect of the Pb surfactant is restricted to the film morphology. It has little influence on the film structure.

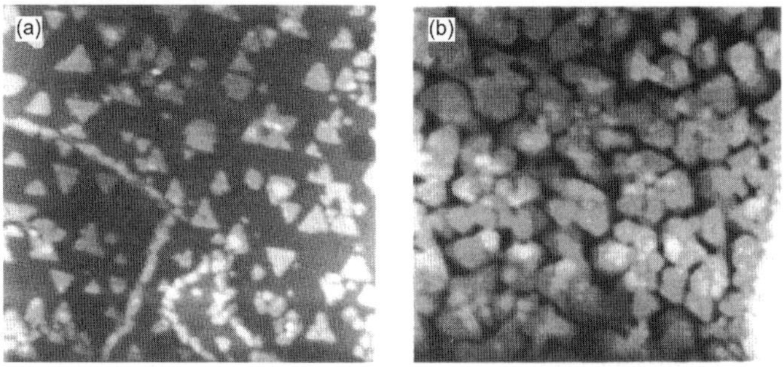

Fig. 5.39. STM images of (**a**) 0.6 ML Co and (**b**) 5 ML Co grown on Cu(111) at room temperature. The image size is 2000×2000 Å^2 for (**a**) and 1200×1200 Å^2 for (**b**). From [149], used with permission

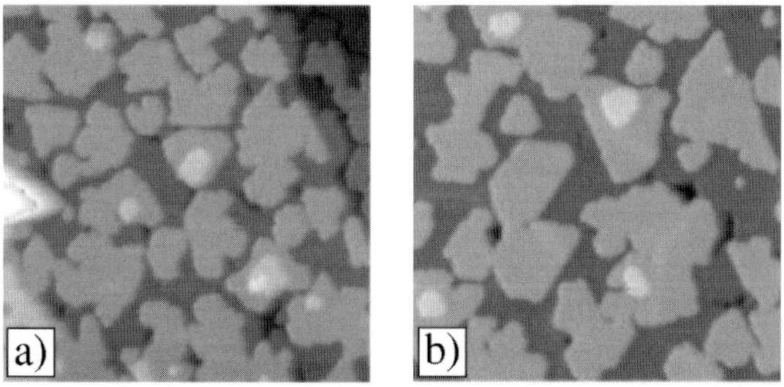

Fig. 5.40. STM images (400×400 Å^2) of (**a**) 1.6 ML Co and (**b**) 2.7 ML Co deposited on (4×4)Pb/Cu(111) at room temperature. Compare with Fig. 5.39. From [154], used with permission

The Curie temperature of Co films prepared on a Cu(111) substrate was found to be higher than that of films of the same thickness prepared on a Cu(001) substrate [146]. MOKE detected both polar and longitudinal hysteresis loops in a 1.5 ML Co/Cu(111) system, showing that the perpendicular anisotropy of Co films is enhanced on Cu(111) compared with the case on Cu(100) [146]. The films grown using a layer of surfactant (Pb) showed a stronger perpendicular anisotropy than did the films grown without application of the surfactant [155].

In contrast to the immediate structural relaxation from fcc to hcp on Cu(111), it has been shown that Co films can be grown with an fcc structure on Cu(110) substrates with a thickness exceeding 1000 Å [156]. This unusual stability of fcc Co was inferred to be due to two cooperating factors. First, very thin Co(110) films are stabilized by the energy cost of the large dislocation network which is required for a complete phase transformation. Second, once the film is thick enough to thermodynamically favor the hcp

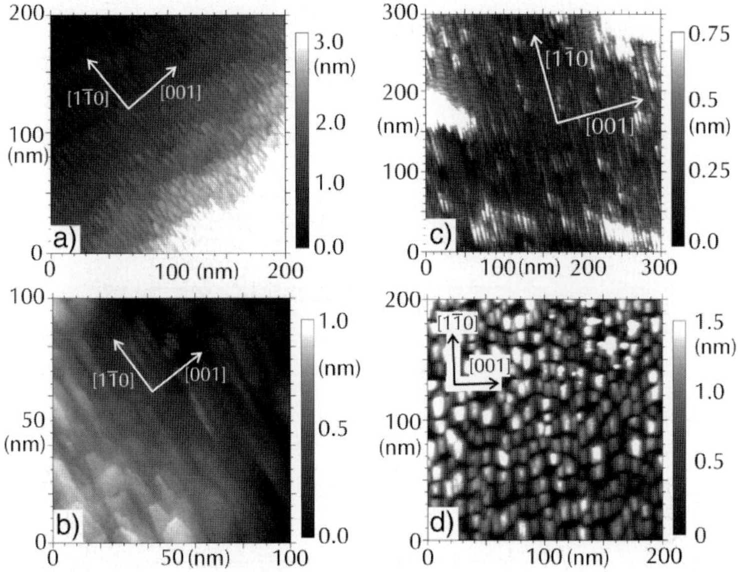

Fig. 5.41. STM images of Co/Cu(110) films with a thickness of (**a**) 3 Å, (**b**) 20 Å, (**c**) 30 Åand (**d**) 130 Å. From [157], used with permission

structure, the time constant for this relaxation is very long. The growth of Co(110) films starts with 2–3 ML high islands near step edges, which show a stripe shape orientated parallel to the $[1\bar{1}0]$ direction(see Fig. 5.41) [157]. The stripe shape of the islands becomes more pronounced with increasing thickness, and regular monatomic step arrays with the step edge parallel to the [001] direction are observed at a thickness of 30 Å. However, with further increase of the Co film thickness, square-like islands develop. The growth below 14 Å is considered to be pseudomorphic. This growth regime is followed by an intermediate thickness range dominated by stripe-shaped islands and anisotropic strain relaxation within the film plane, which is then followed by a gradual transition near 50 Å into growth of square-shaped islands with isotropic in-plane strain [157].

The magnetic anisotropy of Co/Cu(110) films has been investigated by Brillouin light-scattering measurements [157]. The thickness-dependent anisotropy constants, namely, the uniaxial anisotropy constants $K_{in-plane}$ and $K_{out-of-plane}$ and the cubic magnetocrystalline anisotropy constant K_1, are shown in Fig. 5.42; they were obtained by fitting the data by using an expression for the anisotropy energy of the form

$$\begin{aligned} E_{anisotropy} &= K_1(\alpha_{x'}^2\alpha_{y'}^2 + \alpha_{y'}^2\alpha_{z'}^2 + \alpha_{z'}^2\alpha_{x'}^2) \\ &\quad + K_{in-plane}\alpha_x^2 - K_{out-of-plane}\alpha_z^2 \,, \end{aligned} \tag{5.4}$$

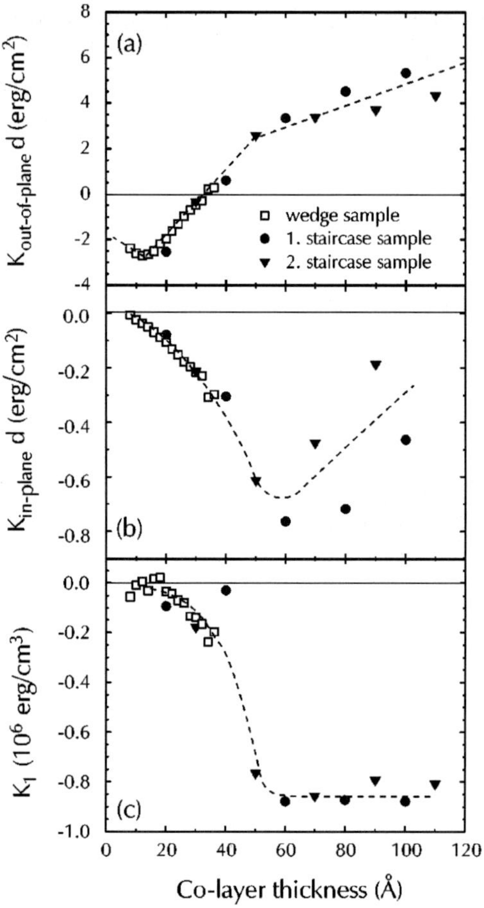

Fig. 5.42. (**a**) Effective out-of-plane anisotropy constant $K_{out-of-plane}$, and (**b**) effective in-plane anisotropy constant $K_{in-plane}$, multiplied by the Co film thickness d as a function of d. (**c**) Magnetocrystalline anisotropy constant K_1 as a function of d. The three different symbols denote three different samples. From [157], used with permission

where $\alpha_{i'}$ and α_i are the direction cosines in the crystallographic reference frame and in the film coordinate system, respectively. A conspicuous change in magnetic anisotropy is found around a film thickness of 50 Å, which includes a change in slope in $K_{out-of-plane}$, the presence of a minimum in $K_{in-plane}$ and the appearance of a bulk magnetocrystalline anisotropy. These changes have been associated with a transition from anisotropic to isotropic strain relaxation [157].

Ultrathin Bcc Co Films. Metastable bcc Co ($a_0 = 2.827$ Å) was first stabilized by epitaxy on a GaAs(110) substrate by Prinz [158]. The bcc structure was found to be stable up to a thickness of 357 Å. The magnetic moment per atom was determined to be $1.53\mu_B$, which is lower than the reported value of $1.71\mu_B$ for hcp Co. The in-plane [001] direction is the magnetic easy axis, and the out-of-plane [110] axis is hard. The bcc metastable structure was first determined from RHEED patterns and x-ray diffraction and was further confirmed by conversion-electron x-ray absorption fine-structure measurements[3] [159].

Ultrathin bcc Co films have also been grown on metal substrates. Li and Tonner found that 3 ML Co films grown on an ultrathin bcc Fe film on an Ag(001) substrate show features in angle-resolved x-ray photoelectron scattering and angle-resolved Auger electron spectroscopy (ARAES) similar to those measured for Fe in a 3.5 ML bcc Fe film on Ag(001) [160]. Since these spectra are found to be similar for different materials, the Co film was suggested to have the same structure as the bcc Fe film. A subsequent study by Zhang et al. by x-ray photoelectron diffraction (XPD) showed that Co films grown on Fe(001) single crystals had a bcc structure for a thickness of up to 20 Å [161]. The Co layers in Co/Fe multilayers on Si(111) also have a bcc structure when the thickness of the Co layer is smaller than 20 Å [162]. However, a mixed fcc–hcp structure is observed when the thickness of the Co layer exceeds 20 Å. Similar conclusions have been drawn from LEED studies [163].

When Co is deposited on a clean Fe(001) single crystal, the (1×1) LEED pattern persists up to a thickness of 15 Å. Quantitative LEED analysis showed that Co films within this thickness range have a tetragonally distorted bcc structure with $a = 2.87$ Å (pseudomorphic with Fe(001)) and $c = 2.792$ Å. When the Co film thickness exceeds 15 Å, a $c(2 \times 2)$ LEED pattern develops, with a background which increases with thickness. The structure of the $c(2 \times 2)$ films was found to be a pseudomorphic distortion of hcp Co with a $(11\bar{2}0)$ orientation, which has a rectangular unit mesh and requires a 6.6% compression in one direction and a 0.4% compression in the other direction in the surface plane to adapt to the Fe(001) unit cell (see Fig. 5.44). The film has a bulk interlayer spacing of 1.29 Å and about a 10% contraction of the first interlayer spacing. The $c(2 \times 2)$ pattern is visible until 30 Å has been deposited. The structural transition from bct at lower thicknesses to hcp at higher thicknesses can be clearly seen from the difference in the LEED $I(V)$ curves shown in Fig. 5.43. It was found that a well-ordered $p(1 \times 1)$ oxygen overlayer on the Fe(001) substrate, which was obtained by exposing the

[3] Conversion-electron x-ray absorption fine structure is a modification of the standard EXAFS technique, in which the electrons emitted by the sample owing to Auger decay of core holes produced during x-ray absorption are detected in a He-gas ionization chamber. The measured current signal directly reflects the x-ray-induced core-hole creation and therefore can be used to determine the local atomic structure by analyzing the variation of the current with photon energy just as is done for EXAFS.

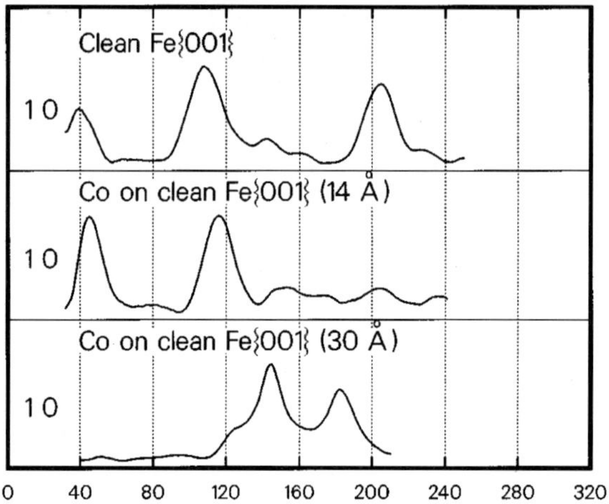

Fig. 5.43. Normal-incidence LEED spectra from clean Fe(001) (*top*), a 14 Å film of Co grown on Fe(001) (*middle*), and a 30 Å film of Co on Fe(001) (*bottom*). The 14 Å Co film obviously has a structure similar to that of Fe(001), but with a contracted interlayer spacing, whereas the 30 Å film has a different structure. From [163], used with permission

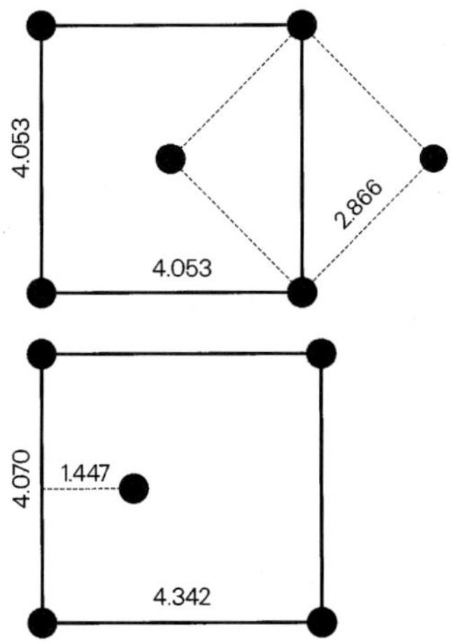

Fig. 5.44. *Top*: face-centered (*solid lines*) and primitive (*dashed lines*) meshes of Fe(001). *Bottom*: unit mesh of hcp Co(11$\bar{2}$0). All distances are in Å. From [163], used with permission

clean Fe(001) substrate to oxygen gas, can increase the critical thickness for the growth in the hcp (11$\bar{2}$0) orientation to beyond 30 Å [163, 164]. This is similar to the case of Fe/Cu(100) ultrathin films, where the use of CO as a surfactant can extend the layer-by-layer growth regime and push the fcc–bcc transition to larger thickness [165].

To clarify whether the bct–hcp structural transition proceeds in a martensitic-like way, i.e., involving the film throughout the whole thickness, or whether it proceeds smoothly, involving only the layers above the critical thickness, the structural evolution of Co layers grown on Fe(001) has been investigated by primary-beam diffraction modulated electron emission (PDMEE)[4] [166] measurements during both film growth by thermal evaporation and erosion by mild ion sputtering [167]. The bct structure was observed for films below 12 ML during both the growth procedure and the erosion procedure. The erosion procedure was started after the completion of the bct–hcp transition during the growth procedure. This result indicates that the structural transition is a layer-depth-dependent crossover rather than an overall martensitic-like transformation. The PDMEE data (Co LMM Auger electrons) were analyzed assuming that the film structure can be separated into two distinct phases above a critical thickness, i.e., the bct phase at the Co/Fe interface and the hcp phase grown on top of it. For a given coverage, the thickness of each phase was determined by fitting the measured PDMEE curves using the following formula:

$$I_{calc}(\theta, t) = (e^{-t/\lambda} - e^{-T/\lambda}) I_{bct}(\theta) + (1 - e^{-t/\lambda}) I_{hcp}(\theta) \,, \tag{5.5}$$

where T is the total Co coverage, λ is the inelastic mean free path of the Co LMM electrons, t is the hcp layer thickness, and $I_{bct}(\theta)$ and $I_{hcp}(\theta)$ are the reference PDMEE curves for the bct and hcp layers. The fitting results are given in Fig. 5.45. The growth curve in Fig. 5.45 suggests that the phase transition, immediately after the critical thickness, occurs continuously. The bct and hcp phases both grow until the total coverage reaches 22 ML. However, after this coverage, the bct phase thickness reduces because the hcp phase begins to grow downwards too, by forcing the phase transition to occur in the topmost part of the bct layer. This accounts for the observed difference between the growth curves and the erosion curves above the critical thickness.

A similar bct–hcp structural transition has been observed for Co films grown on an FeAl(001) substrate [168]. FeAl is a cubic alloy with a CsCl-type structure, whose lattice constant is 2.904 Å, slightly larger than that of bcc Fe (2.87 Å). The critical thickness for the structural transition was found to be 13 Å. The LEED intensity analysis determined interlayer spacings of $d_{bulk} = 1.40 \pm 0.03$ Å and $\triangle d_{12} = -0.15 \pm 0.03$ Å for the (1×1) bct film below the critical thickness, and $d_{bulk} = 1.29 \pm 0.04$ Å and $\triangle d_{12} = -0.125 \pm 0.03$ Å for the $c(2 \times 2)$ strained hcp $(11\bar{2}0)$ phase above the critical thickness.

Another alloy substrate, TiAl(010), has also been exploited for the growth of an ultrathin metastable Co film [169]. TiAl has the tetragonal structure

[4] In PDMEE measurements, the secondary-electron yield and/or Auger electron yield are/is recorded as a function of the incidence angle of the primary electron beam which is usually rotated azimuthally about a high-symmetry direction of the substrate. The yield is enhanced owing to forward-scattering effects when the primary electron beam is aligned along low-index atomic chains, as we have seen in the case of angle-resolved x-ray photoemission scattering (ARXPS).

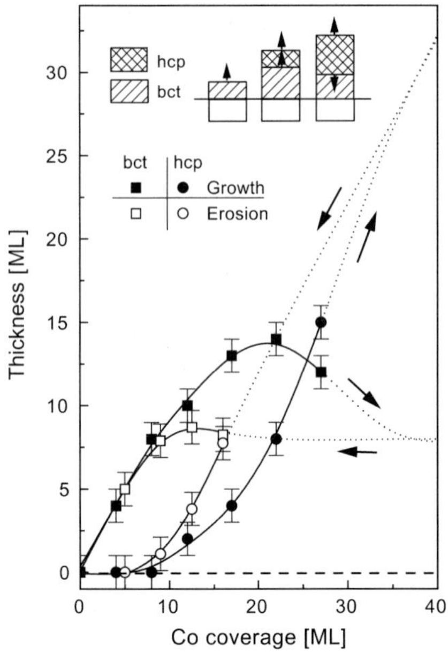

Fig. 5.45. By fitting the PDMEE spectra, the thicknesses of the layers with bct and hcp structures were obtained for both the growth and the erosion processes as a function of the total thickness of the Co film. The *dotted lines* show tentative trends for the thicknesses of the two phases in the high-coverage region. The evolution of the bct and hcp layers during growth is also sketched at the *top*. From [167], used with permission

of AuCu-I. In this structure, the (001) planes contain alternately 100% Ti and 100% Al, whereas the (010) planes are composed of 50% Ti and 50% Al. However, the (010) surface is chemically reconstructed: Ti and Al atoms in the first and second layers change place, so that the first layer is all Al and the second layer is all Ti. The sides of the rectangular TiAl(010) mesh are 4.0707 and 4.005 Å. Pseudomorphic growth of Co has been observed for about 15–20 ML on this substrate, with a bulk interlayer spacing of $d_{bulk} = 1.40 \pm 0.03$ Å and a contracted surface layer ($d_{12} = 1.28 \pm 0.03$ Å). A strain analysis suggested that the Co film has a strained bcc structure [170]. The pseudomorphic growth of Co on TiAl(010) exemplifies a different type of epitaxial growth. In this case, the equilibrium phase, i.e., the unstrained metastable phase, of the film material has an in-plane unit mesh which differs from the substrate in symmetry.

The studies of ultrathin Co films on various surfaces of W (bcc) single crystals [171, 172] were more or less motivated by the requirement for spin-polarized STM tips. The growth of Co on W(100), W(110) and W(111) shows a common trend, in which the growth starts with a two-dimensional or layer-by-layer-grown pseudomorphic layer followed by a structural transition to the close-packed phase. On the W(100) surface [172], Co initially forms two pseudomorphic monolayers. Later, three-dimensional islands grow on the top of this double layer. The close-packed 3D crystals have the $(11\bar{2}0)$ plane parallel to the substrate, with the [0001] and $[1\bar{1}00]$ directions parallel to the $[1\bar{1}0]$ and $[\bar{1}10]$ directions, respectively, of the substrate in two equivalent orientations (see Fig. 5.44). The crystals are long in the $[1\bar{1}00]$ direction and short in the orthogonal [0001] direction. The pseudomorphic double layer was found to be nonmagnetic at room temperature, since it produced no magnetic contrast in spin-polarized low-energy electron microscopy (SPLEEM).

Though the observed RHEED intensity oscillations are indicative of a quasi-layer-by-layer growth mode for Co on W(110), the upper limit for pseudomorphic growth is extremely small (around 0.7 ML) [174]. Upon further deposition, the pseudomorphic monolayer transforms into the hexagonal close-packed structure. The film has a Nishiyama–Wassermann orientation, with (0001)Co‖(110)W and $[11\bar{2}0]$Co‖[001]W [173, 174]. It is only slightly strained in the direction with large misfit ($[11\bar{2}0]$Co or [001]W, 20.8% misfit; see Fig. 5.46), in which the Co atoms are not in register with respect to the substrate lattice, while the film continues the substrate periodicity in the

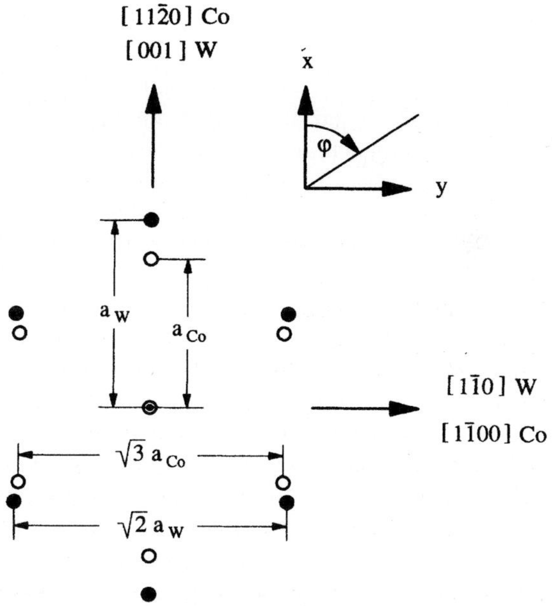

Fig. 5.46. Comparison of atomic positions in bulk W(110) (*solid circles*) and Co(0001) (*open circles*) for the Nishiyama–Wassermann orientation $[11\bar{2}0]$Co‖[001]W. The misfit of Co with respect to W in the x and y directions (labeled in the figure) is −20.8% and −3.0%, respectively. From [173], used with permission

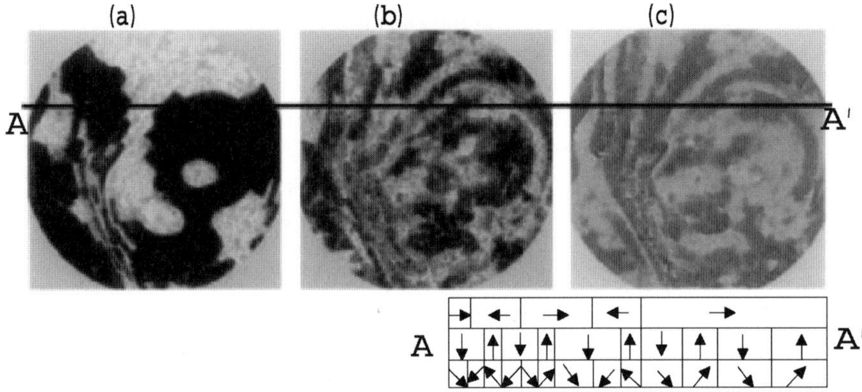

Fig. 5.47. SPLEEM images of a 5 ML hcp Co film on W(110). The electron energy was 2.1 eV, and the field of view 8 μm. (**a**) Image of the in-plane component of the magnetization. (**b**) Image of the out-of-plane component. (**c**) Tilt angle image. Below (**c**) a schematic presentation of the magnetization distribution along the section AA' is shown, with the *top row* corresponding to (**a**), the *middle row* to (**b**), and the *bottom row* to (**c**). From [175], used with permission

low-misfit direction ($[1\bar{1}00]$Co or $[1\bar{1}0]$W), which produces a strain of about 3%. This strain decreases only slowly with thickness, and causes a uniaxial in-plane anisotropy with the easy axis in the $[1\bar{1}00]$Co or $[1\bar{1}0]$W direction [173]. A canted magnetization and a wrinkled domain structure have been observed by SPLEED for hcp Co films on W(110) over a wide thickness range between 3 and 8 ML (see Fig. 5.47) [175]. It is not surprising that an ultra-thin magnetic film shows a canted magnetization. The canted magnetization results from the competition between different anisotropies which favor in-plane and out-of-plane magnetization, respectively. In its simplest form, the anisotropy energy of a (0001) hcp magnetic film can be written in the form

$$E = \frac{1}{2}\mu_0 M_s^2 \cos^2\theta + K_1 \sin^2\theta + K_2 \sin^4\theta \,, \tag{5.6}$$

where θ is the angle between the film normal and the magnetization, and M_s is the saturation magnetization. The second- and fourth-order anisotropy coefficients K_1 and K_2 can be further decomposed into a thickness-dependent and a thickness-independent contribution, i.e., $K_\mathrm{i} = K_\mathrm{ib} + K_\mathrm{is}/t (i = 1, 2)$, with t being the film thickness. It can be derived from (5.6) that there exist two critical thicknesses, $t_{c1} = K_{1s}/(\frac{1}{2}\mu_0 M_s^2 - K_{ib})$ and $t_{c2} = (2K_{2s} + K_{1s})/(1/2\mu_0 M_s^2 - 2K_{2b} - K_{1b})$. In the absence of an applied magnetic field, the magnetization is perpendicular to the film below the critical thickness t_{c1} and in the film plane above the critical thickness t_{c2}. Provided that $t_{c1} < t_{c2}$, a canted magnetization is energetically favored, with

$$\sin^2\theta = \frac{1/2\mu_0 M_s^2 - K_\mathrm{ib} - K_{1s}/t}{2K_{2b} + 2K_{2s}/t} \,. \tag{5.7}$$

The observation that the out-of-plane component has a smaller domain size is understandable from the viewpoint of minimizing the magnetic dipolar energy.

A pseudomorphic bcc-like Co film can persist to a considerable thickness when it is grown on a W(111) substrate [176, 177]. At 380 K, a (1×1) pseudomorphic film can be grown up to at least 15 ML, while at elevated temperatures (for example 685 K or 810 K) the (6×6) close-packed phase starts to nucleate when the pseudomorphic film is thicker than 3 ML [177]. The (6×6) LEED pattern is characterized by sharp satellite spots surrounding integer spots with threefold symmetry. The separation between the satellite and integer spots indicates a sixfold periodicity in the in-plane $\langle 11\overline{2}\rangle$ substrate directions. A (6×6) Co film can be formed by lateral compression of a bulk hcp close-packed plane by 2.8%. This reduces the in-plane lattice constant from the bulk value of $a = 2.507$ to 2.438 Å, and gives rise to a coincidence between 11 Co spacings and 6 W spacings. The magnetic properties have been examined mainly for pseudomorphic Co/W(111) films. These films become ferromagnetic above about 7 ML [172]. Between 7 and 8–9 ML the magnetic signal rises rapidly, and thereafter increases linearly with increasing thickness (see Fig. 5.48). The magnetization is dominated by the in-plane anisotropy. The favored in-plane magnetization directions are $\langle 11\overline{2}\rangle$ and $\langle 1\overline{1}0\rangle$.

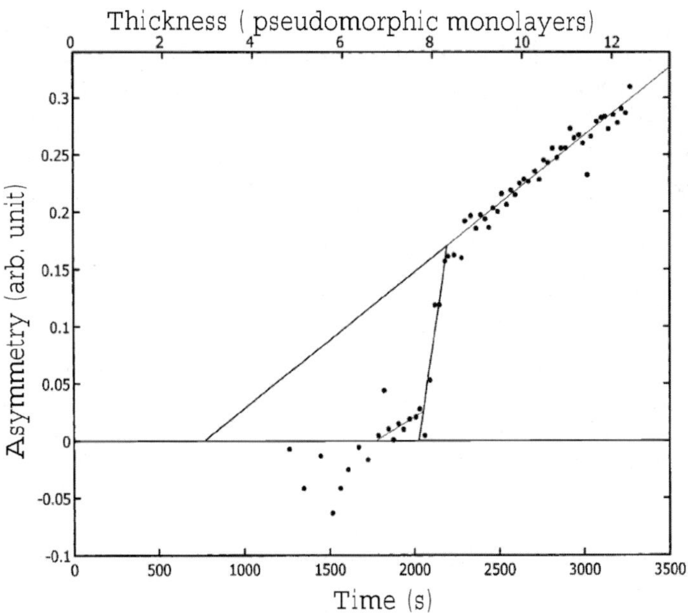

Fig. 5.48. Magnetic SPLEEM asymmetry of Co films on W(111) grown at about 400 K after the growth of three initial pseudomorphic monolayers at about 800 K, as a function of deposition time. From [172], used with permission

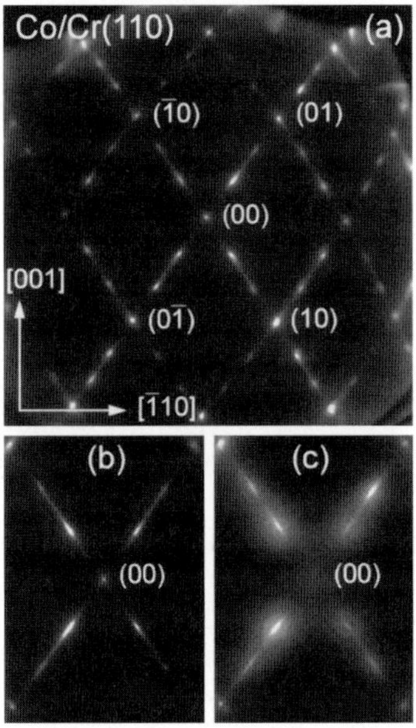

Fig. 5.49. (**a**) LEED pattern of a 5 ML Co/Cr(110) film at 125 eV showing a (3×1) reconstruction. (**b**) and (**c**) show detailed scans at 133 eV for an 8 ML film before and after removal of about 2 ML coverage by Ne^+ sputtering. From [179], used with permission

On bcc Cr(100), pseudomorphic Co growth occurs for the first few monolayers, followed by a continuous relaxation towards the hcp structure [178]. However, pseudomorphic growth takes place over a considerable thickness range (> 30 ML) on a Cr(110) substrate [179]. This situation is similar to the growth of Co on Cu single crystals, where fcc Co is extremely stable on the Cu(110) surface. A (3×1) reconstruction has been observed for Co/Cr(110) films. The LEED patterns shown in Fig. 5.49 result from the coexistence of two equivalent (3×1) reconstruction domains, each of which shows uniaxial superperiodicity along one of the two in-plane close-packed row directions. The persistence of the (3×1) LEED pattern upon removal of 2 ML coverage by moderate sputtering at low temperature implies that the reconstruction is robust and is extended normal to the film surface. It has been suggested that the pseudomorphic strain is responsible for the observed reconstruction and that vertical buckling is involved in the reconstruction, just as in the (5×1) reconstruction in Fe/Cu(100) [180].

5.2.6 Ultrathin Bcc Ni Films

The studies of epitaxy-stabilized Ni structures have so far been limited to bcc Ni on an Fe(100) substrate. It is the general consensus that Ni grows in a pseudomorphic way with a bcc structure for the first 3–6 ML on Fe(100)

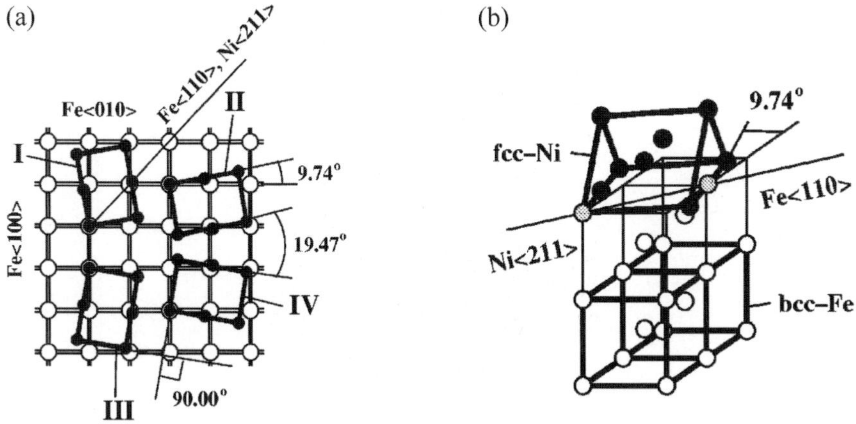

Fig. 5.50. (**a**) Four possible fcc Ni {110} domains (*filled circles*) on the bcc Fe{100} mesh (*open circles*). (**b**) The atomic positions of Ni atoms after the bcc-fcc transition of a Ni layer on bcc Fe{100}. The *gray circles* indicate Ni atoms whose positions remain unaffected during the transition. From [185], used with permission

[181, 182, 183]. The remarkable RHEED oscillations observed, indicate a layer-by-layer-like growth for low coverage. Upon further growth, a $c(2 \times 2)$ reconstruction is observed, and the film gradually changes its structure to fcc with twinned Ni{110} domains rotated by $\pm 9.74°$, with an epitaxial relation of fcc Ni$\{110\}\langle 211\rangle \|$bcc Fe$\{100\}\langle 110\rangle$ (see Fig. 5.50) [184]. The epitaxial relationship has been confirmed by Rutherford backscattering spectroscopy and XRD measurements [185]. The $c(2\times 2)$ reconstruction observed by LEED and RHEED has also been explained as an artifact resulting from the coexistence of four equivalent fcc Ni domains (see Fig. 5.50a) [185].

The magnetic properties of bcc Ni films are still somewhat controversial. Polarized-neutron reflection studies at 4 K of a 3 ML bcc Ni film on 5 ML Fe prepared on an Ag(001) substrate gave a nickel moment of 0.55–0.80μ_B and an iron moment close to the bulk value [186]. Photoemission spectra indicated a long-range ferromagnetic order at room temperature, with a magnetic moment around 0.4μ_B for a 7 ML bcc Ni film on Fe(001) [187]. X-ray magnetic dichroism measurements on Ni/Fe(001) showed that the total magnetic moment of Ni is around 0.3μ_B without any apparent dependence on the layer thickness [188]. Here we lay aside these discrepancies and focus on a recent x-ray magnetic-circular-dichroism study of Ni/Fe(001) multilayers [189]. The multilayers were prepared on 325 Å Cr on an MgO(001) substrate with a structure of Fe12Å[Ni0–30Å/Fe12Å]$_{20}$. The average lattice constants and the magnetic moments of Ni and Fe as a function of Ni layer thickness are shown in Fig. 5.49. When the thickness of the Ni layers increases while the thickness of the Fe layers is kept constant at 12 Å, the multilayers undergo a coherent structural transition from bcc at low Ni thickness to fcc at high Ni thickness.

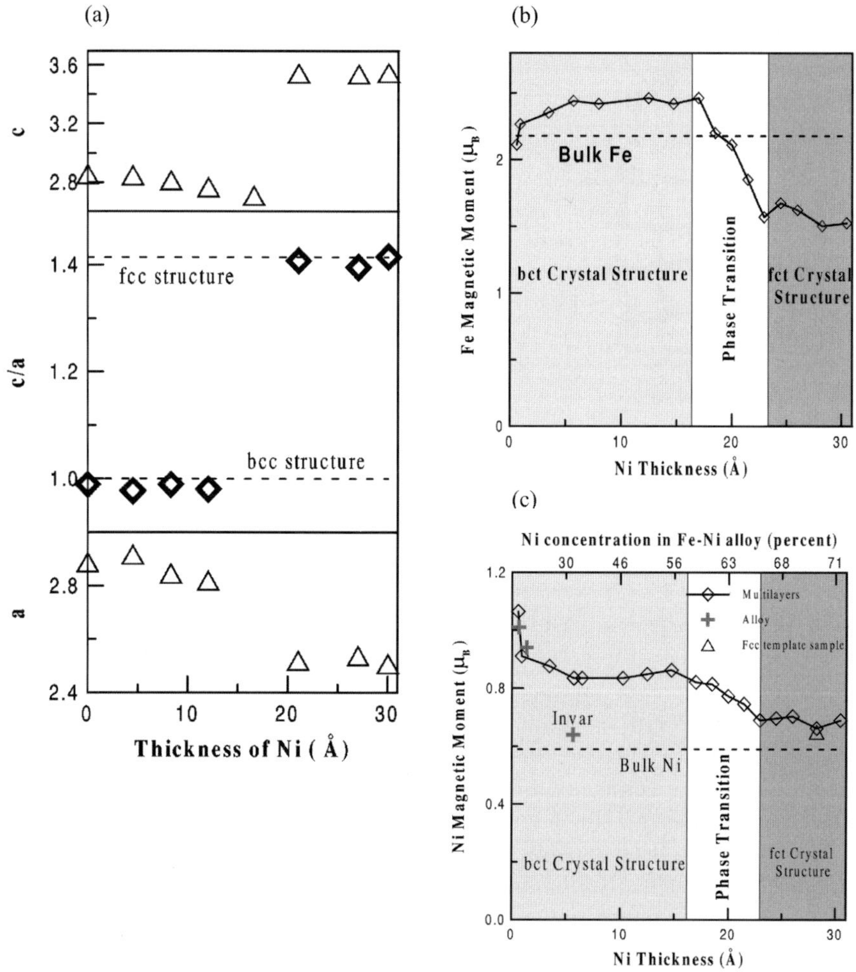

Fig. 5.51. (**a**) The average perpendicular spacing c (*top panel*), in-plane lattice constant a (*bottom panel*) and the c/a ratio (*center*) for Ni/Fe(001) multilayers as a function of Ni layer thickness. The Fe layer thickness is constant at 12 Å. (**b**) Summary of the element-specific magnetic moment of Fe in Ni/Fe(001) multilayers as measured by XMCD. (**c**) Summary of the element-specific magnetic moments of Ni in Ni/Fe(001) multilayers as measured by XMCD. For comparison, the bulk value for fcc Ni is plotted as a *dashed line*, and the Ni magnetic moment in bulk Ni alloys with a corresponding Ni thickness is plotted as *gray crosses*. From [189], used with permission

The multilayers with thin Ni layers (< 16 Å) are bct and ferromagnetic. The average moment of the Ni atoms is enhanced well above that of bulk fcc Ni. It was suggested that this enhancement could be ascribed to the reduced coordination of the bcc structure and to interaction at the Fe/Ni interfaces.

5.3 Epitaxially Stabilized Ultrathin Alloy Films

In the previous section, we have shown with a number of examples that a variety of metastable crystallographic structures of 3d elements can be stabilized under normal conditions via epitaxial growth, by choosing suitable substrates. We shall now show that alloying in systems which are immiscible in their bulk form may occur in ultrathin epitaxial films. In Sect. 5.3.1, a theory will be presented to explain why epitaxial strain can reduce the miscibility gap. Then several examples will be given in Sect. 5.3.2 of the experimental realization of nonequilibrium alloy structures in epitaxial films. The work done in this field is relatively sparse, but we believe that stabilizing metastable compounds or alloys by epitaxial growth will attract more attention in the future owing to its inherent potential.

5.3.1 Theory of Epitaxy-Assisted Alloying

The effects of strain imposed by interface interactions on the binary temperature–composition phase diagram, including the order–disorder transition, have been investigated theoretically from the viewpoint of energy by Flynn [190] and by Wood and Zunger [191]. The basic idea is that, if a substrate with a lattice constant intermediate between those of the crystals of the individual components is chosen, then an epitaxial film of an alloy phase which is phase-separated in the bulk may be energetically favored owing to two effects [192]: (1) a reduced strain energy compared with pseudomorphic single-component films, and (2) the fact that the atoms in the alloy phase can be accommodated with regard to the substrate lattice without forming dislocations. In the following, we shall show how the epitaxial strain can stabilize such an alloy structure, on the basis of continuum elasticity theory [191].

Consider a crystalline material with an fcc structure. Its total energy E_{bk} (per fcc site) for the bulk phase, as a function of its cubic lattice parameter a, can be expanded about the equilibrium zero-pressure value a_{eq} as

$$E_{bk}(a) = E_{eq} + \frac{9}{8} B a_{eq} (a - a_{eq})^2 + \cdots \,, \tag{5.8}$$

where E_{eq} is the equilibrium total energy and $B = (C_{11} + 2C_{12})/3$ is the bulk modulus. C_{ij} is an element of the matrix of elastic constants. Now consider the same material grown as a thin epitaxial film on a substrate with a lattice parameter a_s. For pseudomorphic growth, the film is coherently strained and the lattice parameters of a unit cell parallel to interface plane are determined by a_s, while the cell dimension c normal to the interface plane is free to relax. The equilibrium perpendicular dimension $c_{eq}(a_s)$ can be obtained by minimizing the elastic energy density $U = 1/\varepsilon \cdot C \cdot \varepsilon$ with respect to c:

$$c_{eq}(a_s) = c(a_{eq}) - \left(\frac{3B}{A} - 1 \right) (a_s - a_{eq}) \,, \tag{5.9}$$

where $A = B + 4C_{44}/3$ for the (111) orientation, $A = C_{44} + 1/2(C_{11} + C_{12})$ for the (110) orientation and $A = C_{11}$ for the (001) orientation; ε and C are the strain tensor and the elastic-constant matrix, respectively. $c(a_{eq})$ equals a_{eq} for a [001]-orientated substrate. For substrate orientations other than [001], a geometric factor relates $c(a_{eq})$ to a_{eq} and the strain-induced distortion is not tetragonal. For example, it is trigonal for a (111) substrate. Total energy per fcc site of the coherently strained epitaxial film is

$$E_{ep}(a_s, c) = E_{eq} + \frac{9}{8} q B (a_s - a_{eq})^2 + \frac{1}{8} A a_{eq} [c - c_{eq}(a_s)]^2 + \cdots , \quad (5.10)$$

where $q = 1 - B/A$. The last term on the right-hand side of (5.10) is zero for thin epitaxial films, since c is unstrained ($c = c_{eq}$). To illustrate how the epitaxial constraint can stabilize certain alloys which are immiscible in the bulk, consider a schematic solid-state reaction between the solids α and β to produce an alloy phase $\alpha_x\beta_{1-x}$:

$$x\alpha + (1 - x)\beta \rightleftharpoons \alpha_x\beta_{1-x} . \quad (5.11)$$

If the reaction takes place incoherently in the bulk, all three species α, β and $\alpha_x\beta_{1-x}$ adopt their equilibrium lattice dimensions $a_{eq}[\alpha]$, $a_{eq}[\beta]$ and $a_{eq}[\alpha_x\beta_{1-x}]$, respectively. The bulk formation enthalpy is defined as the change in enthalpy in this reaction:

$$\triangle H_{bk}(x) = E_{eq}[\alpha_x\beta_{1-x}] - xE_{eq}[\alpha] - (1 - x)E_{eq}[\beta] . \quad (5.12)$$

However, if the reaction takes place epitaxially and all the species have the same in-plane lattice constants as the substrate, the change in enthalpy is the epitaxial formation enthalpy

$$\triangle H_{ep}(a_s, x) = \triangle H_{bk}(x) + \triangle E_{ES}(a_s, x) . \quad (5.13)$$

Here $\triangle E_{ES}(a_s, x)$ is the excess epitaxial strain energy and can be written as

$$\triangle E_{ES}(a_s, x) = E_{ES}[\alpha_x\beta_{1-x}] - xE_{ES}[\alpha] - (1 - x)E_{ES}[\beta] . \quad (5.14)$$

The epitaxial strain energy for each species (denoted by the index i in square brackets) is

$$E_{ES}[i] = \frac{9}{8} q[i] B[i] a_{eq}[i] (a_s - a_{eq}[i])^2 , \quad (5.15)$$

as can be obtained from (5.10). The epitaxial stabilization effects can be discussed on the basis of (5.12) and (5.14). An alloy with composition $\alpha_x\beta_{1-x}$ can be formed with assistance from epitaxy if $\triangle H > 0$ but $-\triangle E_{ES} > \triangle H_{bk}$. This is schematically shown in Fig. 5.52a. The right-hand side of Fig. 5.52a shows the thickness dependence of the energies (per site) of such an epitaxially stabilized phase $\alpha_x\beta_{1-x}$ and of its constituents $\alpha + \beta$, with dislocations included. Assuming $\alpha\beta$ is lattice-matched to the substrate, its energy is constant for all film thicknesses h. In contrast, $\alpha + \beta$ is strained and its energy

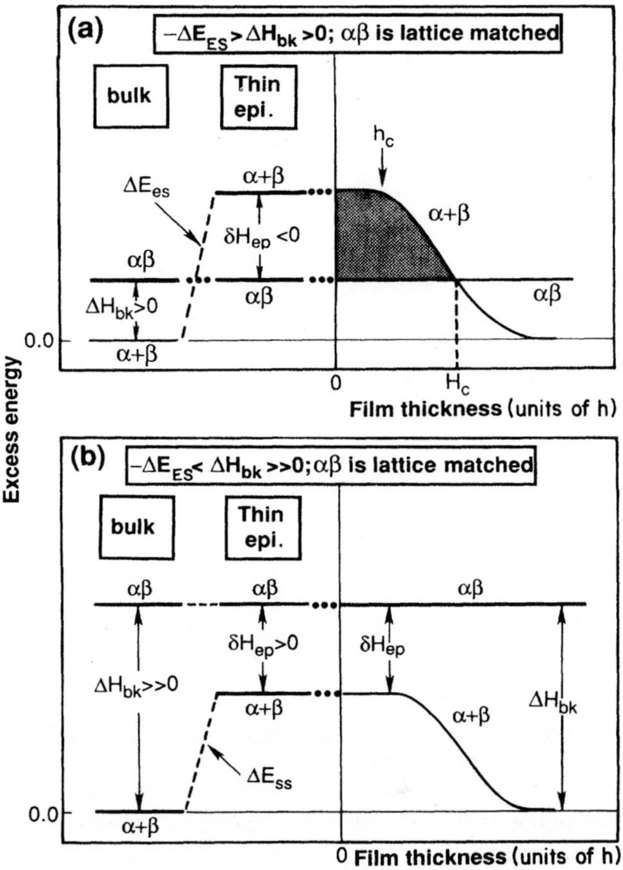

Fig. 5.52. Relative energies of the unstable bulk alloy $\alpha_x\beta_{1-x}$ (denoted by $\alpha\beta$) and of the phase-separated constituents $\alpha+\beta$ in bulk and epitaxial form on a substrate lattice. $\alpha\beta$ is assumed to be lattice-matched to the substrate. (**a**) If $-\triangle E_{ES} > \triangle H_{bk} > 0$, $\alpha\beta$ is epitaxially stabilized. (**b**) When $-\triangle E_{ES} < \triangle H_{bk} \gg 0$, $\alpha\beta$ cannot be epitaxially stabilized. The *right-hand panels* indicate the energy change upon loss of coherence as the film thickness increases. From [191], used with permission

decreases towards its unstrained value after the film exceeds the critical thickness h_c for dislocation formation. Therefore there exists a thickness h below which only the alloy film $\alpha_x\beta_{1-x}$ can be epitaxially stabilized. However, if $\triangle H_{bk}$ is so large that $\triangle H_{bk} + \triangle E_{ES}$ is still positive, the alloy film cannot be stabilized (see Fig. 5.52b).

The thermodynamic functions and phase diagrams of bulk binary alloys $\alpha_x\beta_{1-x}$ have been successfully calculated by the method of expanding the alloy energy as a linear combination of cluster energies [193, 194]. The cluster energies have been either directly calculated [194] or derived from the results

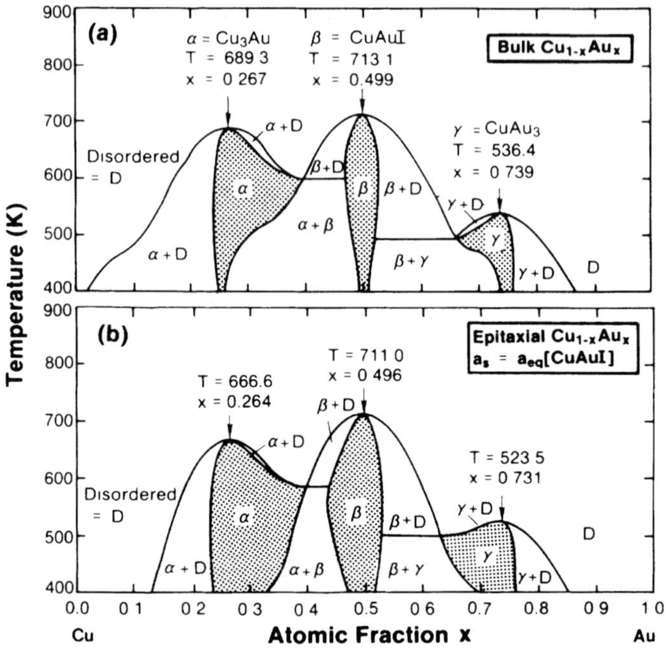

Fig. 5.53. Calculated phase diagrams for $Cu_{1-x}Au_x$ (**a**) in the bulk and (**b**) in epitaxial form on a (001) substrate of the CuAu-I compound. *Shaded areas* indicate single-phase ordered compounds. From [191], used with permission

for ordered periodic structures [191, 195]. Following the same path, epitaxial alloys can be treated just by adding the epitaxial constraint $a_{\|} = a_s$ to the calculations. Figure 5.53b shows a calculated epitaxial phase diagram for a Cu–Au binary alloy with the in-plane lattice matched to the CuAu-I compound. Compared with the bulk phase diagram (Fig. 5.53a), the epitaxial phase diagram shows a wider composition range for the single-phase ordered alloys.

5.3.2 Epitaxial Alloy Thin Films

Perhaps owing to the technical complexity involved in the preparation and characterization of magnetic alloy films, only a few 3d alloy systems have been produced by MBE. However, it can be seen clearly from the limited studies that the structural and magnetic properties of alloy materials can be modified by epitaxial growth to a considerable extent. Metastable fcc $Co_{1-x}Mn_x$ alloy films on Si(100) and Si(111) have been found to be ferromagnetic for x up to 0.7, while bulk alloys show only antiferromagnetic behavior when x exceeds 0.3 [196]. A spin reorientation transition upon variation of the alloy concentration has been observed in Ni_xPd_{1-x} films on $Cu_3Au(100)$ [197] and in Fe_xCo_{1-x} films on Cu(100) [198]. Metastable hcp $Ni_{1-x}Fe_x$ ($0 < x < 20\%$) films have been stabilized on Cr(211) [199]. In the following, two other alloy systems will be discussed in detail. These are $Au_{1-x}Ni_x$ films on Au(001) and $Fe_{1-x}Ni_x$ films on Cu(100). The former system shows that a system can be stabilized by epitaxial growth as an alloy film even though it is immiscible in the bulk. The latter system exemplifies that the structural and magnetic properties of an alloy can be considerably modified in epitaxial films.

Epitaxial-Strain-Stabilized $Au_{1-x}Ni_x$ Alloy Thin Films

Both Au and Ni crystallize with fcc structure. Owing to the large size difference (14%) between these two elements, the bulk equilibrium phase diagram of the Au–Ni binary system is characterized by a strong miscibility gap with a maximum at a temperature of 1083 K and a composition of 70.6 at.% Ni [200]. At temperatures below the miscibility gap, the homogeneous alloy phase is not the stable phase. The room-temperature solubilities of both Au in Ni and Ni in Au are very small (~ 0.01). However, metastable disordered fcc AuNi solid solutions can be obtained over the whole composition range at room temperature by rapid quenching from a temperature above the miscibility gap [201]. At temperatures above 573 K, the quenched Au–Ni alloys decompose by means of a discontinuous precipitation process. At lower temperatures, spinodal decomposition[5] occurs. Modulated structures have been observed in Au–Ni alloys along $\langle 100 \rangle$ and $\langle 110 \rangle$ [202].

Thin films of $Au_{1-x}Ni_x$ solid solutions have been prepared at room temperature by MBE growth on an Au(001) buffer layer on an MgO substrate with an atomic fraction of Ni of up to 0.6 (see [203] and references therein). During the growth, the RHEED intensity of the specular reflection shows strong oscillations, indicating a layer-by-layer growth mode. These oscillations are present up to 30–70 ML, depending on the Ni content and the surface quality of the buffer layer (see Fig. 5.54). The effect of aging at an elevated temperature depends strongly on the composition. For $x = 0.14$ and $x = 0.19$, no composition modulation was observed after 150 hours in a temperature range between 220 and 250 °C. For $0.28 < x < 0.5$, however, modulated structures were formed along the growth direction after annealing. The period of the modulated structures decreases with increasing Ni content. A period of 4.4 ML was observed for $x = 0.28$ and it decreased to

[5] Spinodal decomposition occurs in a temperature and composition range where the second derivative of the free energy with respect to the alloy composition is negative, i.e., $\partial^2 F/\partial c^2 < 0$. Under this condition, the homogeneous alloy phase is unstable against nonlocal fluctuations of composition of infinitesimal amplitude. Spinodal decomposition involves a continuous process: first, composition fluctuations around the average composition develop, then they grow in amplitude with a diffusion flux against the concentration gradient and a negative diffusion coefficient, and finally a state consisting of a two-phase mixture is reached.

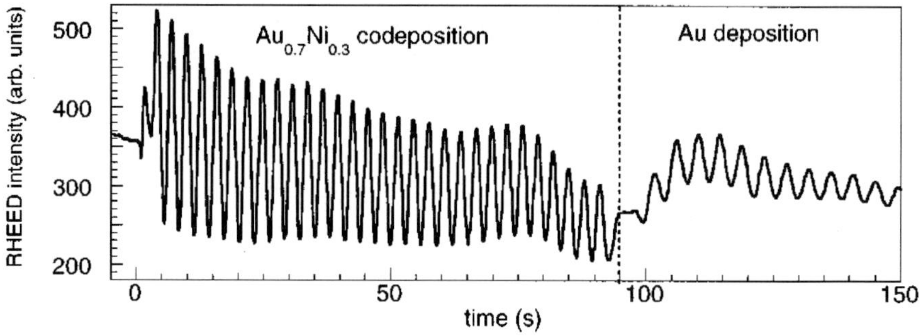

Fig. 5.54. Oscillations in RHEED intensity during the growth of an $Au_{0.7}Ni_{0.3}$ film on an Au(001) buffer layer. The *vertical dashed line* corresponds the closure of the shutter on the Ni source, and afterwards only Au was evaporated. The change in oscillation period after the closure of the shutter on the Ni source allows one to determine the composition of the alloy film. From [203], used with permission

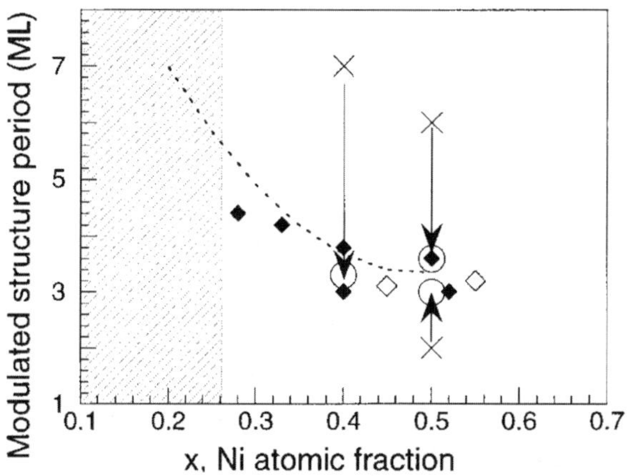

Fig. 5.55. Variation of the period of the modulated structure with Ni content for coherent layers (*solid diamonds*), partially relaxed layers (*open diamonds*) and artificially modulated structures (*crosses* denote the initial state and *open circles* the final state). From [203], used with permission

about 3 ML for $x = 0.5$ (see Fig. 5.53). Differently from the bulk spinodal decomposition, the modulated structures observed in the $Au_{1-x}Ni_x$ alloy films show considerable stability. They do not change with further annealing treatment and their periods show no dependence on the annealing temperature. This stability is more clearly shown by the fact that artificially modulated structures always relax to the same modulation upon annealing, in spite of their initial structure. It has been also found that a certain amount of strain is necessary to form the modulated structure, indicating that the modulation is strain-driven.

Fcc Fe_xNi_{1-x} Films on Cu(100). At room temperature, Fe-rich Fe–Ni bulk alloys show a bcc structure and Ni-rich alloys show an fcc structure. The fcc–bcc martensitic transition takes place around $x = 0.7$ for Fe_xNi_{1-x}. The change in crystal structure is accompanied by a strong downward shift of the magnetic moment from the Slater–Pauling curve [204] and a fast decrease of the Curie temperature. For bulk alloys with a concentration near $x = 0.65$, the thermal expansion coefficient is very small over a wide range of temperatures and the elastic coefficients are almost temperature-independent. This is the Invar effect, which is believed to be correlated with the moment–volume instability [205].

When Fe_xNi_{1-x} alloy films are prepared on Cu(100) substrates, a different behavior is observed. The alloy films on Cu(100) are characterized by an fcc structure pseudomorphic with the substrate over a wide range of concentrations and thicknesses [206, 207, 208]. RHEED oscillations are generally observed during the growth of Fe_xNi_{1-x} alloy films, indicating a layer-by-layer-like growth mode (see Fig. 5.56). In contrast to the corresponding bulk alloys, the fcc–bcc transition is suppressed in the epitaxial film even in the case where the Fe concentration is as high as 90 at% [208]. The fcc local atomic environment of a 30 ML $Fe_{80}Ni_{20}$ alloy film on Cu(100) has been confirmed by x-ray absorption near-edge spectroscopy and extended x-ray absorption fine structure [208]. On the other hand, it has been claimed in a study that $Fe_{80}Ni_{20}$ alloy films show a gradual decrease of the in-plane

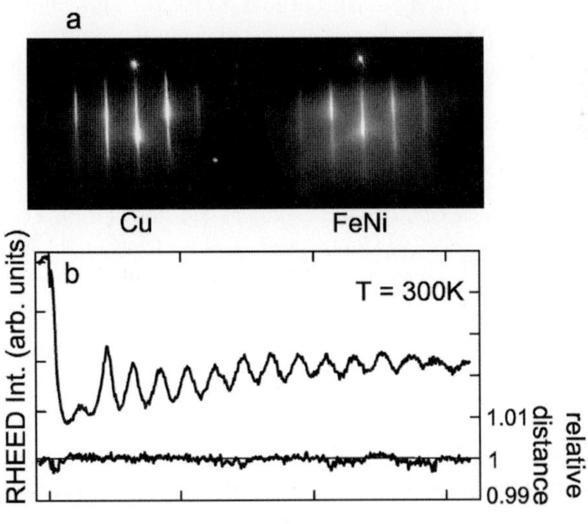

Fig. 5.56. (**a**) RHEED pattern observed for a Cu(100) substrate and for a subsequently deposited 30 ML $Fe_{84}Ni_{16}$ film on Cu(100) along the [011] azimuth. (**b**) RHEED oscillations observed during the growth of an $Fe_{75}Ni_{25}$ alloy film on Cu(100). From [208], used with permission

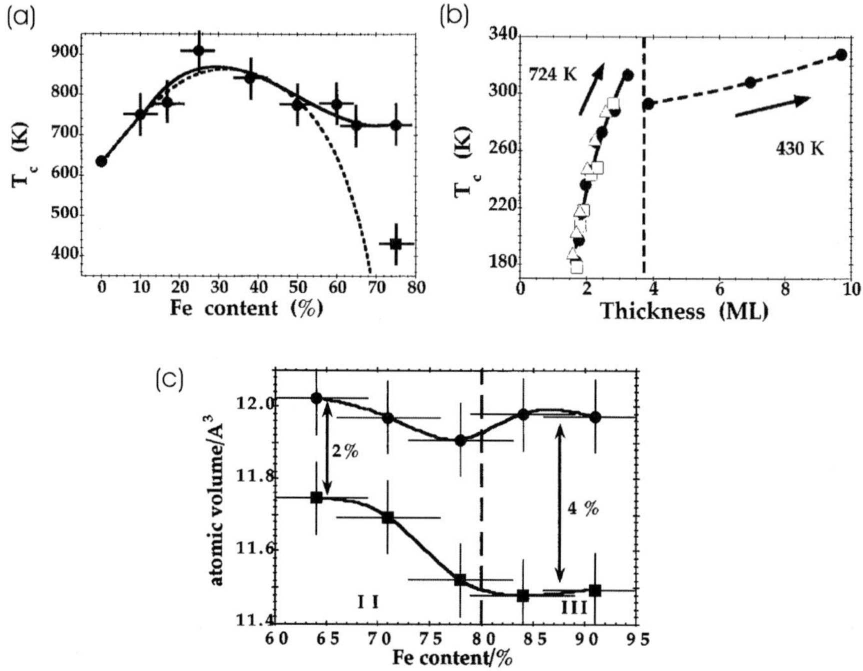

Fig. 5.57. (**a**) Comparison of Curie temperatures between bulk Fe_xNi_{1-x} alloys (*dashed curve*) and epitaxial alloy films on Cu(100). To eliminate the influence of the film thickness on the Curie temperature, the data for the alloy films were obtained by extrapolating the thickness-dependent experimental values to infinitely large thickness using an empirical scaling formula $T_C(n)/T_C(\infty) = 1/[1+[(n-n')/n_0]^{-\gamma}]$ [146], where n' and n_0 are empirical parameters and γ is a fitting parameter. The *solid square* shows T_C for the $Fe_{75}Ni_{25}$ alloy film with a larger thickness. (**b**) Thickness dependence of Curie temperature for an $Fe_{75}Ni_{25}$ alloy film on Cu(100) (*solid symbols*). The *solid line* is a fit to the empirical formula and the *dashed line* is a guide to the eye. The *open squares* and *triangles* are data for an $Fe_{65}Ni_{35}$ film and an $Fe_{60}Ni_{40}$ film, respectively. (**c**) Comparison of the variation of the atomic volume of 2 ML (*circles*) and 6 ML (*squares*) thick alloy films near the Invar concentration. From [205, 206], used with permission

lattice constant of $\sim 0.7\%$ between 1 and 4 ML [205]. The alloy films also show a different concentration dependence for the Curie temperature. In contrast to the bulk material, the rapid drop in Curie temperature around the bulk Invar concentration is suppressed (see Fig. 5.57a), and the alloy films with a concentration beyond this Invar concentration are still ferromagnetic [206]. An interesting thickness dependence of the Curie temperature has been observed for these Fe-rich alloy films [206]. As shown in Fig. 5.57b, in the low-thickness range (< 4 ML), the thickness dependence of the Curie temperature of $Fe_{60}Ni_{40}$ and $Fe_{65}Ni_{35}$, as well as that of $Fe_{75}Ni_{25}$, can be fitted by the same empirical scaling formula [146, 206]. However, a rapid drop in Curie temperature is observed for $Fe_{75}Ni_{25}$ around a film thickness of 4 ML, and the Curie temperature increases only slowly with further deposition. This is in contrast to the $Fe_{60}Ni_{40}$ and $Fe_{65}Ni_{35}$ films, which comply with the scaling law over a wide thickness range. Consistently, the Fe 3p asymmetry in x-ray magnetic linear dichroism (XMLD) measurements on $Fe_{72}Ni_{28}$ and $Fe_{80}Ni_{20}$ films coincides with the value for Fe concentrations below 60% when the film is 2–3 ML thick. However, it decreases gradually with increasing thickness [205]. Correspondingly, a reduction in atomic volume has been observed for alloy films around the Invar concentration when the film thicknesses increased from 2 ML to 6 ML [205]. The reduction in atomic volume is more pronounced for films with a higher Fe concentration (see Fig. 5.57c). A sudden drop in Curie temperature and magnetic moment with thickness has been observed around 4–5 ML for Fe/Cu(100) films, which is due to a magnetic phase transition from the homogeneous ferromagnetic phase at low thickness to a special phase at larger thickness in which only the surface layer is ferromagnetic [165, 209]. However, the mechanism leading to the change in magnetic properties in Fe–Ni alloy films with an Fe concentration above the Invar value is believed to be different from this "magnetic live surface layer" picture [205, 206], since the XMLD asymmetry decreases with increasing film thickness gradually, rather than by a jump. The behavior shown by Fe–Ni alloy films has been associated with a transition from a high-spin state at low thickness to a low-spin state at large thickness [205, 206]. The coexistence of the high-spin state and low-spin state, together with the moment–volume instability, has been proposed as an explanation for the Invar effect [210]. The total magnetic-moments of Fe and Ni in Fe–Ni alloy films have been determined by applying the sum rule to x-ray magnetic-circular-dichroism spectra (see Fig. 5.58) [208]. The most striking result is that both the Fe magnetic moment and the Ni moment decrease with increasing Fe concentration on the Fe-rich side. The decrease in the Fe magnetic moment is more pronounced than that for Ni. Accompanying the decrease in atomic magnetic moment is a reduction in atomic volume (see Fig. 5.58a). These results again show a moment–volume instability for fcc Fe_xNi_{1-x} epitaxial films on Cu(100) when the film has an Fe concentration larger than the critical value for the bulk fcc–bcc transition. The loss of total magnetic moment in Fe_xNi_{1-x} films with increasing Fe concentration has also been ascribed to the emergence of a low-spin antiferromagnetic state in the region of high Fe concentration in coexistence with a high-spin ferromagnetic state [211].

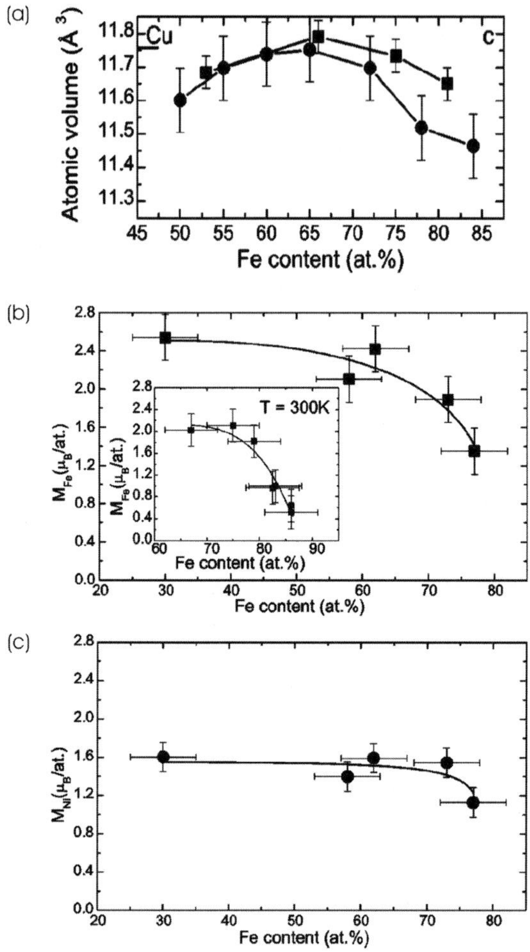

Fig. 5.58. (**a**) Average atomic volume of Fe_xNi_{1-x} films on Cu(100) as a function of Fe concentration: the *black squares* denote the data measured by XRD for 50–ML thick films, and the *black circles* denote data obtained from RHEED and LEED measurements for an alloy thickness of 6 ML. (**b**) and (**c**) show the total magnetic moment of Fe and Ni in Fe_xNi_{1-x} films on Cu(100) as a function of Fe concentration at 10 K. The data are derived from XMCD spectra. The *inset* shows the Fe total magnetic moment versus the Fe concentration at 300 K. From [208], used with permission

References

1. G. A. Prinz: J. Magn. Magn. Mater **100**, 469(1991)
2. M. van Schilfgaarde, I. A. Abrikosov and B. Johansson: Nature **400**, 46(1999)
3. M. Wuttig and B. Feldmann: Surf. Rev. Lett. **3**, 1473(1996)
4. S. C. Abrahams, L. Guttman and J. S. Kasper: Phys. Rev. **127**, 2045(1962)
5. V. L. Moruzzi, P. M. Marcus, K. Schwarz and P. Mohn: Phys. Rev. B **34**, 1784(1986)
6. V. L. Moruzzi and P. M. Marcus: Energy band theory of metallic magnetism in the elements. In: *Handbook of Magnetic Materials*, vol 7. ed by K. H. J. Buschow (North-Holland, Amsterdam London New York Tokyo) pp97-137
7. P. J. Craievich, M. Weinert, J. M. Sanchez and R. E. Watson: Phys. Rev. Lett. **72**, 3076(1994)
8. F. Milstein, H. Fang and J. Marschall: Philos. Mag. A **70**, 621(1994)
9. V. L. Sliwko, P. Mohn, K. Schwarz and P. Blaha: J. Phys. : Condens. Matter **8**, 799(1996)
10. P. Alippi, P. M. Marcus and M. Scheffler: Phys. Rev. Lett. **78**, 3892(1997)
11. P. M. Marcus and P. Alippi: Phys. Rev. B **57**, 1971(1998)
12. Y. Tian, F. Jona and P. M. Marcus: Phys. Rev. B **59**, 12286(1999)
13. P. M. Marcus, V. L. Moruzzi and S. -L. Qiu: Phys. Rev. B **60**, 369(1999) and references therein
14. S. L. Qiu and P. M. Marcus: Phys. Rev. B **60**, 14533(1999)
15. V. L. Moruzzi, P. M. Marcus and P. C. Pattnaik: Phys. Rev. B **37**, 8003(1988)
16. M. Podgórny and J. Goniakowski: Phys. Rev. B **42**, 6683(1990)
17. C. Wolverton, G. Ceder, D. de Fontaine and H. Dreyssé: Phys. Rev. B **48**, 726
18. R.E. Thomas and G.A. Haas: J. Appl. Phys. **43**, 4900(1972)
19. M. B. Brodsky, P. Marikar, R. J. Friddle, L. Singer and C. H. Sowers: Solid State Commun. **42**, 675(1982)
20. M. B. Brodsky and H. C. Hamaker: In *Superconductivity in d- and f-Band Metals*, ed by W. Buckel and W. Weber (Kernforschungszentrum Karlsruhe, Karlsruhe, West Germany, 1982) pp. 291–298
21. J. -H. Xu, A. J. Freeman, T. Jarlborg and M. B. Brodsky: Phys. Rev. B **29**, 1250(1984)
22. J. Goniakowski and M. Podgórny: Phys. Rev. B **44**, 12348(1991)
23. G. Y. Guo and H. H. Wang: Phys. Rev. B **62**, 5136(2000)
24. Q. Xie and P. Chen: Phys. Rev. B **56**, 5223(1997)
25. A. T. Paxton, M. Methfessel and H. M. Polatoglou: Phys. Rev. B **41**, 8127(1990)
26. D.I. Bolef and J. de Klerk: Phys. Rev. **129**, 1063(1963)
27. L. F. Mattheiss and D. R. Hamann: Phys. rev. B **33**, 823(1986)
28. R. W. G. Wyckoff: *Crystal Structure* (Wisley, New York, 1963) Vol. 1, p. 50
29. J. L. Fry, Y. Z. Zhao, N. E. Brener, G. Fuster and J. Callaway: Phys. Rev. B **36**, 868(1987)
30. V. L. Moruzzi and P. M. Marcus: Phys. Rev. B **39**, 6957(1989)
31. T. Asada and K. Terakura: Phys. Rev. B **47**, 15992(1993)
32. T. Oguchi and A. J. Freeman: J. Magn. Magn. Mater. **46**, L1(1984)
33. S. L. Qiu, P. M. Marcus and Hong Ma: Phys. Rev. B **62**, 3292(2000)
34. P. Krüger, O. Elmouhssine, C. Demangeat and J. C. Parlebas: Phys. Rev. B **54**, 6393(1996)

35. V. L. Sliwko, P. Blaha, P. Mohn and K. Schwarz: In *Proceedings of the International Conference on the Physics of Transition Metal*, edited by P. M. Oppeneer and J. Kübler (World Scientific, New York, 1993), P. 614
36. T. Takahashi and W. A. Basset: Science **145**, 483(1964)
37. E. F. Wassermann: In *Magnetismus von Festkörpern und Grenzflächen*, (IFF/KFA Jülich 1993)
38. C. S. Wang, B. M. Klein and H. Krakauer: Phys. Rev. Lett. **54**, 1852(1985)
39. P. Bagno and O. Jepsen and O. Gunnarsson: Phys. Rev. B **40**, 1997(1989)
40. T. Asada and K. Terakura: Phys. Rev. B **46**, 13599(1992)
41. Jing Zhu, X. W. Wang and S. G. Louie: Phys. Rev. B **45**, 8887(1992)
42. J. Häglund: Phys. Rev. B **47**, 566(1993)
43. W. Zhong, G. Overney and D. Tománek: Phys. Rev. B **47**, 95(1993)
44. H. C. Herper, E. Hoffmann and P. Entel: Phys. Rev. B **60**, 3839(1999)
45. D. J. Singh, W. E. Pickett and H. Krakauer: Phys. Rev. B **43**, 11628(1991)
46. M. Wuttig, B. Feldmann and T. Flores: Surf. Sci. **331-333**, 659(1995)
47. S. L. Qiu, P. M. Marcus and Hong Ma: Phys. Rev. B **64**, 104431(2001)
48. T. Asada and S. Blügel: Phys. Rev. Lett. **79**, 507(1997)
49. D. Spišák and J. Hafner: Phys. Rev. B **67**, 134434(2003)
50. Ch. Kittel: *Introduction to Solid State Physics*, 6th edition (John Wiley New York 1986) pp23
51. G. Johanson, M. B. McGirr and D. A. Wheeler: Phys. Rev. B **1**, 3208(1970)
52. C. Amador, W. R. L. Lambrecht and B. Segall: Phys. Rev. B **46**, 1870(1992)
53. O. N. Mryasov. V. A. Gubanov and A. I. Liechtenstein: Phys. Rev. B **45**, 12330(1992)
54. M. Körling and J. Ergon: Phys. Rev. B **54**, R8293(1996)
55. B. Schirmer, B. Feldmann and M. Wutig: Phys. Rev. B **58**, 4984(1998)
56. M. Friák, M. Šob and V. Vitek: Phys. Rev. B **63**, 052405(2001)
57. J. Crangle: Philos. Mag. **46**, 499(1955)
58. B. I. Min, T. Oguchi and A. J. Freeman: Phys. Rev. B **33**, 7852(1986) and references therein
59. D. J. Singh: Phys. Rev. B **45**, 2258(1992)
60. A. Y. Liu and D. J. Singh: Phys. Rev. B **47**, 8515(1993)
61. G. A. Prinz: Phys. Rev. Lett. **54**, 1051(1985)
62. V. L. Moruzzi, P. M. Marcus, K. Schwarz and P. Mohn: J. Magn. Magn. Mater. **54-57**, 955(1986)
63. F. Herman, P. Lambin and O. Jepsen: Phys. Rev. B **31**, 4394(1985)
64. S. Fox and H. J. F. Jansen: Phys. Rev. B **60**, 4397(1999)
65. D. A. Papaconstantopoulos, J. L. Fry and N. E. Brener: Phys. Rev. B **39**, 2526(1989)
66. V. L. Moruzzi and P. M. Marcus: Phys. Rev. B **38**, 1613(1988)
67. H. P. J. Wijn: *Landolt-Börnstein: Magnetic Properties of Metals*, (Springer-Verlag 1986) Vol. 19, subvolume a
68. Y. Tian, F. Jona and P. M. Marcus: Phys. Rev. B **58**, 14051(1998)
69. J. Jandeleit, Y. Gauthier and M. Wuttig: Surf. Sci. **319**, 287(1994)
70. D. Rouyer, C. Krembel, M. C. Hanf, D. Bolmont and G. Gewinner: Surf. Sci. **307-309**, 477(1994)
71. P. Ohresser, F. Scheurer, B. Carrière, J. P. Deville and A. Dobroiu: Surf. Sci. **352-354**, 567(1996)

72. M. Albrecht, M. Maret, J. Köhler, B. Gilles, R. Poinsot, J. L. Hazemann, J. M. Tonnerre, C. Teodorescu and E. Bucher: Phys. Rev. Lett. **85**, 5344(2000); Phys. Rev. B **66**, 205410(2002)
73. S. M. Durbin, L. E. Berman, B. W. Batterman, M. B. Brodsky and C. Hamaker: Phys. Rev. B **37**, 6672(1988)
74. G. Zajac, S. D. Bader and R. J. Friddle: Phys. Rev. B **31**, 4947(1985)
75. M. C. Hanf, C. Pirri, J. C. Peruchetti, D. Bolmont and G. Gewinner: Phys. Rev. B **39**, 1546(1989)
76. M. Wuttig, T. Flores and C. C. Knight: Phys. Rev. B **48**, 12082(1993)
77. M. Wuttig, Y. Gauthier and S. Blügel: Phys. Rev. Lett. **70**, 3619(1993)
78. W. L. O'Brien and B. P. Tonner: Phys. Rev. B **50**, 2963(1994)
79. J. Dresselhaus, D. Spanke, F. U. Hillebrecht, E. Kisker, G. van der Laan, J. B. Goedkoop and N. B. Brookes: Phys Rev. B **56**, 5461(1997)
80. Y. Tian, F. Jona and P. M. Marcur: Phys. Rev. B **59**, 12647(1999)
81. J. Pohl, E. U. Malang, B. Scheele, J. Köhler, M. Ch. Lux-Steiner and E. Bucher: J. Vac. Sci. Technol. B **12**, 3202(1994)
82. S. T. Purcell, M. T. Johnson, N. W. E. McGee, R. Coehoorn and W. Hoving: Phys. Rev. B **45**, 13064(1992)
83. S. K. Kim, Y. Tian, M. Montesano, F. Jona and P. M. Marcus: Phys. Rev. B **54**, 5081(1996)
84. R. Wu and A. J. Freeman: J. Magn. Magn. Mater. **161**, 89(1996)
85. B. T. Jonker, J. J. Krebs and G. A. Prinz: Phys. Rev. B **39**, 1399(1989)
86. D. Tian, S. C. Wu, F. Jona and P. M. Marcus: Solid State Commun. **70**, 199(1989)
87. B. Schirmer, B. Feldmann, A. Sokoll, Y. Gauthier and M. Wuttig: Phys. Rev. B **60**, 5895(1999)
88. A. S. Arrott, B. Heinrich, S. T. Purcell, J. F. Cochran and K. B. Urquhart: J. Appl. Phys. **61**, 3721(1987); B. Heinrich, C. Liu and A. S. Arrott: J. Vac. Sci. Technol. B **3**, 766(1985); B. Heinrich, A. S. Arrott, C. Liu and S. T. Purcell: J. Vac. Sci. Technol. A **5**, 1935(1987)
89. J. H. Van Vleck : Phys. Rev. **45**, 405(1934)
90. F. R. McFeely, S. P. Kowalczyk, L. Ley and D. A: Shirley: Solid State Commun. 15, 1051(1974)
91. P. S. Bagus, A. J. Freeman and F. Sasaki: Phys. Rev. Lett. **30**, 850(1973)
92. B. Hermsmeier, C. S. Fadley, M. O. Krause, J. Jimenez-Mier, P. Gerard and S. T. Manson: Phys. Rev. Lett. **61**, 2592(1988)
93. S. -J. Oh, Gey-Hong Gweon and Je-Geun Park: Phys. Rev. Lett. **68**, 2850(1992)
94. F. U. Hillebrecht, R. Jungblut and E. Kisker: Phys. Rev. Lett. **65**, 2450(1990)
95. G. E. Brodale, R. A. Fisher, N. E. Phillips and K. Matho: J. Magn. Magn. Mater. **54-57**, 194(1986)
96. D. Tian, A. M. Begley and F. Jona: Surf. Sci. Lett. **273**, L393(1992)
97. D. Tian, H. Li, S. C. Wu, F. Jona and P. M. Marcus: Phys. Rev. B **45**, 3749(1992)
98. M. T. Butterfield, M. D. Crapper and V. R. Dhanak: Phys. Rev. B **66**, 165414(2002)
99. M. C. Desjonquères and D. Spanjaard: *Concepts in Surface Physics*, 2nd edn (Springer 1996)
100. W. L. O'Brien and B. T. Tonner: J. Vac. Sci. Technol. A **13**, 1544(1995)

101. B. T. Thole and G. van der Laan: Phys. Rev. B **38**, 3158(1988)
102. For example, B. T. Thole, R. D. Cowan, G. A. Sawatzky, J. Fink and J. C. Fuggle: Phys. Rev. B **31**, 6856(1985)
103. M. Fonin, Yu. S. Dedkov, U. Rüdiger and G. Güntherodt: Surf. Sci. **529**, L275(2003)
104. X. Jin, M. Zhang, G. S. Dong, M. Xu, Y. Chen, Xun Wang, X. G. Zhu and X. L. Shen: Appl. Phys. Lett. **65**, 3078(1994)
105. X. Jin, Yong Chen, X. W. Lin, D. S. Dong, Yan Chen, M. Xu, W. R. Zhu, Xun. Wang, X. L. Shen and L. Li: Appl. Phys. Lett. **70**, 2455(1997)
106. X. Jin, Y. Chen. G. S. Dong, M. Zhang. M. Xu, X. G. Zhu, Xun Wang, E. D. Lu, H. B. Pan, P. S. Xu, X. Y. Zhang and C. Y. Fan: Phys. Rev. B **51**, 9702(1995)
107. R. D. Taylor, G. Cort and J. O. Willis, J. Appl. Phys. **53**, 8199(1982)
108. Chikashi Egawa, Tetsuya Aruga and Yasuhiri Iwasawa: Surf. Sci. Lett. **185**, L506(1987)
109. M. Maurer, J. C. Ousset, M. F. Ravet and M. Piecuch: Europhys. Lett. **9**, 803(1989)
110. C. Liu and S. D. Bader: phys. Rev. B **41**, 553(1990)
111. D. Tian, H. Li, F. Jona and P. M. Marcus: Solid State Commun. **80**, 783(1991)
112. M. De Santis, A. De Andres, D. Raoux, M. Maruer, M. F. Ravet and M. Piecuch: Phys. Rev. B **46**, 15465(1992)
113. F. Baudelet, A. Fontaine, G. Tourillon, D. Guay, M. Maurer, M. Piecuch, M. F. Ravet and V. Dupuis: Phys. Rev. B **47**, 2344(1993)
114. P. A. Lee, P. H. Citrin, P. Eisenberger and B. M. Kincaid: Rev. Mod. Phys. **53**, 770(1981)
115. J. J. Rehr and R. C. Albers: Rev. Mod. Phys. **72**, 621(2000)
116. M. C. Saint-Lager, D. Raoux, M. Brunel, M. Piecuh, E. Elkaim and J. P. Lauriat: Phys. Rev. B **51**, 2446(1995)
117. S. Andrieu, M. Piecuch and J. F. Bobo: phys. Rev. B **46**, 4909(1992)
118. M. Maurer, J. C. Ousset, M. Piecuch, M. F. Ravet and J. P. Sanchez: Mat. Res. Soc. Symp. Proc. **151**, 99(1989)
119. Tao Lin, M. A. Tomaz, M. M. Schwickert and G. R. Harp: Phys. Rev. B **58**, 862(1998)
120. F. Perjeru, M. M. Schwickert, Tao Lin, A. Anderson and G. R. Harp: Phys. Rev. B **61**, 4054(2000)
121. D. Spišák, R. Lorenz and J. Hafner: Phys. Rev. B 63, 94424(2001)
122. Ruqian Wu and A. J. Freeman: Phys. Rev. B **44**, 4449(1991)
123. H. Zenia, S. Bouarab and C. Demangeat: Surf. Sci. **521**, 49(2002)
124. H. L. Luo and P. Duwez, Can. J. Phys. **41**, 758(1963)
125. L. Gonzalez, R. Moranda, M. Salmeron, J. A. Verges and F. Yndurain: Phys. Rev. B **24**, 3245(1981)
126. M. T. Kief and W. F. Egelhoff, Jr.: Phys. Rev. B **47**, 10785(1993)
127. U. Ramsperger, A. Vaterlaus, P. Pfäffli, U. Maier and D. Pescia: Phys. Rev. B **53**, 8001(1996)
128. P. Poulopoulos, P. J. Jensen, A. Ney, J. Lindner and K. Baberschke: Phys. Rev. B **65**, 64431(2002)
129. D. Pescia, G. Zampieri, M. Stampanoni, G. L. Bona, R. F. Wills and F. Meier: Phys. Rev. Lett. **58**, 933(1987)
130. A. Clarke, G. Jennings, R. F. Willis, P. J. Rous and J. B. Pendry: Surf. Sci. **187**, 327(1987)

131. Hong Li and B. P. Tonner: Surf. Sci. **237**, 141(1990)
132. A. K. Schmid and J. Kirschner: Ultramicroscopy **42-44**, 483(1992)
133. J. Fassbender, R. Allenspach and U. Dürig: Surf. Sci. Lett. **383**, L742(1997)
134. F. Nouvertné, U.May, M. Bamming, A. Rampe, U. Korte and G. Güntherodt: Phys. Rev. B **60**, 14382(1999)
135. R. Miranda, D. Chandesris and J. Lecante: Surf. Sci. **130**, 269(1983)
136. O. Hechmann, H. Magnan, P. le Fevre, D. Chandesris and J. J. Rehr: Surf. Sci. **312**, 62(1994)
137. W. Weber, A. Bischof, R. Allenspach, C. H. Back, J. Fassbender, U. May, B. Schirmer, R. M. Jungblut, G. Güntherodt and B. Hillebrands: Phys. Rev. B **54**, 4075(1996)
138. J. A. C. Bland, D. Pescia and R. F. Willis: Phys. Rev. Lett. **58**, 1244(1987)
139. D. Kerkmann and D. Pescia and R. Allenspach: Phys. Rev. Lett **68**, 686(1992)
140. N. D. Mermin and H. Wagner: Phys. Rev. Lett. **17**, 1133(1966)
141. T. Beier, H. Jahrreiss, D. Pescia, Th. Woike and W. Gudat: Phys. Rev. Lett. **61**, 1875(1988)
142. C. M. Schneider, P. Bressler, P. Schuster, J. Kirschner, J. J. de Miguel and R. Miranda: Phys. Rev. Lett. **64**, 1059(1990)
143. P. Krams, F. Lauks, R. L. Stamps, B. Hillebrands, and G.Güntherodt: Phys. Rev. Lett. **69**,3674(1992)
144. F. O. Schumann, M. E. Buckley and J. A. C. Bland: Phys. Rev. B **50**, 16424(1994)
145. U. Bovensiepen, P. Poulopoulos, W. Platow, M. Farle and K. Baberschke: J. Magn. Magn. Mater. **192**, L386(1999)
146. F. Huang, M. T. Kief, G. J. Mankey and R. F. Willis: Phys. Rev. B **49**, 3962(1994)
147. Th. Fauster, G. Rangelov, J. Stober and B. Eisenhut: Phys. Rev. B **48**, 11361(1993)
148. A. Rabe, N. Memmel, A. Steltenpohl and Th. Fauster: Phys. Rev. Lett. **73**, 2728(1994)
149. J. de la Figuera, J. E. Prieto, C. Ocal and R. Miranda: Phys. Phys. B **47**, 13043(1993)
150. M. T. Butterfield, M. D. Crapper, T. C. Q. Noakes, P. Bailey, G. J. Jackson and D. P. Woodruff: Phys. Rev. B **62**, 16984(2000)
151. Ch. Rath, J. E. Prieto, S. Müller, R. Miranda and K. Heinz: Phys. Rev. B **55**, 10791(1997)
152. P. Le Fevre, H. Magnan, O. Heckmann, V. Briois and D. Chandesris: Phys. Rev. B **52**, 11462(1995)
153. B. P. Tonner, Z. -L. Han and J. Zhang: Phys. Rev. B **47**, 9723(1993)
154. J. E. Prieto, Ch. Rath, S. Müller, L. Hammer, K. Heinz and R. Miranda: Phys. Rev. B **62**, 5144(2000)
155. J. Camarero, J. J. de Miguel, R. Miranda, V. Raposo and A. Herando: Phys. Rev. B **64**, 125406(2001)
156. G. R. Harp, R. F. C. Farrow, D. Weller, T. A. Rabedeau and R. F. Marks: Phys. Rev. B **48**, 17538(1993)
157. J. Fassbender et al.: Phys. Rev. B **57**, 5870(1998)
158. G. A. Prinz: Phys. Rev. Lett. **54**, 1051(1985)
159. Y. U. Idzerda, W. T. Elam, B. T. Jonker and G. A. Prinz: Phys. Rev. Lett. **62**, 2480(1989)

160. Hong Li and B. P. Tonner: Phys. Rev. B **40**, 10241(1989)
161. J. Zhang, Z. -L. Han, S. Varma and B. P. Tonner, Surf. Sci. **298**, 351(1993)
162. Ph. Houdy, P. Boher, F. Giron, F. Pierre, C. Chappert, P. Beauvillian, K. Le Dang and E. Velu: J. Appl. Phys. **69**, 5667(1991)
163. S. K. Kim, C. Petersen, F. Jona and P. M. Marcus: Phys. Rev. B **54**, 2184(1996)
164. L. Duò, R. Bertacco, G. Isella and F. Ciccaci: Phys. Rev. B **61**, 15294(2000)
165. J. Thomassen, F. May, B. Feldmann, M. Wuttig and H. Ibach: Phys. Rev. Lett. **69**, 3831(1992)
166. S. Valeri and A. Di Bona: Surf. Rev. Lett. **4**, 141(1997)
167. G. C. Gazzadi and S. Valeri: Europhys. Lett. **45**, 501(1999)
168. C. P. Wang, S. C. Wu, F. Jona and P. M. Marcus: Phys. Rev. B **49**, 17385(1994); ibid **49**, 17391(1994)
169. S. K. Kim, F. Jona and P. M. Marcus: Phys. Rev. B **51**, 5412(1995)
170. P. M. Marcus and F. Jona: Phys. Rev. B **51**, 17081(1995)
171. E. Bauer: J. Phys.: Condens. Matter **11**, 9365(1999)
172. Th. Duden, R. Zdyb, M. S. Altman and E. Bauer: Surf. Sci. **480**, 145(2001)
173. H. Fritzsche, J. Kohlhepp, and U. Gradmann: Phys. Rev. B **51**, 15933(1995)
174. H. Knoppe and E. Bauer: Phys. Rev. B **48**, 1794(1993)
175. T. Duden and E. Bauer: Phsy. Rev. Lett. **77**, 2308(1996)
176. J. Guan, R. A. Campbell and T. E. Madey: Surf. Sci. **341**, 311(1995)
177. K. L. Man, R. Zdyb, S. F. Huang, T. C. Leung, C. T. Chan, E. Bauer and M. S. Altman: Phys. Rev. B **67**, 184402(2003)
178. N. Metoki, W. Donner and H. Zabel: Phys. Rev. B **40**, 17351(1994)
179. S. Fölsch, A. Helms, A. Steidinger and K. H. Rieder: Phys. Rev. B **57**, R4293(1998)
180. S. Müller, P. Bayer, C. Reischl, K. Heinz, B. Feldmann, H. Zillgen and M. Wuttig: Phys. Rev. Lett **74**, 765(1995)
181. Z. Q. Wang, Y. S. Li, F. Jona and P. M. Marcus: Solid State Commun. **61**, 623(1987)
182. B. Heinrich, et al.: J. Cryst. Growth **81**, 562(1987); B. Heinrich et al.: J. Vac. Sci. Technol. **A4**, 1376(1986)
183. B. Heinrich, S. T. Purcell, J. R. Dutcher, K. B. Urquhart, J. F. Cochran and A. S. Arrott: Phys. Rev. B **38**, 12879(1988)
184. Y. Kamada, M. Matsui and T. Asada: J. Phys. Soc. Jpn. **66**, 466(1997); Y. Kamada and M. Matsui: J. Phys. Soc. Jpn **66**, 658(1997)
185. A. V. Mijiritskii, P. J. M. Smulders, V. Ya. Chumanov, O. C. Rogojanu, M. A. James and D. O. Boerma: Phys. Rev. B **58**, 8960(1998)
186. J. A. C. Bland, R. D. Bateson, A. D. Johnsona, B. Heinrichb, Z. Celinskib and H. J. Lauterc: J. Magn. Magn. Mater. **93**, 331(1991)
187. N. B. Brookes, A. Clarke and P. D. Johnson: Phys. Rev. B **46**, 237(1992)
188. J. Vogel and M. Sacchi: Phys. Rev. B **53**, 3409(1996)
189. T. Lin, M. M. Schwickert, M. A. Tomaz, H. Chen and G. R. Harp: Phys. Rev. B **59**, 13911(1999)
190. C. P. Flynn: Phys. Rev. Lett. **57**, 599(1986)
191. D. M. Wood and Alex Zunger: Phys. Rev. Lett. **61**, 1501(1988); Phys. Rev. B **40**, 4062(1989)
192. J. L. Stevens and R. Q. Hwang: Phys. Rev. Lett. **74**, 2078(1995)
193. J. M. Sanchez, F. Ducastelle and D. Gratias: Physica **A 128**, 334(1984)

194. M. Asta, C. Wolverton, D. de Fontaine and H. Dreyssé: Phys. Rev B **44**, 4907(1991); C. Wolverton, M. Asta, D. Gratias and D. de Fontaine: Phys. Rev. B **44**, 4914(1991)
195. J. W. Connolly and A. R. Williams: Phys. Rev. B **27**, 5169(1983)
196. T. H. Kim and J. S. Moodera: Phys. Rev. B **66**, 104436(2002)
197. M. Seider, R. Kaltofen, U. Muschiol, M. -T. Lin and C. M. Schneider: J. Appl. Phys. **87**, 5762(2000)
198. A. Dittschar, M. Zharnikov, W. Kuch, M. -T. Lin, C. M. Schneider and J. Kirschner: Phys. Rev. B **57**, R3209(1998)
199. J. C. A. Huang, Y. M. Hu, C. C. Yu, C. H. Tsao and C. H. Lee: Phys. Rev. B **57**, 11517(1998)
200. H. Okamoto and T. B. Massalski: In: *Binary alloy phase diagram*, ed by T. B. Massalski (American Society for Metals, Metal park, OH, 1986)
201. E. C. Elwood and K. Q. Bagley: J. Inst. Met. **80**, 617(1952)
202. Y. Fukano: J. Phys. Soc. Japan. **16**, 1195(1961)
203. G. Abadias, I. Schuster, A. Marty and B. Gilles: Phys. Rev. B **61**, 6495(2000) and references therein
204. J. C. Slater: J. Appl. Phys. **8**, 385(1937); L. Pauling: Phys. Rev. **54**, 899(1938)
205. F. O. Schumann, R. F. Willis, K. G. Goodman and J. G. Tobin: Phys. Rev. Lett. **79**, 5166(1997)
206. F. O. Schumann, S. Z. Wu, G. J. Mankey and R. F. Willis: Phys. Rev. B **56**, 2668(1997)
207. M. G. Martin, E. Foy, F. Chevrier, G. Krill and M. C. Asensio: Surf. Sci. **433-435**, 88(1999)
208. E. Foy, S. Andrieu, M. Finazzi, R. Poinsot, C. M. Teodorescu, F. Chevrier and G. Krill: Phys. Rev. B **68**, 94414(2003)
209. A. Berger, B. Feldmann, H. Zillgen and M. Wuttig: J. Magn. Magn. Mater. **183**, 35(1998)
210. E. F. Wassermann: J. Magn. Magn. Mater. **100**, 346(1991)
211. J. W. Freeland, I. L. Grigorov and J. C. Walker: Phys. Rev. B **57**, 80(1998)

6 Correlation Between Magnetism, Structure and Growth for Ultrathin Fe Films on Cu(100)

6.1 Introduction

To obtain a comprehensive understanding of the correlation between the magnetism and structure of materials has been a research goal of scientists for a long time. Early in the 1930s, Bethe and Slater found a phenomenological relationship between the direct exchange interaction and the atomic separation [1, 2] (Fig. 6.1). This is the well known Bethe–Slater curve. Recently, the correlation between the structure and the magnetism for 3d metals has been investigated with advanced self-consistent energy-band calculations [4, 5, 6]. These calculations show that the magnetic properties of a 3d metal are closely related to the atomic volume. Normally magnetic transition metals will lose their magnetic moment at a compressed volume and, on the other hand, nonmagnetic transition metals will become ferromagnetic at an expanded volume. Thus a transition from the nonmagnetic to a magnetic state is expected for all 3d transition metals when the atomic volume is increased. In the limit of large volume, the magnetic moment approaches the value determined by Hund's rule for the free-atom configuration. Three types of transition behavior, classified as type I, type II and type III depending on the number (one, two and three, respectively) of critical points, have been predicted (see Fig. 6.2). Bcc Sc, Ti, Fe, Co and Ni have been predicted to show a type I tran-

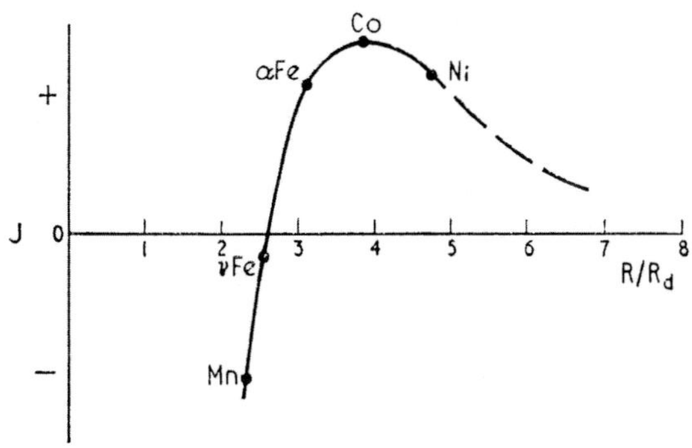

Fig. 6.1. The Bethe–Slater curve relates the exchange constant J to the ratio of the atomic separation R to the radius of the d shell R_d. From [3], used with permission

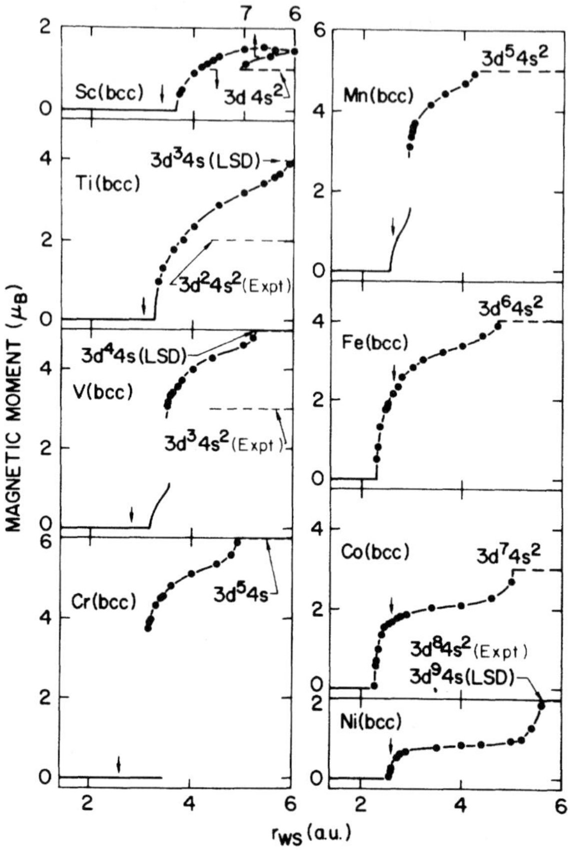

Fig. 6.2. Calculated magnetic moments for 3d transition elements with bcc structure. The Hund's-rule atomic limit is shown as a *horizonal broken line*. From [6], used with permission

sition in which the termination of the nonmagnetic behavior is followed by a ferromagnetic phase. This is a second-order transition, since the magnetic moment changes continuously. If the calculations take only the nonmagnetic and ferromagnetic states into account, bcc Cr is found to undergo a first-order type II transition, which is characterized by a discontinuous jump in magnetic moment at the transition from the nonmagnetic to the ferromagnetic state. Bcc V and Mn, which have a type III transition, exhibit a second-order transition from the nonmagnetic state to a low-spin state, followed by a first-order transition from a low-spin to a high-spin state at larger atomic volume. The type of transition is determined by the details of the nonmagnetic density of states and the location of the Fermi level. In general, if the Fermi level falls in a deep minimum of the nonmagnetic DOS, the system will undergo a first-order transition. On the other hand, if the Fermi level is located at a peak in the nonmagnetic DOS, the system will undergo a second-order transition. Considering the diversity in the structure of the DOS, the transition could be more complicated than that shown in Fig. 6.2. A typical example is provided by fcc iron.

The dependence of the magnetic moment on the lattice spacing in both bcc and fcc iron has been calculated by various groups (e.g. [5]). The resulting relationship is graphed in Fig. 6.3. On the left side, the results for bcc iron are shown. The calculations reproduce the ferromagnetic ground state, with a moment of $2.2\mu_B$ and a lattice spacing of 2.8 Å. Experimentally, a

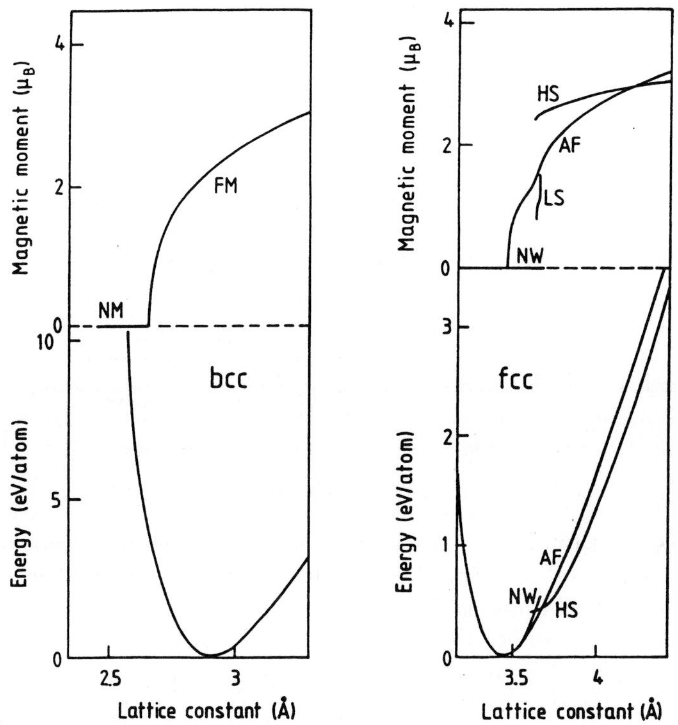

Fig. 6.3. Magnetic moment and total energy vs. lattice spacing for bcc and fcc Fe. The total energy (relative to the minimum energy) and the magnetic moment for fcc and bcc iron are shown as a function of the lattice spacing. For fcc Fe, calculations excluding antiferromagnetic (AF) behavior yield nonmagnetic (NM), lowspin (LS) and high-spin (HS) solutions. With increasing atomic volume, the system undergoes two successive first-order transitions from NM to LS and then to HS. When AF order is included in the calculations, the AF solution has the lowest energy from the equilibrium volume to $a = 3.66$ Å. This volume range covers the range where the LS state exists. Therefore the LS state should not appear as a minimum-energy state for any lattice constant. At $a = 3.66$ Å, a first-order transition from the AF to the HS state occurs. After [4, 5], used with permission

moment of $2.12\mu_B$ and a lattice spacing of 2.86 Å are found. With increasing lattice spacing, the calculations predict a monotonic increase of the magnetic moment.

In the case of fcc iron, the predicted behavior is significantly more complicated. Figure 6.3 shows that nonmagnetic (NM), ferromagnetic (FM) and antiferromagnetic (AF) states are found. For a lattice spacing smaller than 3.44 Å, the nonmagnetic solution is the most stable one. With increasing lattice spacing, the antiferromagnetic solution becomes energetically more favorable. Since for lattice spacings around 3.5 Å the nonmagnetic and antiferromagnetic solutions are almost degenerate, it is not completely clear from such calculations whether the ground state of fcc Fe should be nonmagnetic or antiferromagnetic. This ambiguity has been resolved experimentally by measurements performed on small γ-Fe precipitates in Cu [4]. These precipitates were found to be antiferromagnetic below 68 K. The ferromagnetic solution finally becomes energetically preferred upon further increase of the lattice spacing ($a > 3.66$ Å). This phase exists in a low-spin state (LS) with a moment of $1.1\mu_B$ and a high-spin state (HS) with a moment of $2.5\mu_B$ (for $a > 3.66$ Å). The low-spin state is nearly degenerate with the nonmagnetic solution and is therefore not shown in Fig. 6.3. At a lattice constant of 3.66 Å, a first-order phase transition occurs from the antiferromagnetic state with a moment around $2.5\mu_B$. This discontinuous change of magnetic moment with increasing lattice spacing (atomic volume) is an example of a *moment–volume instability*. It is now believed that moment–volume instabilities are crucial for an understanding of Invar alloys, for example [7]. Such alloys show anomalies in the temperature dependence of their thermal expansion, lattice constant, elastic properties, susceptibility, etc.

To confirm or refute these theoretical predictions, it is necessary to produce fcc iron with different and, preferably, enlarged lattice spacings at low temperature and to determine the corresponding magnetic properties. Unfortunately, such a simple-minded approach cannot easily be realized, since it is experimentally impossible to expand a crystal lattice. Epitaxial growth offers a more plausible means to stabilize expanded fcc iron. In the previous chapters, it was shown that for heteroepitaxial systems with small lattice mismatch, it is possible to grow strained films that adopt the lattice constant of the substrate in the contact plane. A suitable substrate for the growth of expanded fcc iron should be copper. Cu has a lattice spacing of 3.61 Å at 300 K, which is only slightly larger than the extrapolated lattice spacing of fcc iron at room temperature. The resulting small misfit of -0.7% is favorable for the growth of strained, dislocation-free films, a fact that was demonstrated by electron microscopy almost 20 years ago. More importantly, the lattice spacing of Cu is located in the range in which the theoretically predicted energies of the nonmagnetic, antiferromagnetic and ferromagnetic states are very close to each other, so that a subtle interplay between growth, structure and magnetism might be expected. In fact, in our work we have experimentally observed and described a complex correlation between growth, structure and magnetic properties for ultrathin Fe/Cu(100) films. We shall present these results in Sect. 6.2, including growth studies, determinations of the structure and the characterization of the magnetic properties. The complex correlation will be discussed at the end of the section. An extensive discussion involving related topics will be given in Sect. 6.3.

6.2 Ultrathin Fe Films on Cu(100)

6.2.1 Early Work on Iron Films on Cu(100)

Early investigations of the magnetic properties of iron films on Cu(100) concentrated on the characterization of the magnetic ground state and the determination of the magnetic anisotropy. Little attention was given to measurement of the magnetic moment for fcc Fe [8], and a direct correlation between the structure and magnetism of the films had not yet been established [9, 10]. Moreover, even the results that describe the magnetic ground state and the magnetic anisotropy showed surprising discrepancies. Table 6.1 summarizes some early experimental observations. Since a pronounced dependence of the magnetic properties upon film thickness and growth temperature was found, these parameters are also specified. While some researchers found that the ground state of the fcc iron films was antiferromagnetic [11, 12], others claimed that it was ferromagnetic [13, 14, 15, 16, 17] or even found "the absence of ferromagnetism in the usual sense" [9]. Similarly contradictory findings were also reported for the magnetic anisotropy. There is general consensus, however, that for ultrathin iron films the magnetization is initially aligned perpendicular to the surface. Above a certain critical thickness, the magnetic anisotropy changes from perpendicular to in-plane. However, the exact thickness at which the magnetization axis is found to reorient is not without its fair share of confusion: it varies from 6 [15, 16] to 17 ML [9].

Liu et al. were the first to measure the magnetic anisotropy as a function of film thickness and growth temperature [16]. Their results indicate that iron films grown at room temperature are ferromagnetic and have a critical thickness for the rotation of the magnetization axis of 6 ML [16]. On the other hand, Stampanoni [17] observed a critical thickness for room-temperature-grown films of 14 ML. Finally, Pescia et al. [14] and Pappas et al. [15] noted a temperature-dependent rotation of the magnetization from perpendicular at low temperature to in-plane at higher temperatures. The transition temperature, however, differs substantially between these two studies.

To some extent as a consequence of the controversial magnetic properties found, the structure and growth of iron films on Cu(100) have also received considerable interest. It was initially believed that the iron films grew with a strained face-centered cubic structure that adopted the in-plane spacing of the Cu(100) substrate. It was only above 20 Å that the formation of bcc iron

Table 6.1. Summary of various experimental investigations of the magnetic properties of Fe on Cu(100). ⊥ and ‖ denote a magnetization perpendicular or parallel, respectively, to the surface

Reference	Measuring technique	Film thickness, growth temperature	Magnetic ground state (Curie or Néel temperature)	Magnetic anisotropy
[11]	Mössbauer spectroscopy	4 ML, 300 K	Antiferromagnetic (40 K)	–
[12]	Mössbauer spectroscopy	? ML, 420 K	Antiferromagnetic (65 K)	–
[13]	Spin-polarized photoemission, SMOKE	Up to 20 ML, 420–460 K	Ferromagnetic (>420 K)	‖
[14]	Spin-polarized photoemission	1, 3, 5 ML, 300 K	Ferromagnetic (300 K)	1 ML, ‖ 3, 5 ML, ⊥ (for $T < 30$ K)
[15]	Spin-polarized secondary-electron emission	5–7ML, 125 K (annealed at 300 K)	Ferromagnetic (∼350 K)	5 ML, ‖ ($T < 300$ K) 6 ML, ‖($T < 150$ K) 7 ML, ‖
[16]	SMOKE	1–8 ML, 110-350 K	Ferromagnetic	$d \leq 6$ ML, ⊥ $d > 6$ ML, ‖
[17]	Spin-polarized photoemission	Up to 20 ML 300 K	Ferromagnetic ($d < 14$ ML, ∼300 K; $d > 14$ ML, >500 K)	$d < 14$ ML, ⊥ $d > 14$ ML, ‖
[9]	Brillouin scattering	Up to 20 ML 300 K	$d < 6$ ML and $d >17$ ML, ferromagnetic (350 K); 6 ML< $d < 17$ ML, paramagnetic	$d < 6$ ML, ⊥ $d > 17$ ML, ‖

was observed by transmission electron microscopy [18]. The pioneering investigations were later corroborated by low-energy electron diffraction studies [10, 19, 20]. The latter showed that up to a certain critical thickness, fcc iron with the Cu(100) in-plane lattice spacing did indeed grow. Above this thickness, the (1×1) diffraction pattern observed became considerably more diffuse. The coverage at which the diffraction pattern changed from sharp to diffuse was found by different authors to be 5, 10 and 21 ML [14, 20, 21]. More recent studies have shown that the structure of iron films is considerably more complex than initially thought. The first hints came from a full dynamical LEED analysis [10]. These calculations could not satisfactorily reproduce the beam intensities for 1 and 3 ML thick Fe films if it was simply assumed that the iron structure closely resembled the underlying atomic arrangement of the Cu(100) substrate. The authors of [10] discussed these findings in terms of an incomplete wetting of Cu(100) by the iron films. Such an explanation is difficult to accept, since electrons scatter similarly at iron and copper atoms and hence these elements have almost identical phase shifts. Therefore, the I/V curves are not sensitive to the detailed morphology, i.e., they cannot distinguish between three-dimensional and two-dimensional growth. Thus, the inability to reproduce the measured I/V curves by theoretical calculations is indicative of a structural rearrangement of the films rather than incomplete wetting. Support for this statement comes from a recent surface extended x-ray-absorption fine structure (SEXAFS) study [22], which demonstrated that for coverages up to 4 ML, iron films grown at room temperature have an fcc-like structure with considerable vertical disorder. Perpendicular disorder was clearly evident from the broad radial distribution function measured. For films thicker than 4 ML, an indisputably sharper nearest-neighbor distance distribution was found. In these cases, both the atomic spacings and the observed coordination were characteristic of undistorted fcc iron [22]. Such a structural rearrangement is also in accordance with diffraction studies [23, 24, 25]. At low coverages, (4×1), (5×1) and (2×1) phases are observed with increasing film thickness. These superstructures are suggestive of atomic rearrangements. Even though there is general agreement about the sequence in which these superstructures appear, the absolute thicknesses are still a matter of debate. Daum et al. [23], for example, observed the $p2mg(2\times1)$ superstructure at coverages between 2 and 3 ML, while Glatzel et al. [26] found this structure at around 6 ML. Xhonneux and Courtens claim that the $p2mg(2\times1)$ phase exists between 6 and 17 ML [9]. The precise atomic positions for these superstructures have not been determined either.

At coverages above 10 ML, a (2×1) superstructure is found [25, 27]. No structural model has previously been proposed to explain this phase. There is general speculation, however, that with increasing film thickness, a structural transformation into bcc iron should occur. Such a behavior has been suggested in several studies, in which structural transitions at thicknesses varying from 5 to 21 ML were observed [14, 20, 21]. In none of these studies, however, was an analysis performed to determine the film structure after the transformation had occurred. The clearest evidence for the formation of bcc iron comes from a transmission electron microscopy study [18]. This investigation reveals evidence for α- (bcc) iron above a film thickness of 20 Å and suggests the formation of (110) iron in a Nishiyama–Wassermann orientation [28] on fcc (111) facets or the growth of (110) iron on (100) or (110) facets. The latter is known as the Pitsch orientation and is only observed for thin films [29]. XPS forward-scattering results also find evidence for the formation of bcc iron but with the [100] direction almost parallel to the surface normal [30].

The actual growth mode of the iron films is not without controversy either. In particular, the growth mode in the initial stage remains unclear. Earlier studies have claimed layer-by-layer growth, from an observation of linear segments and breakpoints in Auger uptake curves [14, 23]. Recently, a combined Auger and Rutherford backscattering (RBS) analysis concluded, however, that the first two breakpoints are due to the completion of bilayers [26]. X-ray photoelectron diffraction and reflection high-energy electron diffraction [31] provide clear evidence for iron agglomeration even at room temperature. Scanning tunneling microscopy studies confirm that there is considerable deviation from layer-by-layer growth for the first two monolayers [32, 33, 34]. At higher coverages, however, an almost perfect layer-by-layer growth mode is encountered.

The inconsistencies presented above are in part due to wrong film thickness calibrations. Additionally, it has already been pointed out several times [31, 35] that the different properties of the iron films can be related to the growth temperatures chosen. Above 370 K, for example, copper is known to diffuse to the film surface [31, 36]. Nevertheless, there is evidence that not all of the reported discrepancies were caused by different thickness determinations or growth temperatures. Steigerwald and Egelhoff [37] and Xhonneux and Courtens [9] have demonstrated the importance of the growth conditions such as the background pressure during deposition and the preparation of the substrate surface.

Under these circumstances, it is rather difficult, if not impossible, to combine the various experimental results in order to produce a clear and consistent understanding of the magnetism, growth and structure of iron films. Thus the aim of our work was to investigate these properties for iron films deposited on Cu(100) in one chamber. Care was taken to ensure that a precise thickness calibration was achieved through the combination of three different methods, namely the observation of RHEED oscillations, the measurement of Auger uptake curves and a calibration of deposition rates using a quartz crystal microbalance. To relate the magnetic properties to films with a definite structure and morphology, a complete structural analysis of the atomic positions and the film morphology was performed. Most of the measurements were carried out for films grown at room temperature but other temperatures were tested as well. In the following subsections (Sects. 6.2.2–6.2.5), we deal exclusively with films prepared at room temperature (300 ± 2 K). The properties of iron deposited at other temperatures are discussed only briefly in Sect. 6.2.6. In Sect. 6.2.2, a short outline of the magnetic properties that we found will be given.

6.2.2 Magnetic Properties of Fe Films on Cu(100)

The magnetic properties of the iron films were investigated using the magnetooptic Kerr effect. In these studies, hysteresis loops were measured in both the longitudinal and the polar geometry. By employing a modulation technique, the Kerr ellipticity ε_K and the Kerr rotation θ_K were determined. Figure 6.4 shows typical longitudinal and polar Kerr ellipticity curves for a 10 ML thick film at 170 K. While no hysteresis loop is obtained in the longitudinal geometry, a perfectly square loop is obtained in the polar measurement. This clearly reflects the perpendicular magnetic anisotropy of the film.

Such hysteresis loops were determined for a large number of films with different thicknesses. In each case, ellipticities and Kerr rotations were measured between 110 K and the Curie temperature T_c of the films. Figure 6.5

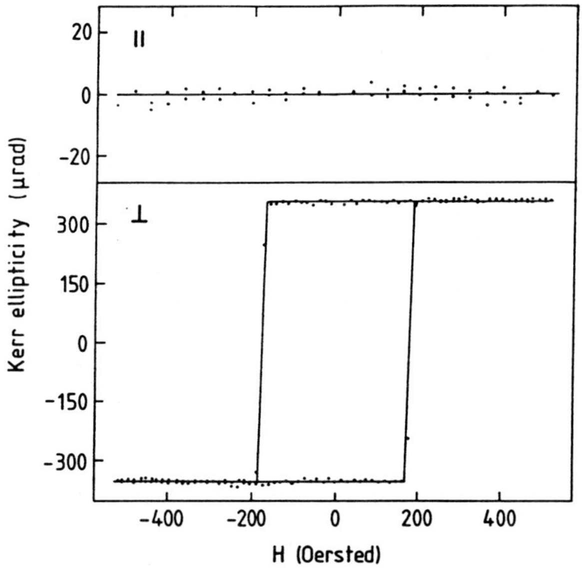

Fig. 6.4. Kerr ellipticity curves for a 10 ML thick film. The curves were recorded at 170 K in the longitudinal (*upper curve*) and the polar (*lower curve*) geometry. The angles of incidence for the longitudinal and polar measurements were 65° and 25°, respectively, from the surface normal. The azimuthal direction of incidence was chosen along the [110] direction

presents such a sequence of hysteresis loops, obtained for a 7.5 ML thick iron film at different temperatures. With increasing temperature, the magnitude of the Kerr ellipticity at saturation ε_K^S decreases. From these curves, ε_K^S was extrapolated to $T = 0$ K. The resulting values of ε_K^S as a function of film thickness are plotted in Fig. 6.6. The broken line separates the region where the films show perpendicular magnetic anisotropy from the thickness region where parallel anisotropy is found. The values were obtained by extrapolating the ellipticities at saturation to $\varepsilon_K^S = 0$. This allows the determination of the Curie temperature to within ±3 K.

Three different regimes of magnetic behavior can be distinguished in Fig. 6.6. Up to about 4 ML, a perpendicular anisotropy is found, with a Kerr ellipticity that increases almost linearly with film thickness. Around 4 ML, the Kerr ellipticity drops sharply and remains constant between 5 ML and approximately 10 ML (region II). At about 11 ML, the magnetization switches from perpendicular to in-plane. In this region, the Kerr ellipticity again increases linearly with film thickness. These changes are also reflected in the behavior of the Curie temperature with increasing thickness. The Curie temperature has a maximum of 370 K around 3 ML. In region II, it is constant and fairly low but rises above 500 K in region III. Here it was impossible to determine more than a lower limit for the Curie temperature, since above

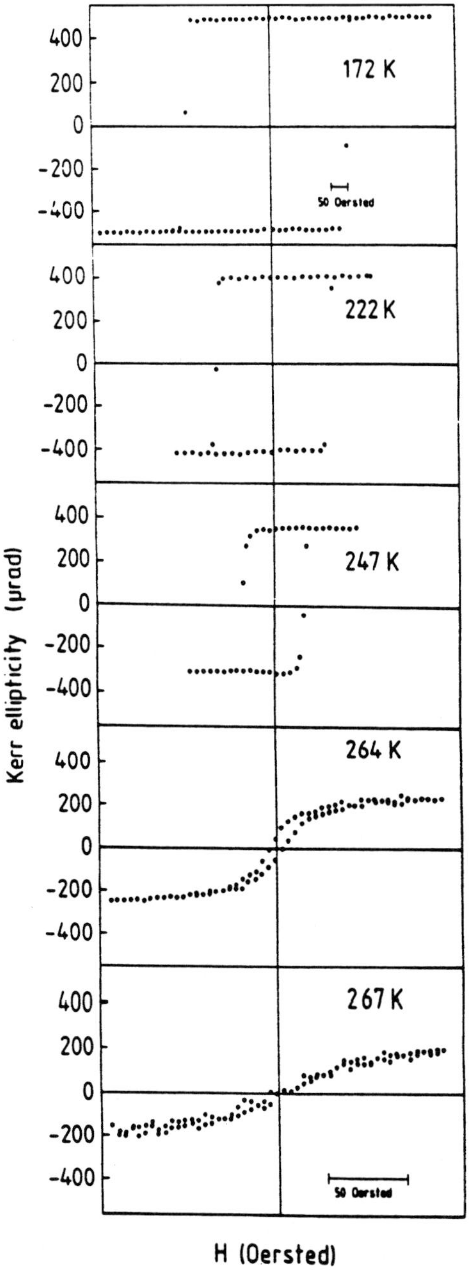

Fig. 6.5. Polar Kerr ellipticity curves for a 7.5 ML thick film. The curves were recorded at different temperatures between 172 and 267 K. The Curie temperature of the film is 273 ± 3 K. The scale of the x axis is enlarged by a factor of 5 for the measurement at 172 K

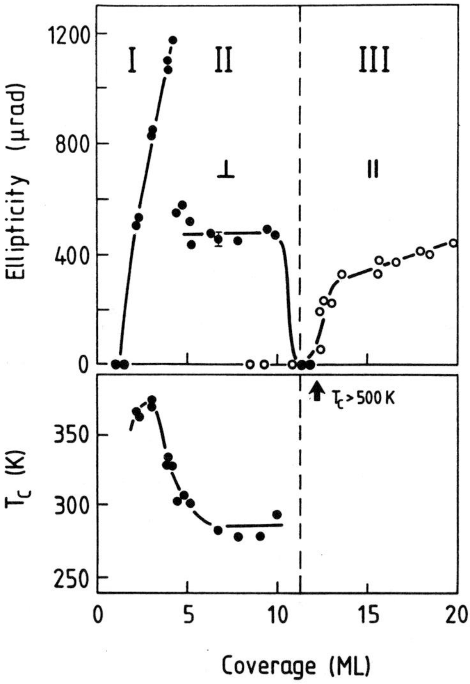

Fig. 6.6. Thickness dependence of Kerr ellipticity and Curie temperature. *Upper curve*: thickness dependence of the Kerr ellipticity at saturation extrapolated to $T = 0$ K. The *solid* and *open circles* denote the ellipticities obtained from hysteresis loops recorded in the polar and the longitudinal geometry, respectively. The error bar is indicated in the figure. The longitudinal ellipticities are smaller than the polar ellipticities owing to the reduced sensitivity of the longitudinal Kerr effect. For the chosen angles of incidence, a value of 7–8 is calculated for the ratio between the magnitudes of the ellipticities. Three regions with different magnetic anisotropy can be distinguished. *Lower curve*: thickness dependence of the Curie temperature T_c. Around 11 ML, the Curie temperature jumps to values above 500 K. The experimental error lies within the size of the symbols

500 K, Cu diffusion to the surface takes place [25]. This irreversibly changes the magnetic properties of the films. One can therefore only state that the Curie temperature is higher than 500 K.

Since the Kerr ellipticity of a homogeneously magnetized film increases linearly with increasing film thickness, further conclusions can be drawn from Fig. 6.6. For example, it can be concluded that in regions I and III the entire film is ferromagnetic. The different magnitudes of the Kerr ellipticity in regions I and III are due to the higher sensitivity of the Kerr effect in the polar geometry. For the chosen angles of incidence, a value of around 7–8 is calculated for the ratio of the magnitudes of the Kerr ellipticities. Taking this into account, one can deduce that the magnetization per iron layer is fairly similar in regions I and III. The slopes of the ellipticities with respect to film thickness are consistent with theoretical calculations of the Kerr ellipticity for bcc Fe. Since the magnetooptic constant for fcc iron is not known, it is not possible to say a priori if the ellipticity curves in regions I and III are also consistent with data expected for fcc iron.

On the other hand, the relatively small and constant Kerr ellipticity found between 5 and 11 ML (region II) indicates that only *a small and constant number of iron layers is ferromagnetic*, while the bulk of the film does not contribute to the ferromagnetic signal. The magnetism must therefore be localized either at the Fe–Cu interface or at the film surface. This is consistent with the decrease of the Curie temperature at the transition from region I to region II. Additionally, it was found that a CO exposure of 0.5 L lowers the Curie temperature by 40 K. Since it is unlikely that CO adsorption will influence the potential magnetism at the interface, this strongly suggests that the magnetism is localized at or near the surface of the film. For the temperature investigated ($T > 110$ K), one can exclude an antiferromagnetic coupling between successive iron layers. This would lead to an oscillation of the Kerr ellipticity with film thickness, as the net magnetic moments of films with even and odd numbers of layers would be different. Thus the bulk of the film is either paramagnetic, or antiferromagnetic with a Néel temperature below 110 K. In the following, attempts will be made to correlate the different regimes of magnetic behavior with the growth and structure of the iron films. More specifically, the goal is to understand the surprising magnetic properties in region II and the transition of the magnetic behavior from region I to II. Furthermore, we shall focus on the switching of the magnetic anisotropy from region II to region III. In the following subsections, the growth and structure of the iron films will be presented.

6.2.3 Growth of Fe Films on Cu(100)

To follow the growth of Fe films, the MEED (medium-energy electron diffraction) intensity was measured during deposition. The intensity changes of a MEED beam reflect changes in film morphology. In particular, layer-by-layer growth is discernible by regular intensity oscillations, as we have discussed in Chap. 2. However, structural changes can influence the intensity of the MEED beam as well. For iron on Cu(100), this is a considerable problem, since previous work has already shown the appearance of several superstructures with increasing film thickness [25, 26, 28]. To partly overcome this difficulty, several different MEED beams were measured simultaneously. While changes in step density (morphology) should have a similar effect upon all beams, a change in film structure should in principle affect different beams differently. Therefore, in our initial analysis, only features common to all beams were attributed to changes in morphology. Figure 6.7 and 6.8 show the MEED intensities during two different growth experiments. The curves were measured using an electron energy of 3 keV and an angle of incidence of $81.1 \pm 0.2°$ from the surface

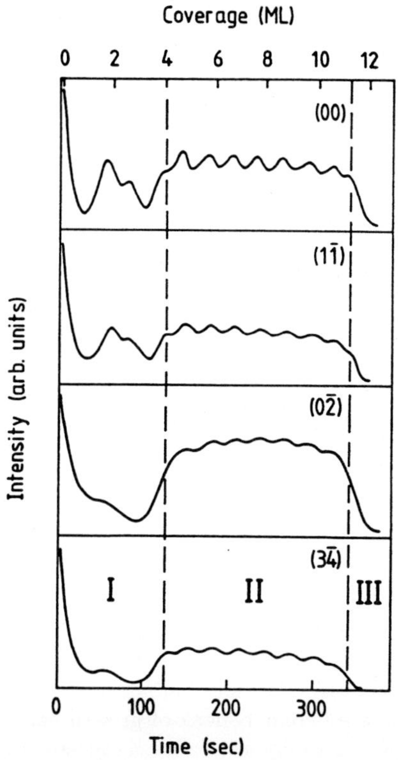

Fig. 6.7. MEED curves for four different beams during Fe deposition. The beam intensities are depicted as a function of deposition time (*lower scale*) and coverage (*upper scale*). The curves were recorded for an electron energy of 3 keV, an angle of incidence of 81.1° from the surface normal (destructive interference) and an azimuthal angle of 4° from the [100] direction. The three different regions are marked

normal. This corresponds to a destructive interference condition of adjacent terraces for the (00) beam (out-of-phase condition). The azimuthal angle of incidence was 4° and 6°, respectively, from the [100] direction.

In Fig. 6.7, the MEED intensities of four different beams are reproduced. Three regions with different growth behavior can be distinguished. The first region, up to 4 ML, is characterized by a rapid decrease in intensity and a deep minimum at 1 ML. At 2 ML, the intensities of all beams show a more or less pronounced maximum. Several beams show a further maximum at 3 ML. A drastic increase of intensity is observed for all beams around 4 ML, connected with a considerable sharpening of the MEED spots. In region II (4 to 11 ML), regular intensity oscillations with a high average intensity are observed. This is a fingerprint of layer-by-layer growth. From the small oscillation amplitude, it can be concluded that rather large two-dimensional islands

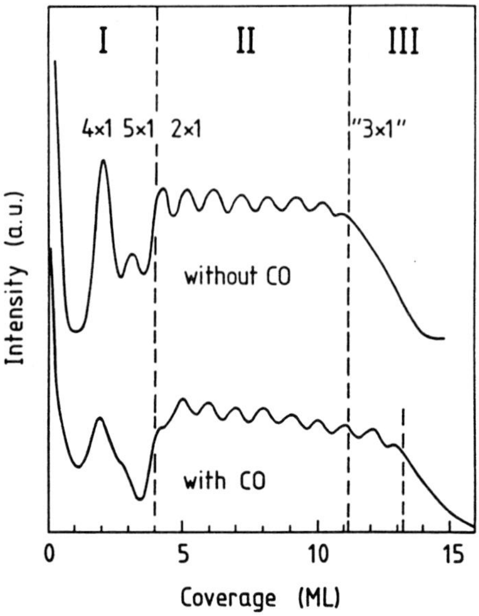

Fig. 6.8. Influence of CO codeposition on Fe film growth. MEED curves of the specular beam for growth under UHV conditions ($p_{TOT} = 1 \times 10^{-8}$ Pa, *upper curve*) and for a CO partial pressure of 7×10^{-8} Pa (*lower curve*). The curves were recorded with an electron energy of 3 keV and at an angle of incidence of 8.9° from the sample surface. Three different regions of growth can be distinguished. With increasing film thickness, four different superstructures evolve

grow on the film surface. Above 11 ML (region III), the intensity decreases exponentially, indicating the growth of three-dimensional islands. Connected with the decreases of intensity, streaking of the diffraction spots in the [011] and $[0\bar{1}1]$ directions is observed in the MEED pattern.

Figure 6.8 shows the influence of CO codeposition on the film growth. The MEED oscillations for the (00) beam displayed in the upper portion were obtained for an iron film grown under a background pressure p_{TOT} of 1×10^{-8} Pa. This curve resembles the data for the (00) beam shown in Fig. 6.7. Above this curve, the superstructures observed in the different stages are indicated. In the lower portion of the figure, the MEED intensity oscillations are depicted for a film grown under a CO partial pressure of 7×10^{-8} Pa ($p_{TOT} = 8 \times 10^{-8}$ Pa). The main difference between the upper and the lower curve is the extension of layer-by-layer growth during CO codeposition. This enlarges the region where intensity oscillations are observed from approximately 11 ML to 13 ML.

Two additional techniques were employed to characterize the morphology of the iron films. These were high-resolution spot-profile analysis LEED (SPALEED) and scanning tunneling microscopy. These experiments were performed through a cooperation with the Professor Neddermeyer's group in Bochum (now at Halle), by A. Brodde and H. Hannemann. In contrast to the RHEED technique, which could be used on-line (during deposition), the SPALEED and STM studies were limited to a post-growth analysis. Figure 6.9 shows SPALEED measurements for a 10 ML thick film in the upper half of the figure and for a 20 ML thick film in the lower half. The full width at half maximum (FWHM) of the (00) beam is displayed as a function of the primary electron energy. By inserting the electron energies at which the (00) beam was broad or sharp into (3.16), the step height of the resulting Fe islands was determined to be 1.74 ± 0.04 Å, in close agreement with the value of 1.77 Å expected for strained fcc iron on Cu(100) [38]. For the 20 ML thick film, a markedly different oscillation frequency is observed. The resulting step height of 2.02 ± 0.04 Å is a clear signal that the films have a different structure in region III. Further support for this finding will be presented in Sect. 6.2.4. Additionally, the form of the modulation of the half width in Fig. 6.9 was analyzed for clues about the film morphology [39]. The maxima of the FWHM of the (00) beam of the 10 ML thick film are rather sharp, while the minima are broad. The reverse is seen in the case of the 20 ML thick film. Here the maxima are broad and the minima are sharp. This can be analyzed by assuming different sequences of up and down steps. The resulting morphology is shown in the two insets of Fig. 6.9. For the 10 ML thick film, basically every up step is followed by a down step. This is characteristic feature of two-dimensional growth. For the 20 ML thick film, on the contrary, up steps follow each other, a situation typical of three-dimensional growth.

This film morphology is directly visible in the STM topographs depicted in Figs. 6.10 and 6.11. In Fig. 6.10, a constant-current topograph (CCT) is depicted for a 950×600 Å^2 area of a film which showed the (1×1) LEED pattern typical of films in region II. The film thickness was approximately 12 ML. The topograph shows the presence of large two-dimensional islands and terraces, as expected for layer-by-layer growth. However, deviations from layer-by-layer growth are also encountered. On the main terrace, for example, three atomic layers are growing simultaneously. Additionally, a 30×600 Å^2 large, protruding island is seen in this figure. Figure 6.11 reproduces an STM image of a 16 ML thick film which exhibited the characteristic "(3×1)" LEED pattern of region III. The large, two-dimensional islands characteristic of a layer-by-layer growth mode are no longer visible; instead, the roughness of the surface has increased. The overall corrugation is typically 10 Å, with a maximum of 15 Å. This is further evidence for the three-dimensional growth that had already been proposed from the exponential decay of the MEED intensities and the SPALEED data for films in region III.

To elucidate the mechanism responsible for the change in morphology, diffusion experiments were performed in regions II and III [25]. The change of growth mode from II to III is also reflected in a drastically different diffusion behavior of Cu to the top of the Fe film. While annealing of a 15 ML thick film at 575 K for 20 minutes leads to considerable Cu diffusion to the top of

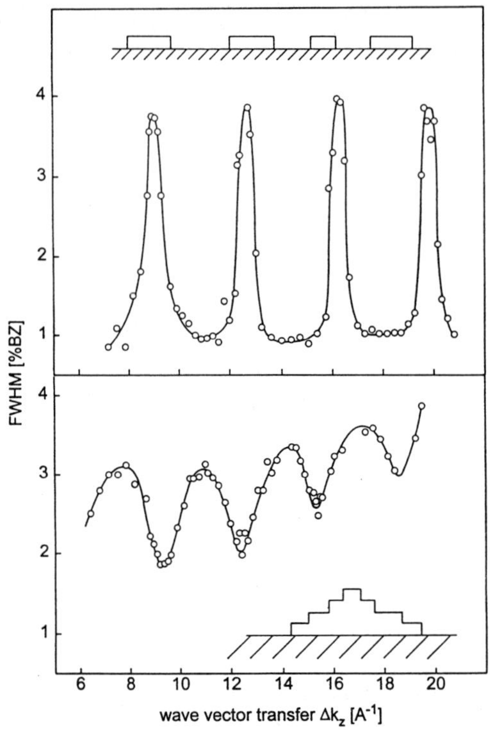

Fig. 6.9. Morphology of 10 and 20 ML thick Fe films on Cu(100). From an analysis of the dependence of the FWHM of the (00) beam on the wave vector transfer, the step height and film morphology can be derived. The *upper portion* shows the resulting curve for a 10 ML thick Fe film deposited at 300 K. The *lower portion* displays the FWHM for a 20.5 ML thick Fe film. 20 ML of Fe was deposited at 300 K. Then the film was annealed for 20 minutes at 470 K. This led to the formation of a sharp "(3 × 1)" LEED pattern. Subsequently, the film was cooled to 170 K and additional 0.5 ML Fe was deposited to produce an increased step density. The resulting morphology is shown as an *inset* for both films

the film, as evidenced by the increase in the Auger intensity ratio of the Cu and Fe transitions, no measurable Cu diffusion is observed for either a 7.5 or a 10 ML thick film after the same annealing procedure [25]. This is remarkable, since the thicker Fe films in region III should form a higher barrier for Cu diffusion to the film surface. Obviously, upon the transition from region II to region III, a new diffusion channel is opened. It has been shown that the new mechanism is diffusion along dislocation lines caused by the misfit between the film and substrate. Evidence for these dislocations is directly visible in Fig. 6.12. This STM topograph was measured after annealing of an 8 ML thick iron film for 40 minutes at 370 K. Again, large two-dimensional islands and terraces are visible. As in Fig. 6.10, protruding islands in the [011] and

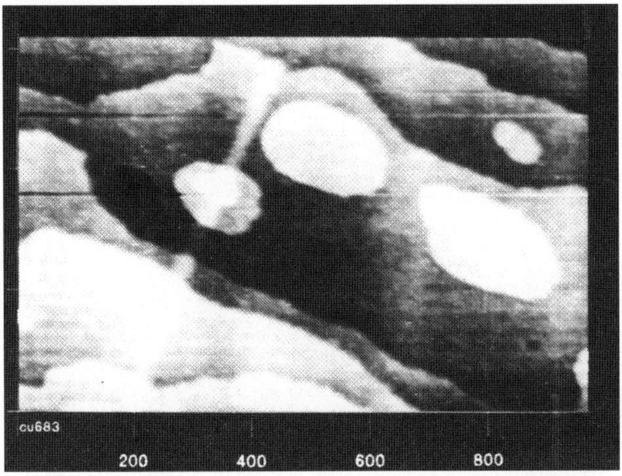

Fig. 6.10. STM image of a film in growth region II. Constant-current topograph from a 12 ML thick iron film on Cu(100) with a (1×1) LEED pattern (region II) after room-temperature deposition. A sample bias voltage of $U = -6$ mV (i.e., probing the filled states of the sample) and a tunneling current of $I = 0.5$ nA were used. The area is approximately $950 \times 600\,\text{Å}^2$. STM images taken by A. Brodde and H. Hannemann

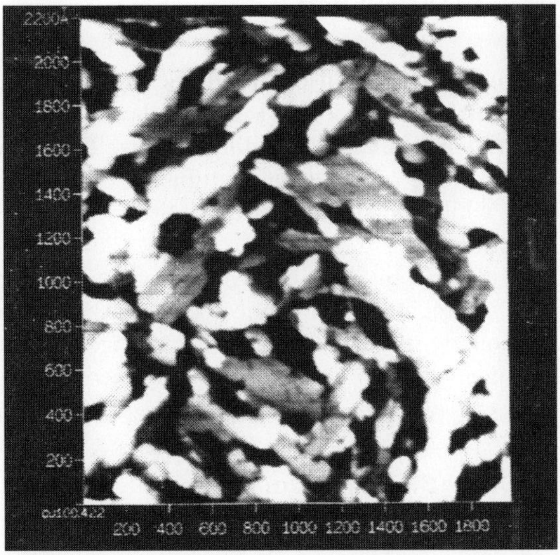

Fig. 6.11. STM image of a film in growth region III. Constant-current topograph from a 16 ML thick iron film with a "(3×1)" LEED pattern (region III) after room-temperature deposition and subsequent heating to 370 K and 500 K. $U = 1$ mV, $I = 0.5$ nA; the area is approximately $1900 \times 2200\,\text{Å}^2$. The gray tone scale covers a range of 15 Å

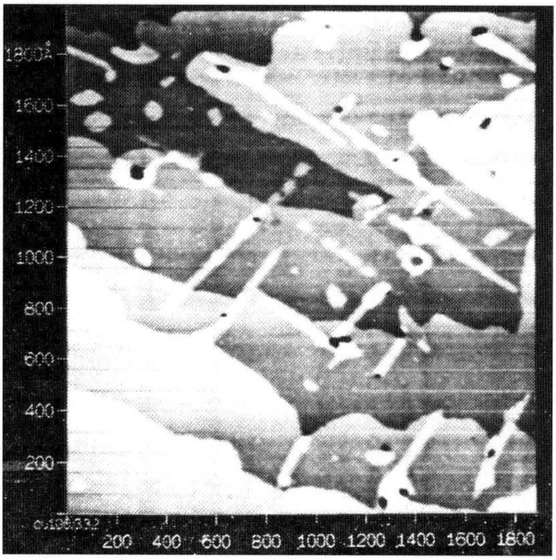

Fig. 6.12. STM image of a film in growth region II after annealing. Constant-current topograph from an 8 ML thick iron film on Cu(100) with a "(1 × 1)" LEED pattern (region II) after room-temperature deposition and subsequent heating to 370 K and 500 K. $U = -10$ mV, $I = 1$ nA; the area is approximately 1850×1900 Å^2. The gray tone scale covers a range of 15 Å

$[0\bar{1}1]$ direction are observed. Additionally, holes with a lateral extension of approximately 30 Å and a depth of at least two layers appear. These holes are seen mostly at the end of the protruding islands. Therefore we attribute these holes to the presence of preferential diffusion channels in the vicinity of dislocation lines, while the stripes are attributed to glided crystal planes between the dislocation lines.

Thus, the transition from layer-by-layer growth in region II to three-dimensional growth in region III is connected with the formation of misfit dislocations. Additionally, the SPALEED data in Fig. 6.9 give evidence for a structural rearrangement in films in region III. Therefore the structure of the films will be described in the next subsection.

6.2.4 Structure of Fe Films on Cu(100)[1]

The iron superstructures that evolve on the Cu(100) surface at 300 K with increasing film thickness have already been listed in Fig. 6.8. A (4×1) and a (5×1) structure are observed in region I. The (4×1) phase forms between 2 and 2.5 ML, and the (5×1) phase exists between approximately 3.5 and 4.5 ML. Connected with the transition from growth region I to growth region II is the formation of the $p2mg(2 \times 1)$ superstructure, which is evident between

[1] See [25, 39, 40, 41].

6 and 7 ML at room temperature. For a coverage above 7 ML, a (1×1) pattern is observed. However, upon cooling films between 5 and 11 ML (region II) to 150 K, the $p2mg(2 \times 1)$ structure is visible over the entire coverage range [42]. Above 11 ML (region III), new but rather streaky diffraction spots manifest themselves. The streaks run in the [011] and $[0\bar{1}1]$ directions and their intensity is centered around positions that are close to the diffraction spots expected for a (3×1) structure. With increasing coverage, the intensity of these spots increases. As will be explained later, this phase does not really have a (3×1) structure, and therefore the label "(3×1)" will be enclosed in quotation marks.

The LEED patterns of the (5×1), $p2mg(2 \times 1)$ and "(3×1)" structures are shown in Fig. 6.13. To decide if these different structures have a similar atomic arrangement, I/V curves for the (01) beam were measured for different film thicknesses. The logic behind this experiment was that electron scattering at Cu and Fe atoms is so similar that the I/V curves should reflect local atomic positions, independently of the nature of the scattering entity or the detailed surface morphology (i.e., 3D or 2D growth).

In the upper portion of Fig. 6.14, the I/V curve for the (10) beam of the clean Cu(100) surface is shown for comparison. Below, the (10) beam from a 3.3 ML thick iron film which displayed a (5×1) diffraction pattern is plotted. The resemblance in the I/V curves indicates that the atomic arrangement of the (5×1) phase is similar to that of the fcc (100) surface of Cu. The peak positions in the I/V curves, however, are shifted to lower energies, and the magnitude of this shift increases with increasing electron energy. Furthermore, the peaks in the I/V curve for the (5×1) structure become increasingly broad at higher energies. This points towards a pronounced variation in the atomic spacing between different layers (vertical disorder). Since similar I/V curves were also measured for thinner films showing a (4×1) diffraction pattern, one can conclude that the local atomic environments of the two phases are quite similar. The I/V curve for a 6.6 ML thick Fe film has a remarkably closer resemblance to the I/V curve for the Cu(100) surface. The peak positions match almost perfectly, demonstrating that the film grows with an fcc structure. There is no broadening of the peaks at high energies, so there is no vertical disorder as for the (5×1) structure. Finally, the I/V curves for the "(3×1)" structure show little similarity to the other spectra. This corroborates the conclusion of the SPALEED analysis that a rather different structure is produced in this thickness region.

To enable a quantitative structure determination, I/V curves for a large number of different beams for several film thicknesses were taken and compared with full dynamical calculations. The data were measured at low temperature for the $p2mg(2 \times 1)$ phase by C. Stuhlmann (IGV/KFA Jülich) and G. Schmidt (University of Erlangen-Nürnberg). The LEED intensity analysis was performed by Landskron et al. [42]. It can be seen from the LEED pattern of the $p2mg(2 \times 1)$ structure that the $(h/2, 0)$ and $(0, k/2)$ spots

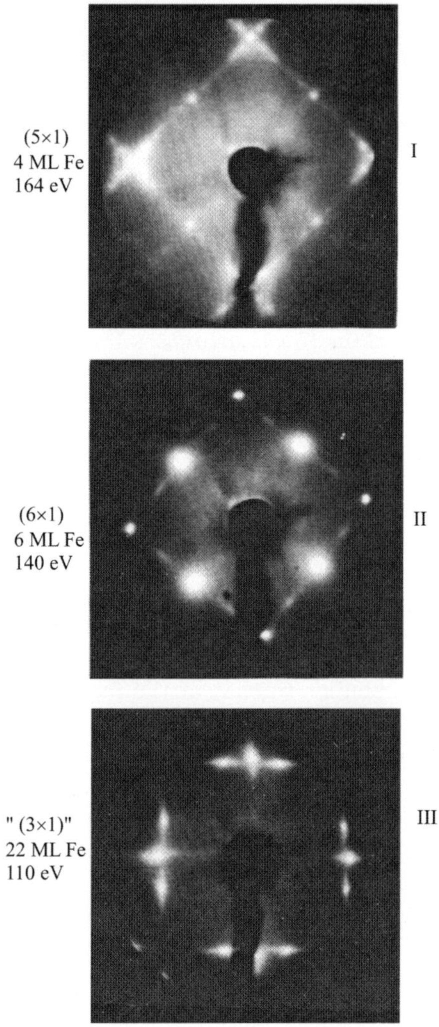

Fig. 6.13. LEED patterns of the (5×1), $p2mg(2 \times 1)$ and "(3×1)" structures. The corresponding phases were formed after deposition of 4 ML Fe, 6 ML Fe and 22 ML Fe, respectively. The LEED patterns displayed were taken at 164 eV, 140 eV and 110 eV, respectively

are systematically extinct. This indicates that a glide plane running in the [011] direction must be present. A second glide plane, running in the $[0\bar{1}1]$ direction, exists for the other domain orientation. Assuming that the reconstruction is localized at the surface layer, only one possible model can explain the existence of this glide plane. This model, which is characterized by an antiparallel displacement of adjacent iron atoms, is depicted in Fig. 6.15, where this displacement is labeled "SD". To simplify the picture, only one of the

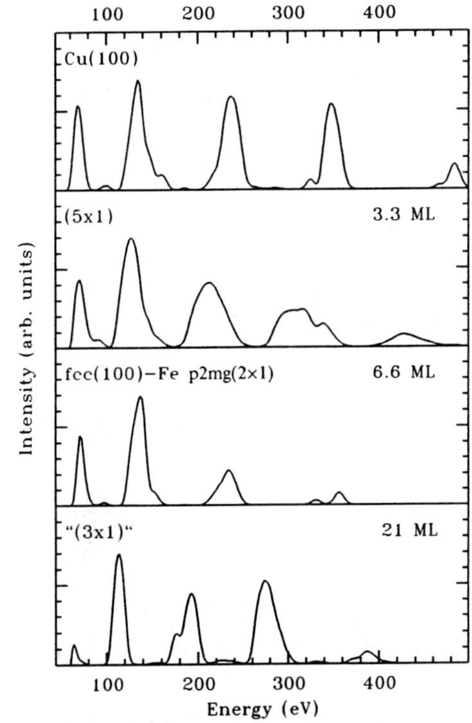

Fig. 6.14. LEED I/V curves from the (10) beam from different Fe films on Cu(100). The corresponding film thicknesses are given as well

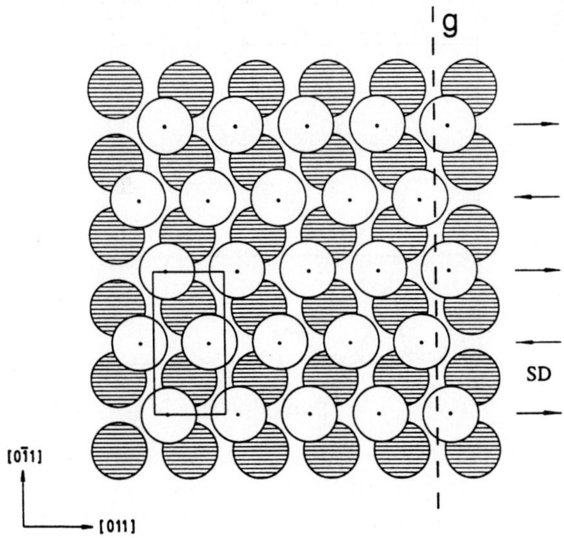

Fig. 6.15. Structural model for the $p2mg(2 \times 1)$ structure [23]

two possible orientations is drawn. In the LEED analysis, the displacement SD and the first three interlayer spacings were optimized. The best overall fit gave a Pendry R-factor of 0.25 with the following structure:

SD= 0.14 ± 0.09 Å, $d_{12} = 1.88 \pm 0.02$ Å, $d_{23} = 1.81 \pm 0.03$ Å, $d_{34} = 1.77 \pm 0.04$ Å.

Thus the top Fe layer not only shows a reconstruction but also has a considerably larger spacing (d_{12}) than that expected for a strained fcc iron film (1.77 Å).

Fig. 6.16 shows the experimental I/V curves for 8 and 11 ML thick iron films that were prepared and measured at room temperature. These films showed the typical growth behavior of region II, i.e., regular MEED intensity oscillations above 5 ML and a sharp (1×1) LEED pattern. The I/V curves for the 8 ML thick film (dotted lines) and the 11 ML thick film (dashed lines) closely resemble each other, implying that they are characteristic of the structure formed in this thickness regime.

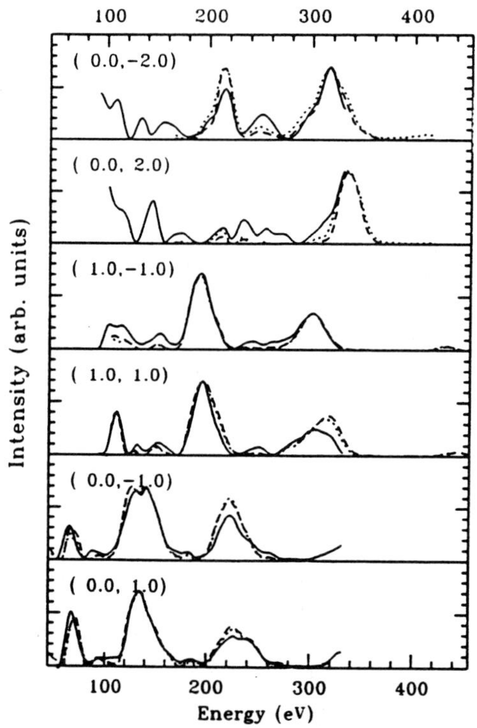

Fig. 6.16. LEED I/V curves for 8 and 11 ML thick films at 300 K. The experimental I/V curves for the 8 ML (*dotted lines*) and the 11 ML (*dashed lines*) thick films were measured at 300 K. The *solid curves* denote the best-fit structure for the 8 ML thick film at 300 K

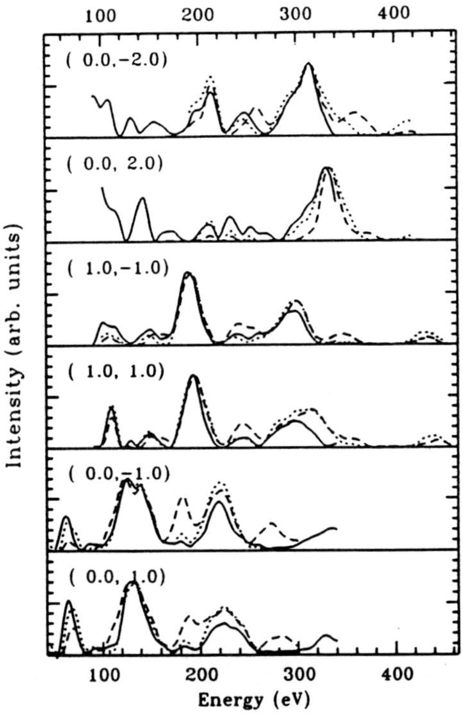

Fig. 6.17. LEED I/V curves for 8 and 11 ML thick films at 170 K. The *dotted lines* refer to the 8 ML film and the *dashed lines* to the 11 ML film as before. The differences between the I/V curves for the 8 ML and the 11 ML thick film are due to a structural rearrangement of the 11 ML thick film upon cooling. The *solid curves* denote the best-fit structure for the 8 ML thick film at 170 K

The I/V curves after cooling to 170 K are shown in Fig. 6.17. While cooling the 8 ML thick film has no significant influence on the I/V curves, additional peaks appear in the I/V curves of the 11 ML thick film. These new features in the I/V curves remain even after warming to 300 K. They are thus indicative of the irreversible formation of a new phase. Consistent with this interpretation is the observation that the LEED pattern now shows streaky "(3×1)" spots.

Finally, in Fig. 6.18, the I/V spectra of a 15 ML thick film measured at 170 K are compared with those of the 8 ML thick film. The LEED pattern for the 15 ML thick film shows the "(3×1)" pattern characteristic of the films in thickness region III. Pronounced differences are visible between the films. The I/V curves for the 15 ML thick film have intense new peaks, in addition to the peaks present for the 8 ML thick film. Therefore, parts of the 15 ML thick film must still have the structure characteristic of the thinner films in region II. For the LEED analysis, the coexistence of two phases in this thickness range poses a severe challenge.

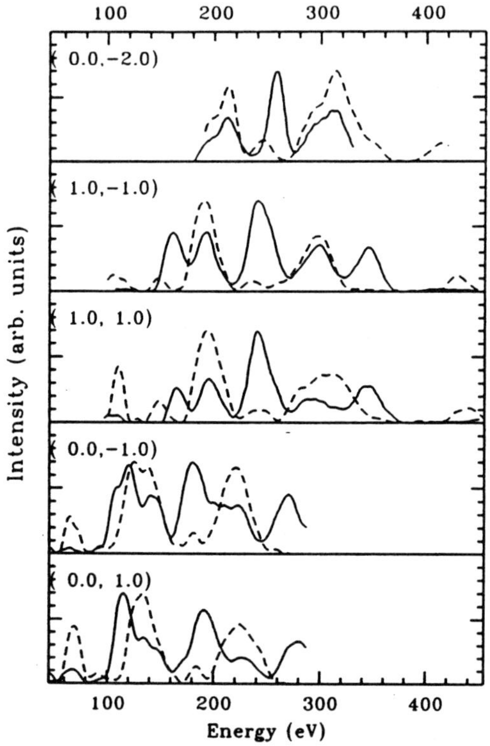

Fig. 6.18. LEED I/V curves for 8 ML (*dashed lines*) and 15 ML (*solid lines*) thick Fe films on Cu(100) at 170 K

Full dynamical LEED calculations for six different films were performed: 8 ML thick films at 300 and 170 K; 11 ML thick films at 300 K, after cooling down to 170 K and after subsequent rewarming to 300 K; and a 15 ML thick film at 170 K. To reproduce the measured I/V curves, the first two interlayer spacings and the bulk interlayer spacing were optimized. All three interlayer spacings were varied between 1.4 and 2.2 Å in steps of 0.05 Å. The comparison between theory and experiment was evaluated using different R-factors: the Pendry R-factor (R_p) [43], the Zanazzi–Jona R-factor (R_{ZJ}) [44] and the R_{DE} R-factor [45], which evaluates the deviation between the normalized experimental and theoretical intensity spectra. Estimates of the error bars in the structure determination were derived from the variance of the R-factor, $\mathrm{var}(R_p) = R_p\sqrt{8V_i/\triangle E}$, where $\triangle E$ is the range of energies where the calculated and measured spectra overlap.

By use of the stress–strain relationship for thin films, one can calculate the expected bulk interlayer spacing for fcc (100) and bcc (100) films [38] growing with an in-plane lattice constant of 2.55 Å. The assumption that the iron films grow with the in-plane lattice constant of Cu is consistent with

Table 6.2. Optimum R-factor for 8, 11 and 15 ML thick films measured at 170 and 300 K and the corresponding best-fit structure. Consistent results are obtained from the different R-factors for the first three films only. For 8 and 11 ML thick films, the optimum bulk interlayer spacing is 1.77 Å. For the 15 ML thick film, bulk spacings (d_b) between 1.65 and 1.77 Å lead to comparable R-factors, showing that the 15 ML thick film cannot be described properly by assuming a single Fe phase in the film. The values listed for this film have been obtained for $d_b = 1.77$ Å

Structure	d_{12} (Å)	d_{23} (Å)	V_0	R-factor
8 ML Fe (300 K)	1.85	1.80	1.0	0.252 (R_p)
	1.83	1.80	0.0	0.087 (R_{ZJ})
	1.85	1.80	−1.0	0.331 (R_{DE})
8 ML Fe (170 K)	1.86(5)	1.80	0.0	0.233 (R_p)
	1.84(5)	1.80	−1.0	0.099 (R_{ZJ})
	1.88	1.80	−1.0	0.262 (R_{DE})
11 ML Fe (300 K) (before cooling)	1.84(5)	1.80	1.0	0.285 (R_p)
	1.83(5)	1.80	1.0	0.100 (R_{ZJ})
	1.84(5)	1.79	0.0	0.320 (R_{DE})
11 ML Fe (170 K)	1.92	1.74	2.0	0.300 (R_p)
	1.83	1.80	−1.0	0.163 (R_{ZJ})
	1.89	1.76	0.0	0.356 (R_{DE})
11 ML Fe (300 K) (after cooling)	1.92	1.72	2.0	0.435 (R_p)
	1.74	1.86	−1.0	0.243 (R_{ZJ})
	1.90	1.74	−2.0	0.472 (R_p)
15 ML Fe (170 K)	1.72	1.92	2.0	0.487 (R_p)
	1.60	1.70	3.0	0.149 (R_{ZJ})
	1.60	1.72	−5.0	0.476 (R_p)

the observed (1 × 1) LEED structure.[1] For bcc films, an interlayer spacing of 1.65 Å is expected, while 1.77 Å is expected for the bulk spacing in fcc films. The variation thus considers the possibility that epitaxial fcc or bcc iron films with a (100) structure could have been formed. Table 6.2 shows the resulting optimum structures for the three different R-factors (R_p, R_{ZJ} and R_{DE}). The calculated intensities for an 8 ML thick film at 300 K and 170 K are shown in Figs. 6.16 and 6.17, respectively. Acceptably small variations in the best-fit structure for different R-factors are obtained only for three of the films, namely the 8 ML films at 170 and 300 K and the 11 ML

[1] A LEED analysis performed at the University of Erlangen-Nürnberg finds evidence for an in-plane lattice constant of 2.52 Å, i.e., relaxed iron rather than strained iron [K. Heinz, private communication]. This surprising result has not yet been confirmed by other techniques. If this result is reproduced by different methods, it would indicate that our present understanding of ultrathin-film growth needs to be modified to account for this observation.

thick film at 300 K before cooling. For these films, low R-factors and minor variations in the optimum structure are found. The bulk interlayer spacing of 1.77 Å shows conclusively that the films have a strained fcc structure and not a bcc structure. The analysis of the 11 ML thick films cooled to 170 K yields unacceptable variations in the best-fit structure. Clearly, this is due to the formation of a second phase upon cooling. The discrepancies become even more pronounced after subsequent rewarming to 300 K, as evidenced by the significantly worse R-factors. Obviously, we cannot model the structure of these films by assuming the presence of only one epitaxial iron phase. This added complexity is also evident for the 15 ML thick film, which also cannot be properly described. In Fig. 6.19, the variation of the R-factor with the interlayer spacing is shown for 8 and 11 ML thick films. While the results for the 8 ML thick film and the 11 ML thick film before cooling are in close agreement, a considerable shift is found after cooling of the 11 ML film.

In an effort to understand the structural transition from region II to III, the structure of the films above 11 ML was characterized using SPALEED. As previously mentioned, these films showed streaky spots close to the positions of a (3×1) structure.

A 22 ML thick Fe film was prepared at 300 K on Cu(100). The film was subsequently annealed for 20 minutes at 470 K to improve the quality of the LEED pattern. Since annealing at 470 K does not induce measurable Cu diffusion to the top of the Fe film, this procedure does not change the film composition but leads only to sharper diffraction spots. In Figure 6.20, a portion of the diffraction pattern close to the position of the (10) beam of the Fe/Cu(100) surface is shown. Three different LEED spots are visible. Two of them are streaked in the [011] direction, while the third is streaked in the $[0\bar{1}1]$ direction. This suggests that the corresponding superstructure exists in at least two different orientations on the surface. Figure 6.21 shows a schematic view of a large portion of $\boldsymbol{k}$-space around the (00) beam. Diffraction spots are denoted by ellipses. The elongation of the ellipses indicates the direction of streakiness. Open and filled ellipses indicate the two different unit cells for each direction. The two reciprocal-lattice vectors $\boldsymbol{b}_1$ and $\boldsymbol{b}_2$ that span the unit cell have been measured. From a number of SPALEED measurements, $|\boldsymbol{b}_1|$ and $|\boldsymbol{b}_2|$ have been determined to be $0.772 \pm 0.042\,\text{Å}^{-1}$ and $0.769 \pm 0.042\,\text{Å}^{-1}$, respectively, and the angle between $\boldsymbol{b}_1$ and $\boldsymbol{b}_2$ to be $72.1 \pm 0.7°$. Since $\boldsymbol{b}_1$ does not coincide with the (10) beam of the clean Cu(100) surface, the resulting superstructure is not an integer multiple of the (1×1) structure of the substrate, and thus is not a (3×1) phase. This is the reason that it has been called the "(3×1)" structure in the preceding text.

From the knowledge of the unit cell in reciprocal space, the unit cell in real space was determined. The real space unit cell is shown in Fig. 6.22, together with a model that describes a possible atomic arrangement on the Cu(100) surface. The two vectors that span the unit cell have lengths of 2.50 ± 0.04 Å and 2.51 ± 0.04 Å for $\boldsymbol{a}_1$ and $\boldsymbol{a}_2$, respectively. The angle between

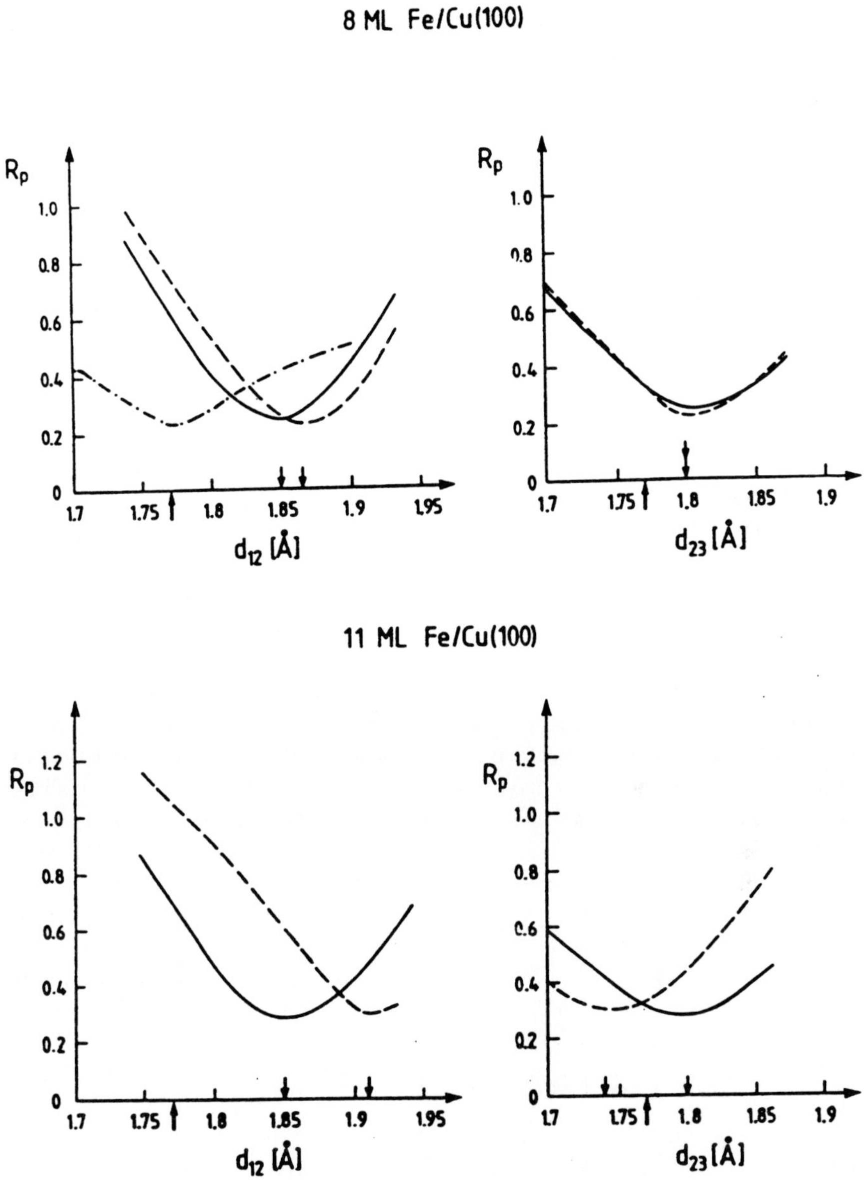

Fig. 6.19. Dependence of the Pendry R-factor on structural modification: the variation of the Pendry R-factor with the first two interlayer spacings d_{12} and d_{23}. The *dashed lines* denote the behavior for the 170 K data. The *solid lines* were obtained for the 300 K data prior to cooling. *Arrows* indicate the position of the R-factor minimum and the bulk interlayer spacing (1.77 Å). The variation of the Pendry R-factor with the bulk layer spacing for the 8 ML thick film at 170 K is shown as a *dashed–dotted curve* in the *upper left diagram*

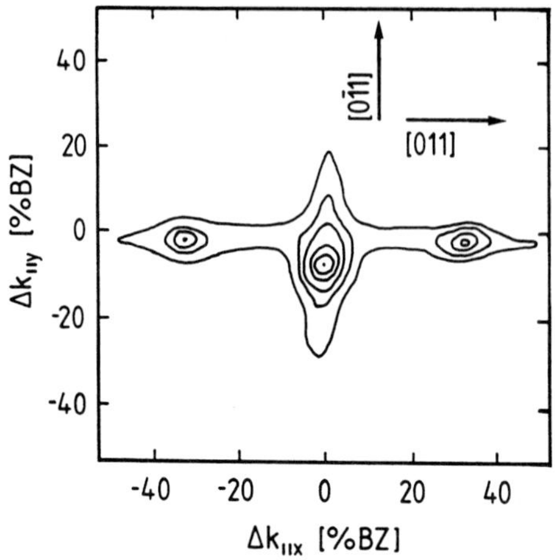

Fig. 6.20. Contour plot of the LEED pattern (E = 180 eV, T = 120 K) around the position of the (10) beam after deposition of 22 ML Fe on Cu(100) at 300 K and subsequently annealing for 20 minutes at 470 K. Only a portion of the reciprocal space is shown. This portion is depicted schematically in Fig. 6.21. The direction of streakiness for the three LEED spots indicates that the beam represents two different domain orientations

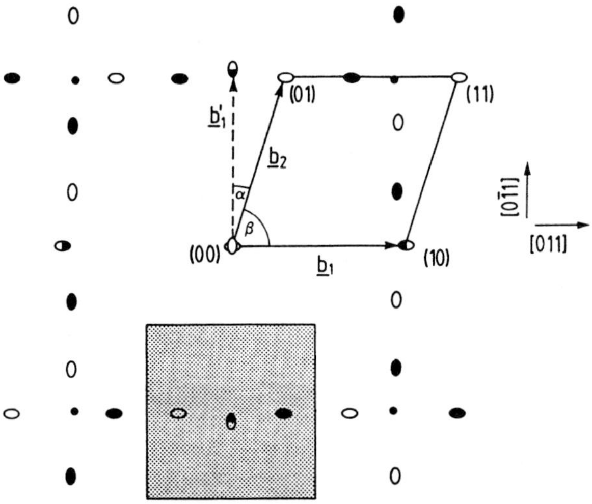

Fig. 6.21. Schematic LEED pattern for thick Fe films (> 15 ML) on Cu(100). *Ellipses* represent the positions of the LEED beams. The elongation of the ellipses indicates the direction of streakiness. *Open* and *filled ellipses* are used to distinguish the two different domain orientations. LEED spots with contributions from two domains are shown as half-filled. One of the four different unit cells is marked and the beams are labeled. The *small filled spots* denote the positions of the (11) beams of the Cu(100) substrate. The directions given are those of the Cu(100) substrate

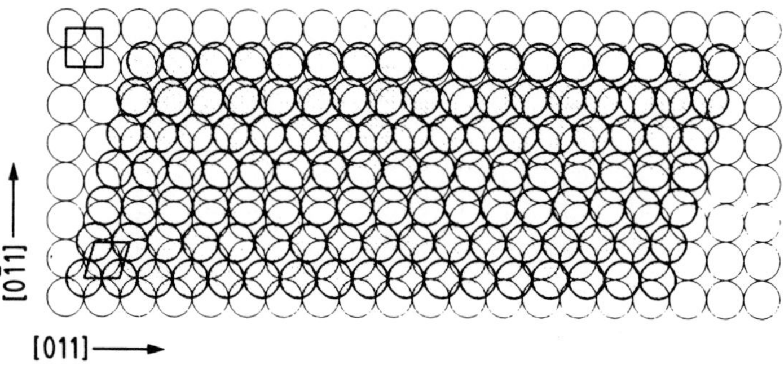

Fig. 6.22. Real-space model of bcc (110)-like iron growing on Cu(100). The *thin circles* denote the Cu atoms of the substrate and the *thick circles* denote Fe atoms. The unit cells of the Cu(100) surface and the bcc (110)-like film are shown, together with crystallographic directions for Cu(100)

$\boldsymbol{a}_1$ and $\boldsymbol{a}_2$ is 70.5°. The resulting geometry of the iron film thus closely resembles the atomic structure of the bcc iron (110) surface.

Three different experiments confirm that the resulting phase is indeed similar to the bcc (110) geometry. As mentioned earlier, SPALEED was used to determine the interlayer spacing, LEED I/V curves were measured and the atomic geometry was investigated with STM. As described in Sect. 6.2.3, the step height of films with the "(3×1)" structure was determined using SPALEED to be 2.02 ± 0.04 Å, in excellent agreement with the step height of 2.03 Å expected for bcc Fe(110). On the other hand, for 10 ML thick films, the step height was found to be 1.74 ± 0.04 Å, close to the expected value of 1.77 Å for fcc iron epitaxially grown with a (100) orientation on Cu(100) [38]. The result of the LEED analysis for 8 and 11 ML thick films also gives an interlayer spacing of 1.77 ± 0.04 Å.

LEED I/V curves were also measured for a 20 ML thick film. Again the sample was annealed for 5 minutes after deposition. Then the sample was cooled to 170 K, which did not change the film structure. Afterwards, LEED I/V curves were measured for several beams, including the (01), (10) and (11) beams. The labeling of these beams is depicted in Fig. 6.21. Figure 6.23 shows the resulting I/V curves. A visual comparison with I/V curves for the (110) surface of bcc iron (Figs. 3 and 4 in [46]) shows close agreement. This supports the result of our SPALEED analysis, which suggests that the atomic arrangement of a 20 ML thick iron film on Cu(100) closely resembles the bcc iron (110) surface. The small discrepancies between the I/V curves for bcc Fe(110) and our results might be due to the fact that our LEED data were not measured at normal incidence but with a small tilt angle of approximately 2°.

Further evidence for the suggested atomic arrangement comes from STM pictures taken with atomic resolution. An atomically resolved image of the clean Cu(100) surface is shown for comparison in Fig. 6.24. The STM topo-

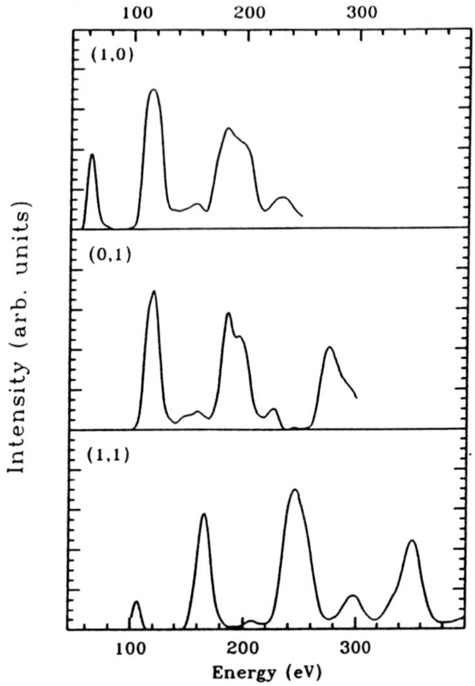

Fig. 6.23. LEED I/V curves for a 20 ML thick iron film on Cu(100): see text for details

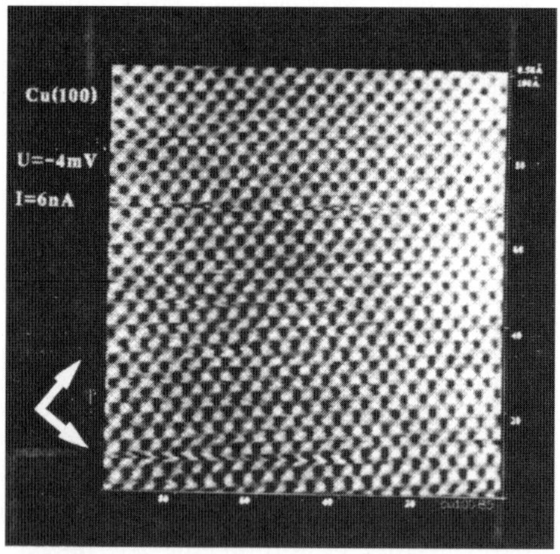

Fig. 6.24. Atomically resolved STM image of Cu(100): constant-current topograph, $U = -4$ mV, $I = 6$ nA. The area is $70\times70\,\text{Å}^2$. The atomic corrugation is typically 0.3 Å

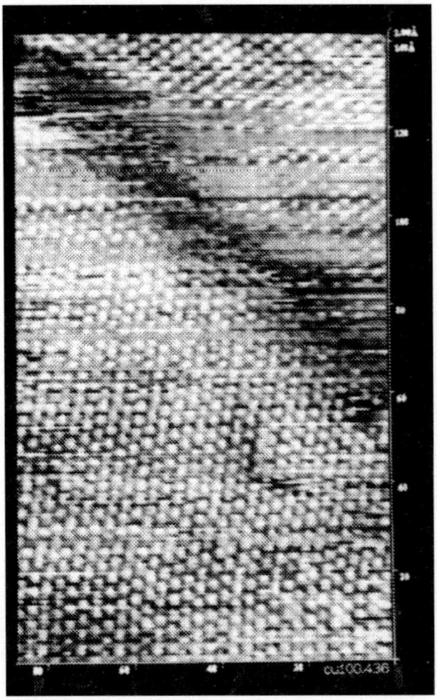

Fig. 6.25. Atomically resolved STM image of bcc-like iron on Cu(100): constant-current topograph from a 16 ML thick iron film with a "(3×1)" LEED pattern after room-temperature deposition and subsequent heating to 370 K and 500 K (the same sample as used for the STM measurement shown in Fig. 6.11). $U = -1$ mV, $I = 15$ nA. The area is 70×70 Å^2. The atomic corrugation is 0.4 Å

graph resolved the fourfold symmetry of the (100) surface. The deviations from the square shape of the unit cell are due to minor deviations of the xy piezoelectric system from orthogonality and possible drift effects. The atomically resolved constant-current topograph obtained from Cu(100) allows the identification of the [011]-like directions, as indicated on Fig. 6.24. In Fig. 6.25, a CCT obtained from one of the larger Fe islands in the 16 ML thick film shown in Fig. 6.11 is displayed. By analyzing the orientation of the atomic rows visible, one can recognize that rows of atomic features follow the [011] but not the $[0\bar{1}1]$ high-symmetry direction. The latter would be expected in addition to the [011] direction for the growth of fcc Fe(100). In Fig. 6.26, a schematic representation of the arrangement of the atoms in Fig. 6.25 is shown, where the [011] high-symmetry direction is indicated. It is quite obvious that the orientations of the unit cell are different on the two sides of the depression seen in the center of Fig. 6.25 and do not correspond to that of either Cu(100) or epitaxial fcc Fe(100). The symmetry of the observed protrusions is more hexagonal-like and is explained by formation of a

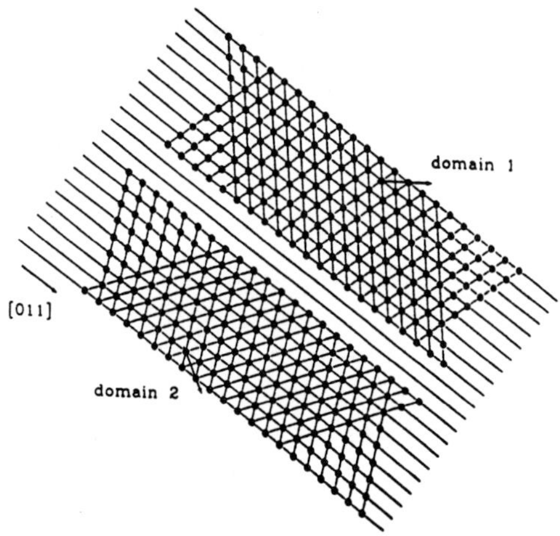

Fig. 6.26. Schematic representation of the STM image shown in Fig. 6.25

bcc Fe(110) surface. A comparison of Figs. 6.22, 6.25 and 6.26 gives further evidence for the formation of bcc iron. This modification grows in four different orientations on the substrate. Two of these orientations are visible in Fig. 6.25.

We have shown that bcc iron with a (110) orientation grows on the Cu(100) surface above a critical thickness of 11 ML. The same structure can also be obtained by cooling or by mild sputtering for 2 minutes with 600 eV Ar ions (sample current 1 μA) of an 11 ML thick film.

6.2.5 Conclusions and Comparison with Other Studies

Before the correlation between structure, growth and magnetism is unraveled in the next subsection, the results of the last two subsections will be summarized and compared with other studies.

The growth mode in the initial stage has been most controversial. Initially, it was believed that the iron films grow in a layer-by-layer fashion from the start [14, 23]. Such a scenario is at odds with thermodynamic arguments, since the surface free energy of iron (2.84 J/m^2) is larger than the surface free energy of Cu (2.36 J/m^2) and the interface energy is positive [47]. As a result, Bauer's criterion for layer-by layer growth is not fulfilled [48]. Photoelectron diffraction and RHEED investigations provide evidence for Fe agglomeration at room temperature [31]. The MEED intensities shown in Figs. 6.7 and 6.8, as well as similar data by Steigerwald and Egelhoff [37], can be explained by agglomeration at small thicknesses followed by coalescence and, finally, flattening of the films at around 4 ML. The latter is the reason for the drastic increase in intensities seen in Figs. 6.7 and 6.8. This interpretation, however, is at variance with recent STM and ion scattering studies (ISS), which find evidence for agglomeration and interdiffusion, but only up to approximately 2 ML [31, 32, 33, 49]. Above this thickness, almost perfect layer-by-layer growth has been found by STM. The ISS and STM results can explain the observation of MEED oscillations at 2 and 3 ML and the absence of an intensity maximum at 1 ML [50]. However, the question still arises as to why the MEED intensity increases around 4 ML. The STM investigation indicates that no dramatic change in film morphology occurs around this coverage even though the island size of the 6 ML films is larger than for the 2 ML films [33]. Therefore, only a structural rearrangement can be the cause of the increase in MEED intensity. Indeed, the transition from growth region I to region II is connected with a structural transition from the (5×1) to the $p2mg(2 \times 1)$ phase. The (5×1) structure is characterized by considerable vertical disorder, as shown by an EXAFS analysis of films in this thickness range [22] and our LEED I/V curves (Fig. 6.14). The (2×1) structure, on the other hand, shows a well-defined interlayer spacing. The vertical disorder of the (5×1) phase, therefore, is responsible for the low average intensity in region I, and the structural rearrangement to the (2×1) phase leads to the increase of the MEED intensity. The EXAFS analysis [22] shows that films in region I have an fcc-like structure. However, detailed LEED measurements and a full dynamical quantitative analysis reveal that the whole film in region I is reconstructed [40, 52, 53]. The iron atoms of all layers are displaced from the ideal fcc positions both vertically and in-plane (see Fig. 6.27). The maximum in-plane displacement is calculated to be 0.5 Å for the (4×1) phase and 0.4 Å for the (5×1) phase, while the maximum vertical displacement is 0.18 Å for the (4×1) and 0.28 Å for the (5×1) phase. As a consequence of these displacements, the interlayer spacings in the Fe film are considerably enlarged compared with the value extrapolated from the stable high-temperature phase of fcc Fe ($d = 1.78$ Å). Equivalently, the atomic volume of Fe in either a (4×1) or a (5×1) reconstructed film increases to 12.1 Å^3.

The behavior in thickness region II is the most easily understood. For film thicknesses between 5 and 11 ML, regular intensity oscillations are observed by MEED. This is evidence for layer-by-layer growth. Further support comes from the STM images presented in the last subsection and the SPALEED analysis of the energy dependence of the full width at half maximum of the (00) beam.

The films in region II have a $p2mg(2 \times 1)$ structure which is characterized by a parallel displacement of adjacent atoms in the top layer of the Fe film. The distance between this surface layer and the next layer is considerably enlarged compared with the bulk interlayer distance. The second interlayer spacing is still slightly enlarged, but deeper interlayers have a bulk-like spacing characteristic of strained fcc iron (1.77 Å). The $p2mg(2 \times 1)$ structure is visible at room temperature for film thicknesses between 5 and approximately 6 ML. For large film thicknesses, up to 11 ML, this structure is only

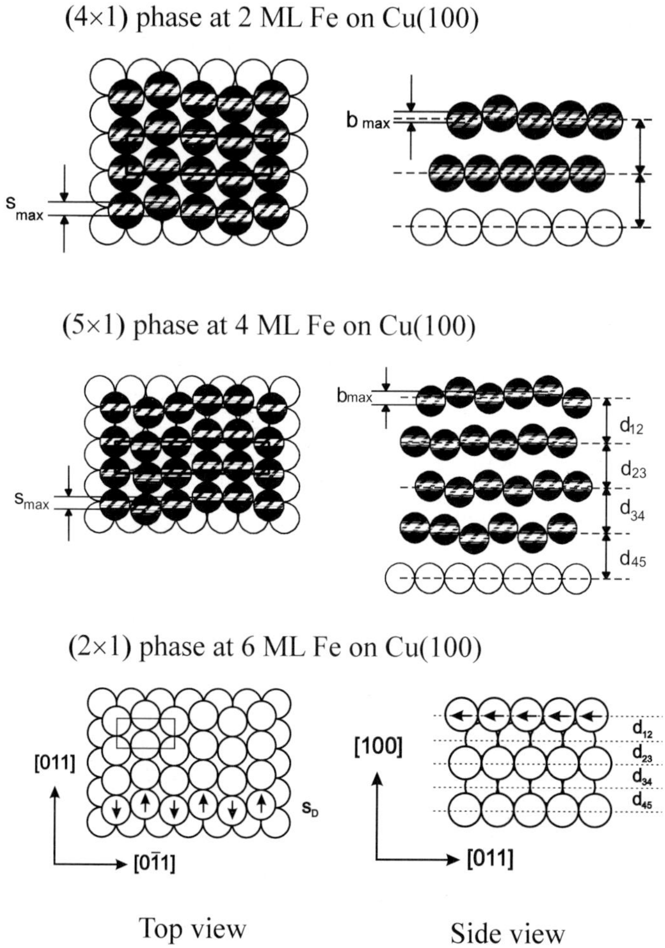

Fig. 6.27. Structural model for a 2 ML, a 4 ML and a 6 ML Fe film on Cu(100) based on the LEED patterns and I/V curves. For the 2 and 4 ML films, the iron atoms are shown as *hatched spheres* and the copper atoms as *open circles*. Only the top Fe layers are shown in the top views, while all layers are displayed in the side views. The maximum amplitudes of buckling (b_{max}) and shift (s_{max}) are exaggerated. The structural parameters have been determined to be $d_{12} = 1.87$ Å, $d_{23} = 1.84$ Å, $b_{max} = 0.18$ Å and $s_{max} = 0.5$ Å for the 2 ML (4×1) Fe film, and $d_{12} = 1.86$ Å, $d_{23} = 1.90$ Å, $d_{34} = 1.85$ Å, $d_{45} = 1.78$ Å, $b_{max} = 0.28$ Å and $s_{max} = 0.4$ Å for the 4 ML (5×1) film. For the 6 ML Fe/Cu(100) film, the first two iron layers are shown, with *arrows* denoting the reconstructive displacement of the iron in the top layer from the fourfold positions. From [51], used with permission

Table 6.3. Comparison of the interlayer spacing for Fe films on Cu(100). All films were grown at room temperature. T_{exp} denotes the temperature at which I/V curves were measured. All films showed a (1×1) pattern, except in [42], where the $p2mg(2 \times 1)$ structure was analyzed

	Structures [reference]						
	5 ML Fe [19]	6 ML Fe [42]	8 ML Fe	8 ML Fe	10 ML Fe [20]	11 ML Fe	12 ML Fe [10]
T_{exp}	300 K	90 K	300 K	170 K	300 K	300 K	300 K
d_{12}(Å)	1.88±0.02	1.88±0.02	1.85±0.02	1.87±0.02	1.81±0.02	1.85±0.02	1.84±0.03
d_{12}(Å)	1.92±0.02	1.81±0.03	1.80±0.03	1.80±0.03	1.78±0.02	1.80±0.03	1.76±0.03
d_b(Å)	1.92±0.02	1.77±0.03	1.77±0.03	1.77±0.03	1.78±0.02	1.77±0.03	1.77±0.0

observed at low temperature (150 K). LEED analysis of films in this coverage region also finds an enlarged first interlayer spacing. Table 6.3 lists the results of several LEED studies of iron films grown at room temperature, together with our data. There is general agreement that the first interlayer spacing is considerably enlarged. This is rather surprising, since on most closed-packed metal surfaces, the first interlayer spacing is contracted (see [54] and references therein). The observation of a $p2mg(2 \times 1)$ phase upon cooling films up to 11 ML thick suggests that, locally, the top-layer atoms in these films show the same reconstruction as observed for 5 and 6 ML thick films [9]. This is confirmed by a recent LEED analysis, which finds a considerable improvement in the R-factor for films which exhibit a (1×1) LEED pattern when, locally, a (2×1) reconstruction is assumed [55, 56]. In contrast to the films in region I, the reconstruction is limited to the top layer for films in region II. Except for the top layer, the film is almost ideally fcc-structured. In Sect. 6.2.6, it will be shown that the structural transition from the entirely reconstructed (5×1) phase to the (2×1) phase where only the top layer is reconstructed is essential for an understanding of the magnetic properties of Fe/Cu(100) ultrathin films.

The transition from region II to region III is characterized by an exponential decay of the MEED intensities (Figs. 6.7 and 6.8). Such a decay is typical of three-dimensional growth, in agreement with the STM topographs and the SPALEED analysis of the film morphology. The diffusion behavior provides additional evidence for the formation of misfit dislocations upon the transition from region II to III.

Such misfit dislocations occur above a critical film thickness d_c because of the lattice mismatch between the substrate and the deposit. Using (3.4), a critical thickness of 20 ML can be calculated for fcc iron on Cu(100), assuming the formation of complete edge dislocations with a Burgers vector $\boldsymbol{b} = (1/2)\langle 110 \rangle$. Experimentally, a smaller critical thickness of 11 ML is observed. A possible explanation is the formation of partial dislocations such as Frank or Shockley dislocations. The energy required to introduce these dislocations is considerably smaller, resulting in a reduced critical thickness. These partial dislocations are connected with the formation of faults in the stacking sequence of (111) planes. Such stacking faults lead to steps with a height of 1/3 and 2/3 of a monatomic step, consistent with the STM observations [39]. These steps run in the [011] and $[0\bar{1}1]$ directions, precisely the directions where the streaking of the diffraction spots is observed. With this information to hand, the mechanism responsible for the transition in the growth mode can be explained.

Above a film thickness of 11 ML, a square network of partial dislocations is formed. These dislocations lead to the formation of steps running in the [011] and $[0\bar{1}1]$ directions on the film surface. Such steps form nucleation centers for adatoms, leading to preferential growth at the step edges. This preferential growth, in turn, leads to the buildup of walls in the vicinity of the dislocations as schematically depicted in Fig. 6.28. Our proposed mechanism not only explains the observed three-dimensional growth but also accounts for the wall-like islands seen in Fig. 6.12.

The transition from growth region II to region III is also accompanied by a structural change. This is evident from the formation of additional diffraction spots close to the positions of a (3×1) structure and the appearance of additional peaks in the I/V curves of 15 ML thick films. The structural analysis presented in the last subsection shows that bcc iron growing with a (110) orientation on the substrate is formed in region III. Thus the transition from region II to region III is connected both with a change of growth mode and with a change in the structure of the iron film. The transition

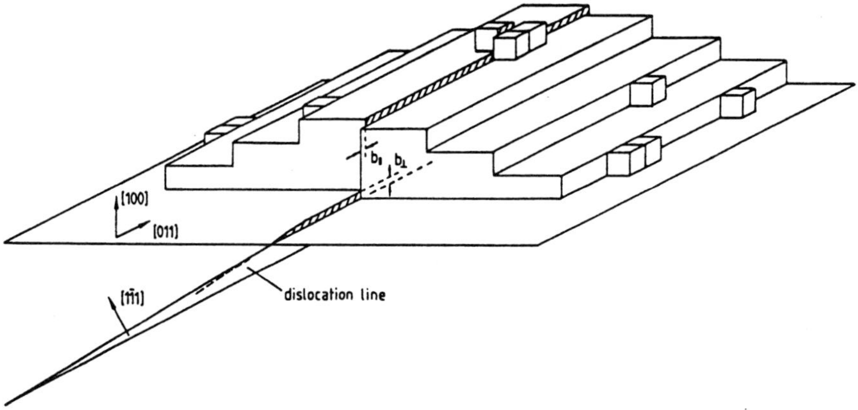

Fig. 6.28. Schematic view of the film surface above the critical thickness. The walls form along a step of 1/3 of the monatomic height. The step is due to a Shockley partial dislocation with a Burgers vector $\boldsymbol{b} = (1/6)[112]$ on a $(1\bar{1}1)$ glide plane. $b_\parallel$ and $b_\perp$ are the projections of the Burgers vector onto the [011] direction and onto the surface normal, respectively

from fcc to bcc is gradual. At 15 ML, the I/V curves still show peaks that are characteristic of fcc iron growing with a (100) orientation on the substrate. Presumably, the bcc iron is initially localized in the vicinity of the dislocations, while the regions between the dislocations and the related walls still have an fcc structure. With increasing film thickness, the walls expand until the whole film has the bcc structure. This scenario is consistent with the metastability of fcc iron films on Cu(100). Prinz [57] has pointed out that these films are metastable in a twofold sense. They are metastable since they grow with the "wrong" structure, namely, with the fcc structure instead of the bcc structure. Additionally, the films are metastable since they grow in a strained condition owing to the 0.7% mismatch between the fcc iron and fcc copper. With each additional layer of fcc iron that grows on the substrate, the energy difference between the fcc iron film and a comparable film with a bcc structure increases. Ultimately this energy difference will exceed the energy necessary to introduce defects, and the system will relax into the lower-energy structure.

For a strained film, the energy associated with the strain distortion increases with increasing film thickness until the critical thickness d_c is reached. Above d_c, the system introduces defects such as misfit dislocations to reduce the strain energy. Our investigation shows that dislocation formation opens the channel for the structural transition to the formation of bcc iron. That dislocation formation is an essential step in the structural transition can be seen from several experiments. The stabilization of fcc iron films by CO co-deposition up to 13 ML shows that a suppression of dislocation formation suppresses the structural transition to bcc iron. Furthermore, we have demonstrated in the LEED I/V analysis of iron films with thicknesses of 8, 11 and 15 ML described above, that cooling an 11 ML thick film results in the same bcc (110) structure as that found for a 15 ML thick film. Mild sputtering of 10 ML thick iron films also leads to the characteristic LEED pattern of bcc iron on Cu(100) [39]. In fact, all processes such as sputtering, cooling or further Fe deposition that cause dislocation formation in the iron films studied are accompanied by the structural transformation. Thus, for these films, dislocation formation initiates the structural transformation.

6.2.6 Correlation Between Magnetism, Structure and Growth

Figure 6.29 shows our most direct evidence for the correlation between magnetism, structure and growth for Fe films on Cu(100). In this figure, the magnetic properties, the growth mode and the structure of the films in different thickness regimes are summarized. This comparison demonstrates that the three different regimes of magnetic behavior are correlated with the three different regimes of growth.

The magnetic behavior of the films in region II is the most peculiar. For films between 5 and 11 ML, only the outermost layers at the surface are ferromagnetic, while the bulk of the film does not contribute to the measured Kerr

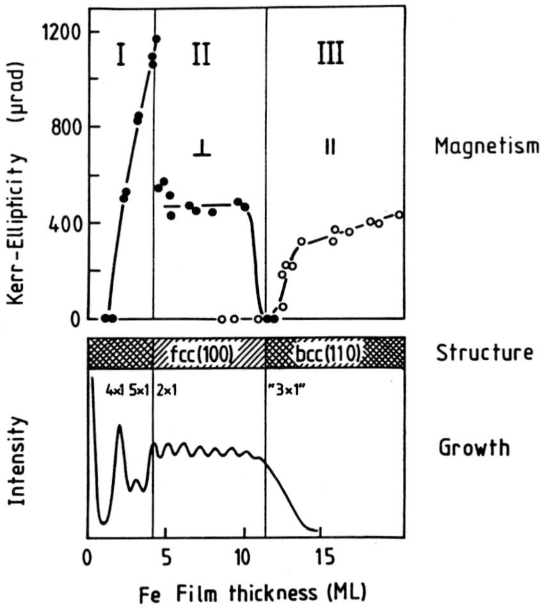

Fig. 6.29. Correlation between magnetism, structure and growth. In the *upper portion* the Kerr ellipticity at $T = 0\,\mathrm{K}$ is depicted for the polar (*filled circles*) and the longitudinal (*open circles*) geometry. In the first two regions, the magnetization is perpendicular to the film surface, and in region III it is parallel to the surface. In the *lower portion*, the intensity of the (00) MEED beam during Fe deposition is depicted showing the same three different regions. The corresponding superstructures are indicated as well. Above 11 ML, the films transform from a strained fcc (100) structure to the bcc (110) structure

signal. The magnetic ground state of this part of the film is either antiferromagnetic and above the Néel temperature, or paramagnetic. The formation of a ferromagnetically ordered ("live") surface layer is a direct consequence of the film structure. Full dynamical LEED calculations for films in region II show that the interlayer spacing is significantly expanded by $0.1 \pm 0.02\,\text{Å}$. While the second interlayer spacing is still slightly expanded to 1.80 Å, deeper interlayer spacings are close to 1.77 Å, the value expected for strained (epitaxially expanded) fcc iron. The correlation between an expanded interlayer spacing and ferromagnetic behavior is in line with theoretical calculations for bulk fcc iron [5]. These calculations predict the existence of a ferromagnetic high-spin state for fcc iron when the lattice constant is larger than 3.6 Å. More to the point, recent theoretical calculations for thin iron films on Cu(100) have reproduced our experimental findings, namely the formation of a ferromagnetic surface layer separated by an enlarged first interlayer spacing from the deeper antiferromagnetic layers [58].

The iron films in this growth and coverage region show a tendency to order, at least locally, into a $p2mg(2 \times 1)$ structure characterized by an antiparallel shift between adjacent atoms (see Fig. 6.27). The displacement pattern can be related to the tendency of neighboring atoms to increase (maximize) their spacing within the surface layer. This is a further indication of an enhanced atomic volume of the ferromagnetic surface atoms. From a comparison of the magnitude of the Kerr ellipticity in region II with that in regions I and III, we can conclude that, at most, two surface layers are ferromagnetic. This can also account for the slightly expanded second interlayer spacing compared with the bulk interlayer spacing.

For most ultrathin magnetic films, it is possible to grow 5–6 layers at most before the magnetization switches from perpendicular to in-plane [59]. This is due to the shape anisotropy of the film, which leads to the reorientation of the magnetization axis. Our findings of a perpendicular anisotropy for films up to 11 ML (region II) are not in contradiction with this, because in our case only the topmost two surface layers are ferromagnetic. Since the magnetization is confined to the surface, the shape anisotropy is independent of the film thickness. In region III (above 11 ML), a parallel magnetic anisotropy is found. The LEED structural analysis confirms the formation of bcc iron, which in turn explains the high Curie temperature of above 500 K. In this thickness region, the entire film is ferromagnetic. The switching from perpendicular to parallel magnetization is therefore only a ramification of the formation of a homogeneously magnetized iron film with a bcc structure and is not controlled by the actual thickness of the film. That the change in the structure of the film over a narrow coverage region is accompanied by a change in the magnetic properties over the same region is further evidence of a correlation between structure and magnetism. In an attempt to confirm that this correlation is not coincidental, the layer-by-layer growth region was extended by codeposition of CO (see Fig. 6.30). For the exposure used, the onset of dislocation formation and hence the structural rearrangement to bcc iron were delayed to above 13 ML. For these films, the observation of layer-by-layer growth and perpendicular magnetization was prolonged to above 13 ML. More significantly, the breakdown of layer-by-layer growth and the switching to parallel magnetization occurred simultaneously. CO deposition after the film had been prepared influenced neither the structure and morphology nor the magnetic anisotropy. Therefore, it can be concluded that the change of the magnetic properties between region II and region III is directly linked to the structural rearrangement to bcc iron accompanying the breakdown of layer-by-layer growth [60]. Sputtering or cooling of a 10 ML thick film also changes the magnetic properties of the film, since these operations lead to the transformation of the film structure into bcc iron as well.

Even for region I (<4 ML), a close correlation between structure and magnetization exists. Films in this thickness region show a (4×1) and then a (5×1) superstructure with increasing film thickness. Both phases are characterized

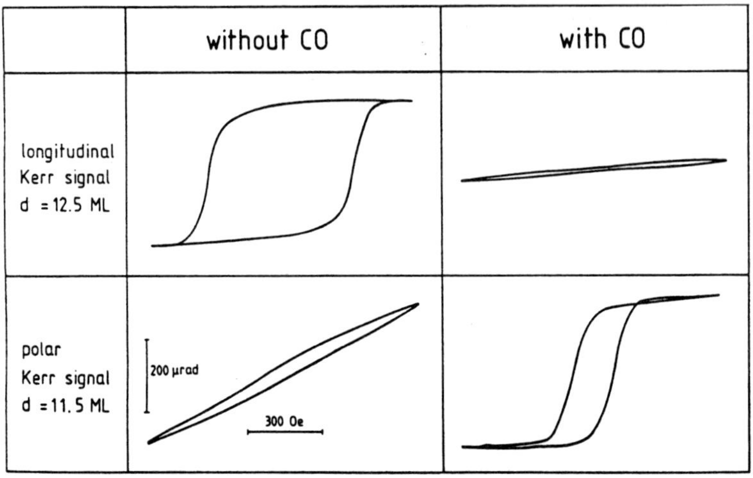

Fig. 6.30. Influence of CO codeposition on the magnetic properties. Kerr ellipticity curves are shown for the polar and longitudinal geometries at 170 K for Fe films grown under optimum UHV conditions or a CO partial pressure of 7×10^{-8} Pa during Fe deposition. Films grown under codeposition of CO grow layer-by-layer up to 13 ML, while the critical thickness for dislocation formation under the optimum UHV conditions is 11 ML (Figs. 6.7 and 6.8). The scales for the x and y axes are shown next to the *lower left curve*. The x axis in the *lower right panel* is expanded by a factor of 5

by an fcc-like structure with considerable vertical disorder and an enlarged interlayer spacing. Total-energy calculations such as those displayed in Fig. 6.3 suggest a high-spin ferromagnetic ground state for an enlarged interatomic spacing (or atomic volume). Most likely, the (4×1) and (5×1) structures with an expanded atomic volume (12.1 Å^3) are intimately related to the predicted ferromagnetic state of bulk iron. This is once again indicative of an interrelationship between enlarged atomic spacings and ferromagnetism for the iron films. Since for both the (4×1) and the (5×1) structure the whole film is reconstructed, it is plausible that for this expanded fcc Fe phase, the mismatch to the Cu(100) surface is so large that this phase cannot be stabilized above approximately 4 ML. Above this coverage, the film rearranges presumably into an antiferromagnetic, low-spin state but with an ferromagnetic surface layer on top that maintains the high-spin moment.

Our results and conclusions are in agreement with experiments by Stampanoni [17], who found a constant polarization of photoemitted electrons for film thicknesses between 5 and about 14 ML. Magnetic live surface layers can also account for the “absence of ferromagnetism in the usual sense” [9]. Similarly to other studies [9, 31, 36, 37], we have also found that the growth and structure of the iron films are dependent upon the growth conditions. This dependence will be discussed in the next section.

6.3 Topics Related to Ultrathin Fe/Cu(100) Films

In the last section, we concentrated on the growth, structure and magnetic properties of Fe films grown on Cu(100) at room temperature, in order to clearly demonstrate the correlation between structure and magnetism. In fact, in the last two decades, numerous studies related to Fe/Cu(100) films have been performed. These studies cover the magnetic ordering and spin structure in Fe/Cu(100) films, the effects of growth conditions and preparation methods on the structure and magnetism of the films, the evolution of the structure and magnetism upon variation of external parameters (temperature, magnetic field, etc.), and the modification of the film properties when the substrate lattice parameter is slightly changed. All of these topics are discussed in this section.

6.3.1 Magnetic Order in Fe/Cu(100) Films and Its Origin

We have seen in the last section that Fe/Cu(100) films show different types of magnetic order with variation of film thickness. Below 4 ML (region I), the films show ferromagnetic order in the entire film, with a Curie point above room temperature. Between 5 and 11 ML (region II), the ferromagnetic order is located in the vicinity of the surface only. Though the picture described above is generally accepted, some aspects concerning the detailed magnetic ordering and spin structure of Fe/Cu(100) films are still unclear. In the following, we shall discuss two issues which need to be investigated in more detail.

Though ferromagnetic order has been well established [60, 61, 62] for Fe/Cu(100) films in region I, the physical explanation is still a disputed issue. In most studies, the ferromagnetism observed in Fe/Cu(100) films with a thickness below 4 ML is connected with an expanded atomic volume. The films in this thickness region are found to be reconstructed throughout the whole film and show an enlarged atomic volume of 12.1 Å^3 as compared with that of fcc Cu (11.76 Å^3) [40]. As a result of the moment–volume instability, an enlarged atomic volume favors a large atomic magnetic moment. At the same time, an enlarged atomic volume also favors a positive exchange constant and thus a ferromagnetic coupling, according to the Bethe–Slater curve (see Fig. 6.1). The magnetic properties of Fe/Cu(100) films in region II can also be explained within this picture. Since only the atoms of the surface layer are reconstructed and have an enlarged atomic volume, only the surface layer shows ferromagnetic order. The interior atoms have an fcc structure with an atomic volume of 11.4 Å^3, so they are supposed to be in the antiferromagnetic low-spin state.

However, the above explanation has been challenged recently by an STM study [63]. Figure 6.31 shows three atomically resolved STM images of Fe/Cu(100) films. All the films imaged have a thickness between 2 and 4 ML, but they were prepared and measured under different conditions. However,

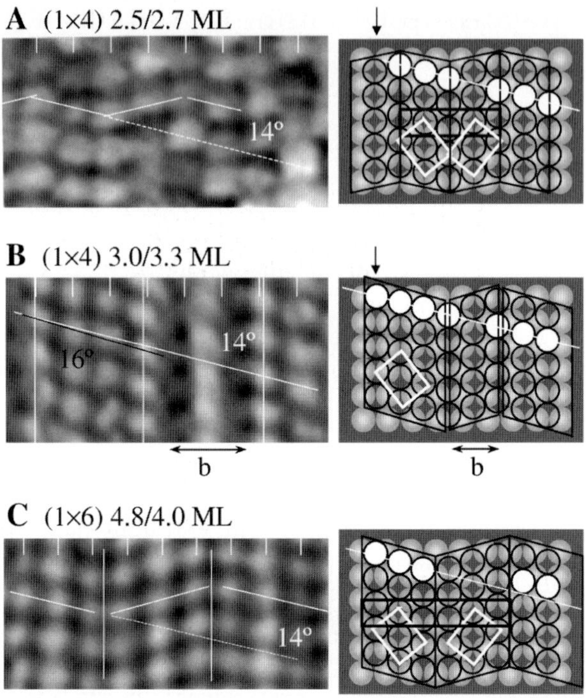

Fig. 6.31. Atomically resolved STM images of 2–4 ML iron films on Cu(100), prepared and measured under different conditions. (A) Film with (1×4) superstructure, grown at 130 K, annealed at 300 K and imaged at 80 K. (B) Film with (1×4) superstructure, grown at 310 K and imaged at 80 K. (C) Film with (1×6) superstructure, grown at 310 K and imaged at 300 K. The three images all show bcc-like stripes sheared by the +14° or −14° with widths varying between 2 and 4 atom rows. The *white rectangles* show the (110) bcc-like cell. The first thickness value given above each image was measured with a quartz crystal microbalance, and the second by AES. From [63], used with permission

all visible structures can be imagined as resulting from a shearing of the fcc lattice by 14°, which makes the local atomic arrangement very similar to that of a (110)-oriented bcc film in the Pitsch orientation [29] with respect to the fcc substrate (Fig. 6.32). The shear instability of fcc iron in ultrathin films has been confirmed by ab initio local-spin-density calculations [64], which predicted a shear angle of 13° in good agreement with the shear angle of 14° deduced from the STM experiment. Since the atomic surface density of the bcc (110) local structure is equal to that of the fcc substrate, these bcc-like structures are considerably strained. Their surface atomic density in the relaxed state would be 12% higher. The films do not show the long-range translation symmetry of a bcc lattice, but consist of alternating sequences of stripes of bcc-like twins, leading to a zigzag deformation of the original

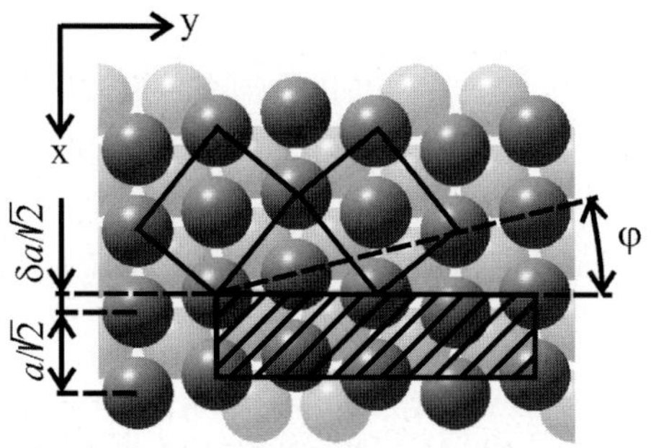

Fig. 6.32. The two topmost layers of the (1×4) striped bcc structure on top of the underlying fcc structure of the substrate. The *hatched rectangle* shows the (1×4) surface cell. The *smaller rectangles* mark the strained (110) bcc-like cell. The shear angle φ of 14° is that measured in the STM experiment. From [64], used with permission

straight rows of atoms. Some aspects of the zigzag-like deformation of the fcc lattic resemble the sinusoidal mode proposed by Müller et al. [40], which was derived from I/V LEED data. However, the multiphase microdomain structure is a new feature revealed by this STM study. An enlarged atomic volume is also expected for this zigzag multidomain structure from linear elastic theory.

On the basis of these STM observations, it has been argued that the ferromagnetism of Fe/Cu(100) films in region I comes from the nanomartensitic bcc structure. The volume (interlayer) expansion, which has been traditionally taken as the key point that motivates one to suggest a ferromagnetic fcc-like phase, is explained in this scenario as an elastic effect due to the strain state of the commensurate bcc-like film. Obviously, this explanation focuses more on the local atomic environment than on the atomic volume. One should note, however, that the interpretation of the STM images as evidence for a bcc (110)-like structure creates a number of open questions. Since the bcc phase is the equilibrium structure at room temperature, why should the film form an fcc structure with increasing film thickness and only form a bcc(110) oriented film above 9 ML? For this bcc (110) phase, interlayer mass transport is very difficult, leading to three-dimensional growth. This is in line with observations for Fe(110) [66], where at room temperature three-dimensional growth is observed. For the supposedly bcc (110)-like thin-film phase (region I), however, interlayer mass transport is possible, leading to smooth films. This is not easily explained under the assumption that the STM image is evidence for a bcc (110)-like structure. Finally, also the observed perpendicular magnetic anisotropy has never been observed for films with a bcc (110) surface. Hence this interpretation of the STM images cannot easily explain a large number of established experimental facts.

Now we turn to the magnetic order in Fe/Cu(100) films in region II (4–11 ML). For these films, an antiferromagnetic ground state has been generally accepted. However, the details of the antiferromagnetic spin configuration and even the Néel temperature are still controversial or unclear issues.

Early neutron diffraction measurements on γ-Fe in a single-crystal copper matrix [65] suggested an antiferromagnetic structure similar to γ-manganese, in which the magnetic moments of atoms in alternate (001) planes are parallel or antiparallel to each other. The spin vector was found to be inclined at about 19° to the normal of the ferromagnetic sheets. An atomic moment of $0.7\mu_B$ and a Néel temperature of 8 K were estimated in the same experiments. Mössbauer effect measurements [67] found a higher Néel temperature for γ-Fe stabilized by the Cu lattice, equal to 55 ± 3 K and 67 ± 2 K for two different sizes of precipitates. Recently, a spin density wave (SDW) antiferromagnetic spin structure with a transition temperature of about 60 K has been proposed on the basis of x-ray and neutron diffraction data from γ-Fe precipitates in Cu [68]. This finding is supported by theoretical calculations [69]. Therefore, this aspect of the spin structure of γ-Fe precipitates is still not settled. Even if the spin structure of γ-Fe precipitates were known, one would have no reason to believe that Fe/Cu(100) films with a thickness in region II would have the same spin structure as γ-Fe precipitates in a copper matrix. The epitaxial constraint and the different film geometry could lead to a different spin configuration for the films. Owing to the experimental difficulties, only limited measurements have been performed to specify the magnetic properties of Fe/Cu(100) films in region II. An early study using ^{57}Fe conversion-electron Mössbauer spectroscopy of fcc Fe/Cu(100) films in UHV by Macedo and Keune [12] showed a drastic broadening of the spectral line below $T_N = 65 \pm 5$ K, indicating the occurrence of magnetic ordering. A similar broadening has been observed for γ-Fe precipitates below their Néel temperature [67]. The measurements also revealed a small average hyperfine field below T_N, indicating a small atomic magnetic moment. The magnetic ordering at low temperature in iron films in region II has been confirmed by Keavney et al. [70], Ellerbrock et al. [61] and Freeland et al. [71] using the same technique, but Freeland et al. measured a lower T_N (~35 K) in a 4 ML Fe/Cu(100) film. On the basis of these Mössbauer results, it seems to be conclusive that Fe/Cu(100) films in region II have an antiferromagnetic ground state and that the magnetic ordering occurs below 70 K. Unfortunately, the Mössbauer effect measurements cannot give any information concerning the spin structure. Up to now, the spin configuration for Fe/Cu(100) films in region II has only been investigated theoretically [75, 58, 72, 73, 74, 76, 77]. The ab initio calculation results of Fu and Freeman [72] and Fernando and Cooper [73] show that in a free-standing Fe layer with the Cu(100) lattice constant, there exists a ferromagnetic coupling between the surface layer (S) and the first subsurface layer ($S-1$), but an antiferromagnetic coupling be-

Table 6.4. Summary for the theoretical results for the ground-state configurations of fcc Fe/Cu(100) films; u = up and d = down indicate the orientation of the magnetic moment in a layer, starting at the free surface

Thickness (ML)	1	2	3	4	5	6	7	8	9
Results of [75]	u	uu	uuu	uudd	uuddd	uudduu			
Results of [77]	u	uu	uuu	uudd	uuddd	uudduu	uuudduu	uudduudd	uudududuu
Results of [76][a]	u	uu	udu	uddu	uuduu	uudduu	uuddduu	udduuddu	uudduddu u

[a] The calculations were based on nFe/1Cu(001) superlattices with an ideal fcc Cu structure.

tween the $S-1$ and $S-2$ layers. Kraft et al. [58] have extended the calculation to a symmetric free-standing 11-layer slab and demonstrated that the antiferromagnetic coupling between neighboring layers continues from the $S-1$ layer to the center of the slab. All calculations agree on a ferromagnetic intralayer spin order and a strong enhancement of the magnetic moment at the surface. These results are consistent with the surface live-layer model. On the other hand, Asada and Blügel [75] have conducted a thorough search of the possible collinear spin configurations for Fe/Cu(100) overlayers and obtained a complicated and thickness-dependent spin structure (see Table 6.4). Calculations including interface effects by Moroni et al. [77] have produced exactly the same spin structures as those given by Asada and Blügel. These results are in agreement with the experimental data showing live surface magnetism for fcc Fe films in region II, but lead to a different explanation of the magnetic live surface. In this picture, the inner Fe layers may also add to the net magnetic moment for these films, besides the surface layer [75].

As to the magnetic properties of Fe/Cu(100) films in region II, we should mention the anomalous magnetic behavior observed for these films around 200 K. This anomalous behavior was first observed by Li et al. [78] in polar MOKE measurements on an Fe wedge grown at 280 K on Cu(100). The measured saturation magnetization shows oscillations with a peak separation of about 2.6 ML in region II (see Fig. 6.33). Since no longitudinal Kerr signal was detected in the same thickness region, a canted spin structure was ruled out as an explanation of the oscillations. The oscillations were thought to result from antiferromagnetic Fe underlayers with alternating sheets of spins. With the total number of these spin sheets changing, the net moment and thus the Kerr signal would oscillate with thickness. A linear extrapolation of the oscillation amplitude with respect to the measurement temperature gives a critical temperature $T \sim 200$ K, above which the oscillations disappear. This temperature has been taken as the Néel temperature of the interior part of 5–11 ML Fe/Cu(100) films by some authors. Similar oscillations have been observed by Qian et al. [79] (see Fig. 6.34). In that study, a spin-density-wave antiferromagnetic configuration was proposed to explain

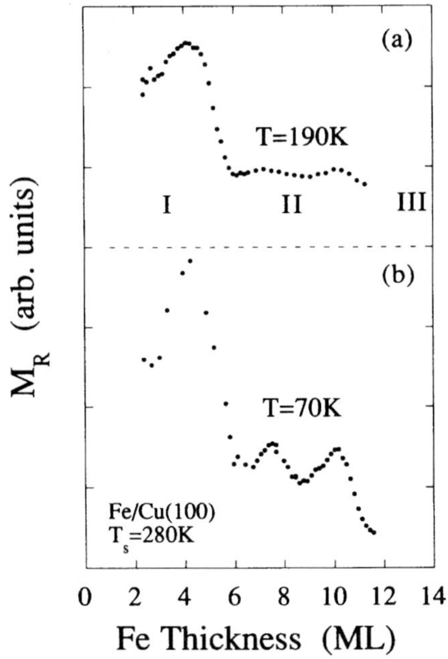

Fig. 6.33. Saturation magnetization M_R measured from polar SMOKE as a function of thickness for an Fe wedge on Cu(100). The measurement was performed at 190 K (**a**) and 70 K (**b**). Oscillations of M_R with thickness appear in region II when the measurement temperature is below 200 K, and they increase in amplitude upon cooling. From [78], used with permission

the oscillations. The z component of the SDW wave vector was determined to be 2.7 ML by fitting the oscillating data; this is in good agreement with the peak separation of 2.6 ML reported by Li et al. Anomalous magnetic behavior has also been observed in bilayer structures which contain an fcc Fe(100) layer [80, 81, 82]. For example, magnetic relaxation behavior has been observed below about 200 K in a 9–ML Fe/15 ML Ni bilayer grown on Cu(100) (Fig. 6.35) [81], and a jump in the longitudinal Kerr signal around 200 K in a 5.3–ML Fe/7–ML Ni bilayer on Cu(100) (Fig. 6.36) [80]. All the results indicate that something happens around 200 K for fcc Fe/Cu(100) films in region II. The question remains of what is the origin of the phenomena at this temperature. An SDW antiferromagnetic transition [78, 79] is one possible answer to this question. However, the SDW model lacks direct evidence, and a theoretical study [83] shows that spin waves are always unstable with respect to collinear configurations. Therefore one cannot exclude other possible mechanisms which could lead to this anomalous behavior. For example, interesting Mössbauer results have been reported by Keavney et al. [70]. Their Mössbauer measurements reveal two types of Fe atomic sites in the interior portion of an fcc Fe/Cu(100) film: low-spin and high-

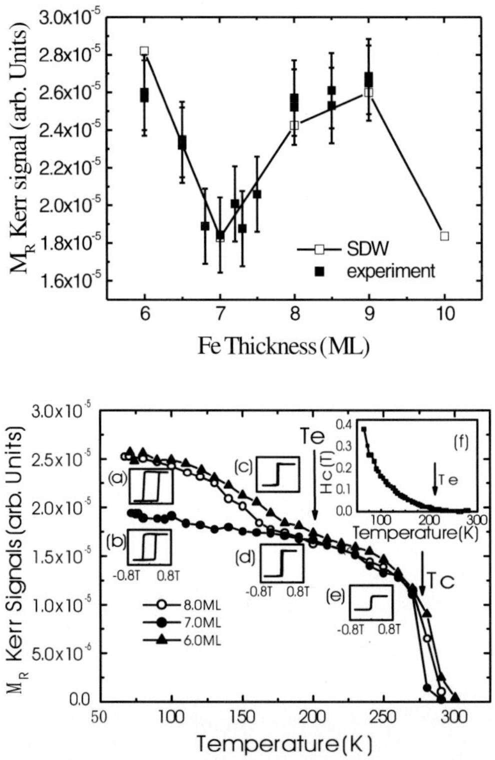

Fig. 6.34. *Top*: saturation magnetization measured from polar MOKE at 70 K as a function of film thickness, together with a fitting curve assuming an incommensurate SDW with $s_z = s_{z0} \cos(qz)$ and $q = 2\pi/2.7d$. *Bottom*: temperature-dependent saturation magnetization as a function of temperature for 6, 7 and 8 ML Fe films on Cu(100). Curves (a)–(e) show representative hysteresis loops at different temperatures; (f) shows the curve of coercivity versus temperature for the 8 ML Fe/Cu(100) film. From [79], used with permission

spin Fe sites (see Sect. 6.3.3). The hyperfine splitting of the sextet lines, which is indicative of the ordering of the high-spin bcc-like Fe atoms, is discernable at 150 K. From this observation, Keavney et al. proposed that there exist bcc-like high-spin clusters in a low-spin matrix. If this is true, the anomalous magnetic behavior described above could be connected to the magnetic ordering of the high-spin Fe clusters. In addition, one cannot rule out the possibility that a slight change in structure occurs below 200 K. To give a definite explanation of the magnetic anomaly around 200 K shown by fcc Fe/Cu(100) films, extensive temperature-dependent measurements are required.

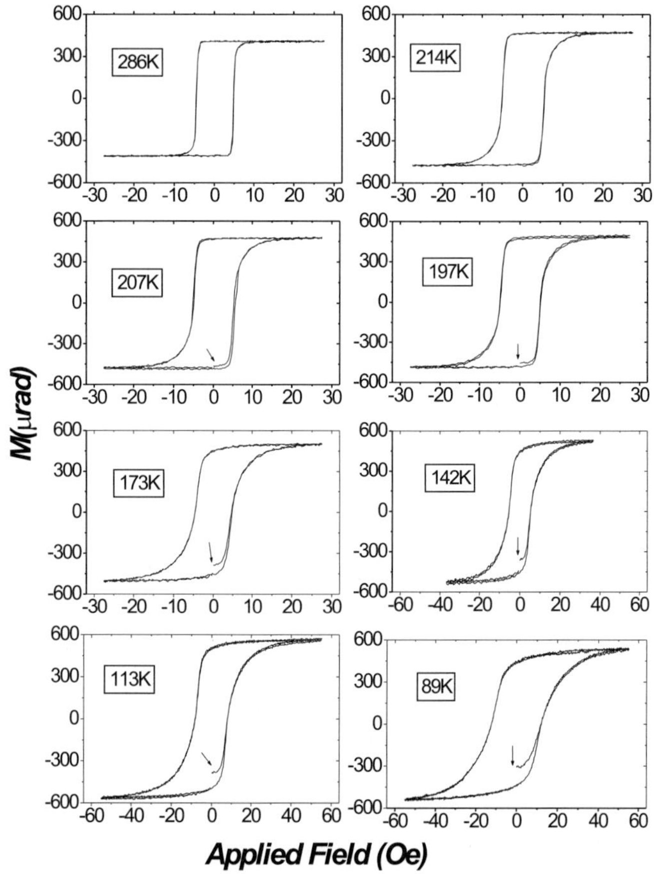

Fig. 6.35. Polar MOKE hysteresis loops for the first two sweeps at different temperatures for a 9–ML Fe/15–ML Ni bilayer on Cu(100). The *arrows* indicate the starting point for the first loop. The deviation of the starting point from the hysteresis loop indicates a magnetic relaxation behavior, which starts to appear around 200 K and becomes more pronounced upon cooling

6.3.2 Structural and Magnetic Properties of Fe/Cu(100) Films Prepared at Low Temperature

It has been found that the deposition temperature has a significant impact on the growth behavior and the magnetic properties for Fe/Cu(100) films. When the films are prepared at low temperature (these films will be referred to as "LT films"), i.e., near 100 K, MEED measurements show no intensity oscillations but an intensity decay with film thickness, indicating a three-dimensional mode for the growth of the LT films [84, 85]. This is in contrast to the nearly layer-by-layer growth of iron films at room temperature ("RT" films). The significant influence of deposition temperature is also reflected in

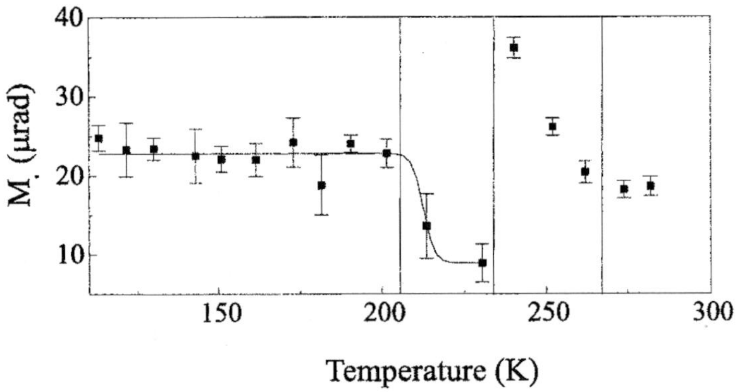

Fig. 6.36. Temperature dependence of the longitudinal Kerr ellipticity at saturation for a 5.3 ML Fe/7 ML Ni bilayer on Cu(100). From [80], used with permission

the magnetic properties of the films. In contrast to the perpendicular magnetic anisotropy persisting up to 11 ML for the RT films, a magnetization reorientation from the perpendicular direction at small thickness to in-plane at large thickness occurs between 5 and 6 ML for LT films [16, 86]. To unravel the reason for this difference and to further clarify the correlation between growth, structure and magnetism, we have also studied the magnetic properties and the structure of iron films deposited on Cu(100) at 100 K and subsequently annealed at room temperature [40, 84, 87].

In Fig. 6.37, the saturation magnetization, the remanent magnetization, the coercive field and the Curie temperature of LT Fe/Cu(100) films are plotted as a function of film thickness. Two regimes with different magnetic properties can clearly be distinguished. Above 5.8 ML (region II), the films show only in-plane magnetization. The saturation magnetization measured in the longitudinal geometry increases linearly with film thickness. This is a characteristic feature of a film which is homogeneously magnetized. On the other hand, films with a thickness below 4.5 ML (region I) show a magnetization only in the perpendicular direction. Again, the ellipticities increase linearly with film thickness, indicating that the entire film is also ferromagnetic in region I. The magnetization direction switches from perpendicular to in-plane between 4.5 and 5.8 ML, in agreement with previous work [16, 86]. The Curie temperature lies above 380 K below 4.3 ML,[2] and then has a minimum of 365 K around 4.6 ML until it begins to increase steeply above 5.5 ML. This minimum of the Curie temperature implies a thickness-dependent structural evolution or a heterogeneously magnetized iron film.

[2] Since considerable Cu diffusion has already set in above 380 K in region I, it is impossible to measure the Curie temperature precisely in this region. The same holds for thicknesses larger than 6.5 ML in region II, because above this thickness the Curie temperature is higher than T_{dif}, the temperature above which Cu diffusion becomes important.

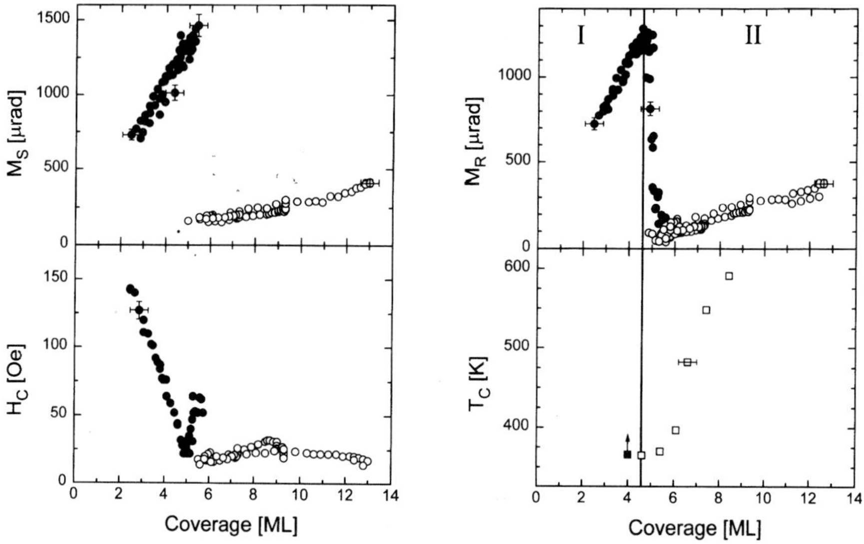

Fig. 6.37. Thickness dependence of the Kerr ellipticity at saturation M_s, the coercive force H_c, the Kerr ellipticity at remanence M_R and the Curie temperature T_c for Fe films deposited at 100 K and subsequently annealed at 300 K. The hysteresis loops were recorded at 166 K. The *solid* and *open circles* denote the values obtained from curves recorded in the polar and the longitudinal geometry, respectively. T_c was determined by extrapolating M_s to zero. The *arrow* indicates that the measured value for T_c has been reduced through the diffusion of Cu to the surface, leading to an irreversible change of the film composition

With increasing film thickness, three superstructures have been identified by LEED. These are a (4×1) and a (5×1) structures and a (3×1)-like phase. The (4×1) and (5×1) superstructures are observed between approximately 2 and 4.5 ML. After Fe deposition of more than 4.5 ML, the diffraction spots become diffuse and the background intensity increases. Streaks running in the [011] and [0$\bar{1}$1] directions are visible in the LEED pattern. Their intensity is centered close to the positions of a (3×1) superstructure. However, closer inspection reveals small deviations from the positions of an ideal (3×1) structure. So we call the structure a (3×1)-like structure. A similar diffraction pattern has been observed for room-temperature-prepared iron films on Cu(100) with a thickness above 11 ML. As has been discussed in the previous section, the (3×1)-like LEED pattern is due to the formation of bcc iron that grows with the (110) face almost parallel to the Cu(100) surface [39]. This gives evidence that Fe films deposited at low temperature transform into a bcc structure as early as around 5 ML. A comparison of the LEED I/V curves for Fe films deposited at 100 K and 300 K (see Fig. 6.38) makes this point clearer. The I/V curve of a 5.8 ML iron film deposited at low temperature shows a sequence of peaks similar to that of a 20 ML iron film prepared

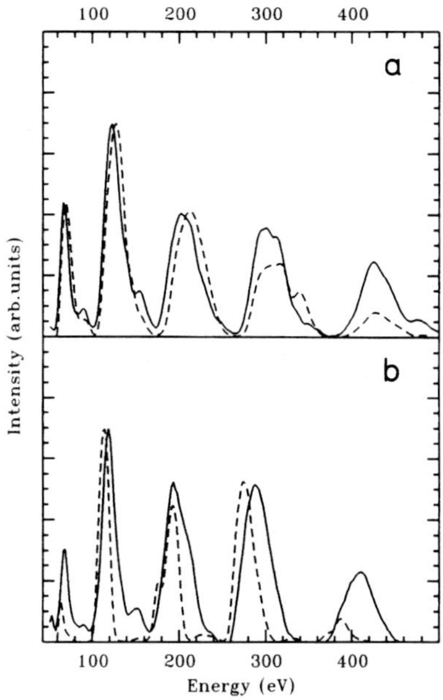

Fig. 6.38. Comparison of LEED I/V curves for Fe films deposited at 100 K and annealed at 300 K (*solid lines*) and for films deposited at 300 K (*dashed lines*). The *upper part* of the figure shows curves for the (5×1) structure at 3.7 ML, while the *lower part* shows the curves for the bcc phase at 20 ML for room-temperature deposition and 5.8 ML for low-temperature deposition

at 300 K, indicative of a similar geometric arrangement for the two films. It is well known that the latter has a bcc structure at this thickness. On the other hand, the 3.7 ML films show similar I/V curves, independent of the deposition temperature. Magnetic-anisotropy measurements provide additional evidence for a direct structural transition from distorted fcc to bcc in LT iron films (see Fig. 6.39) [88]. The two linear segments for the LT films suggest the absence of the intermediate (2×1) phase, which appears in RT Fe/Cu(100) films.

In conclusion, the iron films deposited at room temperature and low temperature have similar structural and magnetic properties up to 5 ML. Both are homogeneously magnetized and show a perpendicular magnetic anisotropy. A (4×1)–(5×1) superstructure is observed by LEED for both deposition temperatures. However, the influence of the deposition temperature on the film structure and magnetism starts to manifest itself when the film thickness exceeds 5 ML. For RT iron films, between the (4×1)–(5×1) distorted thin films ($<$ 4 or 5 ML) and the bcc thick films ($>$ 11 ML), there exists a (2×1) intermediate structure in a thickness range between approximately 5 and 10 ML. These films have a distorted fcc structure [41] and show a $p2mg$(2×1) reconstruction of the surface layer at low temperature [42]. The surface layer(s)

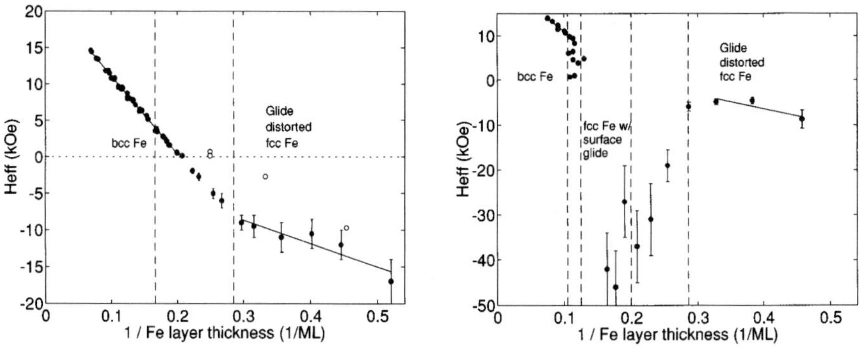

Fig. 6.39. The thickness dependence of the effective anisotropy field H_{eff} for Fe films grown on Cu(100) at 100 K and subsequently annealed at room temperature (*left*) and for films grown at room temperature (*right*). In the left panel, the *open circles* are for Fe films prior to annealing. The *dashed vertical lines* in the left panel mark the onset and completion of the fcc–bcc transformation, while in the right panel they mark the onset and completion of the fcc–bcc transformation and the fcc-like (5×1)–(2×1) transformation. The effective anisotropy field is defined as $H_{eff} = 4\pi M_s - 2K_1/M_s$, where K_1 is the first-order anisotropy constant, and it can be decomposed into thickness-dependent and thickness-independent parts, i.e., $K_1 = K_v + 2K_s/t$, where t is the film thickness. From [88], used with permission

is/are ferromagnetically ordered and referred to as a "magnetic live layer", while the interior of the film shows antiferromagnetic coupling [12, 89]. Only for thicknesses larger than 11 ML does the martensitic transformation from fcc to bcc iron occur via a collective, nondiffusive displacement of many adjacent atomic rows in the fcc [011] direction [39, 90]. In contrast, such a structure is absent for LT films. A direct transition from ferromagnetic fcc iron to ferromagnetic bcc iron occurs around 5 ML. Now the question is why an antiferromagnetic (2×1) Fe phase is stabilized above 5 ML for films grown at room temperature but not for deposition at low temperature. At present, we have no definite answer to this question. We can just discuss several possible causes here.

Independently of deposition temperature, the ferromagnetic fcc phase with characteristic (4×1) and (5×1) LEED patterns can only be stabilized up to slightly above 4 ML. Then this phase becomes unstable. Apparently then two options exist. One is an increase of the shear amplitude and an enlargement of the interlayer spacing, leading to a transformation to bcc Fe. The second possibility is a vanishing of the sinusoidal displacement of the atoms and a decrease of the interlayer spacing in the iron film, resulting in an antiferromagnetic fcc phase. We suggest energetic, kinetic and compositional differences as possible driving forces for the different scenarios. From the viewpoint of energetics, one might argue that the activation barrier for the transition to bcc iron is large for the room-temperature-deposited (5×1) phase compared with low-temperature-deposited (5×1) phase. In contrast, the activation energy for the transition to antiferromagnetic fcc iron is larger for the LT (5×1) Fe phase. These different barriers could be due to small differences in atomic positions which depend on the deposition temperature. In Fig. 6.38, one can see that the peaks of the I/V curves are shifted to somewhat lower energies for LT films than for RT films with equal thickness. This result would indicate an even larger interlayer spacing for the LT films. Medium-energy ion scattering experiments have revealed a more pronounced atomic disorder for LT Fe films than for RT films [91]. These results suggest that the difference in the structural transition between LT and RT iron films could have a structural origin. Another possibility is that the difference could have kinetic reasons. In this case one might argue that an important procedure necessary for the formation of antiferromagnetic fcc iron is suppressed at low temperature and hence the transformation goes towards bcc iron. There is indirect evidence pointing in this direction from STM data by Giergiel et al. [92]. This STM study showed that for room-temperature deposition, a small fraction of the surface already deviates from the ideal fcc structure above 4.6 ML. These features later develop into bcc precipitates. It would be rather instructive to investigate bcc iron formation at low temperatures with the STM as well, to prove that there is a larger fraction of these features. Finally, compositional differences in the film might account for the different activation barriers. It is well known that, at least at room temperature, small amounts of Cu are incorporated into the Fe film up to 2 ML. This incorporation might stabilize the AFM fcc iron Fe phase.

6.3.3 Structural Instability of Ultrathin Fe/Cu(100) Films

The structural instability of ultrathin Fe films on Cu(100) has been manifested by a number of experimental observations. It is indicated by the reconstruction of the entire film for ferromagnetic fcc-like Fe films (<4 ML) [40] and the structural transition with increasing thickness for both RT and LT films. Additionally, it has been found that Fe/Cu(100) films with a thickness close to the structural boundaries, for example a 4 ML film in region I and an 11 ML film in region II, exhibit structural rearrangements upon cooling. These results directly demonstrate the instability of the associated structures.

Zharnikov et al. [93] have found that the loss of the long-range magnetic order above 330 K for a 4 ML Fe/Cu(100) film is actually due to a crystalline restructuring occurring at that temperature and its consequences for the magnetic properties. They have recorded the LEED I/V curves for (00) beam (Fig. 6.40) and the Kerr intensity (Fig. 6.41) for a 4 ML iron film during heating and cooling of the sample. The intensities of the peaks at 145 and 176 eV were used as a measure of the fraction of the reconstructed fct structure and of the unstrained fcc structure, respectively. Upon heating of the sample, the film shows an irreversible change around 300 K both in the intensity ratio of the I/V peaks and in the saturation magnetization. A coincidence

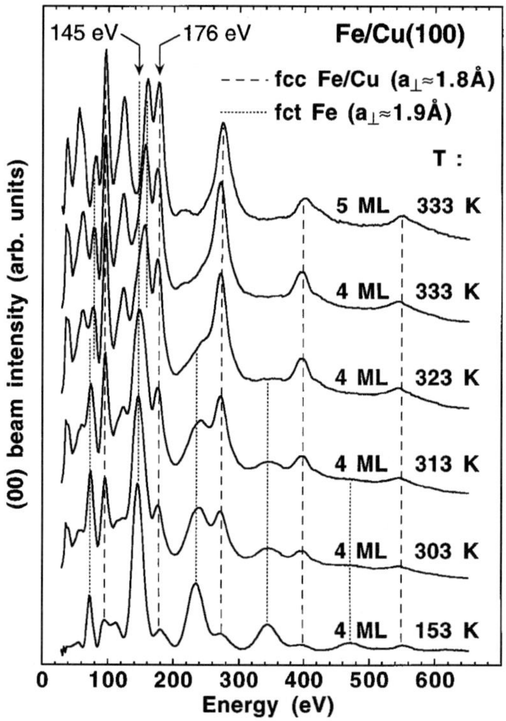

Fig. 6.40. LEED I/V curves of the (00) beam for 4 ML Fe on Cu(100) measured at various temperatures during heating, and for 5 ML Fe measured at 333 K (*topmost curve*). Two periodic sequences of maxima ($S_{fcc}(E)$ and $S_{fct}(E)$), corresponding to the fcc and fct structures of Fe on Cu(100), are traced by *vertical dashed* and *dotted lines*, respectively. From [93], used with permission

between the temperature-dependent behavior of the saturation magnetization and the ratio $I(\text{fct})/I(\text{fcc})$ indicates that the ferromagnetic long-range order in the 4 ML Fe/Cu(100) film is not destroyed by the usual Curie transition, but by the temperature-driven structural transformation. The new structure after heating shows ferromagnetic long-range order at 270–280 K, which is the typical Curie temperature for an Fe film with an fcc structure expanded only in the topmost layer. Atomic interdiffusion can be excluded as the cause of this structural transition, since such irreversible behavior does not appear in 3 ML Fe/Cu(100) films, where the onset of interdiffusion occurs at an even lower temperature. The ratio $I(\text{fct})/I(\text{fcc})$ for a 3 ML film shows a slow and reversible change with temperature (up to 400 K), while the Kerr measurement reveals a reversible Curie transition around 365 K.

Temperature-driven structural rearrangements have also been observed for films in region II which have an fcc structure. Wuttig and Thomassen first found [41], by a LEED I/V investigation, that a room-temperature-

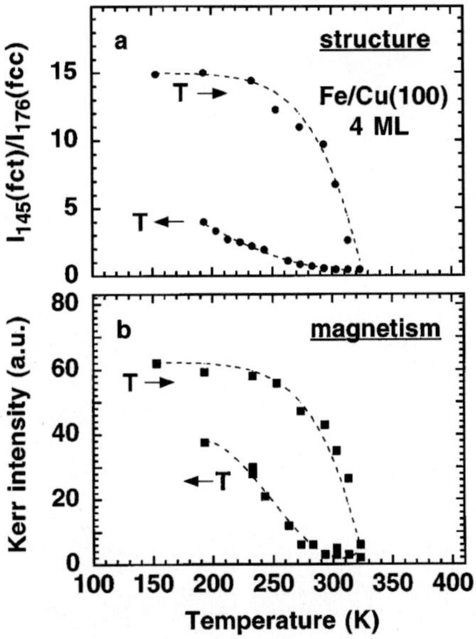

Fig. 6.41. The ratio of the intensities of the peaks at 145 and 176 eV (marked in Fig. 6.40) in the I/V curves for the (00) beam (**a**), and the saturation magnetization (**b**) for a 4 ML Fe/Cu(100) film during heating and cooling of the samples (indicated by *arrows*). From [93], used with permission

deposited 11 ML Fe/Cu(100) film experiences an irreversible structural transition upon cooling. On the other hand, the I/V curves for an 8 ML thick film hardly change upon cooling. This finding has been confirmed by a recent STM study (see Fig. 6.42) [94], which showed that Fe/Cu(100) films with a thickness of 9–11 ML have an almost fcc-like morphology before cooling but a bcc-like morphology after cooling (see Figs. 6.10 and 6.11). The observed structural rearrangement upon cooling is consistent with the fact that the fcc–bcc transition occurs at a much smaller thickness for LT films than for RT films.

Now that we have discussed the structural instability of Fe/Cu(100) films upon variation of temperature, we would like to end this section with a short comment on the structural and magnetic stability of these films in a magnetic field. A field-induced metastable transition has been observed in a room-temperature-grown 3.5 ML Fe/Cu(100) film [95]. After the sample has been subjected to a 9.0 kOe polar magnetic field, the perpendicular component of the remanent magnetization disappears and only the in-plane component is left. A slight heating of the sample can restore the original state. The disappearance of the perpendicular anisotropy was connected to a structural transition with a magnetoelastic origin. To our knowledge, this is the only report concerning the field-induced instability of Fe/Cu(100) films. Obviously this phenomenon deserves further investigation. Identifying if the film struc-

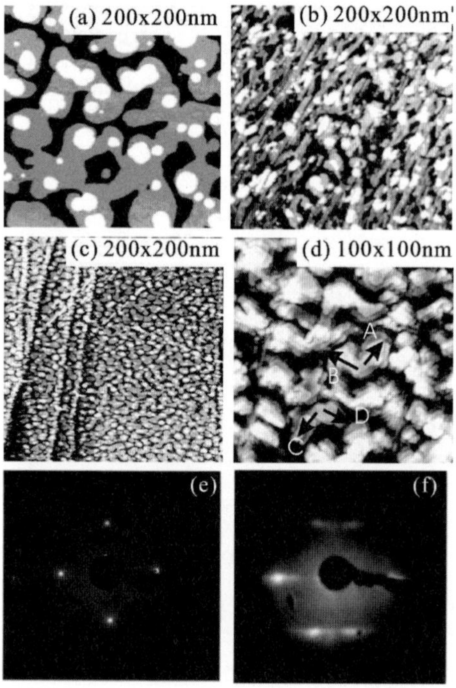

Fig. 6.42. STM images of a 9 ML Fe/Cu(100) film before (**a**) and after (**b**) a cooling–warming procedure, a 10 ML Fe/Cu(100) film as grown at room temperature (**c**), and a 12 ML bcc Fe/Cu(100) film (**d**). (**e**) and (**f**) show the LEED patterns for the 9 ML Fe/Cu(100) film before and after the cooling–warming procedure. From [94], used with permission

ture changes with an applied magnetic field and how it changes if it does will contribute to a complete understanding of the correlation between structure and magnetism for Fe/Cu(100) ultrathin films.

6.3.4 Structure and Magnetism of Pulsed-Laser-Deposited Ultrathin Fe Films on Cu(100)

Pulsed-laser deposition (PLD) is usually employed for the preparation of thin films with a thickness of at least a few nanometers, and is used especially for complex oxide films owing to the advantage of obtaining a stoichiometry consistent with the target. Only very recently, PLD was used to prepare ultrathin metal films in a UHV chamber [96]. To do so, a low laser power (slightly above the ablation threshold) and a large target–substrate distance were employed in order to conveniently control the film thickness and to avoid droplet formation. The PLD method has also been applied to prepare ultrathin iron films on Cu(100) in the expectation of modifying the growth behavior and film structure in comparison with the extensively investigated thermally deposited (TD) films. This could bring about a further clarification of the correlation between structure and magnetism in this system [97, 98, 99]. Compared with the TD Fe/Cu(100) films discussed in the previous sections, PLD films show different features regarding growth, structure and magnetism.

In contrast to the significant deviation from layer-by-layer mode for the growth of the first two monolayers of TD Fe/Cu(100) films, layer-by-layer growth is achieved in the initial stages when the film is prepared by the PLD method. The layer-by-layer growth is characterized by clear, regular RHEED oscillations (see Fig. 6.43) and further confirmed by STM observations [97]. Consistent with the layer-by-layer growth mode, PLD preparation leads to an improved film quality up to 5 ML coverage, as is revealed by the higher LEED intensity [99]. However, above 6 ML, thermal deposition leads to a better film quality [99]. The LEED patterns are very similar for both deposition methods. That is to say, superstructures of type (4×1), (5×1) and (2×1) are

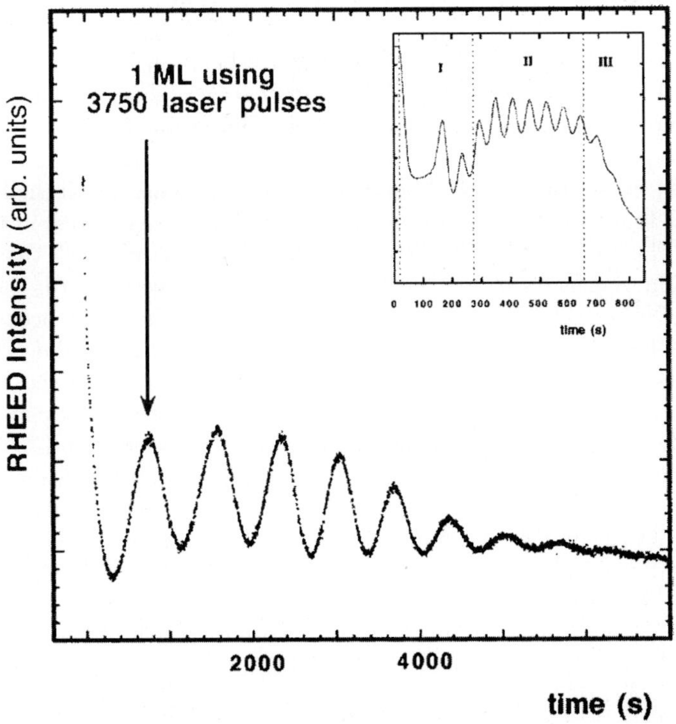

Fig. 6.43. RHEED oscillations of a 10 ML pulsed-laser-deposited iron film grown on Cu(100) at room temperature. Clear, regular oscillations are observed for the first two monolayers, where oscillations are usually absent for thermally deposited films. The *inset* shows irregular MEED oscillations in this thickness range obtained during the thermal deposition of an Fe/Cu(100) film. From [97], used with permission

Table 6.5. Comparison of the best-fit structural parameters for 4 ML Fe films prepared by thermal deposition and pulsed-laser deposition. The vertical layer spacing is denoted by d_{ij}, and the maximum of the vertical buckling amplitude is b_{imax} (see Fig. 6.27). From [99]

Parameter	TD	PLD
b_{1max} (Å)	0.24±0.02	0.41±0.03
b_{2max} (Å)	0.11±0.02	0.19±0.04
b_{3max} (Å)	0.12±0.03	0.07±0.05
b_{4max} (Å)	0.21±0.05	≈0
d_{12} (Å)	1.86±0.02	1.91±0.03
d_{23} (Å)	1.90±0.02	1.82±0.03
d_{34} (Å)	1.85±0.03	1.76±0.05
d_{45} (Å)	1.78±0.06	≈1.78

observed with increasing thickness for both methods. However, the superstructure spots appear to be slightly weaker for PLD films than for TD films. Though the LEED patterns show a similarity between PLD and TD films, a Tensor LEED analysis of the 4 ML films reveals a considerable difference in structure between the PLD and TD films [99]. Table 6.5 lists the best-fit structural parameters for both films. The differences can be summarized as follows: (i) In contrast to the expanded interlayer spacing for all the Fe layers in the TD film, only the top two interlayer spacings are expanded in the PLD film. (ii) The buckling amplitudes of the PLD film are larger than those of the the TD film by a factor of 2 at the surface, but are reduced towards the Fe–Cu interface. Neither buckling nor lateral displacement of the atoms occurs at the substrate–film interface, i.e., in the first layer of the PLD Fe film. The PLD films also show remarkably different magnetic properties from the TD films (see Fig. 6.44) [97, 98]. In contrast to the perpendicular magnetic anisotropy in the TD films, the PLD films with thicknesses between 2 and 5 ML show an in-plane easy magnetization axis. A spin reorientation transition from in-plane to perpendicular occurs above 5 ML and a well-defined perpendicular easy magnetization axis is established above 7 ML. The preferred magnetization direction returns to in-plane above 10 ML, at which thickness the film structure switches from fcc to bcc, similarly to the behavior of TD films. The saturation magnetization (M_s) also shows distinctly different behavior between the PLD and TD methods. For PLD films, M_s increases linearly up to a thickness of 4 ML, and then drops by 20% and remains constant up to 7 ML, from where it slowly decreases to a value corresponding to a 2 ML film at 10 ML.

The different magnetic behavior between the TD and PLD Fe/Cu(100) films reflects the complex magnetic structure of fcc Fe and the subtle correlation between growth, structure and magnetism for this system. Though a definite explanation for these differences is not available at present, we would

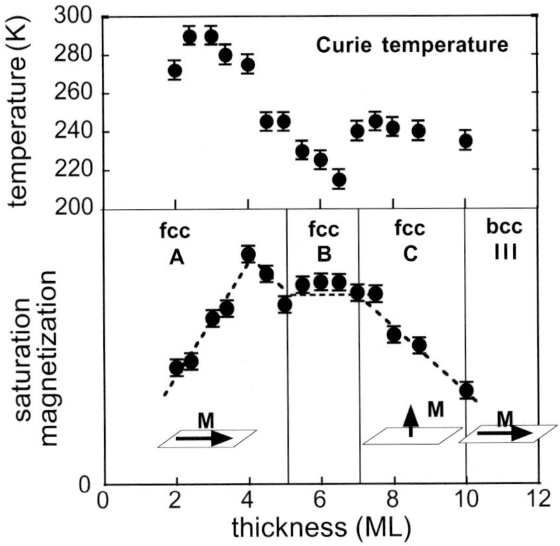

Fig. 6.44. Plots of Curie temperature (*upper panel*) and saturation magnetization M_s (*lower panel*) as a function of thickness for pulsed-laser-deposited Fe/Cu(100) films. M_s was measured at 145 K. The magnetic easy axis is also depicted for different thickness regions. From [98], used with permission

like here to introduce a model proposed by Weinelt et al. [99]. In this model, it is believed that pulsed-laser ablation generates ions with nonthermal kinetic energies even though the laser power is just slightly above the ablation threshold. These high-kinetic-energy ions and their resputtering effect have been assumed to account for the observed differences. These nonthermal kinetic energies, if converted into lateral mobility upon adsorption, will help to overcome the Ehrlich–Schwoebel barrier at island edges, and may in general improve layer-by-layer growth. Furthermore, defects produced by mild sputtering can act as nucleation centers. A large density of nuclei also favors layer-by-layer growth. The high-kinetic-energy particles also lead to the incorporation of Cu atoms into the first iron layer grown by sputter-exchange processes. In this way, the epitaxial stress is reduced and surface wetting is favored. The reduced stress also explains the fact that the iron layer at the interface of a 4 ML PLD film is not buckled but exhibits the structure of a Cu(100) bulk layer. Pronounced atomic mixing at the interface for PLD films could also be responsible for the in-plane easy axis observed for films with a thickness between 2 and 5 ML. According to theoretical results [100, 101], the main contribution to the perpendicular anisotropy stems from the interface and surface layers, and for thin films, contributions from the interface would be strongly reduced when Cu atoms were incorporated into the interface layer. The drop of the magnetization at higher coverages (above 7 ML) is attributed to the decrease of surface quality caused by sputtering in PLD. A

LEED intensity analysis has already shown that the quality of the PLD films is worse than that of TD films when the film thickness is larger than 6 ML. One thing still left to explain is the fact that the saturation magnetization observed in the thickness region of 5–7 ML for PLD films is twice as large as that for TD films with the same thickness. Weinelt et al. suggested that the larger saturation magnetization may stem from a different antiferromagnetic coupling scheme introduced by a changed interface layer. However, the existence of antiferromagnetic order at the measurement temperature (145 K) has not yet been proven unequivocally. To give a clear explanation for this difference, a precise determination of the structure of films in this thickness region is needed.

6.3.5 Fe Ultrathin Films on Various Substrates with a Lattice Constant Close to That of Cu(100)

Since different types of magnetic order and a complicated correlation between structure and magnetism have been found for Fe/Cu(100) films, it is naturally also interesting to investigate iron films on other fcc substrates with a lattice parameter close to that of Cu. These substrates include diamond (100), which has a smaller lattice constant than Cu(100), and Cu–Al and Cu–Au alloys, which have a larger lattice constant than Cu(100). Fe films on diamond show completely different magnetic properties from Fe/Cu(100) films, while the magnetic properties of iron films on Cu–Al and Cu–Au alloys resemble those of Fe/Cu(100) films more or less. This reflects the fact that the nature of the epitaxial strain (in-plane compression or expansion) has significant effects on the film magnetism. Iron films on a $Cu_3Au(100)$ substrate, which has the largest lattice constant among the substrates investigated, show a strained bcc structure over the whole thickness range. This result provides an upper limit for the metastable epitaxy of fcc Fe. The detailed results for iron films on various substrates are summarized in Table 6.6.

Table 6.6. Summary of the growth, structure and magnetic properties of ultrathin Fe films grown on substrates with lattice constants close to that of Cu(100). The lattice constant of fcc Fe extrapolated to room temperature is 3.59 Å. The lattice constant of Cu(100) is 3.61 Å. Continued on next page

Substrate (lattice constant in Å) [Reference(s)]	Growth, structure and magnetic properties
Diamond (100) (3.57) [102]	*Growth.* 2D epitaxial growth for films thinner than 5 ML; 3D growth for films thicker than 5 ML. *Structure.* Fcc pseudomorphic films with 1×1 RHEED pattern for films thinner than 5 ML; bcc structure with its [111] direction matching the fcc [110] direction, as occurs in the fcc-to-bcc transformation of Fe/Cu(100) for films thicker than 5 ML. *Magnetism.* (i) The bcc films above 5.5 ML are ferromagnetic with an in-plane easy-axis; (ii) no magnetic signal is detected for films thinner than 5 ML, but Mössbauer spectra for a 4 ML ^{57}Fe film exhibit a spectral line typical of paramagnetic fcc Fe at 300 and 70 K. However, a line broadening is observed at 40 and 35 K, indicating a low-moment antiferromagnetic spin structure.
$Cu_{84}Al_{16}(100)$ (3.65) [103]	*Growth and structure.* The films were prepared at 150 K. Thickness-dependent valence band photoelectron spectroscopy indicates a layer-by-layer growth mode. The $p(1 \times 1)$ LEED pattern suggests epitaxial growth. *Magnetism.* (i) In-plane magnetization starts to appear at 2.5 ML and reaches a maximum value for 3–4 ML films, as probed by linear magnetic dichroism in the angular distribution of photoelectron intensity (LMDAD). Above 4 ML, the LMDAD asymmetry is reduced by 35%. (ii) A Curie temperature of 288 K and a critical exponent $\beta = 0.212$ are derived from a temperature-dependent Fe 3p LMDAD measurement on a 4 ML $Fe/Cu_{84}Al_{16}(100)$ film. The average magnetic moment is estimated to be $2.5\mu_B$ for 3 and 4ML films.

Table 6.6. Continued

Substrate (lattice constant in Å) [Reference(s)]	Growth, structure and magnetic properties
$Cu_{90}Au_{10}(100)$ (3.66) [104]	*Growth and structure.* (i) The growth of the film is characterized by a MEED curve of the (00) spot similar to that observed for Fe/Cu(100) films. Oscillations up to a film thickness of about 10 ML indicate a layer-by-layer growth mode. (ii) Unlike Fe/Cu(100) films, the Fe/$Cu_{90}Au_{10}(100)$ films always have a $p(1 \times 1)$ LEED pattern. No superstructure due to a reconstruction of the Fe layer is observed at any thickness (1–10 ML). (iii) Kinematical analysis of the LEED I/V curves reveals a thickness-dependent interlayer spacing (see left panel in Fig. 6.45). Up to 4 ML (region I), the films have an expanded vertical interlayer distance compared with the substrate. Between 4 and 10 ML (region II), a reduced interlayer spacing is observed. Above 10 ML (region III), a distinctly higher vertical spacing consistent with a (110)-oriented bcc Fe phase is obtained. *Magnetism.* (i) Below 2 ML, the films show a perpendicular easy axis, while for films with a thickness of 3.3 ML and higher, in-plane magnetization is observed. (ii) Corresponding to the three different structural regions, the longitudinal MOKE signal increases linearly with thickness up to 3.3 ML, then becomes almost constant in region II, and finally increases again rapidly.
$Cu_{1-x}Au_x(100)$ (3.606–3.704) [70]	*Sample preparation.* The alloy substrates were codeposited in situ on NaCl(100) wafers by MBE. On about 5000 Å of a $Cu_{1-x}Au_x$ substrate, 6.5 ML Fe(100) was deposited at 0° C, followed by 40 ML $Cu_{1-x}Au_x(100)$ with the same composition as the substrate. The bilayer structure was repeated four times to increase the Mössbauer signal. Within the 6.5 ML Fe layer, the 2 ML regions close to the $Cu_{1-x}Au_x$ layer were composed of natural Fe, while the interior region is composed of ^{57}Fe. In this way, the Mössbauer spectra measured only the interior region of the Fe layer. *Mössbauer results.* The Mössbauer spectra indicate two kinds of Fe sites. One is a low-spin site, reflected by the broadening of the singlet line at low temperature. The other Fe site is a high-spin site, characterized by the appearance of the sextet lines upon cooling. The intensity of the sextet lines increases with increasing substrate lattice parameter, indicating that the fraction of high-spin Fe sites increases with increasing substrate lattice parameter (see Fig. 6.46). It was inferred that there are two metastable phases in the interior portion of the face-centered Fe films and that an in-plane expansion drives the system closer to one of the two states.

Table 6.6. Continued

Substrate (lattice constant in Å) [Reference(s)]	Growth, structure and magnetic properties
$Cu_3Au(100)$ (3.745) [105, 106, 107]	The most conspicuous magnetic behavior observed for iron films on $Cu_3Au(100)$ is a spin reorientation transition upon variation of the film thickness. The preferred direction of magnetization changes from perpendicular for small thicknesses to in-plane for large thicknesses. The transition thickness depends on the deposition temperature and varies from group to group. For films deposited at room temperature (RT), the transition thickness has been reported to be 2.3 ± 0.2 ML [105] or 3.5 ML [107]. If the films are prepared at low temperature (150–160 K) (LT), the transition thickness increases to 3.2 ± 0.2 ML [105] or 5.5 ML [107], respectively. Two kinds of explanations have been proposed for the spin reorientation transition. Lin et al. [107] associated this transition with a structural transformation from fcc at small thickness to bcc at high thickness. Through kinematical analysis of the LEED (00) I/V curves, they found that the vertical interlayer distance jumps for both RT and LT films at the thicknesses at which the spin reorientation transition occurs. This scenario is supported by STM observations by these authors. On the other hand, we have examined the structural properties of $Fe/Cu_3Au(100)$ by full dynamical LEED I/V calculations and reach a different conclusion [105, 106]. It was found that the structural parameters obtained by Lin et al. can reproduce the I/V curve of only the (00) beam and not the full data set. Furthermore, it was found that the film structure derived from the full dynamical analysis is excellently located on the epitaxial line of bcc Fe (see Fig. 6.47) and, on the other hand, clearly deviates from that of FM or AFM fcc Fe. This is to say, the films over the whole thickness range studied have a strained bcc structure. Hence the spin reorientation transition is explained as resulting from the competition of various magnetic anisotropies.

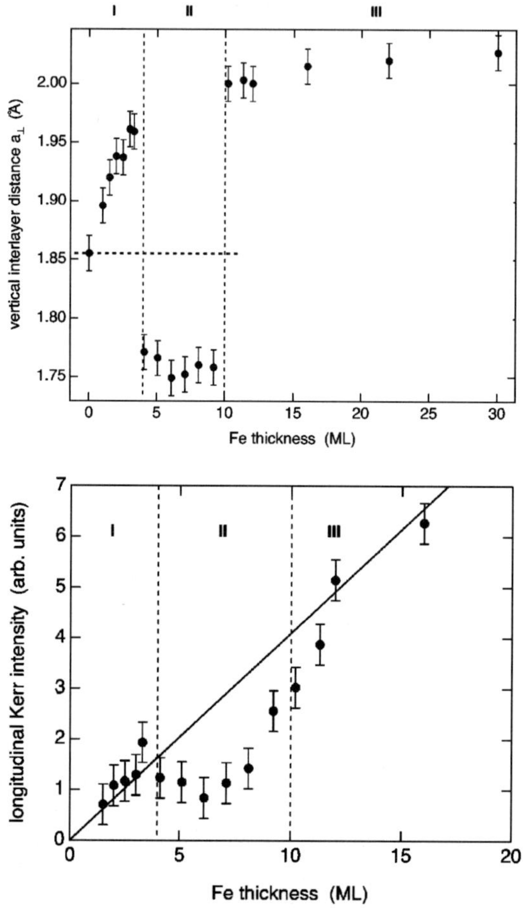

Fig. 6.45. *Top*: average vertical interlayer spacings $a_\perp$ of Fe/$Cu_{90}Au_{10}$(100) films extracted from a kinematical analysis of peak positions in LEED I/V curves. The value of $a_\perp$ measured for the $Cu_{90}Au_{10}$(100) substrate is highlighted by the *dashed horizontal line*. Three different structural regions can be distinguished and are labeled I, II and III. *Bottom*: thickness dependence of the longitudinal MOKE intensity. From [104], used with permission

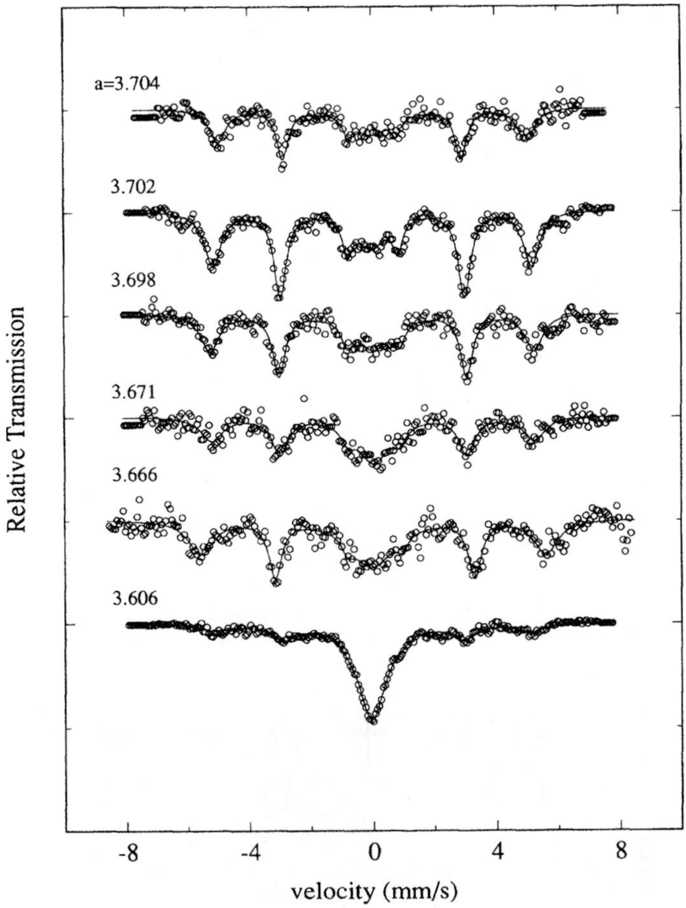

Fig. 6.46. Low-temperature (~15 K) Mössbauer spectra for 6.5 ML fcc Fe(100) films grown on $Cu_{1-x}Au_x$ substrates with six different lattice parameters in the range a = 3.063 to 3.704 Å. The ratio of the intensity of the sextet lines to that of the singlet line increases with increasing lattice parameter. From [70], used with permission

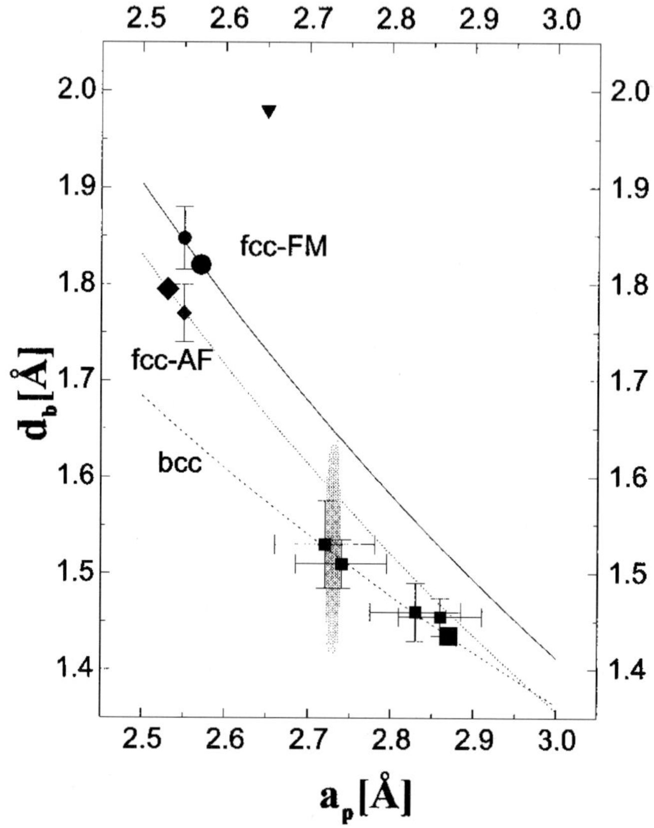

Fig. 6.47. The structures of various bulk phases of iron: FM bcc Fe ($d_b = 1.435$ Å, $a_p = 2.87$ Å) (*large filled square*), FM fcc Fe ($d_b = 1.82$ Å, $a_p = 2.57$ Å) (*large filled circle*) and AF fcc Fe ($d_b = 1.795$ Å, $a_p = 2.53$ Å) (*large filled diamond*). The three curves show the epitaxial lines for FM bcc Fe (*dashed line*), FM fcc Fe (*solid line*) and AF fcc Fe (*dotted line*). The epitaxial lines were calculated using the formula $d/d_{eq} = (a/a_{eq})^{-\gamma}$. In the calculation, a γ value of 1.16 was used for the bcc structure, and 1.64 for the fcc structure. Here d_b and a_p denote the interlayer spacing and the in-plane lattice constant of the bulk phase of iron, respectively. The structures determined by a full dynamical analysis of the LEED I/V curves for 3.3, 4.8, 18, and 53 ML Fe/$Cu_3Au(100)$ are displayed as *small filled squares*. Clearly, the bcc structure gives an excellent fit to the film structure for the entire thickness range studied. The *shaded area* depicts the in-plane nearest-neighbor spacing derived from an analysis of the LEED spot positions. The *triangle* denotes a structure determined in [107] for 5.5 ML Fe grown at low temperature on $Cu_3Au(100)$. The *small diamond* and the *small circle* show experimental data for Fe films on Cu(100)

References

1. H. Bethe: Handbuch der Physik, **24/2**, 595(1933)
2. J. C. Slater: Phys. Rev. **36**, 57(1930)
3. D. H. Martin: *Magnetism in Solids* (The M.I.T. Press, Cambridge, Massachusetts 1967) p326
4. V. L. Moruzzi, P. M. Marcus and P. C. Pattnaik: Phys. Rev. B **37**, 8003(1988)
5. V. L. Moruzzi and P. M. Marcus: Phys. Rev. B **39**, 6957(1989)
6. V. L. Moruzzi and P. M. Marcus: Phys. Rev. B **38**, 1613(1988)
7. E. F. Wassermann: In *Magnetismus von Festkörpern und Grenzflächen*, (IFF/KFA Jülich 1993)
8. J. A. C. Bland, D. Pescia and R. F. Willis: Phys. Rev. Lett. **58**, 1244(1987)
9. P. Xhonneux and E. Courtens: Phys. Rev. B **46**, 556(1992)
10. S. H. Lu, J. Quinn, D. Tian, F. Jona, and P. M. Marcus: Surf. Sci. **209**, 364(1989)
11. R. Hallbauer and U. Gonser: J. Magn. Magn. Mat. **35**, 55(1983)
12. W. A. A. Macedo and W. Keune: Phys. Rev. Lett. **61**, 475(1988)
13. P. A. Montano, G. W. Fernando, B. R. Cooper, E. R. Moog, H. M. Naik, S. D. Bader, Y. C. Lee, Y. N. Darici, H. Min and S. Marcano: Phys. Rev. Lett. **59**, 1041(1987)
14. D. Pescia, M. Stampanoni, G. L. Bona, A. Veterlaus, R. F. Willis and F. Meier: Phys. Rev. Lett. **58**, 2126(1987)
15. D. P. Pappas, K. P. Kämper, B. P. Miller, H. Hopster, D. E. Fowler, A. C. Luntz, C. R. Brundle and Z. X. Shen: J. Appl. Phys. **69**, 5209(1991)
16. C. Liu, E. R. Moog and S. D. Bader: Phys. Rev. Lett. **60**, 2422(1988)
17. M. Stampanoni: Appl. Phys. A **49**, 449(1989)
18. W. A. Jesser and J. W. Matthews: Philos. Mag. **17**, 595(1968)
19. A. Clarke, P. J. Rous, M. Arnott, G. Jennings and R. F. Willis: Surf. Sci. **192**, L483(1987)
20. Y. Darici, J. Marcano, H. Min and Montano: Surf. Sci. **182**, 477(1987)
21. M. T. Kief and W. F. Egelhoff, Jr.: Phys. Rev. B **47**, 10785(1993)
22. H. Magnan, D. Chandesris, B. Villette, O. Heckmann and J. Lecante: Phys. Rev. Lett. 67, 859(1991)
23. W. Daum, C. Stuhlmann and H. Ibach: Phys. Rev. Lett. **60**, 2741(1988)
24. C. Egawa, E. M. McCash and R. F. Willis: Surf. Sci. 215, L271(1991)
25. J. Thomassen, B. Feldmann and M. Wuttig: Surf. Sci. 264, 406(1992)
26. H. Glatzel, Th. Fauster, B. M. U. Scherzer and V. Dose: Surf. Sci. **254**, 58(1991)
27. Chikashi Egawa, Tetsuya Aruga and Yasuhiri Iwasawa: Surf. Sci. Lett. **185**, L506(1987)
28. Z. Nishiyama, K. Shimizu and K. Sugino: Acta metall. **9**, 620(1961)
29. W. Pitsch: Philos. Mag. **4**, 477(1959)
30. S. A. Chambers, T. J. Wagener and J. H. Weaver: Phys. Rev. B **36**, 8992(1987)
31. D. A. Steigerwald, I. Jacob and Jr. W. F. Egelhoff: Surf. Sci. **202**, 472(1988)
32. D. D. Chambliss, K. E. Johnson, R. J. Wilson and S. Chiang: J. Magn. Magn. Mat **121**, 1(1993)
33. D. D. Chambliss, R. J. Wilson and S. Chiang: J. Vac. Sci. Technol. B **10**, 1993(1992)
34. A. Brodde and H. Neddermeyer: Surf. Sci. **287/288**, 988(1993)

35. F. Jona and P. M. Marcus: Surf. Sci. **223**, L897(1989)
36. W. Daum: PhD Thesis, RWTH Aachen (1988)
37. D. A. Steigerwald and W. F. Egelhoff, Jt.: Surf. Sci. **192**, L887(1987)
38. F. Jona and P. M. Marcus: In: *Surface Physics and Related Topics*, ed by Fu-Jia Yang et al., (World Scientific, Singapore, 1991) p213
39. M. Wuttig, B. Feldmann, J. Thomassen, F. May, H. Zillgen, A. Brodde, H. Hannemann and H. Neddermeyer: Surf. Sci. **291**, 14(1993)
40. S. Müller, P. Bayer, C. Reischl, K. Heinz, B. Feldmann, H. Zillgen and M. Wuttig: Phys. Rev. Lett **74**, 765(1995)
41. M. Wuttig and J. Thomassen: Surf. Sci. **282**, 237(1993)
42. H. Landskron, G. Schmidt, K. Heinz, K. Müller, C. Stuhlmann, U. Beckers, M. Wuttig and H. Ibach: Surface Sci. **256**, 115(1991)
43. J. B. Pendry: J. Phys. C **13**, 937(1980)
44. E. Zanazzi and F. Jona: Surf. Sci. **62**, 61(1977)
45. G. Kleinle, W. Moritz and G. Ertl: Surf. Sci. **238**, 119(1990)
46. H. D. Shih, F. Jona, U. Bardi and P. M. Marcus: J. Phys. C **13**, 3801(1980)
47. W. R. Tyson: Surf. Sci. **62**, 267(1977)
48. E. Bauer: Zeitschr. f. Kristallographie **110**, 372(1958)
49. Th. Detzel, N. Memmel and Th. Fauster: Surf. Sci. **293**, 227(1993)
50. D. D. Chambliss and K. E. Johnson: Surf. sci. **313**, 215(1994)
51. K. Heinz, S. Müller and L. Hammer: J. Phys: Condens. Matter **11**, 9437(1999)
52. K. Heinz, S. Müller and P. Bayer: Surf. Sci. **337**, 215(1995)
53. K. Heinz, P. Bayer and S. Müller: Surf. Rev. Lett. **2**, 89(1995)
54. H. L. Davis, J. B. Hannon, K. B. Ray and E. W. Plummer: Phys. Rev. Lett. **68**, 2632(1992) and reference therein
55. S. Müller, P. Bayer, A. Kinne, P. Schmailzl and K. Heinz: Surf. Sci. **322**, 21(1995)
56. K. Heinz, S. Müller and P. Bayer: Surf. Sci. **352-354**, 942(1996)
57. G. A. Prinz: J. Magn. Magn. Mater **100**, 469(1991)
58. T. Kraft, P. M. Marcus and M. Scheffler: Phys. Rev. B **49**, 11511(1994)
59. S. D. Bader: J. Magn. Magn. Mater. **100**, 440(1991)
60. J. Thomassen, F. May, B. Feldmann, M. Wuttig and H. Ibach: Phys. Rev. Lett. **69**, 3831(1992)
61. R. D. Ellerbrock, A. Fuest, A. Schatz, W. Keune and R. A. Brand: Phys. Rev. Lett. **74**, 3053(1995)
62. D. Schmitz, C.Charton, A. Scholl, C. Carbone and W. Eberhardt: Phys. Rev. B **59**, 4237(1999)
63. A. Biedermann, R. Tscheließnig, M. Schmid and P. Varga: Phys. Rev. Lett. **87**, 86103(2001)
64. D. Spisak and J. Hafner: Phys. Rev. Lett. **88**, 56101(2002)
65. S. C. Abrahams, L. Guttman and J. S. Kasper: Phys. Rev. **127**, 2045(1962)
66. U. Köhler, C. Jensen, C. Wolf, A. C. Schindler, L. Brendel and D. E. Wolf: Surf. Sci. **454-456**, 676(2000)
67. U. Gonser, C. J. Meechan, A. H. Muir and H. Wiedersich: J. Appl. Phys. **34**, 2373(1963)
68. Y. Tsunoda: J. Phys.: Condens. Matter **1**, 10427(1989)
69. Y. Kakehashi, O. Jepsen and N. Kimura: Phys. Rev. B **65**, 134418(2002)
70. D. J. Keavney, D. F. Storm, J. W. Freeland, I. L. Grigorov and J. C. Walker: Phys. Rev. Lett. **74**, 4531(1995)

71. J. W. Freeland, I. L. Grigorov and J. C. Walker: Phys. Rev. B **57**, 80(1998)
72. C. L. Fu and A. J. Freeman: Phys. Rev. B **35**, 925(1987)
73. G. W. Fernando and B. R. Cooper: Phys. Rev. B **38**, 3016(1988)
74. R. Lorenz and J. Hafner: Phys. Rev. B **54**, 15937(1996)
75. T. Asada and S. Blügel: Phys. Rev. Lett. **79**, 507(1997)
76. Yu-mei Zhou, Wentong Geng and Ding-sheng Wang: Phys. Rev. B **57**, 5029(1998)
77. E. G. Moroni, G. Kress and J. Hafner: J. Phys.: Condens. Mater **11**, L35(1999)
78. Dongqi Li, M. Freitag, J. Pearson, Z. Q. Qiu and S. D. Bader: Phys. Rev. Lett **72**, 3112(1994)
79. D. Qian, X. F. Jin, J. Barthel, M. Klaua and J. Kirschner: Phys. Rev. Lett **87**, 227204(2001)
80. B. Schirmer and M. Wuttig: Phys. Rev. B **60**, 12945(1999)
81. Xiangdong Liu and Mattias Wuttig: Phys. Rev. B **64**, 104408(2001)
82. X. Liu, B. Schirmer and M. Wuttig: Phys. Rev. B **65**, 224413(2002)
83. D. Spišák and J. Hafner: Phys. Rev. B **66**, 052417(2002)
84. H. Zillgen, B. Feldmann and M. Wuttig: Surf. Sci. **321**, 32(1994)
85. W. F. Egelhoff, Jr. et al.: J. Appl. Phys. **80**, 5183(1996)
86. R. Allenspach and A. Bischof: Phys. Rev. Lett. **69**, 3385(1992)
87. A. Berger, B. Feldmann, H. Zillgen and M. Wuttig: J. Magn. Magn. Mater. **183**, 35(1998)
88. D. E. Fowler1 and J. V. Barth: Phys. Rev. B **53**, 5563(1996)
89. W. Keune, A. Schatz, R. D. Ellerbrock, A. Fuest, K. Wilmers, and R. A. Brand: J. Appl. Phys. **79**,4265(1996)
90. K. Kalki, D. D. Chambliss, K. E. Johnson, R. J. Wilson and S. Chiang: Phys. Rev. B **48**, 18344(1993)
91. J. V. Barth and D. E. Fowler: Phys. Rev. B **52**, 11432(1995)
92. J. Giergiel, J. Kirshner, J. Landgraf, J. Shen and J. Woltersdorf: Surf. Sci. **310**, 1(1994)
93. M. Zharnikov, A. Dittschar, W. Kuch, C. M. Schneider and J. Kirschner: Phys. Rev. Lett **76**, 4620(1996)
94. D. Qian, X. F. Jin, J. Barthel, M. Klaua and J. Kirschner: Phys. Rev. B **66**, 172406(2002)
95. G. G. Hembree, J. Drucker, S. D. Healy, K. R. Heim, Z. J. Yang and M. R. Scheinfein: Appl. Phys. Lett. **64**, 1036(1994)
96. H. Jenniches, J. Shen, Ch. V. Mohan, S. S. Manoharan, J. Barthel, P. Ohresser, M. Klaua and J. kirschner: Appl. Phys. Lett. **69**, 3339(1996)
97. H. Jenniches, J. Shen, Ch. V. Mohan, S. S. Manoharan, J. Barthel, P. Ohresser, M. Klaua and J. kirschner: Phys. Rev. B **59**, 1196(1999)
98. J. Shen, H. Jenniches, Ch. V. Mohan, J. Barthel, M. Klaua, P. Ohresser and J. Kirschner: Europhys. Lett. **43**, 349(1998)
99. M. Weinelt, S. Schwarz, H. Baier, S. Müller, L. Mammer, K. Heinz and Th. Fauster: Phys. Rev. B **63**, 205413(2001)
100. B. Újfalussy, L. Szunyogh and P. Weinberger: Phys. Rev. B **54**, 9883(1996)
101. L. Szunyogh, B. Újfalussy and P. Weinberger: Phys. Rev. B **55**, 14392(1997)
102. Dongqi Li, D. J. Keavney, J. Pearson and S. D. Bader: Phys. Rev. B **57**, 10044(1998)
103. W. A. A. Macedo, F. Sirotti, G. Panaccione, A. Schatz, W. Keune, W. N. Rodrigues and G. Rossi: Phys. Rev. B **58**, 11534(1998)

104. S. S. Kang, W. Kuch and J. Kirschner: Phys. Rev. B **63**, 024401(2000)
105. B. Feldmann, B. Schirmer, A. Sokoll and M. Wuttig: Phys. Rev. B **57**, 1014(1998)
106. B. Schirmer, B. Feldmann and M. Wuttig: Phys. Rev. B **58**, 4984(1998)
107. M. -T. Lin, J. Shen, W. Kuch, H. Jenniches, M. Klaua, C. M. Schneider and J. Kirschner: Phys. Rev. B **55**, 5886(1997)

7 Manganese Films on Cu(100) and Ni(100)

7.1 Introduction

In the previous chapter, a correlation between the structure, growth and magnetism was established for the workhorse of thin-film magnetism, Fe/Cu(100). In this chapter, our attention is turned to a less studied, but potentially equally rewarding system: Mn on Cu(100). To facilitate the discovery of systematic trends, Mn/Ni(100) has been studied for comparison. Bulk manganese has one of the richest varieties of crystallographic and magnetic phases known for metallic elements. This makes Mn an interesting candidate for epitaxial-growth studies, since different crystallographic and magnetic modifications might be stabilized on different substrates. Table 7.1 lists the phases for bulk Mn. As a function of temperature, four different modifications are observed. The complex cubic α phase is stable up to approximately 1000 K. This phase has 58 atoms in the unit cell and is characterized by hexatetrahedral building blocks with nearest-neighbor spacings that vary between 2.24 and 3.0 Å. While α-Mn is antiferromagnetic below 100 K, the other phases are paramagnetic at the temperatures at which they exist. The complex cubic β phase, with 20 atoms per unit cell, is stable between 1000 and 1368 K. The γ and δ modifications are face-centered cubic and body-centered cubic with a nearest-neighbor spacing of 2.73 Å (at 1373 K) and 2.67 Å (at 1413 K), respectively. γ-Mn is stable between 1368 and 1406 K, while δ-Mn is stable between 1406 and 1517 K.

Table 7.1. Crystal structure and magnetic ground state of bulk Mn at different temperatures at a pressure of 10^5 Pa (after [2])

Phase	Structure	Temperature range (K)	Lattice constant (Å)	Magnetic ground state
α	Complex cubic	<1000	8.89	Antiferromagnetic
β	Complex cubic	1000–1368	6.36	Paramagnetic
γ	fcc	1368–1406	3.86	Paramagnetic
δ	bcc	>1406	3.08	Paramagnetic

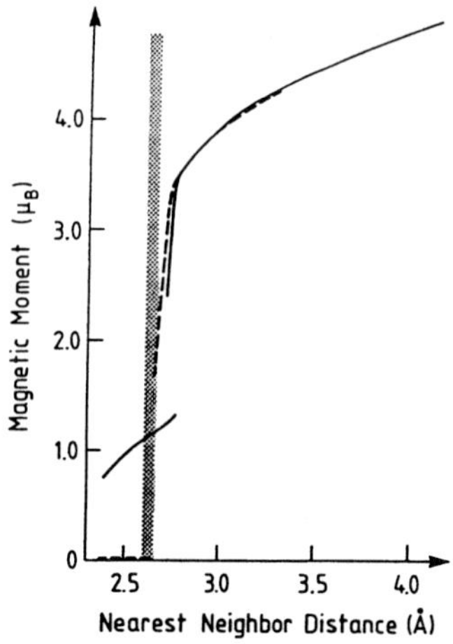

Fig. 7.1. Dependence of the magnetic moment of fcc and bcc Mn upon nearest-neighbor spacing. The *shaded bar* denotes the experimentally observed nearest-neighbor distances. Fcc Mn is denoted by the *broken line*. The *solid line* describes bcc Mn. After [3], used with permission

It is unclear what the magnetic structure and the corresponding moment of bcc and fcc Mn would be if these phases could be stabilized at room temperature or below. Total-energy calculations have addressed this problem for $T = 0$ K. A typical result is shown in Fig. 7.1, where the dependence of the magnetic moment of fcc and bcc Mn upon the nearest-neighbor distance is plotted [3]. Similar calculations have also been performed by other groups [1, 4]. Figure 7.1 shows that the magnetic moment of bcc and fcc Mn has a drastic increase close to, but slightly above the experimentally observed nearest-neighbor spacing. A magnetic-ground-state analysis [3] suggests that bcc Mn should be ferromagnetic at 0 K for nearest-neighbor distances between 2.37 and 2.47 Å. For larger lattice spacings, an antiferromagnetic ground state is possibly energetically favorable [1]. Face-centered cubic Mn should show antiferromagnetism for nearest-neighbor spacings above 2.71 Å [3].

A prerequisite for an investigation of these predictions is the stabilization of fcc and bcc Mn at low temperatures. In principle, this can be realized through epitaxial growth of Mn on a suitable substrate. So far, only very few structural studies have been performed on epitaxially grown Mn films (see Sect. 3.2.3). Complete structural analyses have been reported only for Mn on Pd(100) and Ag(100). LEED I/V measurements of a 21 ML thick Mn film on Pd(100) revealed that the film showed an fcc structure with a tetragonal strain due to the mismatch [5]. A 14 ML Mn film on Ag(100) exhibited a body-centered tetragonal structure with a c/a ratio of 1.13 ± 0.06, as shown by an x-ray photoelectron diffraction (XPD) investigation [6, 7]. Mn films on Fe(100) also show a body-centered tetragonal structure with an expanded interlayer spacing [8, 9]. None of these studies, however, gave any evidence for a magnetic ground state for the films formed.

We have examined the growth, morphology, structure and magnetism of Mn on Cu(100) and Ni(100). Three previous experimental studies have concentrated on these systems. Binns and Norris investigated the growth of Mn on Cu(100) at room temperature using LEED, AES and UPS (ultraviolet photoelectron spectroscopy) [10]. They observed a $c(2\times2)$ structure in the submonolayer region, associated with a breakpoint in the Auger uptake curves. Further breakpoints were attributed to layer-by-layer growth of the Mn films. Using RHEED, Egelhoff et al. [11] also found evidence for layer-by-layer growth of Mn on Cu(100). However, their studies were performed at 77 K. For 4 ML thick films, a (1×1) LEED pattern with very diffuse spots was observed. XPD showed that the interlayer spacing of the Mn films was expanded as compared with the Cu(100) interlayer spacing of 1.8 Å. On the basis of the assumption of epitaxial growth in a (1×1) structure, Blügel et al. calculated an antiferromagnetic ground state for a single Mn layer on Cu(100) [12]. Interestingly enough, such an antiferromagnetic order should give rise to extremely weak superstructure beams in a LEED experiment, enabling the direct observation of antiferromagnetism even in a diffraction experiment using an unpolarized electron beam [13]. The growth of Mn on Ni(100) has not been studied in detail before. Heinrich et al. [14] reported that Mn grows with a complex structure, which they were unable to determine.

Figure 7.2 shows a highly schematic diagram of the superstructures observed in our investigation as a function of growth temperature (i.e. sample temperature during deposition) and coverage for growth on Cu(100). Regions

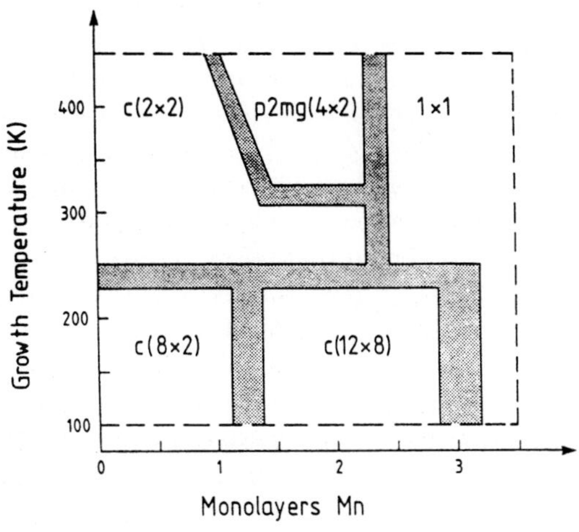

Fig. 7.2. Growth diagram of Mn on Cu(100): schematic diagram of the film structure as a function of growth temperature and coverage. Regions in which the film structure is unknown or a phase transition occurs are shown in *gray*

in which the structure is not known or where structural transitions occur are denoted by gray shading. A more detailed compilation of the superstructures formed as a function of the Mn deposition and growth temperature on Cu(100) and Ni(100) can be found in Sect. 7.2.1–7.3.4. It must be stressed that the schematic diagram in Fig. 7.2 is not a phase diagram. Rather, it shows the superstructures which are observed for a given growth temperature. Not all of the structures depicted are equilibrium structures. Some are metastable configurations which transform upon prolonged annealing to the equilibrium structure. As will be discussed later, kinetics play a dominant role in the formation of Mn superstructures on Ni(100). Because of this delicate relationship between the structure and the temperature during growth and postannealing, it is not feasible to draw such a simplified schematic diagram in the case of Ni(100).

On the one hand, at a given coverage, different superstructures are formed for Mn deposition above and below 270 K. On the other hand, a comparison of the superstructures formed on Cu(100) and Ni(100) shows close similarities. The main differences between Cu and Ni are observed for thick deposits at high temperatures, where on Ni(100) a richer variety of ordered films can be grown. In Figs. 7.3 and 7.4, the LEED patterns for several Mn superstructures observed on Cu(100) and Ni(100), respectively, are shown.

The growth and structure of Mn on Cu(100) and Ni(100) is the first topic presented in this chapter. A number of surface-sensitive probes, including Auger, MEED, LEED, SPALEED and STM, have been applied to characterize the manganese films. Owing to the pronounced dependence of the film structure and the growth mode on the deposition temperature, Mn deposition below 270 K will be discussed separately in Sect. 7.2 (Sec. 7.2.1 for Cu(100) and Sect. 7.2.3 for Ni(100)). At the end of that section, the growth of Mn on Cu(100) is compared with that on Ni(100) and with other related systems. The growth of Mn above 270 K will be described in Sect. 7.3 (Sect. 7.3.1 for Cu(100) and Sect. 7.3.4 for Ni(100)). A comparison of both cases can be found at the end of Sect. 7.3.

Our study of the growth and structure of Mn on Cu(100) and Ni(100) reveals a striking feature of these systems, i.e., the existence of a stable surface alloy phase. We have observed that 0.5 ML of Mn evaporated on Cu(100) or Ni(100) at room temperature results in substitutional adsorption of manganese atoms, which leads to the formation of a stable, ordered surface alloy with a $c(2 \times 2)$ superstructure. Surface alloying after deposition of one metal onto another is gaining more and more interest. This is partly motivated by the desire to tailor the properties of metallic surfaces through intermixing in the near-surface region. Furthermore, understanding the mechanisms that are involved in surface alloying might help to achieve better control of the chemical composition profile of metallic multilayers produced by molecular-beam epitaxy or related techniques. The Cu(100)–$c(2 \times 2)$-Mn surface alloy is of special interest for several reasons: (i) it represents a new class of material,

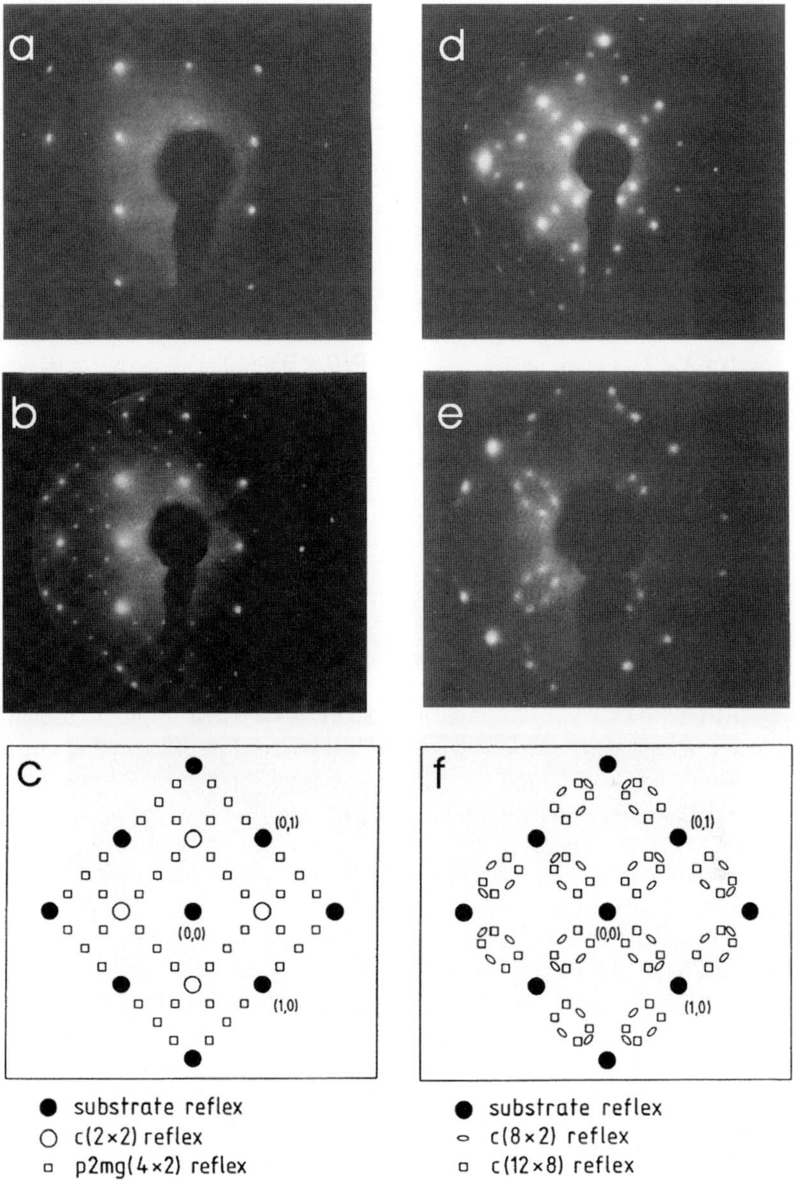

Fig. 7.3. LEED patterns of various Mn superstructures on Cu(100): (**a**) c(2×2) structure (coverage 0.5 ML Mn, growth temperature 325 K); (**b**) *p*4*g*(4×2) structure (coverage 1.4 ML Mn, growth temperature 425 K); (**c**) schematic illustration of *c*(2×2) and *p*4*g*(4×2) structures; (**d**) *c*(8×2) structure (coverage 0.5 ML Mn, growth temperature 200 K); (*e*) Coexistence of *c*(8×2) structure and *c*(12×8) structure (coverage 1.5 ML Mn, growth temperature 200 K); (**f**) schematic illustration of *c*(8×2) and *c*(12×8) structures

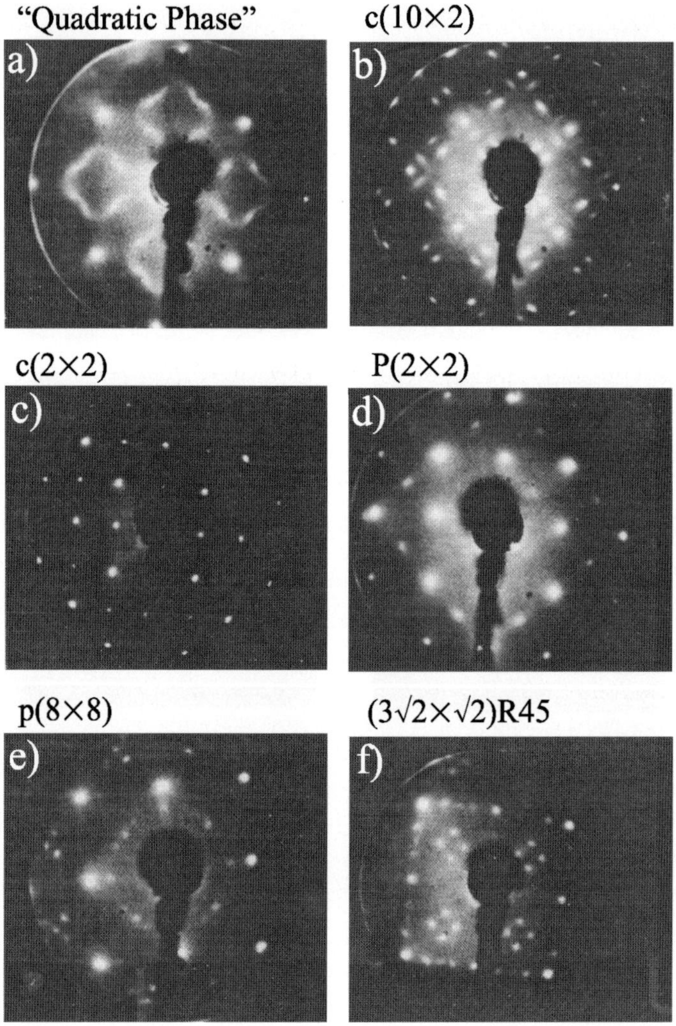

Fig. 7.4. LEED patterns of various Mn superstructures on Ni(100): (**a**) "quadratic phase"(coverage 0.6 ML Mn, growth temperature 200 K) at 85 eV; (**b**) $c(10{\times}2)$ structure (coverage 1.0 ML Mn, growth temperature 200 K) at 152 eV; (**c**) $c(2{\times}2)$ structure (coverage 0.5 ML Mn, growth temperature 475 K) at 398 eV; (**d**) $p(2{\times}2)$ structure (coverage 1.5 ML Mn, growth temperature 425 K) at 127 eV; (**e**) $p(8{\times}8)$ structure (coverage 2.75 ML Mn, growth temperature 550 K) at 44 eV; (**f**) $(3\sqrt{2} \times 2\sqrt{2})R45$ structure (coverage 4.0 ML Mn, growth temperature 550 K) at 112 eV. The second $(3\sqrt{2} \times 2\sqrt{2})R45$ structure, which exists for coverages around 3.5 ML Mn has four additional reflections around the position of the $(1/2, 0)$ beam, rotated by 45° relative to the four beams depicted in (**f**)

i.e., an ordered two-dimensional magnetic alloy; (ii) this ordered alloy phase which exists only at the surface, has no bulk analogue; (iii) the Cu(100)–$c(2 \times 2)$-Mn surface alloy shows an interesting correlation between structure and magnetism.

Therefore, in the later parts of this chapter, we shall concentrate on the surface alloy phases. The surface morphology of this alloy will be described in Sect. 7.4. In Sect. 7.5, total-energy calculations will be set forth to explain the structure and stability of the Cu(100)–c(2×2)Mn phase. The atomic processes involved in the formation of the surface alloy will be discussed in Sect. 7.6.

7.2 Growth and Structure of Mn on Cu(100) and Ni(100) Below 270 K

7.2.1 Growth of Ultrathin Mn Films on Cu(100) Below 270 K

Deposition of Mn on Cu(100) between 110 K and 220 K leads to the formation of a $c(8{\times}2)$ superstructure (see Fig. 7.1). This phase can be detected between 0.2 and 1.6 ML Mn. Above 1.5 ML Mn, a second superstructure, the $c(8{\times}12)$, phase appears. This structure exists in two perpendicular domain orientations up to approximately 2.8 ML Mn. For this phase, a considerably higher background intensity is observed than for the $c(8{\times}2)$ structure. For a more detailed description of the films, it is essential to characterize the film morphology. AES, MEED, LEED, SPALEED and STM have been implemented to investigate the growth of Mn at a variety of sample temperatures. Measurements of the MEED intensity of the (00) beam during Mn deposition at 200 K displayed a single intensity oscillation, which had a maximum intensity after deposition of 0.90 $\pm$ 0.05 ML Mn. This is indicative of the completion of the first Mn layer. Above 1.3 $\pm$ 0.2 ML Mn, the intensity of all MEED beams decreases strongly and the background intensity increases, indicating an increase of disorder. Almost identical curves have been observed for deposition temperatures between 110 and 220 K. A similar behavior is shown in the LEED experiment (Fig. 7.5). The (12) substrate beam shows a pronounced decrease in intensity above 1.1 ML Mn, in agreement with what is observed in the MEED experiment. Clearly, thicker films do not grow in a layer-by-layer fashion. Figure 7.5 also shows the intensity of a $c(8{\times}2)$ superstructure beam (the (7/8 3/2) beam). This beam can be detected between 0.2 and 1.6 ML Mn, but has a clear maximum at 0.90 $\pm$ 0.05 ML Mn. This coverage is identical to the coverage needed to complete the first Mn layer.

That the first layer of Mn atoms is indeed completed after deposition of 0.9 ML Mn on Cu(100) is also shown by high-resolution diffraction (SPALEED) studies [15]. Figure 7.6 shows the intensity distribution around the (00) beam for different primary energies after Mn deposition at 200 K. After deposition of 0.5 ML Mn, a periodic oscillation of the shape of the (00) beam is found with the wave vector transfer Δk_z (Fig. 7.6.a,b). This

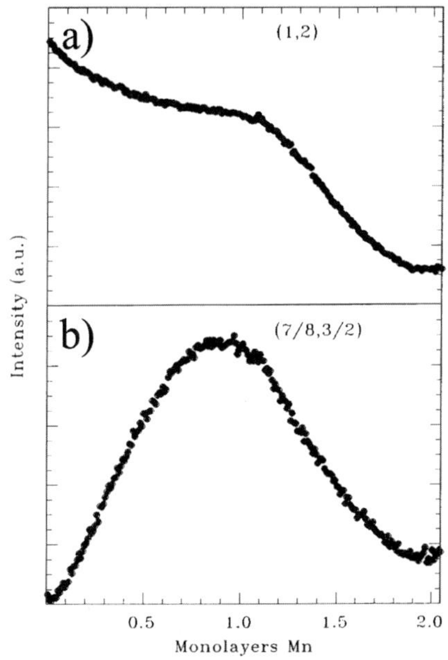

Fig. 7.5. LEED intensities during Mn deposition on Cu(100) at 200 K. The normalized LEED intensities of the (**a**) (12) and (**b**) (7/8 3/2) beams, taken at an angle of incidence of 45° and an electron energy of 155 eV, are plotted as a function of Mn coverage

is clear evidence for the existence of steps on the surface. From the energies at which the (00) beam has the most intense sidebands, the step height can be determined to be 2.08 ± 0.04 Å, assuming that the atomic form factors of Mn and Cu are identical. This value is considerably larger than the step height of the clean Cu(100) surface, which is 1.805 Å. Therefore, this height is associated with the average spacing of the Mn atoms on the Cu(100) substrate. From the position of the sidebands, one can determine the preferential step–step distance. Such an analysis indicates that at a coverage of 0.5 ML, islands form with a preferential size and a preferential separation of 80 Å. In Fig. 7.6(c), the beam profile for the (00) beam is shown for a destructive interference condition after deposition of 0.9 ML Mn. Sidebands are no longer visible; instead, a sharp (00) beam is observed, indicative of the formation of a rather flat manganese film at this coverage. Support for these findings comes from STM observations, which reveal that around 0.9 ML, a flat, continuous film is formed. The STM investigations also show that above 1 ML, the films become rougher and 3D growth starts to evolve. This is consistent with the LEED observation that during the growth of the second monolayer, the diffuse elastic background increases. To determine the structure of the $c(8\times2)$ phase, LEED I/V data have been measured and compared with full dynamical calculations.

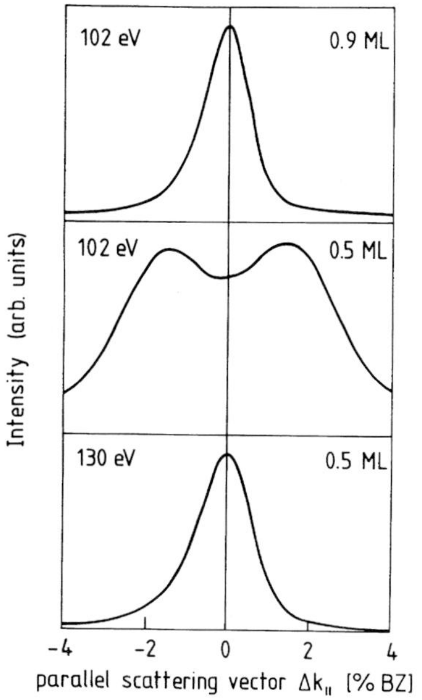

Fig. 7.6. Profiles through the (00) beam after Mn deposition at 200 K. LEED diffraction scans through the specular beam after deposition of 0.5 and 0.9 ML Mn are shown. At a primary energy of 102 eV, electrons scattered from neighboring terraces interfere destructively (out-of-phase condition). The in-phase condition is met at 130 eV

7.2.2 LEED Structure Determination of the $c(8\times2)$ Phase on Cu(100)

The following procedure was adopted to prepare the $c(8\times2)$ overlayer. Mn was deposited on a clean Cu(100) surface held at 200 K. During deposition, the LEED intensity curve for the (1/8, 1/2) beam, which shows a maximum after deposition of 0.9 ± 0.05 ML, was measured. At this maximum, the deposition was stopped and the sample cooled to 120 K. Thereafter, I/V spectra were measured. The $c(8\times2)$ Mn phase formed at 0.9 ML Mn was prepared three times and almost identical I/V curves were obtained each time.

Care was taken to ensure normal incidence in the system. Magnetic fields were strongly reduced by double mu-metal shielding of the entire UHV chamber. Normal incidence was defined as the angle where symmetry-equivalent beams were as similar as possible. The tilt and polar angle were varied until these beams had almost identical curves (Pendry R-factor $R_\mathrm{p} < 0.05$, typically $R_\mathrm{p} \approx 0.03$), showing that normal incidence had been closely approached. After an automatic background subtraction, symmetry-equivalent beams were averaged and normalized to constant incident current.

The quantitative analysis was performed using the standard layer-doubling scheme to describe the scattering of electrons between the layers. 125 non-equivalent beams were employed in this scheme at 320 eV. Owing to the large interlayer spacing between the (100) layers (1.81 Å), no convergence problems were expected for the Cu substrate. The top layer (Mn film) was treated in angular-momentum space as a composite layer allowing, in particular, vertical movements of the atoms inside the layer. The phase shifts were obtained from superposition of atomic densities, and the muffin-tin radii were taken to be equal to half of the bulk interatomic spacing. Up to nine phase shifts were used at the highest energy of 320 eV. The real part of the inner potential was taken to be energy-dependent, V_0 (eV) $= -3 - 0.88\sqrt{E+12}$, where E is the energy in eV, while the absorptive part is assumed to be equal to $0.85E^{1/3}$. Lattice thermal vibrations were accounted for by Debye temperatures of 400 and 360 K for Mn and Cu, respectively.

Both the usual grid search (5 eV steps) and an automatic fit procedure (10 eV steps), which has been added to Moritz's LEED codes [16, 17], were employed. Close to the minimum, the latter precedure was used, whereas the former precedure was used over wide parameter ranges to avoid being trapped in local minima. The quality of the agreement between experiment and theory was assessed by means of various R-factors: R_{DE} [18], R_{P} [19], R_{ZJ} [20]. The fit program was controlled by a minimization algorithm for either R_{DE} or R_{P}. A rigid shift of the inner potential was allowed in that part of the analysis.

For comparison with model calculations, six LEED spots were considered in the energy range from 40 to 320 eV covering a total range of 1210 eV. The integer-order beams were averaged to account for the existence of two domains.

There is no evidence for the incorporation of Mn into the Cu(100) surface below 270 K. Therefore the $c(8\times2)$ structure must be caused by a particular ordering of the Mn atoms on the Cu(100) surface. A quasi-hexagonal arrangement of Mn atoms, as depicted in Fig. 7.7, has been suggested [21]. The unit cell contains 14 atoms, with the following positions in the unit cell:

$$\boldsymbol{r}_j = \left[\frac{8}{7}(j-1)\right]\boldsymbol{a}_1\,, \qquad j = 1,\ldots,7 \tag{7.1}$$

and

$$\boldsymbol{r}_j = \left[\frac{8}{7}(j-8)+\frac{4}{7}\right]\boldsymbol{a}_1 + \frac{1}{2}\boldsymbol{a}_2\,, \qquad j = 8,\ldots,14\,, \tag{7.2}$$

where $\boldsymbol{a}_1$ and $\boldsymbol{a}_2$ are the lattice vectors of the Cu(100) surface. This model predicts a coverage of 0.875 ML Mn, in excellent agreement with our LEED investigation, which shows a maximum intensity for the $c(8\times2)$ beams around 0.9 ML Mn. In a kinematical analysis, this model also reproduces the observed LEED pattern and can qualitatively explain the relative intensities of the superstructure beams. Kinematical calculations, however, are unable to

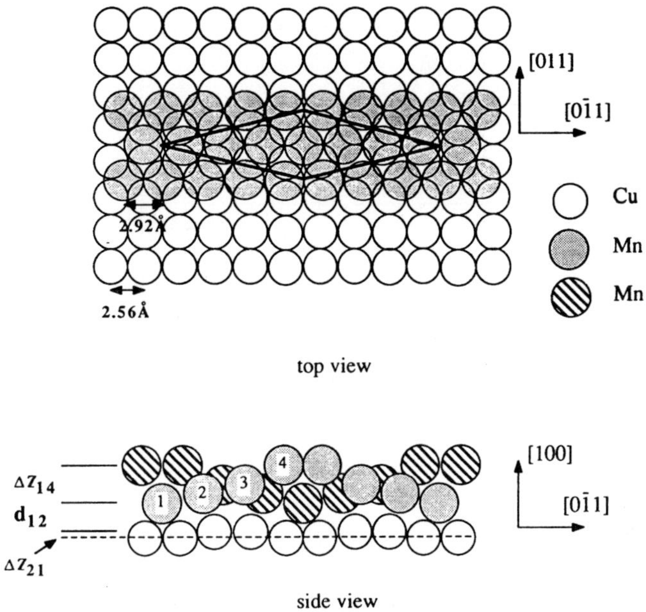

Fig. 7.7. Structural model for the Cu(100) $c(8\times2)$Mn phase: ball model of a pseudohexagonal Mn surface layer on a weakly corrugated Cu substrate. In the side view, Mn atoms in the foreground and background are indicated by different gray scales. The vertical distances are not drawn to scale. The central Mn atom (no. 1) is in the fourfold-hollow position. (**a**) Top view; (**b**) side view

determine the precise surface structure and the registry with respect to the substrate.

There are three main types of registry with respect to the substrate: in each of these, one of the atoms (at the center of the unit cell, for instance) is located in a high-coordination site: (a) the fourfold-hollow site, (b) the bridge site or (c) the top site. The first two cases are rather similar, since the Mn atoms lie in troughs and a small parallel displacement of 0.18 Å in the [0$\bar{1}$1] direction shifts the fourfold hollow site coordination to the bridge coordination. These two bonding configurations seem energetically favorable over the third case, where atoms are in top or bridge-like positions, since in the former two situations, Mn has a higher coordination number. Assuming a rigid substrate and representing the Mn and Cu atoms by hard spheres results in a corrugation of 0.42 Å in the top layer when the center atom is located in the hollow site. Therefore, to include all reasonable possibilities, corrugations of up to 0.65 Å were considered in the LEED analysis. Variations of interlayer distances were performed for the three outer layers, and atomic displacements were allowed in the first and the second layer. All three different registries were investigated. The top layer was assumed to be pure Mn, but some tests were performed to control the composition of the first layer. For

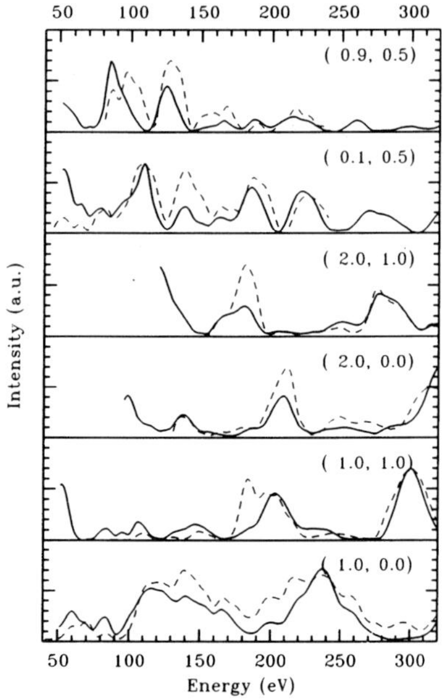

Fig. 7.8. Experimental I/V curves for the $c(8\times2)$ structure (*dashed lines*) and intensities calculated for the best-fit structure, specified in Table 7.2 (*solid lines*)

this purpose, the averaged-t-matrix approximation was used to describe the random distribution of Cu and Mn sites in the first layer.

The final atomic positions for the best-fit solution are shown in Table 7.2. This structure, which has a buckled layer with one atom in the hollow site, is shown in Fig. 7.7. In Fig. 7.8, the I/V curves for the best-fit solution are compared with the experimental spectra. Visual comparison shows that the experimental curves are reproduced reasonably well. This is consistent with the values of the various R-factors, which are 0.112 for R_{ZJ}, 0.309 for R_{P} and 0.355 for R_{DE}. All other adlayer positions yielded much worse agreement, as assessed consistently by the various numerical criteria. A flat layer, for example, produced a Pendry R-factor of 0.548 and an R_{DE} of 0.502, an buckling with half of the optimum amplitude produced an R_{DE} of 0.426, and

Table 7.2. Optimum structure of $c(8\times2)$Mn on Cu(100) with the center of the unit cell in the fourfold-hollow site. The final parameters are compared with the values estimated from a hard-sphere model. See Fig. 7.7 for a definition of the parameters

R_{DE} R_P R_{ZJ}	d_{12} (Å)	d_{23} (Å)	d_{34} (Å)	ΔZ_{14} (Å)	ΔZ_{13} (Å)	ΔZ_{12} (Å)	ΔZ_{21} (Å)
0.112 0.309 0.355	1.77	1.81	1.82	0.53	0.26	0.17	0.03
Hard sphere	1.805	–	–	0.38	0.18	0.07	–

for 2/3 of the optimum amplitude, the R_{DE} was 0.386. The top site can also be excluded, since the R-factors are considerably higher ($R_{DE} > 0.7$) for that site.

In the best-fit model, the highest atoms are those at the corners of the unit cell: they are 0.53 Å higher than the lowest atom, which is above the hollow site (center of the unit cell). This is a larger corrugation than expected from the hard-sphere model. In the second layer, some rippling is also visible: the four atoms at the center of the unit cell, i.e., those in contact with the deepest Mn atom, are repelled inwards. The second layer is thus composed of pairs of [011] rows alternatively up and down, displaced vertically by about 0.03 Å. The Mn layer is separated from the Cu substrate by $d_{12} = 1.77$ Å. d_{12} describes the distance from the lowest atom in the top layer to the highest atom in layer 2 and is slightly smaller than the bulk spacing. The buckling in the copper layer partly explains the shortening of the vertical Mn–Cu spacing. The actual distance from the Mn atom in the hollow site to the neighboring Cu atoms is 1.8 Å (1.77 + 0.03 Å), i.e., close to the bulk value of 1.805 Å. If the distances are referred to the center-of-mass planes, the interlayer spacing is obviously much larger (2.03 Å). This value is in good agreement with the measured step height of 2.08 ± 0.04 Å. Little or no relaxation is found for deeper layers. Finally, the analysis of the composition of the first layer shows the best R-factor for a pure Mn top layer.

7.2.3 Growth and Structure of Mn on Ni(100) Below 270 K

Similarly to the results established for Mn deposition on Cu(100), there is a pronounced influence of the growth temperature on the structure of Mn films on Ni(100). For growth temperatures between 125 and 250 K, three different superstructures have been found. Although the first superstructure is observed between 0.25 and 0.8 ML Mn, the most intense superstructure beams are seen for coverages between 0.45 and 0.8 ML Mn. Independent of growth temperature and coverage in this regime, the superstructure beams are very streaky. The diffraction spots are positioned in an almost square arrangement around reciprocal-lattice points of a $c(2\times2)$ superstructure. For most primary energies, the group of beams consists of four spots, which are streaked in the [011] and $[0\bar{1}1]$ directions. However, at some electron energies, a more complex pattern is observed, which consists of eight spots streaked in directions that deviate slightly from the [011] and $[0\bar{1}1]$ directions. In the following, this structure will be referred to as the "quadratic phase" owing to the particular appearance of its diffraction pattern.

Above 0.8 ML Mn, a $c(10\times2)$ superstructure evolves, which is visible between 0.9 and 1.7 ML Mn. The transition between the quadratic phase and the $c(10\times2)$ structure occurs in a narrow coverage range between 0.8 and 0.9 ML Mn. In this region, the beams of the quadratic phase move with increasing coverage until the beam positions of the $c(10\times2)$ structure are reached. This structure is ordered best at the lowest coverage at which it appears (0.9 ML Mn). Nevertheless, even at this coverage, only streaky superstructure spots that are elongated in the [011] and [$0\bar{1}1$] directions (see Fig. 7.4b) are observed. The degree of "streakiness" depends upon the coverage and, especially, upon the growth temperature. For growth temperatures below 160 K, the films show a pronounced increase of disorder, indicating insufficient mobility of Mn atoms for ordering into larger islands.

Above 1.5 ML Mn, a faint $c(12\times8)$ superstructure is observed. In this coverage region, the superstructure beams are weak and broad. The substrate beams also become broad, but their width depends upon the primary electron energy, indicating the presence of steps strongly on the surface. Above approximately 3.9 ML Mn, only faint substrate beams are visible, on a background that increases continuously with increasing coverage.

To get a better handle on the low-temperature growth of Mn films on Ni(100), the intensity of several substrate beams was measured during deposition in a MEED experiment. Since the intensity of the superstructure beams was very weak at the angles of incidence used, those beams could not be followed. Figure 7.9 shows the intensity of the (00) beam during Mn deposition at 200 K. The curve shows one oscillation, with a maximum of intensity after deposition of 0.9 ML Mn. Above this coverage, the intensity drops quickly until 1.1 ML has been deposited. Thereafter, the decrease is less pronounced. The increase of intensity above 0.4 ML Mn is clearly linked with the formation of the quadratic phase, and the intensity maximum is connected with the well-defined structural transition to the better-ordered $c(10\times2)$ structure between 0.8 and 0.9 ML Mn.

There is no evidence for any Mn incorporation into the Ni(100) surface below 270 K. Therefore the quadratic phase, the $c(10\times2)$ structure and the

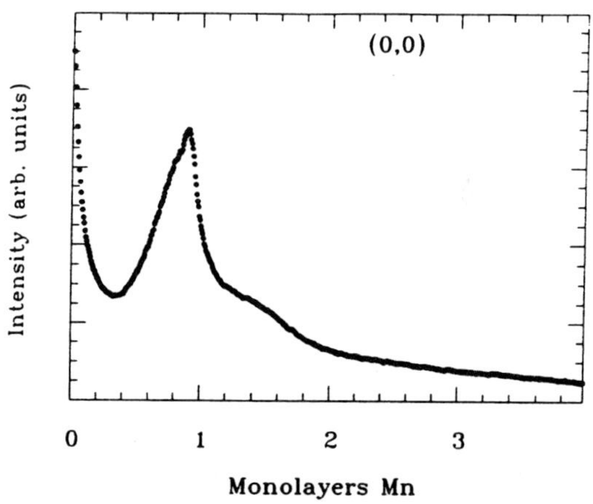

Fig. 7.9. MEED intensity during Mn deposition on Ni(100) at 200 K. The normalized MEED intensity of the (00) beam for an electron energy of 3 keV and an angle of incidence of 3.2° is shown

$c(12\times8)$ structure are all due to particular arrangements of Mn atoms *on* the Ni(100) surface. With this information as a foundation, structural models for the different phases can be discussed.

In Sect. 7.2.1, it was shown that the $c(8\times2)$ structure formed for a monolayer-thick Mn film on Cu(100) is due to a hexagonal arrangement of Mn atoms. This model not only accounted for the observed MEED curves during Mn deposition but also reproduced the measured I/V curves, with the geometrical arrangement shown in Fig. 7.7 and Table 7.2. On the basis of the outcome of this analysis, we propose that Mn atoms on Ni(100) are also arranged in a similar fashion at coverages of around 0.9 ML Mn. To reproduce the $c(10\times2)$ LEED pattern, the arrangement shown in Fig. 7.10 is suggested. The surface unit cell drawn contains 18 atoms, with the following positions:

$$\boldsymbol{r}_j = \left[\frac{10}{9}(j-1)\right]\boldsymbol{a}_1\,, \quad j = 1,\ldots,9\,, \tag{7.3}$$

and

$$\boldsymbol{r}_j = \left[\frac{10}{9}(j-10)+\frac{5}{9}\right]\boldsymbol{a}_1 + \frac{1}{2}\boldsymbol{a}_2\,, \quad j = 10,\ldots,18\,, \tag{7.4}$$

where $\boldsymbol{a}_1$ and $\boldsymbol{a}_2$ are the lattice vectors of the Ni(100) surface. For this model, the first layer is expected to be completed after deposition of 0.9 ML Mn on the Ni(100) surface. This is in excellent agreement with the intensity maximum observed for the (00) beam after deposition of 0.9 ML Mn (Fig. 7.9), lending further support to our proposed model.

Most likely, the Mn atoms show a corrugation as sketched in Fig. 7.10b. This corrugation can be rationalized in a hard-sphere model where Mn atoms

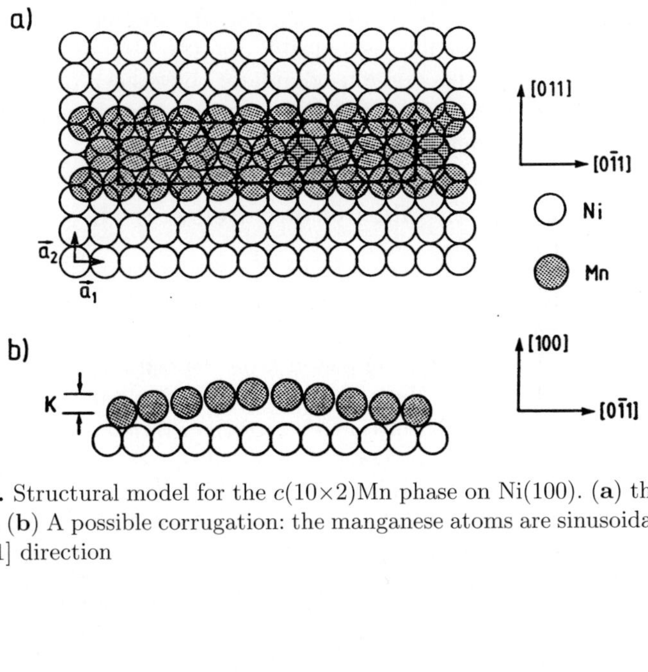

Fig. 7.10. Structural model for the $c(10\times2)$Mn phase on Ni(100). (**a**) the unit cell is marked. (**b**) A possible corrugation: the manganese atoms are sinusoidally shifted in the $[0\bar{1}1]$ direction

in bridge sites have a larger distance from the first substrate layer than do Mn atoms in hollow sites. Using the atomic diameters of 2.73 Å for Mn and 2.49 Å for Ni, one obtains a corrugation of 0.36 Å. For this value, the beam intensities calculated within a kinematical approximation show reasonable agreement with the intensities of the observed diffraction pattern.

On the other hand, it is obvious from the observed LEED pattern that the long-range order of the $c(10 \times 2)$ structure is far from perfect. As stated previously, the superstructure beams are elongated in the [011] and $[0\bar{1}0]$ directions. This indicates either that the hexagonal films grow in rectangular islands whose correlation in the higher-order commensurate direction is better than in the second-order commensurate direction or that there are fewer domain boundaries in this direction. A possible explanation for this behavior can be found in the misfit between Mn and Ni. The atomic radii of fcc manganese and fcc nickel are 2.73 Å and 2.49 Å, respectively [22]. In the direction with tenfold periodicity, the Mn–Mn distance is 2.74 Å, i.e., there is an expansion of less than 0.5%. On the other hand, the distance between neighboring atoms in hollow and bridge sites is 2.84 Å. Thus, there is a much larger expansion, of more than 4%, in that direction. It appears that this strain hinders the formation of long-range order in the twofold direction, leading to the streaky superstructure beams.

Presumably, Mn atoms in the higher-coverage $c(12 \times 8)$ structure also show a hexagonal arrangement similar to the quasi-hexagonal structure of the $c(10 \times 2)$ phase. Owing to the large size of the unit cell, it is difficult to construct a structural model for this phase.

The peculiar diffraction pattern of the quadratic phase observed for Mn films on Ni(100) at around 0.6 ML Mn indicates an ordering of the film with a large number of defects or domain walls. Very similar diffraction patterns have been reported for H on W(100) [23, 24]. A large density of irregular domain boundaries is probably responsible for the streaky superstructure spots [25]. Indeed, maze-like domain boundaries have been observed for the (2×1) structure of oxygen on W(100) [26], a system that is characterized by streaky or split superstructure spots [27].

7.2.4 Comparison of Mn Films Grown Below 270 K on Cu(100) and Ni(100)

The low-temperature ($T < 270$ K) growth of Mn films on Cu(100) and Ni(100) shows close similarities. In both cases, only the first layer wets the substrate surface. This layer is completed after deposition of 0.90 ± 0.05 ML Mn. Thicker Mn films form 3D islands. This behavior can be qualitatively understood within the context of Bauer's growth criterion. According to this rule, layer-by-layer growth should be observed if $\gamma_f + \gamma_i < \gamma_s$. Here γ_f and γ_s are the surface free energies of the film and substrate, respectively, and γ_i is the interface energy. The surface free energy of 1.54 J/m^2 for Mn is smaller than the surface free energies of 1.79 J/m^2 for Cu and 2.38 J/m^2 for Ni [28].

Consequently, both systems fall into the category where layer-by-layer growth is expected in the initial stages of growth. With increasing film thickness, the contribution of the misfit becomes increasingly more important. For a precise calculation of the contribution of the misfit to the interface energy, the strain in the film has to be determined. The associated energy of the strained film can be determined if one knows the structure of the Mn film (see (3.3)). For the first layer, the contribution of this component is of the order of 0.2 J/m^2. This is less than the difference between the surface free energies for Mn–Cu and Mn–Ni. Thus the first layer should indeed wet the substrate. For thicknesses above 1 ML, the misfit contribution becomes increasingly more important and will lead to 3D growth (Stranski–Krastanov growth), as observed in the experiments. The thickness above which 3D growth is encountered will also depend upon the chemical contribution to the interfacial energy. This chemical component should be small (and positive) for the Mn–Cu system since this system does not form ordered bulk alloys. The tendency of the Mn–Ni system to form ordered alloys gives a negative contribution, i.e., there is an enhanced tendency to wet the substrate.

Not only the growth but also the structure of the Mn films on Cu(100) and Ni(100) show pronounced similarities. On both substrates, a faint $c(12 \times 8)$ diffraction pattern is observed above approximately 1.5 ML Mn. Around 0.9 ML Mn, the films form a well-ordered $c(8 \times 2)$ structure on Cu(100) and a $c(10 \times 2)$ structure on Ni(100). The only noteworthy difference occurs at lower coverages, where a "quadratic phase "characterized by two-dimensional disorder is observed on Ni(100) between 0.25 and 0.8 ML Mn. At comparable coverages, a $c(8 \times 2)$ structure is formed on Cu(100).

Both the $c(8 \times 2)$ and the $c(10 \times 2)$ structure are characterized by a quasi-hexagonal arrangement of Mn atoms on the substrate. Such hexagonal top layers have been found for several different classes of systems, including pure metals such as Pt, Au and Ir [29, 30]. Similar arrangements have also been discovered for substitutionally disordered Pt alloys such as $Pt_{80}Fe_{20}(100)$, $Pt_{80}Co_{20}(100)$ [31] and $Pt_{50}Ni_{50}(100)$ [32], and a hexagonal arrangement has been discussed for Ag and Au films on Cu(100) [33] and Ag films on Ni(100) [34]. A feature common to all these examples is that the top layer forms a dense and stable (111)-like layer above the (100) substrate.

For Ir(100), for example, a quasi-hexagonal (5×1) reconstruction is observed, which leads to a local arrangement that resembles the atomic positions expected for the unreconstructed Ir(111) surface [35]. This also holds for $Pt_{50}Ni_{50}(100)$ [32], where a (19×1) structure is found that closely matches the Pt(111) surface. These findings are summarized in Table 7.3, where the ratio of the areal density of some quasi-hexagonal top layers on fcc(100) substrates to the atomic density in the unreconstructed (111) surface of the corresponding element is given. Table 7.3 shows that very often the ratio is close to 1.0. Thus the areal density closely resembles the density of the (111) surface, and the reconstruction can be attributed to the tendency of the sys-

Table 7.3. Atomic densities of various quasi-hexagonal films and surfaces on fcc(100) substrates. $R = F_{\mathrm{Vol}}/F_{\mathrm{Lay}}$ is defined as the quotient of the areal density of the (111) plane in the bulk of the deposited metal, F_{Vol}, and the areal density of the film material on the surface, F_{Lay}. While Ag films on Cu(100) and Ni(100) have an atomic density that closely resembles the atomic density of fcc(111) surfaces, this does not hold for Mn films

System [Reference]	R
Ir(100)[168]	0.97
$Ni_{50}Pt_{50}$(100) [166]	0.99
Ni(100)$c(8 \times 2)$ Ag [127]	0.98
Cu(100)$c(10 \times 2)$ Ag [44]	1.01
Cu(100)$c(8 \times 2)$ Mn [154]	1.16
Ni(100)$c(10 \times 2)$ Mn [169]l	1.07

tem to minimize the surface free energy by formation of a quasi-hexagonal (111)-like top layer. Mn also forms a hexagonal-like structure on Cu(100) and Ni(100), but the areal density differs considerably from the value expected for the corresponding (111) surface. In this sense, Mn behaves anomalously. The Mn in-plane spacing of 2.92 Å is considerably larger than 2.73 Å, the value expected for the γ-Mn(111) surface.

To shed more light on this, the sizes of the unit cell of various quasihexagonal $c(n \times 2)$ superstructures on Cu(100) and Ni(100) are plotted in Fig. 7.11

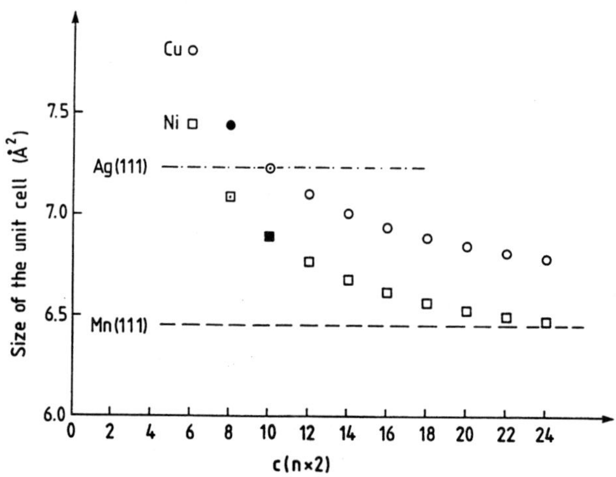

Fig. 7.11. The densities of various quasi-hexagonal structures. The sizes of the unit cell of various quasi-hexagonal superstructures on Cu(100) (*open circles*) and Ni(100) (*open squares*) are shown. For comparison, the sizes of the unit cell of the quasi-hexagonal Ag(111) and Mn(111) surfaces are also shown. *Solid symbols* denote the Mn phases. See text for further explanation

and compared with the values expected for the γ-Mn(111) and Ag(111) surfaces. While Ag films on Cu(100) form a $c(10 \times 2)$ superstructure, a $c(8 \times 2)$ structure is found on Ni(100). These phases are denoted by a dot in Fig. 7.11. Both structures thus have an atomic density that closely approaches the density of the Ag(111) surface. From Fig. 7.11, it is obvious that there are other hexagonal arrangements of Mn atoms on Cu(100) and Ni(100) where the areal density approaches more closely the density of the γ-Mn(111) surface. The deviation is particularly pronounced for the Mn film on Cu(100) (see also Table 7.3). Therefore, the experimentally determined enlarged Mn in-plane spacing must be an intrinsic property of hexagonal Mn films on Cu(100) and Ni(100). This arrangement is not imposed by the structure of the surface. Such a rather surprising finding clearly needs an explanation. In Sect. 7.4, this result is discussed further and an explanation is offered.

7.3 Growth and Structure of Mn on Cu(100) and Ni(100) Above 270 K

7.3.1 Growth and Structure of Mn on Cu(100) above 270 K

The topic of this section is the growth of Mn on Cu(100) above 270 K. Figure 7.12 depicts the Auger signal intensity of the Cu KLL transition at 60 eV and the Mn KLL transition at 40 eV as a function of coverage for a sample temperature of 310 K. In the measurement of such an Auger uptake curve, layer-by-layer growth is characterized by regularly spaced breakpoints. While no prominent breakpoints are observed for the Mn (40 eV) signal, the Cu (60 eV) Auger intensity shows linear segments with breakpoints at coverages Θ_1 and Θ_2. The ratio of the coverages is $\Theta_1/\Theta_2 = 1/3$. Our microbalance calibration shows that the first breakpoint occurs at a coverage of 0.5 ML Mn and the second breakpoint at 1.5 ML Mn. This is inconsistent with a layer-by-layer growth mechanism, as was assumed by Binns and Norris [10], and shows that it is potentially misleading to attribute the observation of a breakpoint in an Auger uptake curve to the completion of an adsorbed monolayer. Towards the goal of understanding the origin of these breakpoints, MEED and LEED were used to unveil the details of the growth of Mn films.

Figure 7.13a shows the MEED intensity of the (00) beam as a function of coverage for a sample temperature of 340 K. The (00) beam is characterized by a rapid decrease in intensity up to a coverage of 0.3 ML Mn. Around 0.3 ML Mn, the intensity increases strongly and reaches a maximum at 0.5 ML Mn. Above 0.5 ML Mn, the intensity decreases again. The absence of regular intensity oscillations strengthens our earlier conclusion that Mn does not grow in a layer-by-layer mode on Cu(100) above 270 K. This holds only for the experiment at 340 K but for all similar growth studies between 270 K and 450 K. On the other hand, the pronounced intensity variation indicates that the structure and/or morphology of the film changes with coverage. Indeed,

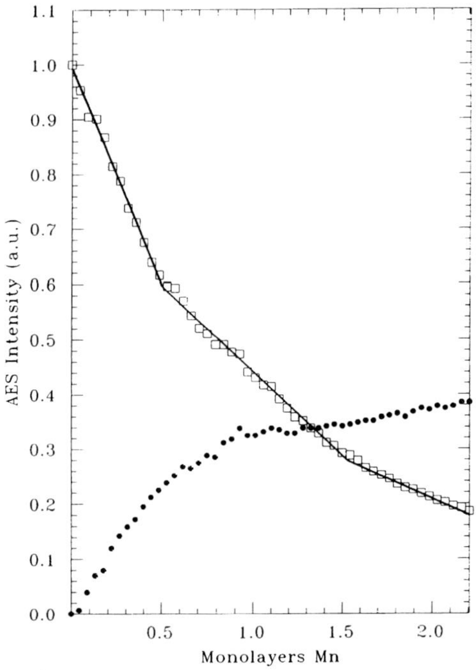

Fig. 7.12. Auger signal intensities of the Cu (60 eV) transition (◇) and the Mn (40 eV) transition (•) as a function of the manganese coverage on Cu(100). The substrate temperature was 310 K

after deposition of 0.5 ML Mn at 300 K, broad $c(2 \times 2)$ superstructure beams are seen in the LEED pattern. The appearance of these additional beams is indicative of the formation of a new structure. The intensity of these diffracted beams was also measured by MEED during deposition. Figure 7.13b depicts the MEED intensity of the $(\overline{1}/2\ \overline{1}/2)$ beam. Above a coverage of 0.3 ML Mn, the intensities of the (00) and the $\overline{1}/2\ \overline{1}/2$ beams behave similarly: both increase strongly until a coverage of 0.5 ML Mn is reached. Above 0.5 ML Mn, their intensity decreases. This shows that the breakpoint in the Auger curve at 0.5 ML Mn corresponds to the coverage where the $c(2 \times 2)$ structure is best ordered.

These changes in surface structure were also followed by the measurement of the LEED intensity during deposition. The result is depicted in Fig. 7.14, where the intensities of three different beams ((3/2 3/2), (1 2) and (2 7/4)) measured at an electron energy of 100 eV and an angle of 45° are shown. The (2 7/4) beam is due to the formation of a $p4g(4 \times 2)$ structure above 0.5 ML Mn. This structure grows at the expense of the $c(2 \times 2)$ structure as can be seen from the simultaneous decrease of the intensity of the $c(2 \times 2)$ beam and the increase of the $p4g(4 \times 2)$ beam above 0.5 ML Mn. These and other

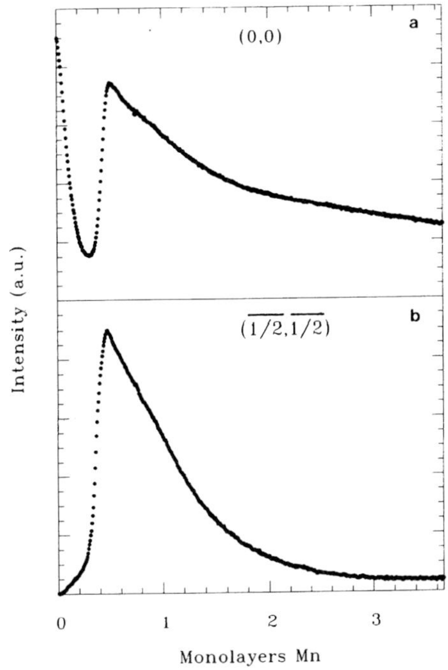

Fig. 7.13. MEED intensities during Mn deposition on Cu(100) at 340 K: The normalized MEED intensities of the (**a**) (00) and (**b**) $(\overline{1/2}\,\overline{1/2})$ beams for an electron energy of 3 keV and an angle of incidence of 3.2° are displayed

similar measurements have been used to determine the stability of a given superstructure with respect to growth temperature and coverage.

The $c(2 \times 2)$ structure has already been observed by Binns and Norris [10] after deposition of submonolayer amounts of Mn on Cu(100) at room temperature. These authors suggested two different structural models for this phase. Either Mn is adsorbed in every second fourfold-hollow site *on* the Cu(100) surface or Mn is incorporated *into* the Cu(100) surface to form an ordered surface alloy. We have taken I/V data for this structure and performed a dynamical intensity analysis [36] to determine the surface structure. These investigations are discussed below.

7.3.2 Structure Determination of the Cu(100) $c(2 \times 2)$-Mn Phase

In order to determine the structure of the $c(2 \times 2)$ phase, LEED I/V curves were recorded for seven nonequivalent beams ((10), (11), (20), (21), (1/2 1/2), (3/2 1/2) and (3/2 3/2)) from 40 eV to around 350 or 400 eV in steps of 0.5 eV. The data sets had an energy range of approximately 1600 eV.

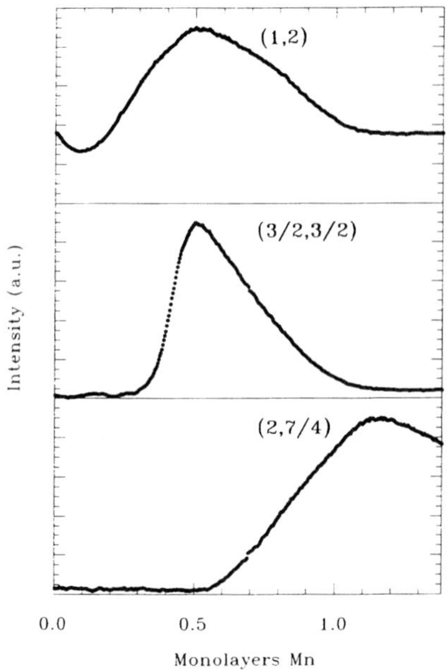

Fig. 7.14. LEED intensities during Mn deposition on Cu(100) at 300 K. The normalized LEED intensities of the (1, 2), (3/2, 3/2) and (2, 7/4) beams measured at an angle of incidence of 45° and an electron energy of 100 eV are reproduced

The $c(2 \times 2)$ structure was produced by deposition of 0.5 ML Mn on surfaces held at 375 K. The coverage calibration was performed using the MEED intensity maximum of the (1/2 1/2) beam as described above. This allowed a precise coverage determination to within ±0.05 ML Mn. I/V spectra measured for three different $c(2\times2)$ structures prepared *ex nuovo* were practically identical. In Sect. 7.4, the morphology of the c(2×2) phase formed on Cu(100) will be presented. The results of that analysis are of relevance for the I/V analysis since they allow a measure of the perfection of the $c(2\times2)$ phase and enable a characterization of the main defects. As will be discussed later, the main source of disorder is steps, while antiphase domain boundaries are practically absent. At a growth temperature of 375 K, the average terrace width is considerably larger than 100 Å. Thus, steps should have only a marginal effect on the LEED I/V measurement.

Two completely different models for the $c(2\times2)$ structure were considered: a Mn overlayer in which Mn atoms are located in every other fourfold-hollow site on the Cu(100) surface, and a surface alloy in which Mn atoms replace every other Cu atom in the topmost plane of the substrate (see Fig. 7.15). In a preliminary analysis, both models were tested using a large variation of the first two interlayer spacings. The corresponding parameters for the surface alloy model are depicted in Fig. 7.16. In the overlayer model, the distance between the Mn overlayer and the topmost atomic plane of the Cu(100)

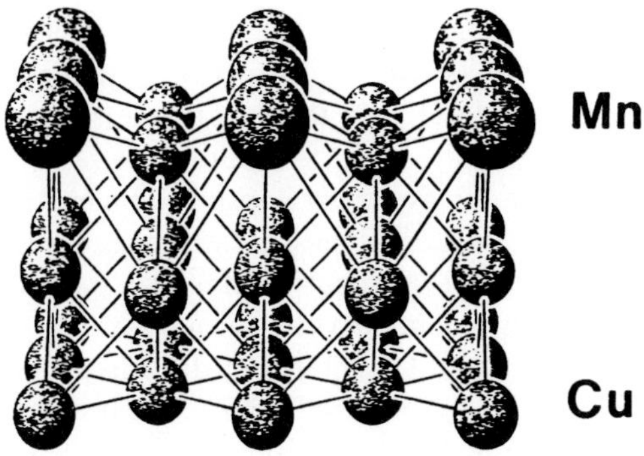

Fig. 7.15. 3D model for the $c(2 \times 2)$ surface alloy

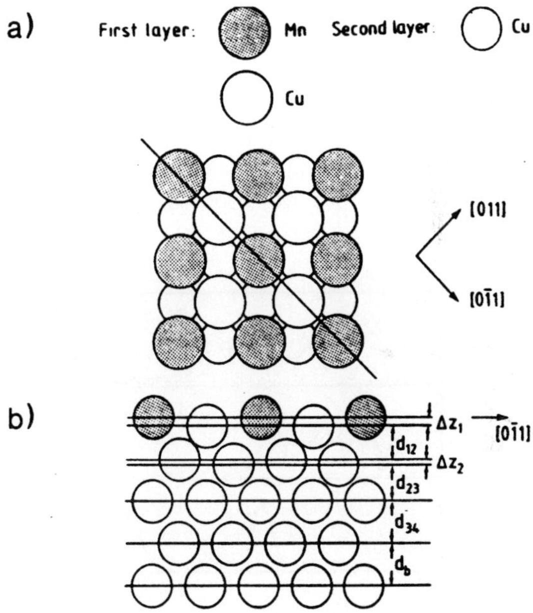

Fig. 7.16. Structural model for the $c(2 \times 2)$ surface alloy: (**a**) top view; (**b**) side view with a definition of the structural parameters used in the LEED analysis

substrate was varied from 1.4 Å to 2.5 Å in steps of 0.1 Å. The corrugation in the alloy layer was varied from +0.5 Å (Mn displaced outwards) to −0.5 Å in steps of 0.1 Å. Additionally, the first interlayer distance was varied in steps of 0.05 Å from 1.6 Å to 2.0 Å. Using only these structural parameters, the best-fit Pendry R-factor R_p was 0.72 for the hollow site and 0.31 for the surface alloy. The difference in the R-factors between the two models is considerably larger than the reliability figure (var R) of the R-factor. For this calculation, an absorptive potential V_{oi} of 4 eV was used and an energy range ΔE of 1200 eV was employed, giving a value of var $R = 0.06$. Therefore it is clear that the fourfold-hollow site model can be discarded. A refinement of structural parameters was thus carried out only for the surface alloy model.

In the next step of the analysis, the first three interlayer distances and the corrugations in the first three layers were optimized using both a grid search and an automated fit procedure. The best-fit structure for the Cu(100)–$c(2\times 2)$-Mn phase has a considerable corrugation of 0.30 ± 0.02 Å in the first layer. Practically no corrugation is found in the second layer (0.02 ± 0.03 Å) and third layer (0.00 ± 0.04 Å), and the interlayer distances are close to the bulk value ($d_{12} = 1.79 \pm 0.02$ Å, $d_{23} = 1.80 \pm 0.03$ Å and $d_{34} = 1.75 \pm 0.04$ Å). The error bar was determined from the reliability figure of the R-factor as described in Sect. 3.2. The R-factors for the optimum structure are:

$$\begin{aligned} \mathrm{Cu}(100)-c(2\times 2)-\mathrm{Mn}: R_\mathrm{p} &= 0.287\ , \\ R_\mathrm{ZJ} &= 0.106\ , \\ R_\mathrm{DE} &= 0.305\ . \end{aligned}$$

In Fig. 7.17, the spectra calculated for the best-fit $c(2 \times 2)$ structure are compared with the measured I/V curves. Visual inspection shows that the peak positions are fairly described well, consistent with the satisfactory R-

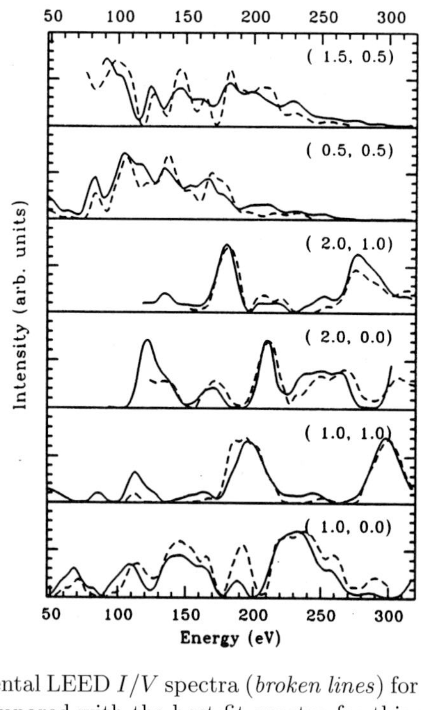

Fig. 7.17. Experimental LEED I/V spectra (*broken lines*) for the Cu(100)–$c(2\times2)$-Mn surface alloy, compared with the best-fit spectra for this phase (*solid lines*)

factors obtained. Additionally, the minima for the different R-factors give the same optimum structure to within ±0.01 Å.

To make the analysis more complete, the composition and ordering of the surface alloy were exhaustively checked, in spite of the fact that the onset of bulk Mn diffusion, as evidenced by changes in the Auger intensity ratio, is found only above 500 K. Therefore, for growth temperatures below 450 K, as used in the LEED study, bulk diffusion of Mn should be negligible. Nonetheless, the mobility might be large enough to allow some Mn atoms to segregate to deeper layers and form additional, dilute Mn–Cu alloy layers. As stated in the previous section, the ATA approximation permits one to optimize the chemical composition layer by layer or, even better, for different sublattices within a layer.

Variation of the Mn coverage in the first layer led to an optimum concentration of 47% Mn. Subsequently, chemical ordering was considered in the top layer. In this step, the two inequivalent sites defined by their different distances to the second layer were assigned different concentrations (c_{11} and c_{12}), which were then optimized. This resulted in an almost perfectly ordered alloy, with $c_{11} = 99\%$ Mn and $c_{12} = 10\%$ Mn. When the composition in the second layer was optimized separately, a Mn concentration c_2 of 12% was obtained. Thus the top layer forms an almost perfectly ordered alloy film above a copper layer containing only a few percent of Mn. The error bar for the determination of the composition in the vicinity of the surface is ≈10% for the first layer and ≈ 20% for the second layer in this case. After optimizing the structure and composition of the Cu(100)–c(2×2)-Mn surface alloy, a metric distance [37] D1 of 16.6 was obtained. The rather low value for this R-factor confirms the quality of the agreement between theory and experiment. The results presented above therefore clearly demonstrate that the surface alloy model is preferred over the Mn overlayer.

The structure of the Cu(100) $c(2\times2)$ Mn surface alloy has also been investigated using photoelectron diffraction by Toomes et al. [38] and using medium-energy ion scattering (MEIS) by Brown et al. [39]. The structural parameters determined by the various techniques are summarized in Table 7.4. Though different d_{12} values are obtained by different techniques, the large corrugation in the outermost layer is confirmed in all three measurements and

Table 7.4. Summary of the values of the structural parameters (see Fig. 7.16 for their definition) determined by LEED (this work), photoelectron diffraction [38] and MEIS [39]

Parameter (Å)	LEED	Photoelectron diffraction	MEIS
d_{12}	1.79 ± 0.02	1.63 ± 0.08	1.72 ± 0.04
d_{23}	1.80 ± 0.03	1.83 ± 0.08	1.83 ± 0.02
$\triangle Z_1$	0.30 ± 0.02	0.39 ± 0.08	0.37 ± 0.06

the corrugation amplitudes determined agree with each other within their error bars. The corrugation is believed to arise from the large atomic volume of magnetic Mn, as is shown by theoretical calculations [40, 41] described later in this chapter.

7.3.3 Structural Transitions upon Annealing for Ultrathin Mn/Cu(100) Films

The experiments described in this section were designed to unveil the relationship between the growth temperature and the film structure. Depending upon the growth temperature, deposition of 0.5 ML Mn leads to the formation of a $c(8 \times 2)$ phase (below 270 K) or a $c(2 \times 2)$ structure (above 270 K). The question arises as to whether both structures are temperature-dependent equilibrium structures or whether one of the structures is metastable upon heating or cooling.

To discriminate between these two choices, a $c(8 \times 2)$ superstructure was produced by evaporating 0.5 ML Mn on Cu(100) at 200 K. Then the sample was cooled to 110 K. This did not change the film structure. Figure 7.18 shows the measured intensities of the $(\overline{1}0)$ substrate beam, the $(\overline{1}/2\ 1/2)$ $c(2\times2)$ beam and the $(\overline{1}/8\overline{1}/2)$ $c(8\times2)$ beam upon annealing and subsequent

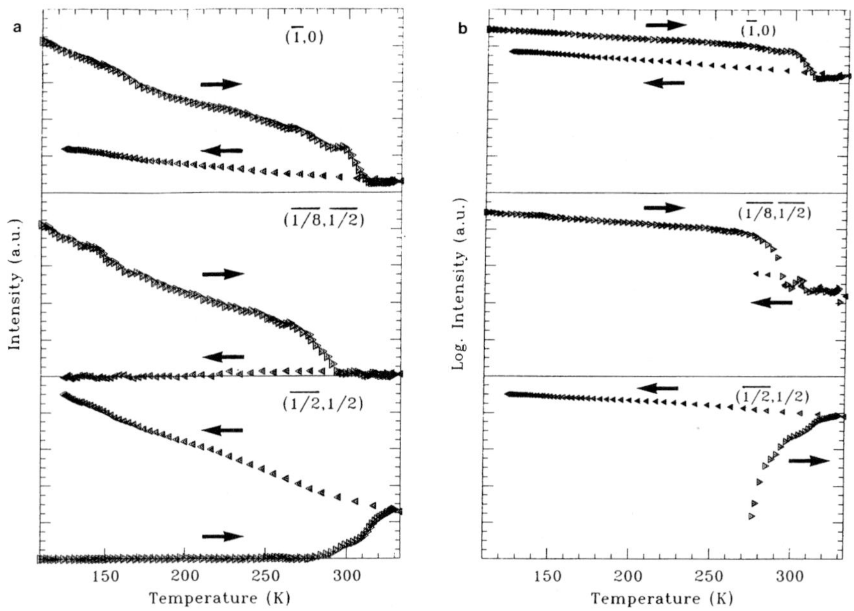

Fig. 7.18. The intensities of the $(\overline{1}0)$, $(\overline{1/8}\ \overline{1/2})$ and $(\overline{1/2}\ 1/2)$ beams are depicted as a function of the substrate temperature on (**a**) linear and (**b**) logarithmic scales for a coverage of 0.5 ML Mn deposited at 200 K. *Arrows* indicate the direction of increasing or decreasing temperature

cooling, on both linear and logarithmic scales. Arrows point in the direction of increasing and decreasing sample temperature during the measurement. The curves can be divided into three regions. Between 110 K and 260 K, the intensities of the $(\bar{1}0)$ and $\bar{1}/8\,\bar{1}/2)$ beams show an exponential decrease, as evidenced by the linear decrease in the logarithmic plot. In this temperature region, no $c(2\times2)$ spots are observed. Between 260 K and 295 K, the intensities of the $(\bar{1}0)$ and $(\bar{1}/8\,\bar{1}2)$ beams drastically decrease. At 280 K, $c(2\times2)$ spots start to appear, while the $c(8\times2)$ beams vanish around 295 K. In this region, the film clearly rearranges from the $c(8\times2)$ to a $c(2\times2)$ structure. Between 295 K and 330 K, the intensities of the $c(2\times2)$ beams increase with increasing temperature, indicating that the film order improves upon heating. In the cooling phase, the $c(8\times2)$ beams do not reappear, and the intensities of the $c(2\times2)$ beams increase further, showing that the $c(8\times2)$ structure is metastable and that the phase transition from the $c(8\times2)$ to the $c(2\times2)$ structure is irreversible. This change of structure upon annealing is also evident from the change in intensity of the $(\bar{1}0)$ beam.

For a more detailed investigation of the rearrangement mechanism, the phase transition was followed by AES. To do so, the sample was prepared in the same way as in the LEED experiment presented above. Figure 7.19a shows the Auger signal ratio of the Cu (60 eV) and Mn (40 eV) Auger transitions as a function of the annealing temperature. The Auger signal ratio hardly changes upon annealing below 240 K and above 300 K. A pronounced increase in the Cu/Mn ratio is observed between 240 K and 300 K. This is the same temperature window in which the phase transition from the $c(8\times2)$ to the $c(2\times2)$ structure is observed. In accordance with the results of the LEED experiment, the signal ratio does not return to its initial value upon cooling. The irreversibility of the phase transition is thus also evident in the Auger measurements. Since Auger electrons at 40 eV and 60 eV are very surface-sensitive, the change in Auger ratio upon annealing indicates that mass transport occurs in the direction normal to the surface. The process responsible is most likely the incorporation of Mn into the near-surface region and not the bulk diffusion of Mn in Cu. This can be seen from a similar measurement performed over a larger temperature range (Fig. 7.19b), which shows both an increase around 270 K corresponding to the $c(8\times2) \longrightarrow c(2\times2)$ phase transition and the onset of volume diffusion, as signaled by the steady increase in the Cu/Mn ratio above 450 K. This interpretation is further supported by measurement of the higher-energy LMM Auger transitions of Cu and Mn. These transitions are less sensitive to incorporation of Mn into the near-surface region, since Auger electrons with higher kinetic energies have a larger information depth. Therefore, the incorporation of Mn into the first layer around 270 K is almost invisible in the Cu/Mn ratio for the LMM Auger transition. The volume diffusion, on the other hand, is clearly reflected in an increase of the Cu/Mn ratio for the LMM Auger transitions.

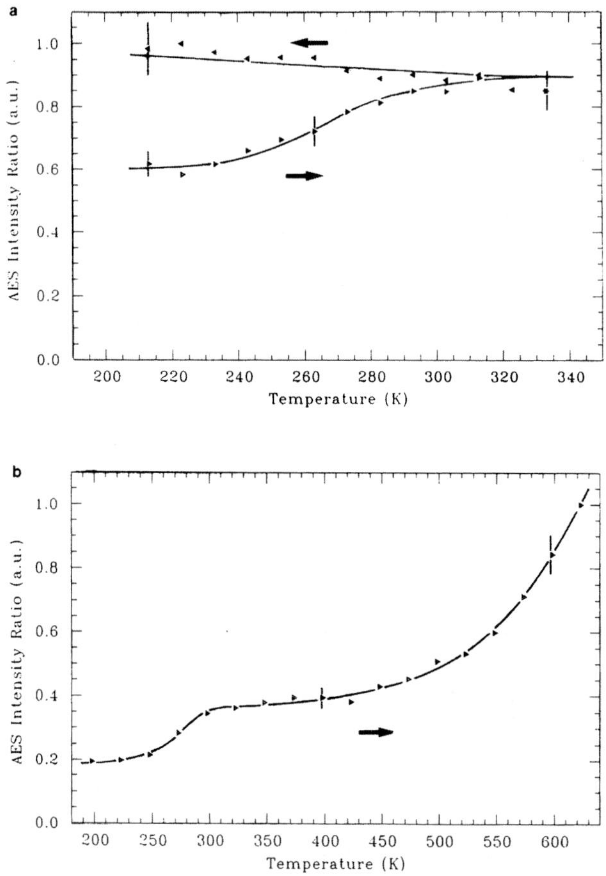

Fig. 7.19. The Auger intensity ratio of the Cu (60 eV) and the Mn (40 eV) Auger transitions as a function of the substrate temperature for a coverage of 0.5 ML Mn deposited at 200 K. (**a**) The sample was heated up to 330 K and then cooled down. (**b**) The sample was heated up to 620 K. *Arrows* indicate the direction of increasing or decreasing temperature

A second phase transition is observed at higher coverages. The $c(12 \times 8)$ structure also rearranges at 400 K into a $c(2\times2)$ structure. No phase transition to the $p4g(4\times2)$ structure is observed. Nevertheless, it could be demonstrated that the latter can also be produced via the incorporation of Mn into the surface. This is shown by the Auger intensity ratio of the Cu transition at 60 eV to the Mn transition at 40 eV. This ratio is (4 ± 1) times larger after deposition of 1.25 ML Mn at 350 K than after adsorption of the same amount of Mn below 270 K. The optimum $p4g(4\times2)$ structure is formed at a coverage of 1.25 ML Mn. Therefore, at least two layers contain Mn. We have explored the possibility that the upper layer consists of a pure Mn layer growing on an alloyed MnCu layer. To do so, 0.5 ML Mn was deposited at 325 K, leading to an ordered surface alloy with a $c(2 \times 2)$ structure. Subsequently, 0.75 ML Mn was deposited at 200 K, a temperature at which Mn incorporation is suppressed. The LEED pattern showed a diffuse $c(2 \times 2)$ pattern with weak $c(8 \times 2)$ beams. Only after annealing at 375 K did we observe the $p4g(4 \times 2)$ structure. This annealing process leads to an incorporation of Mn, as seen from an increase in the Cu/Mn Auger ratio. This confirms that incorporation of Mn into at least two layers is essential for forming the alloyed $p4g(4 \times 2)$ phase. Annealing the $p4g(4 \times 2)$ structure above 470 K leads to the formation of a $c(2 \times 2)$ structure, thereby indicating that the $p4g(4 \times 2)$ phase is only metastable.

It is not trivial to derive a structural model for this phase, since the precise Mn concentrations in the two layers are not known. It is, however, plausible to assume that the Mn concentration in the first layer is at least as high as in the $c(2 \times 2)$ phase and most likely even higher. It would clearly be helpful to determine the precise layer concentration with other techniques, such as ion scattering, as well as to use scanning tunneling microscopy to study this interesting alloy phase.

7.3.4 Growth and Structure of Mn on Ni(100) Above 270 K

An even larger variety of superstructures is observed for growth of Mn on Ni(100) above 270 K. These superstructures not only are dependent upon film thickness but also show a pronounced variation with growth temperature. This, together with the strong impact of postannealing on the film structure, points to the influence of kinetics on the formation of superstructures. This will be discussed in more detail after the growth and structure of various phases has been analyzed.

With the goal of understanding the growth of the Mn films on Ni(100), MEED intensity curves have been measured. Figure 7.20a shows the MEED intensity of the (00) beam during manganese deposition at 370 K. The curve is characterized by two intensity minima after adsorption of 1.0 and 2.1 ML Mn, two maxima at 1.45 and 2.5 ML Mn, and an asymmetric maximum around 0.25 ML Mn. At low coverages, a $c(2 \times 2)$ superstructure appears upon Mn deposition. To follow the formation and decay of the $c(2 \times 2)$ structure, the intensity of a superstructure beam was measured during deposition. Figure 7.20b shows the intensity of the $(\overline{1}/2\ 1/2)$ beam. In striking contrast to what is seen on Cu(100) (Fig. 7.6), two clear maxima are observed, at around 0.45 and at 1.0 ML Mn. It is only after deposition of 1.0 ML Mn that the intensity of the $(\overline{1}/2\ \overline{1}/2)$ beam decreases strongly.

To unveil the temperature dependence of the growth of the $c(2 \times 2)$ structure, the intensity of the $(\overline{1}/2\ \overline{1}/2)$ beam was measured for different growth temperatures. The results are shown in Fig. 7.21. For comparison, the curve obtained for deposition at 370 K is reproduced. An almost identical behavior is observed at 420 K (not plotted). The only minor differences are that

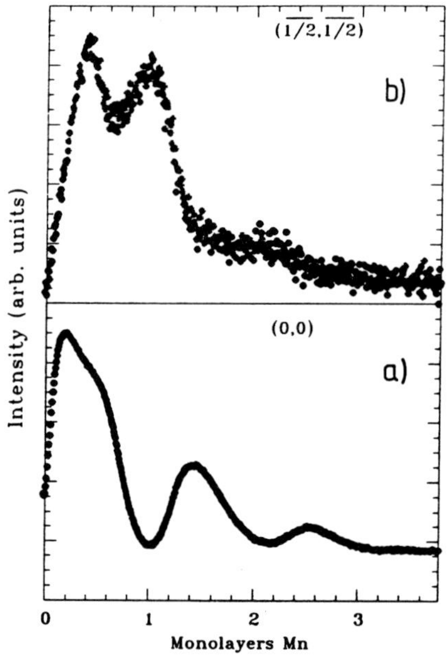

Fig. 7.20. MEED intensities during Mn deposition on Ni(100) at 370 K. The normalized MEED intensities of the (**a**) (00) and (**b**) ($\overline{1}/2$ 1/2) beams during Mn deposition, measured at an angle of incidence of 3.2° and an electron energy of 3 keV, are shown

the maxima are less pronounced, i.e., the minimum around 0.75 ML is shallower and the decrease of intensity above a deposition of 1 ML is slower. At a growth temperature of 470 K, a qualitatively different behavior is apparent. Again, the first maximum appears at around 0.5 ML. Subsequently, the intensity remains almost constant up to approximately 1.0 ± 0.1 ML Mn and then decreases linearly with a breakpoint at approximately 1.5 ± 0.1 ML Mn. At 510 K (not shown), the first intensity maximum is reached only after deposition of 0.65 ± 0.1 ML Mn, even though the first breakpoint exists around 0.45 ± 0.1 ML Mn. At this temperature, the intensity of the $c(2 \times 2)$ beam drops only above 1.5 ±0.1 ML Mn. At higher temperatures (the results for 550 K is reproduced), a similar behavior is found, but with slower decrease in intensity. This trend becomes even more evident at 600 K, where the intensity drops only very slowly above 2.5 ML Mn and the first maximum is reached only after deposition of 1.0 ± 0.1 ML Mn.

The curves in Fig. 7.21 show that Mn forms a well-ordered $c(2 \times 2)$ structure up to coverages ranging from 1.0 ML at 370 K to more than 2.5 ML Mn at 600 K. With postannealing at 600 K, it is even possible to form a well-ordered $c(2 \times 2)$ structure after deposition of 4 ML Mn. To determine the structure of the $c(2\times2)$ phase at different thicknesses, LEED I/V curves have been measured and compared with full dynamical calculations. In the following

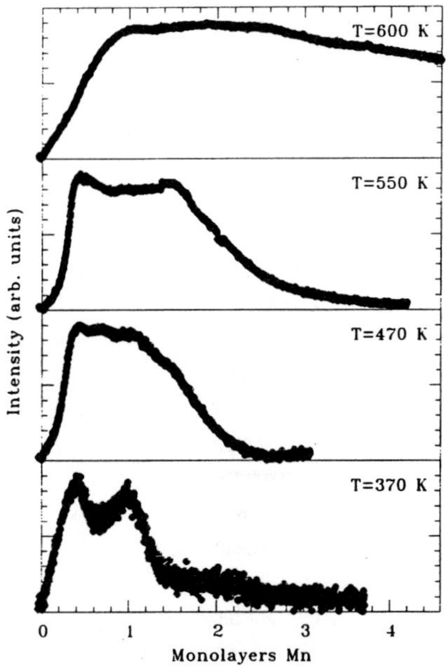

Fig. 7.21. MEED intensities during Mn deposition on Ni(100) at different temperatures. The normalized MEED intensities of the $(\overline{1}/2\ 1/2)$ beam are displayed for different substrate temperatures. The same scattering conditions as in Fig. 7.20 were used

two subsections, we shall concentrate on the determination of the structure of the $c(2 \times 2)$ phases produced by deposition of 0.5 ML Mn (at 450 K) and 3–4 ML Mn (at 600 K).

7.3.5 Structure Determination of the Ni(100)–$c(2 \times 2)$ 0.5 ML Mn Phase

The $c(2 \times 2)$ structure was produced by deposition of 0.5 ML Mn on Ni(100) at 450 K. The coverage calibration was performed using the first MEED intensity maximum of the $(\overline{1}/2\,\overline{1}/2)$ beam. To check the influence of the sample temperature during Mn deposition, growth temperatures ranging from 330 K to 470 K were tested. The LEED spots were considerably broader and the background was more pronounced for films grown at lower temperatures (e.g. around 330 K). The LEED intensity spectra, however, did not differ from those of the well-ordered phase produced at around 425 K or above within the error of the measurement. During the LEED measurements, the sample was held at 110 K.

LEED I/V curves were measured from 40 eV to between 350 and 400 eV after a careful alignment of the LEED system to normal incidence. Seven nonequivalent beams ((10), (11), (20), (21), (1/2,1/2), (3/2,1/2) and

(3/2,3/2)) were measured, so that the entire data set had an energy range of 1600 eV.

Details of the LEED calculations have been described in Sect. 3.2 and 7.2.2. As in the analysis of the Cu(100)–$c(2 \times 2)$-Mn superstructure, two completely different models for the $c(2 \times 2)$ structure were considered: a Mn overlayer in which Mn atoms are located in every other fourfold-hollow site on the Ni(100) surface, and a surface alloy in which Mn atoms replace every other Ni atom in the topmost plane of the substrate (see Figs. 7.15 and 7.16).

In the first step of the analysis, both models were tested using a large variation of the first two interlayer spacings, and of the corrugation for the alloy model. These three parameters were varied in steps of 0.1 Å. For the hollow site model, a minimum R-factor of 0.69 was obtained, while an R-factor of 0.34 was attained for the surface alloy model. The reliability figure of the R-factor (var R) defined in Sect. 3.2 had a value of 0.07. Consequently, the fourfold-hollow site model can be discarded. A refinement of structural parameters was thus carried only out for the surface alloy model. For the best-fit structure of the Ni(100)–$c(2 \times 2)$-Mn surface alloy phase, no corrugation is found in the second and third layers, while a pronounced corrugation of 0.25 ± 0.02 Å is observed in the first layer. The first three interlayer distances are all close to the bulk value of 1.76 Å. In Fig. 7.22, the dependence of the R-factor on the structural parameters is shown, demonstrating that the R-factor has a deep minimum at the optimum structure. The different R-factors for the optimum structure are:

$$\begin{aligned} \mathrm{Ni(100)} - c(2 \times 2) - \mathrm{Mn}: \quad R_{\mathrm{p}} &= 0.283\,, \\ R_{\mathrm{ZJ}} &= 0.086 \\ R_{\mathrm{DE}} &= 0.395\,. \end{aligned}$$

From the variation of the R-factor around the best-fit structure, the error bars for the structural parameters were obtained. In Fig. 7.23, the spectra calculated for the best-fit $c(2 \times 2)$ structure are compared with the measured I/V curves.

7.3.6 Structure Determination of the Ni(100)–c(2×2) 4 ML Mn Phase

To determine the atomic coordinates of the $c(2\times2)$ phase formed after deposition of 3–4 ML Mn, a second LEED analysis was performed. The structure was prepared by deposition of 4 ML Mn at 600 K, followed by further annealing at 600 K for 5 additional minutes to increase the order of the film. This treatment leads to sharp, intense $c(2\times2)$ superstructure beams on a low background. I/V curves were measured after cooling the sample to 150 K. The intensities of six nonequivalent beams were measured from 40 eV to around 350 eV at normal incidence. After an automatic background subtraction, symmetry-equivalent beams were averaged and normalized to constant

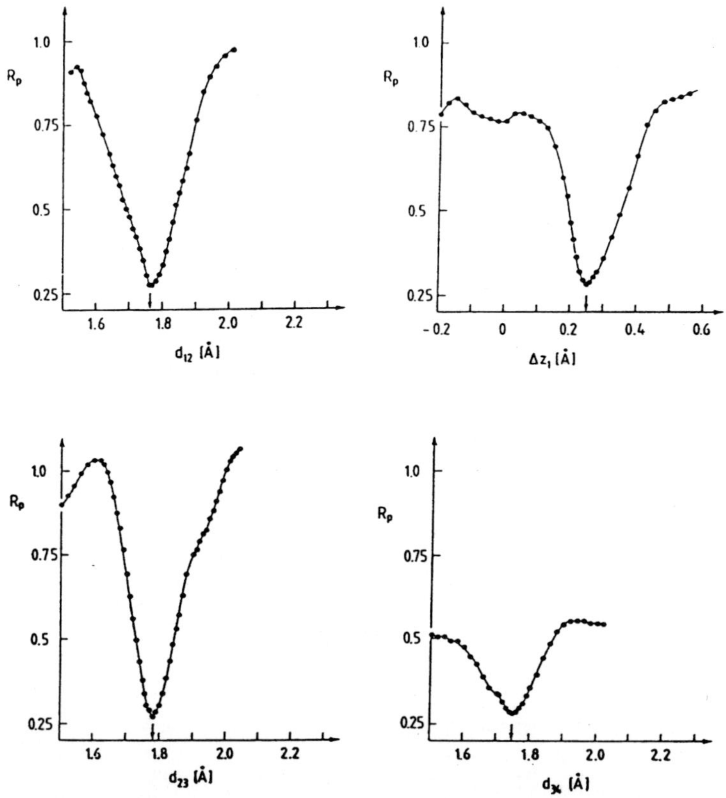

Fig. 7.22. Variation of the R-factor variation around the optimum solution for the MnNi surface alloy. The influence of variation of various structural parameters on the comparison between experiment and theory as estimated from the Pendry R-factor is shown for the Ni(100)–$c(2 \times 2)$-Mn surface alloy. The results of variation of the first (d_{12}), second (d_{23}) and third (d_{34}) interlayer distances and of the first-layer corrugation (Δz_1) around the best-fit structure are displayed. *Arrows* show the position of the R-factor minimum

incident current. The whole data set had an energy range of approximately 1400 eV. To check the reproducibility, the $c(2\times2)$ structure was prepared three times and practically identical I/V curves were obtained. Additionally, the influence of film thickness and deposition temperature was checked. For films between 3 and 4 ML, the I/V curves were basically indistinguishable, provided that the deposition temperature was above 550 K. In general, the background was lower at higher growth temperatures, but the positions of the peaks and their relative intensities did not change. In Figure 7.24, the I/V curves obtained after depositing 4 ML at 600 K are shown (solid lines). For comparison, the I/V curves obtained for the $c(2\times2)$ surface alloy formed after deposition of 0.5 ML Mn at 425 K are shown (*broken lines*). Figure 7.24

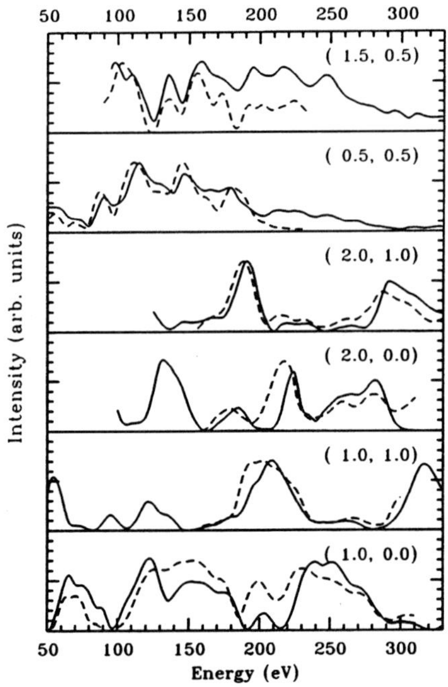

Fig. 7.23. Experimental spectra (*broken lines*) for the Ni(100)–$c(2 \times 2)$-Mn structure, compared with the best-fit spectra for the surface alloy (*solid lines*)

shows pronounced differences in the I/V curves, especially for the substrate beams. This indicates that the structure of the surface alloy is considerably different from that of the 4 ML thick film.

The Auger intensities of the Mn transition at 589 eV and the Ni transition at 848 eV were measured in the first-derivative mode with a cylindrical-mirror analyzer using a primary energy of 3 keV. For the 4 ML thick films, the intensity ratio of these transitions ($I_{\mathrm{Mn}}/I_{\mathrm{Ni}}$) was determined to be 0.92±0.08. Using the known relative sensitivity of 0.88 [42], a composition of approximately 50% Mn and 50% Ni is obtained.

Theoretical I/V curves were calculated for a variety of models, including pure Mn overlayers and a variety of $\mathrm{Mn}_x\mathrm{Ni}_{1-x}$ alloys with different structures and compositions, even though the Auger analysis indicated a composition of 50% Mn and 50% Ni. For the various models, the atomic corrugation at the surface, the first three interlayer distances at the surface d_{12}, d_{23}, d_{34} and the bulk interlayer distance d_{b} were optimized to determine the best-fit structure. The interlayer distances were varied between 1.5 and 2.1 Å in steps of 0.1 Å, and the corrugation of the first two layers was varied between −0.5 and 0.5 Å in steps of 0.1 Å.

It was assumed throughout the calculations that the films grew epitaxially on the Ni(100) substrate. This is supported by the observation that

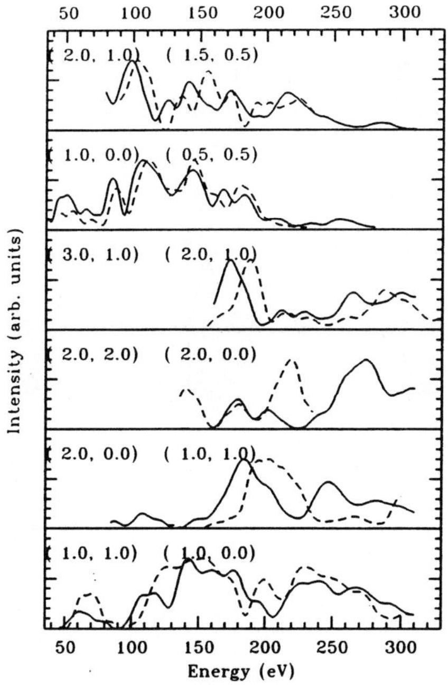

Fig. 7.24. Comparison of I/V spectra for the surface and the bulk alloys. Experimental spectra obtained after deposition of 4 ML Mn at 600 K (*solid lines*) and after deposition of 0.5 ML Mn at 400 K (*broken lines*). Equivalent beams are compared and are labeled using two different notations. The labeling of the beams on the *right* is for the unit cell shown in Fig. 7.26, while the labeling on the *left* is for the primitive unit cell of the Ni(100) surface

the positions of the integer-order LEED beams did not vary with increasing film thickness. As a consequence, the MnNi films have the in-plane lattice constant of 3.52 Å characteristic of the Ni(100) surface.

Only one model gave a reasonable fit to the experimental data. This model, which is shown in Figs. 7.25 and 7.26, is characterized by an ordered arrangement of Mn and Ni atoms where each layer has a Mn concentration of 50%. After the optimum structure was determined on a rough grid, the search was continued on a finer grid, where a step width of 0.01 or 0.02 Å was used. Table 7.5 shows the optimum structure obtained for the three different R-factors (R_P, R_{ZJ} and R_{DE}). The variations in the best-fit structure for the different R-factors are rather small and the values for the R-factor minima are quite low, indicating a satisfactory agreement between theory and experiment.

Visual inspection of the theoretical I/V curves obtained for the optimum structure and the experimental spectra shown in Fig. 7.27 confirm the agreement. Both the peak positions and the intensities of the peaks are satisfactorily described, with the exception of the (20) beam around 180 eV. In

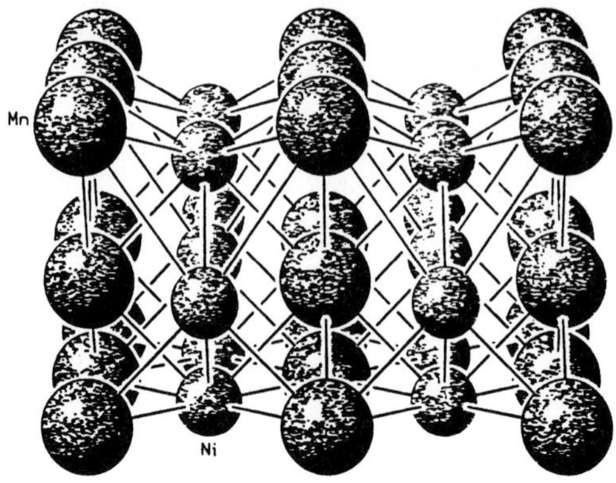

Fig. 7.25. 3D model for the MnNi bulk alloy

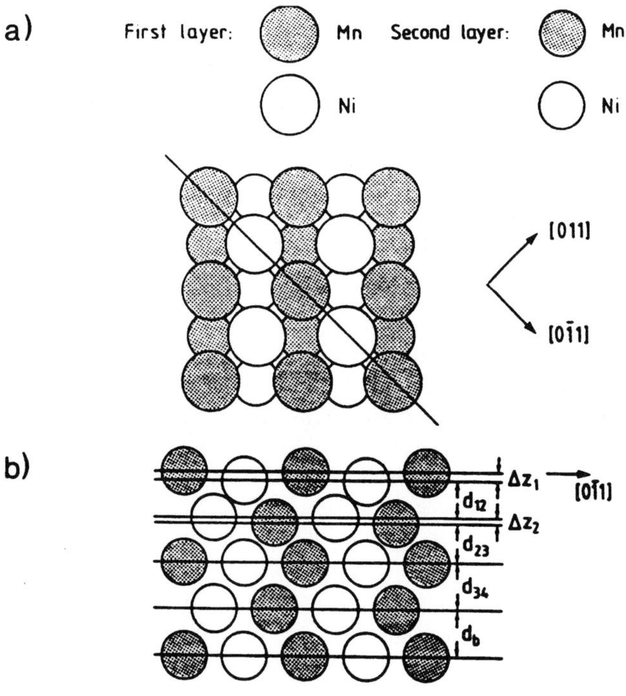

Fig. 7.26. Structural model for the $c(2{\times}2)$ bulk alloy. *(a)* Top view, where the *solid line* denotes the section used for the side view. *(b)* Side view, with a definition of the structural parameters used in the LEED analysis

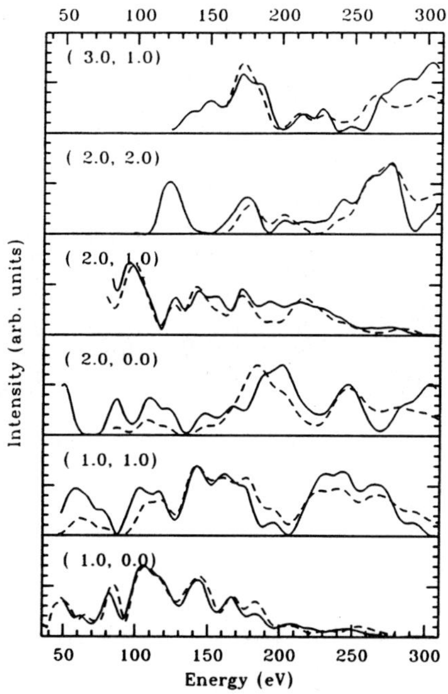

Fig. 7.27. Best-fit spectra for the bulk alloy (*solid lines*) compared with the experimental spectra for the Ni(100)–$c(2\times2)$ Mn structure obtained after deposition of 4 ML Mn at 600 K (*broken lines*)

Fig. 7.28, the variation of the Pendry R-factor with surface corrugation and interlayer spacing is shown. This figure shows that the R-factor minima are deep, thus allowing a precise determination of the corresponding interatomic distances.

In a further stage of the structural analysis, atomic concentrations of different subplanes were varied to see if a unique solution had been found. Several different compositions were checked, and in each case the optimum structure was determined. As limiting cases, two structures were considered in which the second and deeper layers contained pure Mn or pure Ni and only the first layer was composed of an ordered alloy. For both models, the

Table 7.5. Optimum R-factors for the MnNi bulk alloy produced by deposition of 4 ML Mn on Ni(100) at 600K. Consistent results for the different R-factors are obtained

d_{12} (Å)	d_{23} (Å)	d_{34} (Å)	d_b (Å)	ΔZ_1 (Å)	ΔZ_2 (Å)	R-factor
1.76	1.86	1.87	1.88	0.30	0.00	0.327 (R_p)
1.76	1.88	1.87	1.88	0.30	0.00	0.072 (R_{ZJ})
1.76	1.86	1.87	1.88	0.32	0.00	0.299 (R_{DE})

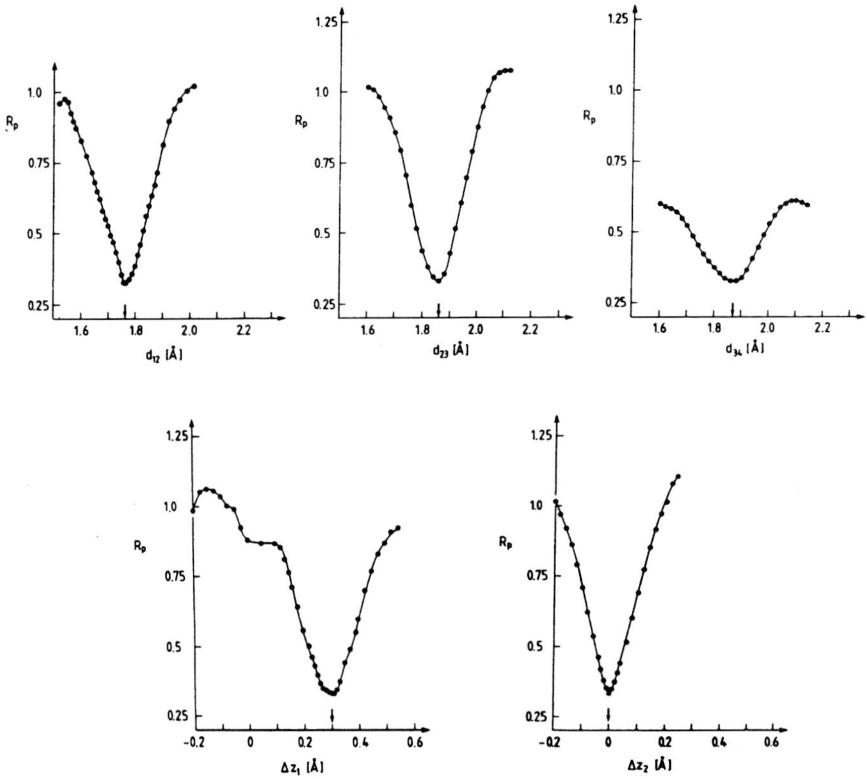

Fig. 7.28. Variation of the R-factor around the optimum solution for the MnNi bulk alloy. The influence of variation of various structural parameters on the comparison between experiment and theory, as estimated from the Pendry R-factor is shown. The results of variation of the first (d_{12}), second (d_{23}) and third (d_{34}) interlayer distances and of the corrugation of the first layer (Δz_1) and the second layer (Δz_2) around the best-fit structure are depicted. *Arrows* show the position of the R-factor minimum. The corresponding values of the structural parameters are listed in the first row of Table 7.5

interlayer distances found after optimization were identical to those of the best-fit structure of the ordered bulk alloy listed in Table 7.5. This shows that the interlayer distances found in the R-factor analysis are independent of the composition of deeper layers. All three R-factors found for these two limiting cases (pure Ni and pure Mn in deeper layers) are worse than the three minimum R-factors for the ordered bulk alloy. Nevertheless, the variations of the R-factors with composition are only of the order of 3%, which is still within the error bars of the structural analysis. For the first layer, on the other hand, the composition can be uniquely determined. Satisfactory agreement between theory and experiment is only achieved for a Mn concentration close to 50%. Basically, all Mn atoms occupy one sublattice and are displaced outwards by 0.30 ± 0.02 Å. An outward displacement of Ni atoms by the same amount leads to a considerably worse R-factor ($R_P = 0.46$) after optimizing the other interlayer distances.

Thus, at present one can conclude that upon deposition of 3 to 4 ML Mn on Ni(100) at elevated temperatures (550–600 K), a film with the structure listed in Table 7.5 is obtained. An ordered alloy is found for the surface plane, in which Mn atoms are displaced outward by 0.30 ± 0.02 Å. For deeper layers, a composition of 50% Mn and 50% Ni for each plane is favored but other concentrations cannot be excluded. On the other hand, the interlayer distances can be uniquely determined, independently of the composition of the deeper layers. This is in agreement with LEED studies by Gauthier et al. (see [31] and, for example, [43]), who found that for various disordered PtNi, PtFe and PtCo alloys, the composition and the interlayer distances in the surface region could be optimized independently of each other. Similarly to their findings, we have observed that the dependence of the R-factor on the interlayer distances is more pronounced than that on the composition. These conclusions are further supported by structural analyses for MoRe alloys [44]. In the studies mentioned above, the relatively large difference in atomic number of the constituent elements enabled a determination of the composition profile in the near-surface region. For the alloy studied here, the atomic numbers of the elements differ only by 3. Hence, the phase shifts of Mn and Ni are quite similar. This has the drawback that it becomes difficult to uniquely determine the composition of the deeper layers. Fortunately, the interlayer distances for the MnNi films, which are summarized in the first row of Table 7.6, are rather different from those of both the clean Ni(100) surface and the $c(2{\times}2)$ structure obtained after deposition of 0.5 ML Mn on Ni(100). The interlayer distances can thus be used as a fingerprint of the structure that was grown.

In the bulk phase diagram, a variety of Mn_xNi_{1-x} alloy phases exist [46]. The question is, therefore, whether any MnNi alloys have interlayer distances close to the values found in our analysis and could account for the observed diffraction pattern. Disordered alloys are formed above 900 K. Since the alloy films in our study were stable at much lower temperatures, disordered MnNi films are excluded. In the bulk phase diagram, there are three ordered alloys which could possibly account for the $c(2{\times}2)$ structure and the observed interlayer spacings. These are the tetragonal and the cubic phase of MnNi and the cubic phase of $MnNi_3$ [2]. The cubic phase of MnNi has the CsCl structure with a lattice constant of 2.98 Å [2]. Only an unreasonably high strain would allow this structure to match the in-plane lattice constant of 3.52 Å of the Ni(100) surface. Therefore this phase can be excluded. Cubic $MnNi_3$ has the Cu_3Au structure, with a lattice spacing of 3.589 Å [2]. The epitaxial growth of $MnNi_3$ would result in a misfit of 2%. As a consequence, the critical thickness for dislocation formation should be approximately 8 Å [47]. In the present work, no evidence for misfit dislocations was found for films up to 15 Å thick.

Table 7.6. Best-fit structures obtained from LEED analyses of the Ni(100) surface [45], the surface alloy formed upon deposition of 0.5 ML Mn on Ni(100) and the bulk alloy produced by deposition of 4 ML Mn on Ni(100). The error bars have been determined from the variation of the Pendry R-factor. For comparison, the expected structure of a tetragonal MnNi(100) film epitaxially grown on Ni(100) is listed. The in-plane lattice constant is denoted by **a**

	Ni(100)	Ni(100)–$c(2\times2)$-Mn 0.5 ML Mn	Ni(100)–$c(2\times2)$-Mn 4 ML Mn	Tetragonal MnNi on Ni(100)
Δz_1 (Å)	–	0.25 ± 0.02	0.30 ± 0.02	–
Δz_2 (Å)	–	0.00 ± 0.03	0.00 ± 0.03	–
d_{12} (Å)	1.74 ± 0.02	1.76 ± 0.02	1.76 ± 0.02	–
d_{23} (Å)	1.76	1.78 ± 0.03	1.88 ± 0.03	–
d_{34} (Å)	1.76	1.75 ± 0.04	1.87 ± 0.04	–
d_b (Å)	1.76	1.76	1.88 ± 0.04	1.87
a (Å)	3.52	3.52	3.52	3.52

Additionally, one would expect an expansion of the interlayer distance to 1.93 Å due to the misfit-induced strain. Since this distance is larger than the observed interlayer spacing of a thick MnNi film, we have also abandoned this possibility. The final choice is the tetragonal phase of MnNi. This structure has a bulk lattice constant of 3.53 Å in the plane and 3.72 Å in the normal direction [2]. Therefore, the misfit between the MnNi(100) film and the Ni(100) surface is only 0.1%. The resulting interlayer spacing of 1.87 Å is in excellent agreement with the observed interlayer distance of 1.88 Å. The observed structure thus corresponds very nicely to a slightly strained tetragonal MnNi film. Each alloy layer has a Mn concentration of 50%, in agreement with the Auger and MEED data. The tetragonal MnNi phase is the most stable bulk phase of all Mn_xNi_{1-x} alloys, i.e., it has the highest heat of formation [46]. This stability, together with the small misfit to the Ni(100) surface, provides the thermodynamic driving force for the growth of a tetragonal MnNi film and not any other Mn_xNi_{1-x} alloy phase on Ni(100).

Finally, it is worthwhile to address the surface structure of a tetragonal MnNi film on Ni(100). While the second and third interlayer spacings are close to the bulk value, the first interlayer spacing is considerably smaller. As can be seen from the definition of the parameters in Fig. 7.26, d_{12} describes the distance from the second layer to the lower-lying Ni atoms in the first layer. The Mn atoms, on the other hand, are displaced outwards by 0.3 Å. This leads to an average interlayer spacing of 1.92 Å, which is similar to the bulk interlayer spacing of the film. It is interesting to speculate on the origin of the strong corrugation of the surface layer. This will be done in Sect. 7.5.

7.3.7 Comparison with Other Systems

The most profound difference in the behavior of Mn deposited on Cu(100) and Ni(100) above 270 K occurs for thicknesses above 1 ML. While thick ordered bulk alloys with the tetragonal MnNi structure can be grown on Ni(100), this is impossible on Cu(100). When Mn is deposited on Cu(100) at 550 K, the LEED pattern shows a $c(2\times2)$ superstructure between 0.3 ML Mn and 4 ML. AES measurements show an increase of the Mn/Cu ratio up to 0.5 ML Mn. This ratio then remains constant up to at least 4 ML. This shows that for Mn on Cu(100), only the top layer can be alloyed, while additional manganese disappears into the bulk at 550 K. Thus, when one deposits Mn on Cu(100), one is unable to grow ordered bulk alloys. The thermodynamic data for bulk MnCu and MnNi alloys explain this striking difference. In Fig. 7.29, the change of entropy, energy and free energy is depicted as a function of alloy composition. For both systems, there is a pronounced entropy gain upon alloy formation. The difference between the Mn–Ni and Mn–Cu systems lies in the energy balance. While the energy is increased upon alloy formation in the Mn–Cu system, it is reduced in the Mn–Ni system. This energetic balance describes the tendency of a system to form AB bonds in preference to AA and BB bonds. Thus, in the case of the Mn–Ni system, MnNi bonds are energetically favorable and hence the system forms ordered alloys. For MnCu,

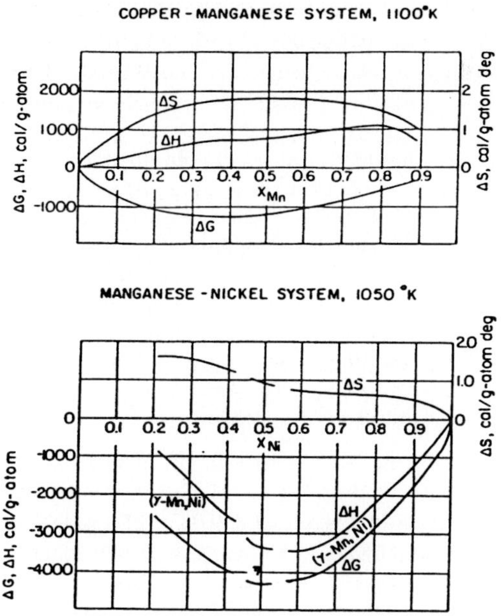

Fig. 7.29. Thermodynamic data for MnNi and MnCu bulk alloys. The change of the entropy (ΔS), enthalpy (Δ) and free energy (ΔG) upon alloy formation are displayed as a function of Mn concentration. After [46], used with permission

this contribution is negative. Hence MnMn and CuCu bonds are preferred over MnCu bonds. On the other hand, because of the pronounced entropy gain, the system forms disordered alloys over the entire composition range (see the sign of ΔG). This information is also evident from the phase diagrams of MnCu and MnNi.

From the structural and thermodynamic data of the various MnNi bulk alloys, it is also clear why ordered tetragonal MnNi films are grown on Ni(100). This structure is energetically favored, since it has the highest heat of formation and only a marginal misfit of 0.1% to the Ni(100) substrate. The small mismatch also identifies the reason why films can be grown for up to eight layers after deposition of 4 ML Mn without any evidence for the formation of misfit dislocations. The films in this thickness range are, instead, controlled by growth kinetics, i.e., mainly by the diffusion of Ni through the alloy film to the film surface, where reaction with Mn takes place.

In view of the thermodynamic data presented here, it is rather surprising that an ordered CuMn surface alloy is formed on the Cu(100) surface. Prompted by the desire to understand this process, a cooperation with a theoretical group was established that enabled the determination of the driving force for the stabilization of the CuMn surface alloy. For the Ni(100)–$c(2\times2)$ Mn system, ordered-alloy formation is less puzzling, since the bulk phase diagram suggests a pronounced tendency to form ordered alloys.

Surface alloy formation has emerged as a topic of current research interest. Ordered surface alloys have even been encountered for alkali metals adsorbed on Al(111) [48, 49]. Earlier examples of ordered surface alloys with a $c(2\times2)$ structure were obtained after deposition of 0.5 ML Au [33, 50] or Pd [51, 52]

Table 7.7. Summary of the structural results obtained for various ordered surface alloys. All structural values, except the ones in parentheses in column 2, which come from EAM calculations [63], have been obtained from LEED analyses. A dash or the symbol d_b indicates that the corresponding value has not been optimized. For comparison, the atomic radii of Au, Pd and Mn are shown. The corresponding values for Cu and Ni are 1.276 and 1.245 Å, respectively. See text for details

	Cu(100)–$c(2\times2)$-Au [180,183]	Cu(100)–$c(2\times2)$-Pd [181]	Cu(100)–$c(2\times2)$-Mn	Ni(100)–$c(2\times2)$-Mn
Δz_1 [Å]	0.1 (0.18)	0.02 ± 0.03	0.30 ± 0.02	0.25 ± 0.02
Δz_2 [Å]	–	–	0.02 ± 0.03	0.00 ± 0.03
d_{12} [Å]	1.88 (1.79)	1.81 ± 0.03	1.79 ± 0.02	1.76 ± 0.02
d_{23} [Å]	d_b (1.81)	d_b	1.80 ± 0.03	1.78 ± 0.03
d_{34} [Å]	d_b (-)	d_b	1.80 ± 0.04	1.75 ± 0.04
d_b [Å]	1.807	1.807	1.807	1.760
Element	Au	Pd	Mn	Mn
Atomic radius	1.442 Å	1.375 Å	1.365 Å	1.365 Å

on Cu(100). In Table 7.7, structural results are compiled for ordered surface alloys formed on Cu(100) and Ni(100). A comparison of the corrugation for the surface layer shows that the two alloys containing Mn have a much larger corrugation, of 0.30 Å (MnCu) and 0.25 Å (MnNi), than have the CuAu and CuPd surface alloys. The earlier trend in the corrugation of the CuAu and CuPd alloys shaped the argument of Wu et al. [51] that the corrugation of the surface alloy layer increases with increasing difference in the atomic radii of the alloy constituents. This statement is consistent with the smaller buckling of the CuPd surface alloy as compared with the CuAu surface alloy. If this tendency also held for Mn, one would expect an even smaller corrugation than for the CuPd surface alloy, since the atomic radius of elemental Mn is smaller than that of Pd. On the contrary, the corrugation found for the Mn surface alloy on Cu(100) and Ni(100) is considerably larger than the corrugation of the CuAu surface alloy. This shows that simple size arguments are insufficient to explain the buckling of surface alloys. Further evidence for this statement comes from a comparison of the corrugation for Mn on Cu(100) and on Ni(100). One would expect that the corrugation would be slightly larger on the Ni(100) surface, since the difference in the radii between Mn and Ni is larger than between Mn and Cu. On the contrary, a larger corrugation is found for the $c(2\times2)$ surface alloy of Mn on Cu(100).

7.4 Morphology of the Cu(100)– and Ni(100)–$c(2\times2)$-Mn Surface Alloys

The last few sections have reported on the growth and structure of ordered alloys formed upon Mn deposition on Cu(100) and Ni(100) above 270 K. Little attention was devoted to a description of the atomic processes that lead to alloy formation. Clearly, the formation of these surface alloys requires incorporation of Mn atoms *into* the topmost substrate layer. This *must* be accompanied by the simultaneous ejection of Cu or Ni atoms onto the surface. In this section, the morphology of the alloy films resulting from these microscopic processes is addressed.

As shown in Sect. 7.2.1 and 7.2.3, Mn forms quasi-hexagonal films with long-range order when deposited on a Cu(100) or Ni(100) surface between approximately 150 and 250 K. This implies that at these temperatures there is sufficient mobility of Mn atoms on both surfaces. Formation of the energetically more favorable surface alloy is only realized above 270 K, suggesting that the activation energy for the incorporation mechanism is only overcome above this temperature. This information is less trivial than it appears at first glance. Recently, it has been suggested that adatom diffusion on various fcc (100) surfaces takes place via an exchange mechanism between the adatom and an atom in the surface layer [53, 54, 55, 56]. This process is schematically represented in Fig. 7.30. The corresponding activation barrier should be lower than the activation energy for diffusion *on* the surface. Obviously, this

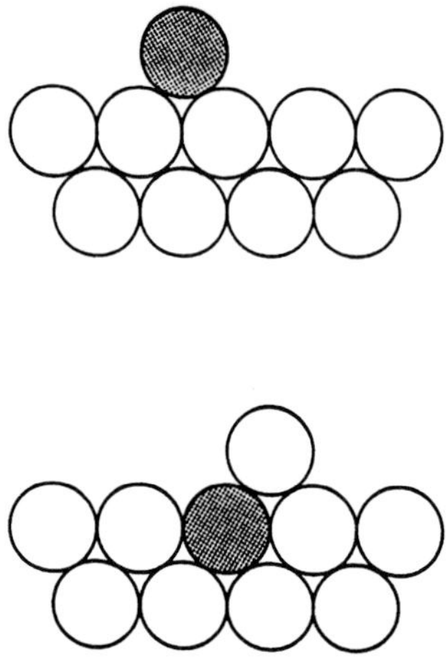

Fig. 7.30. Atomic exchange mechanism leading to surface alloying. A Mn atom (*shaded*) is incorporated into the top layer by simultaneous ejection of a surface-layer atom

does not hold for Mn on Cu(100) and Ni(100), where the activation energy is larger for incorporation than for the diffusion of Mn adatoms.

The growing interest in surface alloy formation stems from the fact that it involves a mass transport mechanism into the surface [57]. Every deposited atom and every released substrate atom must find a sink during the growth process. Two types of sinks generally exist. These are the original steps present on the substrate surface, and the step edges of the islands generated during the growth process.

Step and island formation after Mn deposition on Cu(100) and Ni(100) were investigated from two different perspectives. High-resolution electron diffraction (SPALEED) provided detailed information on the average morphology of the surface, offering a transfer width of 300 Å. Through STM investigations, direct imaging of the film morphology over larger areas (5000 Å × 5000 Å) was realized.[1] Additionally, the effect of defects could be directly monitored.

Figure 7.31 shows the full width at half maximum of the (00) beam for the clean Cu(100) surface. With increasing electron energy, the FWHM increases monotonically owing to the mosaic spread of the sample. A least-squares fit to the SPALEED data gives a value of $0.16 \pm 0.02°$ for the mosaic spread. From the absence of oscillations in the FWHM for the clean Cu surface, it can be concluded that the terrace width of the clean surface is larger than 300 Å. After deposition of 0.5 ML Mn at 350 K on the Cu(100) surface, a

[1] These experiments were carried out in a joint effort by C.C. Knight (Ni(100)) and M. Poensgen (Cu(100)). The investigations were performed in a chamber equipped with a low-temperature STM described elsewhere [58].

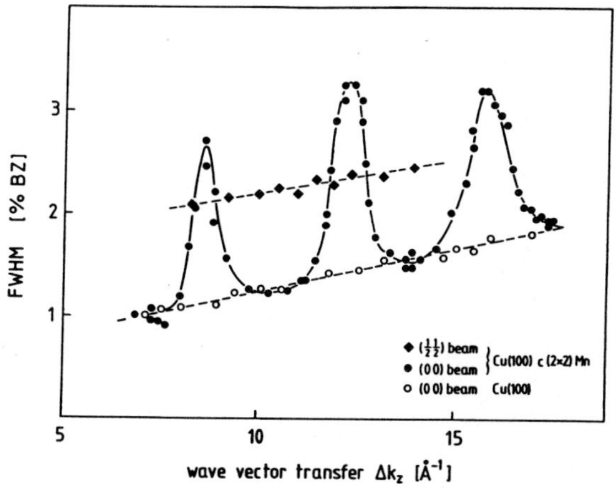

Fig. 7.31. FWHM of the (00) and (1/2, 1/2) beams. The FWHM of the Cu(100) (00) and (1/2, 1/2) beams is shown as a function of the momentum transfer Δk_z, before (*open circles*) and after (*filled circles* and *filled diamonds*) Mn deposition (0.5 ML Mn at 350 K). Strong oscillations due to steps 1.81 ± 0.02 Å in height are visible. The (1/2, 1/2) beam shows no oscillations

$c(2\times2)$ superstructure is formed (see Sect. 7.3.1). Additionally, new features are observed in the beam profiles. One observation is that pronounced oscillations in the FWHM are recorded (Fig. 7.31). The FWHM of the (00) beam is broad for out-of-phase conditions, where electrons scattered from atoms on neighboring terraces interfere destructively, and sharp for in-phase conditions, where constructive-interference conditions are met. This is a clear signature of a drastic increase in step density, which shows that the growth of the surface alloy is connected with the formation of steps. From the oscillation frequency, the step height is calculated to be 1.81 ± 0.02 Å. This agrees well with the value of 1.805 Å expected for monatomic steps on the Cu(100) surface. The absence of higher frequency oscillations confirms the predominance of single-hight, rather than multiple-height, steps. Such an increased surface roughness indicates that the deposited manganese atoms and the released copper atoms do not reach the original step edges. Instead, clusters are formed on the terraces and grow further by the capture of both deposited manganese and ejected copper atoms. This leads to the development of alloyed islands and terraces. Verification of this claim comes from atomically resolved STM images [59].

Using out-of-phase conditions, the profile of the (00) beam was analyzed. This diffraction spot is circular in shape for the Cu(100)–$c(2\times2)$-Mn surface alloy, indicating that the islands are round. Steps (island boundaries) following a preferred direction would render the diffraction pattern anisotropic. Again, this is in agreement with the STM observations [59].

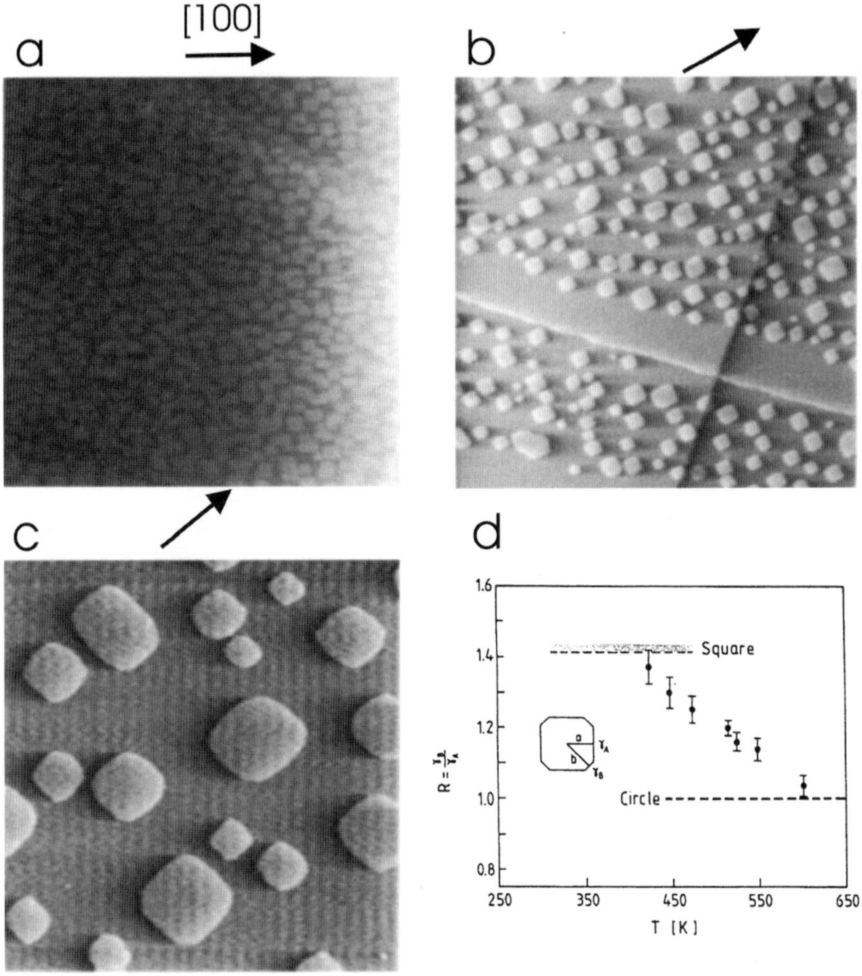

Fig. 7.32. STM images of the Ni(100)–$c(2\times2)$-Mn surface alloy for different growth temperatures. The effect of growth temperature on the surface morphology of alloy formed by deposition of 0.5 ML Mn is shown. (**a**) Scan area $1400\times1400\,\text{Å}^2$, T_{gr} = 420 K, T_{STM} = 420 K; (**b**) Scan area $5600\times5600\,\text{Å}^2$, T_{gr} = 510 K; (**c**) Scan area $2800\times2800\,\text{Å}^2$, T_{gr} = 550 K. Here T_{gr} denotes the growth temperature during deposition, while T_{STM} describes the temperature of the STM measurement. Except for (**a**), the STM images were taken at 300 K. (**d**) Temperature dependence of the ratio R of the energies for [110] and [100] steps

A markedly different behavior of the (00) beam is observed for the Ni(100)–$c(2\times2)$-Mn surface alloy. Under out-of-phase conditions, the (00) beam is elongated in the [100] and [010] direction, pointing to a preferred step orientation in [010] and [100] directions. Intriguingly enough, these are the less close-packed directions of the surface. The existence of preferred step directions is directly visible in STM images. Figure 7.32 shows tunneling images taken on Ni(100)–$c(2\times2)$-Mn alloy surfaces produced at different temperatures. In each case, 0.5 ML was deposited on the Ni(100) surface, and with the exception of the first image shown, the sample was held at the growth temperature only during deposition. Owing to the prolonged annealing of the sample in case (a), the islands are considerably larger than directly after deposition. It should be pointed out that the images were recorded using different scan directions and widths. Arrows label the [100] direction. All of the images show that the island edges run predominantly in two perpendicular directions. Furthermore, the images show the absence of three-dimensional growth. Even on top of the large islands ($> 600 \times 600$ Å^2) formed at 550 K, there is no additional nucleation. This indicates a mobility of manganese adatoms on the MnNi alloy sufficiently high to allow the manganese atoms deposited on these islands to eventually reach the terraces, where they can react with nickel atoms. From this series of pictures, the impression emerges that with increasing temperature the islands become more square-like (compare Figs. 7.32a and 7.32c). Measurement of the island lengths in Fig. 7.32a reveals that the alloy islands grown at 420 K are rectangular-like, with an aspect ratio of 1.6 ± 0.4. This shape must reflect a growth shape, since the [001] and [010] directions of an fcc (100) surface are equivalent. We shall return to this point later.

Additionally, rounding of the island corners at higher growth temperatures is also visible in the STM images (Fig. 7.32d). At a growth temperature of 420 K, the island corners are sharper than at 550 K. This rounding effect can be attributed to the formation of [110] step edges and is indicative of a temperature dependence of the ratio of the [100] and [110] step energies of the alloy [60]. This is even more evident from the appearance of the steps in Figs. 7.32a and 7.32b. In both cases, the macroscopic orientation of the step is the close-packed [110] direction of the substrate. However, for the film grown at 420 K, the steps are faceted in the [001] and [010] directions. This suggests that (1) the steps are alloyed, i.e., the substrate steps have served as sinks for the diffusing manganese and nickel atoms, and (2) for the alloy system, the [001] and [010] step directions are energetically favored by more than $\sqrt{2}$ over the [110] step direction at 420 K. Otherwise, this step faceting, which leads to a $\sqrt{2}$ increase in the step length, would not occur. When the alloy is produced at higher temperatures, the macroscopic steps have no facets in the [010] direction. They run as expected in the [110] direction.

Over a temperature span of 130 K (from 420 to 550 K), there is a two orders of magnitude decrease in the island density (from $\approx 4\times10^{12}$ islands/cm^2 to $\approx 3 \times 10^{10}$ islands/cm^2). The island size increases from $\approx 2 \times 10^3$ Å^2 to $\approx 2\times10^5$ Å^2. From a plot of island density against $1/T$, the activation energy for growth was determined from (2.29) to be 0.75 eV.

The effect of increasing manganese coverage for a growth temperature of 550 K is examined in Fig. 7.33. It should be remarked that below 0.5 ML,

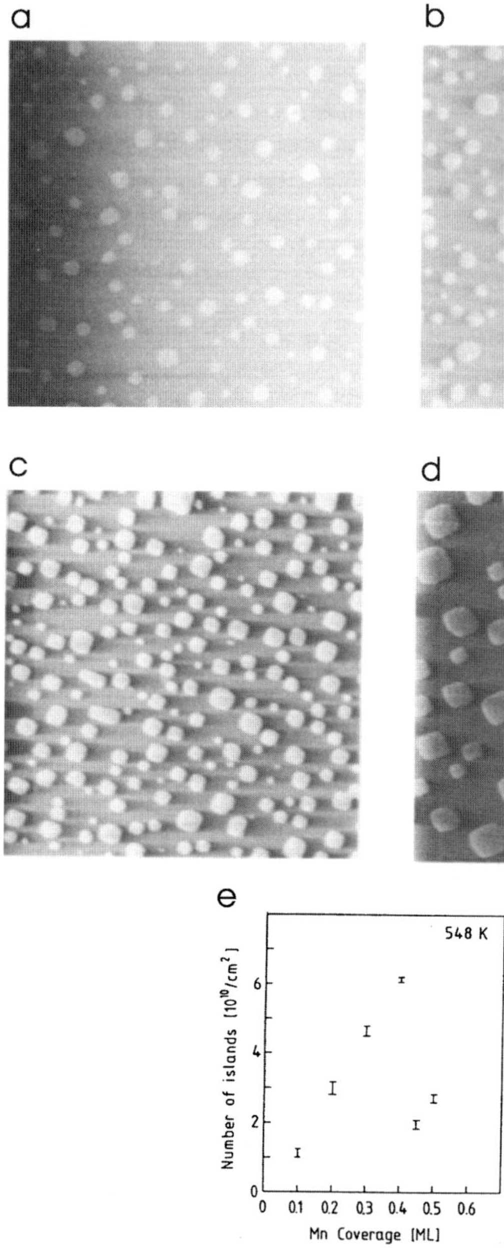

Fig. 7.33. STM images of the Ni(100)–$c(2\times2)$-Mn surface alloy for different coverages. The dependence of the island density on the coverage of the Ni(100)–$c(2\times2)$ Mn alloy formed at 548 K is shown. (**a**) 0.2 ML Mn, (**b**) 0.3 ML Mn, (**c**) 0.4 ML Mn, d) 0.5 ML Mn. (**e**) Island density vs. coverage

the terraces are not completely alloyed. The STM images show that with increasing coverage, the size of the alloy islands does not increase significantly. Rather, it is the number of islands that rapidly grows. Above 0.4 ML, a drastic reduction in the island density is achieved through coalescence. Most conventional theories [61, 62] predict that nucleation of new islands dominates only at low coverages (below 0.1–0.2 ML), while growth of already existing islands takes precedence at higher coverages. Clearly, the coverage dependence of the island density seen in this case is quite different. Equally noteworthy is the dependence of the island shape as a function of coverage. While at 550 K the alloy islands formed at lower coverages are rounded, they become more square-like at higher coverages. The coverage dependence of the island density and shape are signals of the complexity of the alloying process.

We must momentarily return to the STM image (Fig. 7.32a) which shows that the $c(2\times2)$ alloy islands grown at 420 K are rectangular. Since annealing this surface at higher temperatures results in larger, square-like islands, this rectangular form reflects a growth, rather than an equilibrium, shape. A knowledge of this fact is essential for the understanding of a further observation in the LEED pattern. On the Ni(100)–$c(2\times2)$-Mn surface, the half-order beams exhibit pronounced shape variations as a function of electron energy. The intensity distribution oscillates with energy between a circular and a streaked appearance. More surprisingly, the streaks even rotate, i.e., switch their direction by 90° with variation of the energy. Since this variation is observed over three zones in reciprocal space, a kinematical (single-scattering) explanation must apply to it. The energy dependence has to be linked with steps on the surface, since a planar atomic arrangement of defects such as domain boundaries or dislocations cannot be responsible for the changing of the streak direction with energy. The direction of the streaks shows that the steps run along [100] and [010], as already concluded.

Two different step types, denoted as A and B steps (see Fig. 7.34) in the following, are possible. The A step binds Mn atoms in the upper terrace to Ni atoms in the lower terrace, while the B step binds Mn to Mn (or Ni to Ni) across the step. The phase shifts caused by A and B steps are conveniently described by the step vectors $\boldsymbol{g}_{\mathrm{A}}$ and $\boldsymbol{g}_{\mathrm{B}}$ which connect atoms in corresponding positions across the step. Note that $\boldsymbol{g}_{\mathrm{A}}$ and $\boldsymbol{g}_{\mathrm{B}}$ differ by half a lattice constant in the [100] and [010] directions. Thus, A and B steps have opposite influences on the shape of the half-order beams as function of electron energy. This can be rationalized by the structure factor $F(h,k,l)$ which reads for the steps running in the [100] direction:

$$F_{\mathrm{A}}(h,k,l) = (1 + e^{2\pi i(h/2+k+1/2)}) \times (f_{\mathrm{Mn}} e^{2\pi i(h/2+k/2)} + f_{\mathrm{Ni}}) \, , \qquad (7.5)$$

$$F_{\mathrm{B}}(h,k,l) = (1 + e^{2\pi i(h/2+k+1/2)}) \times (f_{\mathrm{Mn}} e^{2\pi i(h/2+k/2)} + f_{\mathrm{Ni}}) \, , \qquad (7.6)$$

where f_{Mn} and f_{Ni} are atomic form factors. A corresponding expression holds for the rotated steps in the [010] direction. From these structure factors, one obtains different elongations for different step types in different step

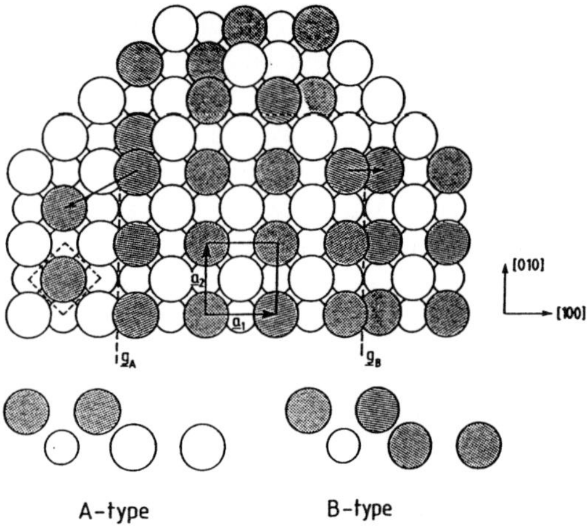

Fig. 7.34. The two different step vectors of the $c(2\times2)$ surface alloy: model for the Ni(100)–$c(2\times2)$-Mn structure (top view). *Shaded circles* represent Mn atoms, while *large* and *small open circles* denote first- and second-layer Ni atoms. The primitive unit cell of the ordered alloy is marked by *full lines*. The two possible types of step on this surface are depicted, in top and side view, with their $\boldsymbol{g}$ vectors. The *broken lines* give the step direction. Steps with Ni–Ni bonding at the step and Ni–Mn bonding at the step lead to beam profiles characteristic of those for the A and B steps, respectively

directions. The relevant message here is that one can distinguish between A and B steps even though the main step directions are equivalent. The selection rule for elongation of the (hk) beam is $h+l=$ odd for A [100] steps, but $k+l=$ odd for B [100] steps. For a given half-order beam, the elongation directions for A and B steps differ by 90° when both [100] and [010] steps are present. An equal distribution of A and B steps will simply lead to a superposition of both patterns and thus to cross-like half-order reflections. Experimentally, such cross-like beams are never observed at 420 K. Instead, only a single elongation is found for each beam, which can only be explained by a preferred type of step on the surface. We observe that the (1/2, 1/2) beam is streaked in the [100] direction for odd l ($E_{\text{exp}} = 115\,\text{eV}$ $(1 = 7)$ and $E_{\text{exp}} = 200\,\text{eV}$ $(l = 9)$), while it is streaked in the [010]-direction for even 1 ($E_{\text{exp}} = 82\,\text{eV}$ $(l = 6)$, $E_{\text{exp}} = 154\,\text{eV}$ $(l = 8)$) and $E_{\text{exp}} = 251\,\text{eV}$ $(l = 10)$), in excellent agreement with the calculated values. We can thus conclude that there is a majority for B steps on the surface. This preference of B steps is an attractive explanation of why the STM image of the surface alloy grown at 420 K shows rectangular islands. The longer step length corresponds to the direction with steps of type B, while the shorter step length is in the direction where the steps of type A are found.

7.5 The Role of Magnetic Energy in Stabilizing the Cu(100)–$c(2 \times 2)$-Mn Surface Alloy

The atomic coordinates derived from the structural optimizations described in the last section show that the $c(2\times2)$ phases of Mn on Cu(100) and Ni(100) have a pronounced buckling in the surface alloy layer. This large corrugation cannot be accounted for by "traditional" arguments based on atomic radii. In the CuPd surface alloy, the Pd atoms, whose radius (1.375 Å) is even larger than that of Mn (1.365 Å), are almost coplanar with Cu atoms [51], and in the CuAu surface alloy, the Au atoms (1.442 Å) are displaced outwards by only 0.1 Å relative to the Cu atoms [50]. Additionally, the question arose as to what mechanism stabilizes the Cu(100–$c(2\times2)$ Mn surface alloy, which has no bulk analogue. To understand the large corrugation of the Mn and to shine more light onto the structural stability of this surface alloy, we performed a structural optimization of paramagnetic (P) and ferromagnetic (F) Cu(100–$c(2\times2)$ Mn surface alloys (S) by ab initio electronic-structure calculations [36]. These calculations, which were performed by S. Blügel (IFF/KFA Jülich), were based on density functional theory in the local-spin-density approximation (see Sect. 4.1.1 and [64]). The equations were solved using the full-potential linearized augmented-plane-wave (FLAPW) method for a thin-film geometry [65]. As a structural model, a seven layer film was chosen, consisting of five layers of Cu with two Cu atoms per layer simulating the Cu(100–$c(2\times2)$ substrate and, at each surface, one Mn and one Cu atom simulating the Cu(100–$c(2\times2)$ surface alloy, as shown in Fig. 7.16. In the structural optimization, which was based on minimizing the total energy of the system $E_S\{\mathrm{P/F}; \Delta z_{Mn,F/P}; \Delta z_{Cu,F/P}\}$, two degrees of freedom were included for both the paramagnetic and the ferromagnetic case. These were the relaxations Δz of the Mn and of the top Cu atom along the surface normal. As confirmed by the LEED analysis presented in Sect. 7.3.2, the Cu(100–$c(2\times2)$ surface alloy structure can be adequately characterized using only these relaxations. Prior to this study, the bulk Cu lattice constant ($a_{Cu} = 3.52$ Å) and the interlayer spacing ($d_{Cu} = a_{Cu/2} = 1.76$ Å) had been optimized by a bulk FLAPW method [66]. The theoretical data, given in absolute units are 3–4% smaller than the experimental results. This is typical for ab-initio calculations in the local-spin-density approximation.

To study the structural stability of the surface alloy against the formation of a similar Cu_3Mn bulk alloy, we repeated the calculation, treating the two $c(2\times2)$ MnCu layers as interlayers (I) separated by a Cu layer and covered by two layers of Cu in an otherwise unchanged environment of a seven-layer slab. Since the direct overlap of the Mn d electrons is of minor importance, this interlayer calculation also serves as a first approximation to the energetic stability of single Mn atoms against interdiffusion.

Two striking points are to be seen upon comparing the results for the paramagnetic solution with those for the ferromagnetic solution (see Fig. 7.35). In the structure of the most stable paramagnetic surface alloy, the position

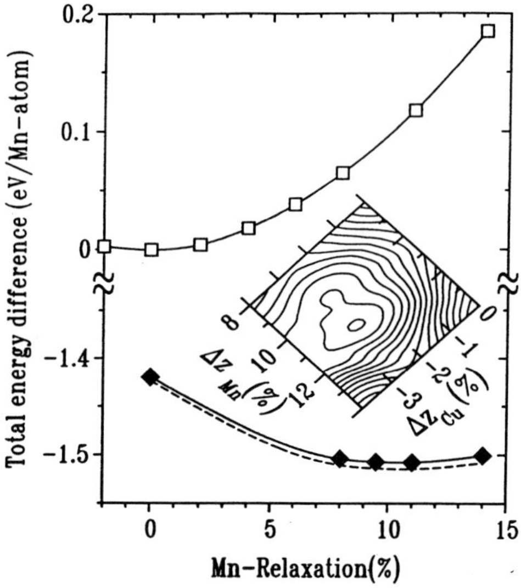

Fig. 7.35. Theoretical total-energy difference per Mn atom of Cu(100)—$c(2\times2)$ Mn vs. the buckling relaxation Δz_{Mn} of Mn in relative units with respect to the theoretical interlayer spacing of Cu, $d_{Cu} = 1.76\,\text{Å}$. The *open squares* represent the paramagnetic and the *solid diamonds* the ferromagnetic results. The *solid lines* (for Cu atoms fixed at the ideally terminated positions $\Delta z_{Cu} = 0$) and *dashed line* (where the top Cu atom is always at its optimally relaxed position) show the fitting polynomials. The inset shows a contour plot of the ferromagnetic total-energy difference with respect to the buckling of Mn and Cu. The minimum, which determines the optimal structure, is found in the inner circle. The contour interval is 1 meV

of the Mn atom remains nearly at the ideally terminated Cu site, i.e., the Mn is displaced inwards by only 0.5% (with respect to the Cu interlayer distance). This is in line with the trend expected on the basis of the experimentally observed corrugation of the CuAu and CuPd surface alloys. On the other hand, this result is in contradiction with the pronounced buckling observed experimentally for the CuMn surface alloy. (2) The ferromagnetic solution is energetically more favorable. In this case, the energy is lower by about 1.4 eV/Mn atom. Thus the CuMn alloy is magnetic. A large outward displacement of 11.5% is found for the Mn atom, while the Cu atoms of the surface alloy layer show a small inward motion of 2.5%. This leads to an overall buckling of 14%. This is in excellent agreement with the experimental value of $16.5 \pm 2.0\%$. Thus we can conclude at this point that (i) the CuMn surface alloy is magnetic, and (ii) the buckling motion of the Mn is caused by its magnetism. Equally convincing is the fact that the small corrugation found for the paramagnetic solution is in line with the corrugation observed for the nonmagnetic CuAu and CuPd surface alloys.

These findings can be understood by considering the bonding behavior of the Mn d electrons. In the paramagnetic case, the d bands are half-filled. Therefore, all bonding states are occupied, while the antibonding states are practically empty. Under these circumstances, Mn favors an environment with a large number of bonding neighbors. Indeed, when we perform a calculation for the paramagnetic c(2×2) MnCu layer as an interlayer in Cu, we find that owing to the increased number of Cu neighbors, the total energy is reduced by 0.56 eV/Mn atom. This means that a paramagnetic MnCu surface alloy would be highly unstable against interdiffusion. When we remove the constraint of paramagnetism, the system becomes magnetic and the Mn atom at the surface of the alloy obtains a magnetic moment of $M_{\mathrm{S}}(\Delta z_{\mathrm{Mn}} = 0) = 3.64\,\mu_{\mathrm{B}}$, which is accompanied by a large gain of magnetic energy. In fact, a magnetic moment of that size means that the majority Mn d orbitals are nearly completely occupied and the minority d bands are nearly empty. Consequently, there are nearly as many Mn bonding d states occupied as antibonding d-states. As a result, the d cohesion becomes less important and is, energetically, more than compensated by the gain of magnetic energy. In the language of atomic radii used above to explain the buckling, the ferromagnetic Mn atom is much bigger than the paramagnetic atom and there is a strong mutual interdependence between the size and the magnetic moment itself. The minor importance of the Mn d electron cohesion also becomes evident when we compare the energy of the ferromagnetic interlayer system $E_{\mathrm{I}}(F,0) = E_{\mathrm{I}}(F,\Delta z_{\mathrm{CuMn}} = 0)$ with that of the ferromagnetic surface layer $E_{\mathrm{S}}(F,0) = E_{\mathrm{S}}(F,\Delta z_{\mathrm{Mn}} = \Delta z_{\mathrm{Cu}} = 0)$. Although the environment of the Mn atom is very different, the energies are nearly degenerate ($E_{\mathrm{S}}(F,0) - E_{\mathrm{I}}(F,0) = -0.04\,\mathrm{eV/Mn}$). The magnetic moment of the Mn overlayer atom is larger than that of the interlayer atom, $M_{\mathrm{I}}(\Delta z_{\mathrm{CuMn}} = 0) = 3.23\,\mu_{\mathrm{B}}$, and it is this additional gain of magnetic energy which stabilizes the surface alloy. The outward displacement of the Mn surface atom is very efficient in releasing the pressure on the Mn atom. Owing to the relaxation, the magnetic moment increases to $M_{\mathrm{S}}(\Delta z_{\mathrm{Mn}} = +11.5, \Delta z_{\mathrm{Cu}} = -2.5) = 3.75\mu_{\mathrm{B}}$ and the total energy is further reduced by 0.1 eV/Mn atom, stabilizing the surface alloy. This relaxational release of compression cannot be matched by a Mn interface atom.

The magnetic origin of the buckling of the Mn atoms in the $c(2\times2)$ alloy is in line with our experimental finding of a slightly reduced corrugation for Ni(100)–$c(2\times2)$-Mn. Electronic-structure calculations show that the magnetic moment M for a Mn impurity in Cu ($M = 3.16\mu_{\mathrm{B}}$ [67]) is larger than for Mn in Ni ($M = 3.02\mu_{\mathrm{B}}$ [68]). The same magnetic trend is expected for Mn atoms on Cu and Ni substrates, which explains the reduced buckling for the MnNi alloy.

No theoretical effort was made to determine the long-range magnetic order, which was assumed to be ferromagnetic. It is known that for Mn monolayers on various substrates, a $c(2\times2)$ antiferromagnetic order is favored [69] owing to a direct in-plane Mn d–d hybridization. However, in the case of the CuMn surface alloy, the magnetic order is determined by an indirect, in-plane RKKY-type interaction resulting from the hybridization of the Mn d electrons with the Cu sp electrons. In such a case, (i) the long-range magnetic order has practically no influence on the structural properties, and (ii) from the exchange coupling of Mn dimers on next-nearest-neighbor sites in Cu [67], the ferromagnetic spin arrangement is expected to be the magnetic ground state.

Finally, we would like to speculate on the possibility of additional magnetic surface alloys by relating this work to the attraction or repulsion between a vacancy and a 3d impurity atom when they are nearest neighbors in bulk Cu or Ni. Ab initio results [68] show that the interaction energy of a vacancy with Cr or Mn in Cu or of a vacancy with V, Cr, Mn, Fe or Co in Ni is nearly degenerate between attraction and repulsion. Similarly to the situation for atoms close to a vacancy, atoms at the surface have a reduced number of nearest neighbors. This reduction in the number of nearest neighbors enhances magnetism. At the surface, there are additional degrees of freedom; i.e., unlike atoms in the bulk, surface atoms are free to relax. Both the potential gain in magnetic energy and the possibility of surface relaxation foster supportive conditions for the stabilization of magnetic surface alloys. Since the buckling relaxation depends on the size of the magnetic moment, on the basis of results for single magnetic impurities in Cu [66] and Ni [67], the largest buckling is expected for Cu-based surface alloys with Cr and Mn and a larger buckling is expected for Cu-based surface alloys than for Ni based alloys. This is in line with our experimental results for the Ni(100)–$c(2\times2)$-Mn surface alloy, where a corrugation of 0.25 ± 0.02 Å has been found [36]. We conclude that Cu(100) c(2×2)Mn is just a prototype of a much wider class of new magnetic materials, namely the 2D ordered surface alloy.

With the information presented in this section, one can also understand the structure of the quasi-hexagonal Mn films on Cu(100) and Ni(100). To prove this, we shall once again come back to the fundamental question of the relationship between magnetism (or more precisely here, magnetic moments) and structure. It has been demonstrated several times that a lattice expansion should be accompanied by an increase in magnetic moment. Now we would like to reverse this argument and say something about interatomic spacings from a knowledge of the magnetic moments. An essential ingredient of such a step is visualized in Fig. 7.36. There, the spin-split density of d electrons is sketched for two different cases, a nonmagnetic solid and a magnetic one. For the ferromagnetic crystal, a large spin splitting has been assumed for simplicity. In the case of Mn (5 d electrons) considered here, half of the d states are filled. The lower-lying states are bonding (hatched), while the higher-lying states (gray) are antibonding. In the nonmagnetic case, both spin-up and spin-down states are occupied with equal weight, thus leading to a filling of only bonding states. In the magnetic case, both bonding and antibond-

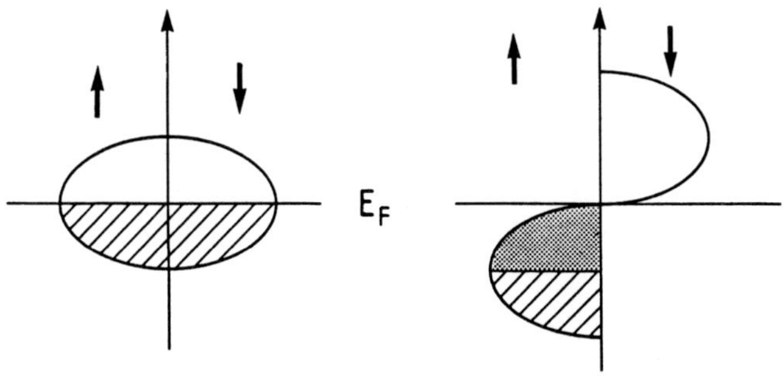

Fig. 7.36. Density of states for spin-up and spin-down electrons. A highly schematic diagram of the density of states for the d electrons in a nonmagnetic and a magnetic crystal is shown. E_F denotes the Fermi energy and the (*hatched* and *gray areas*) describe bonding and antibonding states, respectively

ing states are occupied. The consequence is a smaller cohesive energy and a larger interatomic spacing. In the nonmagnetic case, the cohesive energy is larger. The interatomic spacings are decreased to maximize the overlap between the localized d orbitals of nearest neighbors. This once more illustrates the relationship between magnetic moments and interatomic spacings.

In Sect. 7.2.4, the quasi-hexagonal Mn films formed on Cu(100) and Ni(100) were compared. At a first glance, it is somewhat bewildering that in both cases the atomic spacings between neighboring Mn atoms are considerably larger than the value expected for bulk Mn. This effect is more pronounced on the Cu(100) surface (see Table 7.3). By merging the insight extracted from the theoretical calculations in this section with the discussion of thin-film magnetism presented in Sect. 4.2, such an observation can be understood. When films of magnetic materials are grown epitaxially, a considerable enhancement of the magnetic moment can be expected, especially when the substrate is a noble metal. The low-lying d bands of these metals hybridize only weakly with the d bands of the magnetic film. Therefore, a larger magnetic moment is expected for Mn on Cu(100) than on Ni(100). As stated above, this is consistent with the results of the total-energy calculations for Mn impurities in Cu [66] and Ni [67]. Since increased magnetic moments lead to increased atomic volumes, the Mn atoms occupy more space on Cu(100) and Ni(100) than in bulk Mn, with this effect being more pronounced on the Cu(100) surface. For larger film thicknesses, considerable hybridization between the Mn d bands occurs, since the environment approaches that of bulk Mn. The moments are thus reduced and consequently the interatomic distances are decreased. Hence the reason why above 1 ML, the presumably more densely packed $c(12{\times}8)$ structure is found on Cu(100) and Ni(100). The formation of these complex phases and their surprising structure can be envisioned as a manifestation of the correlation between structure and magnetism in ultrathin films. In line with the arguments presented above, Mn spacings of 2.7 Å, as expected for bulk Mn, have been determined in *thick* films on Cu(111) [70] and Pd(111) [70].

7.6 Atomic Mechanism for the Formation of the Cu(100)–$c(2 \times 2)$-Mn Surface Alloy

In this section, we turn to the atomic processes involved in the formation of the Cu(100)–$c(2 \times 2)$-Mn surface alloy. An in-depth understanding of these processes could help one to enhance alloying if it is required or suppress it if detrimental effects are expected. This could allow a tailoring of surface and interface properties according to our needs. To achieve such a goal, we have employed STM to investigate the formation of the Cu(100)–$c(2 \times 2)$-Mn surface alloy. We shall first present the STM results, as well as the conclusions drawn from those experiments. Then we shall present a scenario for the formation of the surface alloy by incorporating various STM results. Lastly, the detailed atomistic processes involved in alloying will be discussed.

7.6.1 STM Observations

Identification of Incorporated Mn Atoms in STM Images

In our STM study, we always observed small protrusions. The protrusions are created after deposition of Mn. They are already visible after deposition of 0.0033 ± 0.0005 ML Mn, which is the smallest coverage, that we have investigated. The density of the protrusions increases with Mn coverage, and the distribution of the protrusions is strongly influenced by islands and substrate steps. The mobility of the protrusions is dependent on the Mn coverage. Therefore investigating the evolution of the protrusions with coverage and time could provide useful information about the atomic processes involved in the formation of the Cu(100)–$c(2 \times 2)$-Mn surface alloy. To achieve such a goal, we must first identify these protrusions. Figure 7.37 shows a representative STM image for a Mn coverage between 0.005 and 0.3 ML. The image includes both a monatomic step[2] and a number of protrusions. Since the step height is known, this allows us to determine the height of the protrusions by a line scan through two protrusions and the step. The result is 0.32 ± 0.06 Å. This height is very close to the outward corrugation of the Mn atoms in the Cu(100)–$c(2 \times 2)$-Mn surface alloy (~ 0.3 Å), which has been determined by other techniques. Therefore we have reason to believe that *the protrusions are incorporated Mn atoms*.

[2] We have determined the step heights of clean and Mn-covered Cu(100) surfaces, and found a pronounced preference for monatomic steps with a step height of 1.8 Å over double and multiple steps by STM and high-resolution LEED.

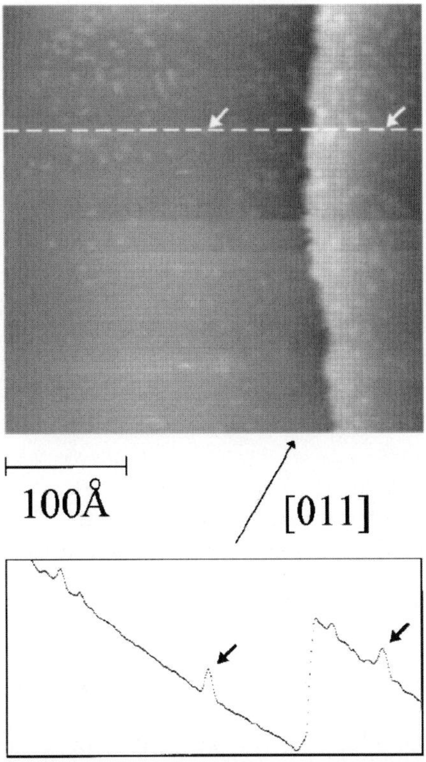

Fig. 7.37. STM image taken after deposition of 0.02 ML Mn at 300 K. The image has a scan range of 360×360 Å^2. Below, a line scan is shown, which was performed along the *dashed line* depicted in the STM image. The *white arrows* in the image denote the deposition of incorporated atoms that lie on the line scan. From [71], used with permission

We have noticed that the protrusions have a larger apparent size in many STM images than expected for a single atom, although the size in the best-resolved images corresponds to a single atom. This is probably due to a convolution of the tip shape and the protrusion. A model proposed by Tersoff and Hamann [72] has been employed to determine theoretically the STM images for the Cu(100)–$c(2 \times 2)$-Mn surface alloy [73]. In this model, the tunnel current is proportional to the local density of states of the sample at the location of the outermost tip atom. This first-principles calculation has confirmed that an incorporated Mn atom is imaged as a protrusion when it exists as an impurity (at low coverage). However, the same work also predicts that, when the Cu(100)–$c(2 \times 2)$-Mn surface alloy is formed, Cu is imaged as a protrusion and Mn as a depression. Figure 7.38 shows an atomically resolved STM image for the Cu(100)–$c(2 \times 2)$-Mn surface alloy, which was obtained after deposition of 0.47 ± 0.02 ML of Mn at 300 K. The resolved

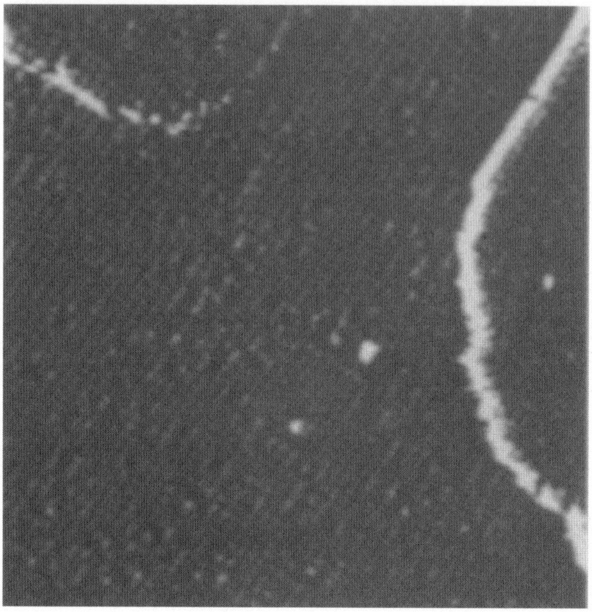

Fig. 7.38. STM image of Cu(100)–$c(2 \times 2)$-Mn structure taken after deposition of 0.47 ML Mn at 300 K. Scan width 180×180 Å^2. From [71], used with permission

atoms have a nearest-neighbor distance of 3.56 ± 0.07 Å along the [010] and [001] directions, which is larger than the distance of 2.55 Å between nearest-neighbor Cu atoms in the Cu(100) surface along the [011] and $[0\bar{1}1]$ directions. This implies that only one atomic species of the surface alloy layer is imaged by STM. In this respect, our STM observation is in agreement with the theoretical prediction. However, the calculations [73] show that the states localized at the Mn sites also spill out into the vacuum and stretch above the positions of the Cu atoms to the next Mn atoms, resulting in a larger STM current at the Cu positions. This prediction is different from our STM interpretation based on a structural corrugation. In our interpretation, the protrusions at higher coverage are considered to be the Mn atoms corrugated outwards. In the following, we shall present a model for the atomic mechanism of the formation of the Cu(100)–$c(2 \times 2)$-Mn surface alloy which is based on the STM observations at low Mn coverage. Under this condition, both the theoretical calculation and the experimental measurement agree that the protrusions are incorporated Mn atoms.

Mobility of Incorporated Mn Atoms. Figure 7.39 shows a superimposed picture composed of two consecutive STM images of a Cu(100) surface after deposition of 0.020 ± 0.007 ML of Mn. The time interval between the two images was 180 s. A substrate step and two-dimensional islands of monatomic height are readily discernible. Additionally, small protrusions (incorporated

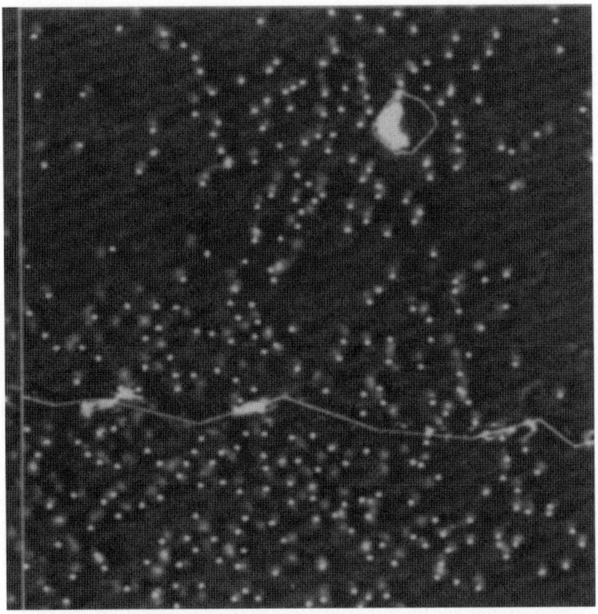

Fig. 7.39. STM images of a Cu(100) substrate after deposition of Mn at 300 K. Coverage 0.020 ± 0.007 ML Mn, scan width 3505 Å×350 Å. In order to visualize the mobility of the incorporated Mn atoms, two consecutively recorded STM images are superimposed. The *white dots* mark the positions of the incorporated Mn atoms 180 s before the image shown was recorded. The *white frame* indicates the margin of the first recorded image with respect to the second. The images do not match exactly, because of thermal drift. From [74], used with permission

Mn atoms) are visible. The islands and protrusions are new features created after the deposition of Mn. By comparing the relative positions of atoms in the two consecutive images, a conclusion can be drawn that *the incorporated Mn atoms are mobile.* The jump frequency Γ can be estimated from the STM images. For instance, it has been observed for a coverage of 0.1 ML Mn that each incorporated Mn atom performs, on average, at least one jump in the time interval of $\triangle t_1 = 180$ s. This gives a lower limit $\Gamma \geq 1/\triangle t_1$ on the jump frequency. The fact that single incorporated Mn atoms are clearly imaged by STM indicates that the majority of Mn atoms do not jump during the scanning of a single Mn atom ($\triangle t_2$). This gives an upper limit on the jump frequency as $\Gamma \leq 1/\triangle t_2$. The diffusion coefficient D can be derived from the jump frequency from $D = a^2 \Gamma z$, where z is the number of nearest neighbors and a is the surface lattice constant. In this way, the diffusion coefficient for the incorporated Mn atom at low coverage has been estimated to be between 1.8×10^{-18} and 6.4×10^{-17} cm^2 s^{-1}. At very small coverage, one can follow the motion of the incorporated Mn atoms. In this case, a direct determination of the diffusion coefficient is possible by evaluating the

Einstein relation $\langle \triangle X^2 \rangle = 4D\triangle t$, where $\triangle X$ is the change of the distance between two Mn atoms projected on to a fixed axis in the time $\triangle t$. In this way, a diffusion coefficient of $(5.3 \pm 0.3) \times 10^{-18}$ has been determined for small coverage; this value falls into the range estimated above.

Using the same methodology as described for the case of small coverages, a jump frequency between 1.1×10^{-2} and 1.3 s^{-1} has been derived at a coverage of approximately 0.45 ML Mn. These values are comparable with the observed jump rate of incorporated Mn atoms at low coverages. For a Mn coverage around 0.3 ML, however, we always observed "fuzzy" images as in Fig. 7.45c, indicating a high mobility. Since both Cu and Mn adatoms move too fast on Cu(100) to be imaged at room temperature, the fuzziness cannot be caused by Mn or Cu adatoms. Therefore, the fuzziness indicates a higher mobility of the incorporated Mn atoms at a coverage around 0.3 ML Mn. This is in contrast to our findings for coverages below 0.2 ML Mn and above 0.4 ML Mn, where most of the incorporated atoms are completely imaged (see Figs. 7.45a, b, d). Since for 0.3 ML Mn, most incorporated Mn atoms are immobile during the time it takes to scan their width, but move within the time it takes to image them completely, an upper and a lower limit on the mobility can be estimated. This leads to a jump rate between 2.2 and 10^2 s^{-1} for 0.3 ML Mn. Such "fuzzy" regions have also been observed in the vicinity of steps or island edges for higher coverages, e.g. around 0.45 ML Mn, showing that the mobility is enhanced in those regions. At coverages around 0.3 ML Mn and above, the diffusion coefficient cannot be determined by use of the Einstein relation, since we are unable to follow the path of individual incorporated Mn atoms with time. At these coverages, however, we can exploit the pronounced mobility of the incorporated Mn atoms. Using a statistical analysis based on time correlation functions (see Appendix B in [74]), we have determined the probability that an incorporated Mn atom makes at least one jump in the time interval $\triangle t$. This leads to a more precise determination of the lower limit on the total jump rate. For a coverage of 0.31 ML Mn, we obtain $\Gamma_{tot} \geq 3\,s^{-1}$.

The coverage dependence of the diffusion coefficient is summarized in Fig. 7.40. The single data point represents the diffusion coefficient in the limit of very small Mn coverages, determined by the evaluation of the Einstein relation. The horizontal bars describe the estimated diffusion coefficient derived from the jump frequency, where only nearest-neighbor jumps are considered and correlation effects are neglected.

Distribution of Incorporated Mn Atoms. At small coverage, an inhomogeneous distribution of incorporated Mn atoms has been observed around islands, as is clearly shown in Fig. 7.41. The STM image presented in Fig. 7.41 was taken after deposition of 0.032 ± 0.004 ML Mn. One can see that most of the Mn atoms are incorporated in the vicinity of island edges, resulting in "halos" of incorporated Mn surrounding the islands. Far away from an island, practically no incorporated Mn atoms are observed. This situation can be clearly demonstrated by plotting the average local coverage of incorpo-

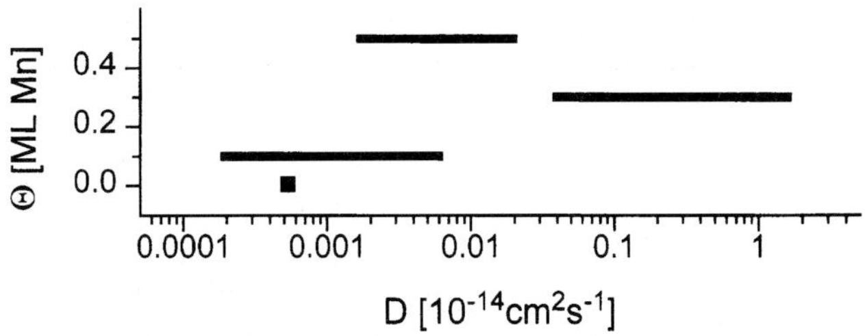

Fig. 7.40. Estimates of the diffusion coefficient D at different Mn coverages Θ. For each coverage, the diffusion coefficient falls in the range marked by the *bars*. The diffusion coefficient was estimated from the total jump frequency, neglecting correlation effects and assuming nearest-neighbor jumps. The single data point is the diffusion coefficient as derived from the Einstein relation. From [74], used with permission

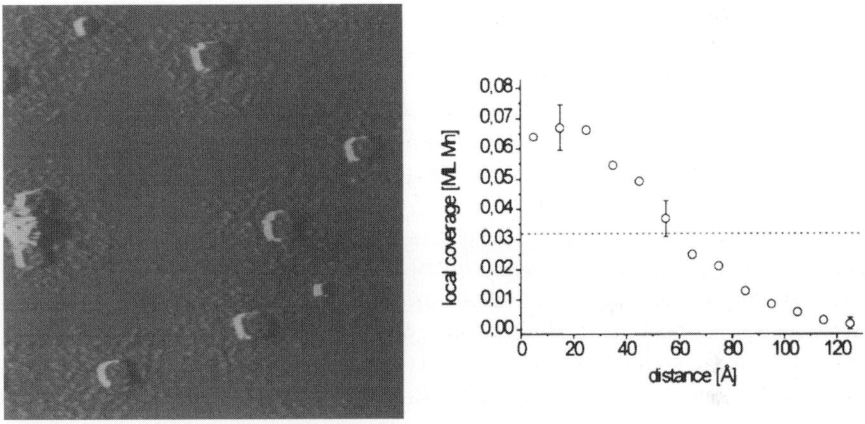

Fig. 7.41. *Left*: STM image of a Cu(100) surface after deposition of 0.032 ± 0.004 ML Mn at 300 K. Scan width 700×700 Å^2. *Right*: the average local coverage of incorporated Mn is depicted as a function of the distance from the island edge for the lower terrace, for a coverage of 0.032 ± 0.004 ML Mn. The local Mn coverage drops from 0.065 ML Mn to approximately zero over a distance of 120 Å. From [75], used with permission

rated Mn as a function of the distance from the island edge (see Fig. 7.41). Furthermore, STM images with a smaller scan width reveal that the density of incorporated Mn in an island is comparable to that in the halo. The number of atoms in an island is also found to be comparable to the number of Mn atoms in the halo. A ratio of 1.3 ± 0.3 has been determined for the average number of atoms in the island to the Mn atoms in the halo.

It has been shown that, for deposition of 0.5 ML of Mn on Cu(100) at temperatures below 270 K, a quasi-hexagonal $c(8 \times 2)$ Mn film is formed instead of the Cu(100)–$c(2 \times 2)$-Mn surface alloy. The $c(8 \times 2)$ structure, with considerable island size, is observed even at temperatures as low as 110 K [21]. Therefore it can be derived that the diffusion of Mn atoms adsorbed (not incorporated) on the Cu(100) surface is fast at room temperature. Accordingly, the islands shown in Fig. 7.41 will initially consist of Mn atoms. In fact, monatomic-height Mn islands have observed after Mn adsorption at 150 K on Cu(100) by SPALEED [59], and on the (100) terraces of a Cu(1 1 141) substrate by STM [15]. Keeping this in mind, we can conclude from Fig. 7.41 that *there must be mechanisms for interlayer mass transport which incorporate the Mn atoms in the islands into the lower terrace and bring an almost equal number of Cu atoms into the the upper terrace.* With such a mechanism, the development of the islands can be described as follows. At the beginning of alloy formation, islands are formed by Mn adatoms. Upon the subsequent incorporation of Mn atoms into the lower terrace, the same number of Cu atoms are brought into the upper terrace. This interlayer mass transport leads to an enhancement of the Cu concentration in the islands. After a sufficiently long time, the Cu concentration in the islands will equal the Cu concentration in the lower terrace. Therefore, at low Mn coverages, the number of Mn atoms in the surrounding lower terrace is comparable to the total number of atoms in the island. In this model, the inhomogeneous distribution around an island can be easily explained as a result of diffusion of the incorporated Mn atoms from their sources, i.e., the islands.

Incorporation in the Vicinity of Substrate Steps. Our STM study has already revealed a seam of Mn atoms at the upper terraces of substrate steps on the vicinal Cu(100) surface [59]. The Mn seam is formed by the attachment of adsorbed Mn atoms, since the substrate steps are preferential sinks for adsorbed Mn atoms. The attachment is found to be independent of the step orientation. Similarly to the situation for island edges, a strong Mn incorporation is also observed in the vicinity of substrate step edges. However, in contrast to the step-orientation-independent attachment of Mn atoms, the incorporation of Mn near a step edge strongly depends on the orientation of the step. As shown in Figs. 7.42a, b, STM images including a step running close to the [010] and the [011] direction, respectively, were recorded after deposition of 0.028 ± 0.007 ML Mn. The main features can be qualitatively described as follows. On the lower terrace, significantly more Mn is incorporated close to the step running in the [010] direction than for the [011] direction. In the case of the upper terrace, more Mn is found in the vicinity of the step running in the [011] direction than for the [010] direction. However, the difference is not as pronounced as on the lower terrace. For both step directions, more Mn is found on the upper than on the lower terrace. This is in striking contrast to the situation for island edges, where the majority of incorporated Mn atoms are found in the lower terrace. The step-orientation-dependent inhomogeneous distribution of incorporated Mn atoms is more clearly shown in Figs. 7.43a, b, where the local Mn coverage on

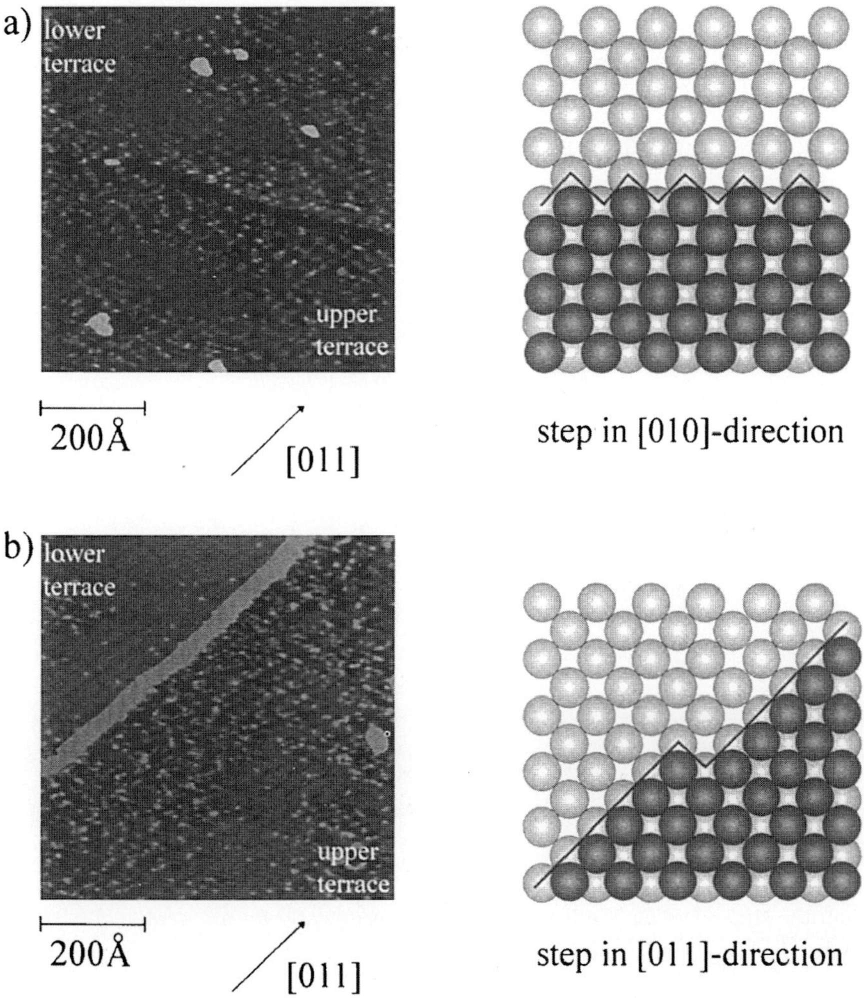

Fig. 7.42. STM images of a Cu(100) surface after deposition of Mn at 300 K. (**a**) Coverage 0.028 ± 0.007 ML Mn, scan width 700×700 Å^2. A step running close to the [010] direction is depicted. (**b**) Coverage 0.028 ± 0.007 ML Mn, scan width 700×700 Å^2. The image shows a step in the [011] direction. From [75], used with permission

the lower and upper terraces for steps in the [011] and the [010] directions, is plotted as a function of the distance from the step. The influence of the step orientation on the distribution of the Mn indicates that, at least for the lower terrace, *kinks play an important role in alloying.* Different step orientations are characterized by different kink concentrations. The kink concentration of a [011] step has been determined to be 2% at room temperature [76]. For

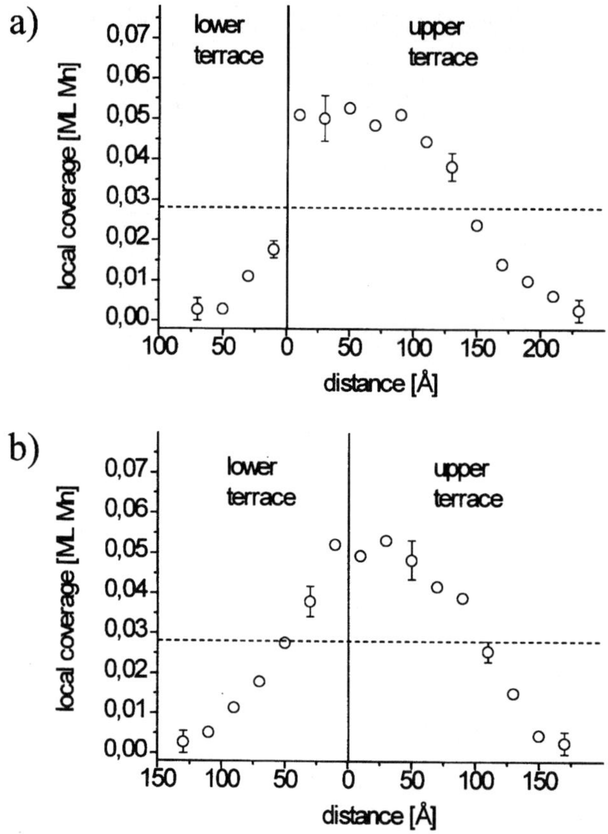

Fig. 7.43. For a [011] step (**a**) and for a step close to the [010] direction (**b**) the average local coverage of incorporated Mn is shown as a function of the distance from the step edge, for the lower and the upper terrace, for a coverage of 0.028 ± 0.007 ML Mn. From [75], used with permission

other directions, the concentration is enhanced owing to forced kinks. In the absence of microfaceting, the [010] step has the largest kink concentration, which is of the order of $100/\sqrt{2} \approx 71\%$. This clearly reveals that for the lower terrace, the degree of Mn incorporation increases with the kink concentration.

Dependence of the Mn Distribution on Annealing and Coverage. We have shown that the incorporated Mn atoms are not homogeneously distributed around islands and substrate steps. These inhomogeneities can be removed by annealing the sample. Figure 7.44 shows a surface after deposition of 0.032 ± 0.004 ML Mn at room temperature and subsequent annealing at 340 K for 10 min. The distribution of incorporated Mn atoms is completely homogeneous. Moreover, islands are no longer observed. This clearly gives

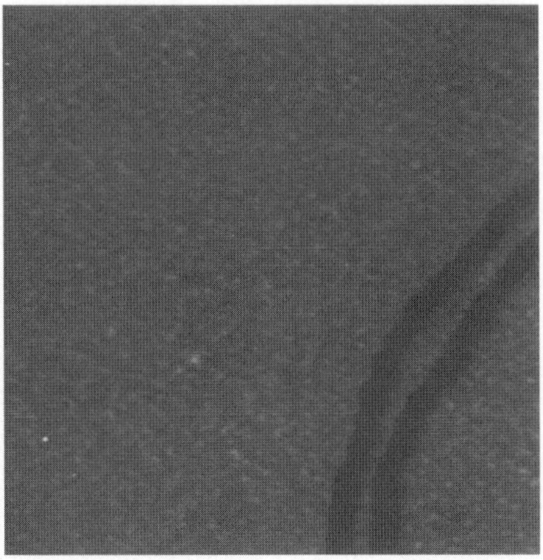

Fig. 7.44. STM image of a Cu(100) surface after deposition of Mn at 300 K and subsequent annealing at 340 K for 10 min. Coverage 0.032 ± 0.004 ML Mn, scan width $700 \times 700\,\text{Å}^2$. From [75], used with permission

evidence that *the inhomogeneous distribution is caused by kinetic limitations rather than energetic constraints.*

Overview of Surface Morphology as a Function of Mn Coverage. In Fig. 7.45, an overview is given of the distribution of incorporated Mn atoms at various coverages. At extremely small coverages, no islands are observed and the incorporated Mn atoms are homogeneously distributed (Fig. 7.45a). With increasing Mn coverage, islands are formed and an inhomogeneous distribution of incorporated Mn atoms is found around the islands and near the substrate steps (Fig. 7.45b). As the number of incorporated Mn atoms increases with Mn coverage, the mobility of the incorporated Mn atoms becomes fast. As a result, it is difficult to capture them at a coverage of 0.3 ML (Fig. 7.45c). With further increase of Mn coverage, the mobility of the incorporated Mn atoms slows down and they gradually become ordered. At a coverage of 0.45 ML, a $c(2 \times 2)$ long-range order can be clearly seen by STM (Fig. 7.45d).

The observation of a homogeneous distribution of incorporated Mn atoms after annealing and the $c(2 \times 2)$ ordering transition at high coverage allow some interesting conclusions about the Mn–Mn interactions. If the Mn–Mn interactions were attractive for next-nearest-neighbor spacings, for example, one would expect to observe the formation of a superstructure in the vicinity of steps or in the islands, and zones less dense in Mn further away from steps. However, such an "energetic" phase separation has never been observed experimentally. This implies that the interaction between incorporated Mn atoms is repulsive, at least at short distances. We have confirmed this by an evaluation of the nearest-neighbor distances between incorporated Mn

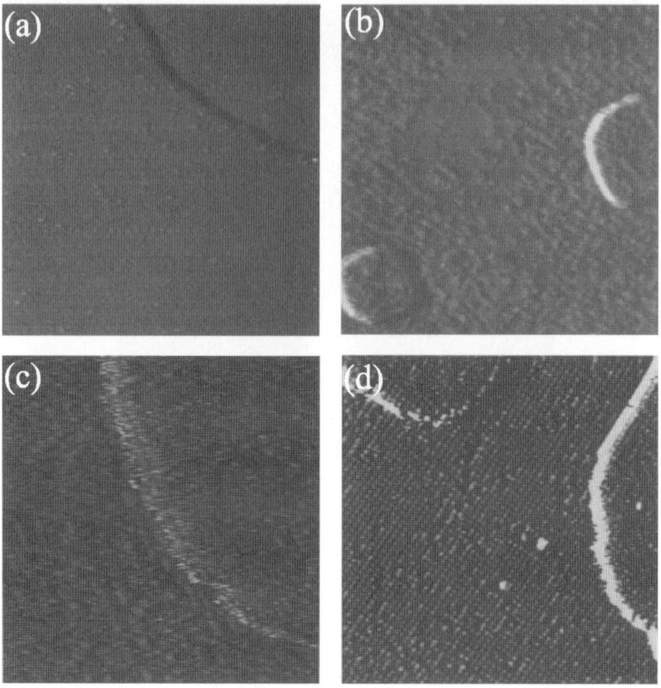

Fig. 7.45. STM images of a Cu(100) surface after deposition of Mn at 300 K. (**a**) Coverage 0.0033 ± 0.0005 ML Mn, scan width 350×350 Å^2. (**b**) Coverage 0.09 ± 0.01 ML Mn, scan width 350×350 Å^2. (**c**) Coverage 0.31 ± 0.03 ML Mn, scan width 90×90 Å^2. (**d**) Coverage 0.45 ± 0.02 ML Mn, scan width 180×180 Å^2. From [74, 75], used with permission

atoms. This analysis shows that the Mn atoms are essentially distributed as noninteracting particles, except at small distances. A nearest-neighbor distance equal to the nearest-neighbor distance of the Cu(100) surface (2.56 Å) was never found. This indicates that *a repulsive Mn–Mn interaction exists, at least over short distances.*

7.6.2 Scenario for Surface Alloy Formation

On the basis of the above STM observations, we shall now develop a scenario that describes the formation of the Cu(100)–$c(2 \times 2)$-Mn surface alloy. This scenario includes three essential processes. These are the diffusion of adsorbed Mn atoms on the Cu(100) surface, the incorporation of Mn into the Cu(100) surface and the subsequent diffusion of the incorporated Mn atoms.

This scenario is schematically described in Fig. 7.46. After Mn adsorption on the surface, Mn adatoms rapidly diffuse on Cu(100) at room temperature until they either reach a substrate step and attach to it or form Mn islands on the Cu(100) surface (phase I). Islands should be easy to form in regions

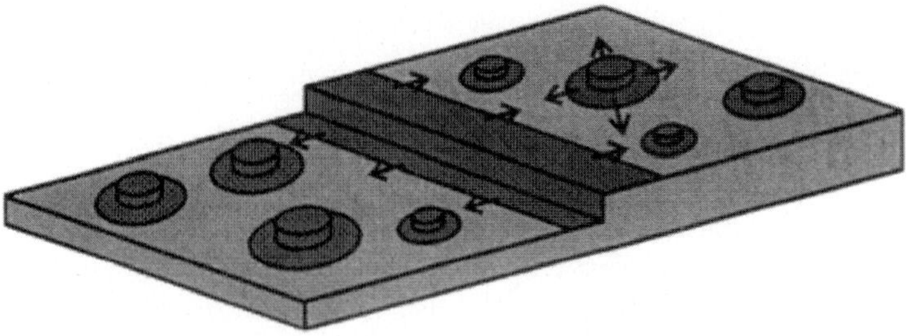

Fig. 7.46. Schematic illustration of alloy formation for the Mn/Cu(100) system. After Mn adsorption on the surface, Mn adatoms rapidly diffuse on Cu(100) at room temperature. The Mn adatoms either reach a substrate step and attach to it or form Mn islands on the Cu(100). Subsequently, Mn atoms are incorporated both at the substrate steps and at island edges. The incorporated Mn atoms diffuse in the Cu(100) surface and finally reach a homogeneous distribution. Owing to the repulsive Mn–Mn nearest-neighbor interaction, a stable $c(2 \times 2)$ surface alloy is formed around a Mn coverage of 0.5 ML. From [75], used with permission

far away from substrate steps, where a high Mn adatom concentration can build up. In a second step, which proceeds on a slower timescale, Mn atoms are incorporated. This occurs both at the substrate steps and at the island edges (phase II). The atomic processes responsible for the incorporation will be discussed in the next subsection. Once incorporated, the Mn atoms diffuse in the Cu(100) surface (phase III). This process is rather slow, and therefore the preferential incorporation close to substrate steps and island edges is still visible in the STM images hours after Mn deposition. Annealing the surface at 340 K for 10 min drastically increases the jump rate of incorporated Mn atoms and leads to a homogeneous Mn distribution. This shows that the inhomogeneous distribution of Mn after incorporation is due to kinetic limitations, i.e., the slow diffusion of incorporated Mn atoms. Owing to the repulsive interaction between Mn atoms at short distances, the incorporated Mn atoms can diffuse sufficiently without phase separation, leading to the final formation of the energetically favored $c(2 \times 2)$ surface alloy.

7.6.3 Atomic Mechanisms for Surface Alloy Formation

After discussing the overall scenario and the timescales relevant to surface alloying, we shall now discuss the atomic mechanisms involved in the formation of the surface alloy. We shall deal first with the atomic mechanisms for the diffusion of incorporated Mn atoms in the Cu(100) surface. Then we shall turn to the mechanisms that can account for the observed incorporation of Mn at substrate steps and island edges. In this discussion, we have to distinguish between incorporation by intralayer mass transport, as could be effective for Mn incorporation into the upper terrace, and Mn incorporation into the lower terrace, which can only proceed by interlayer mass transport.

Furthermore, it might prove necessary to distinguish between alloy formation at substrate steps and at island edges.

Atomic Mechanisms for the Diffusion of Incorporated Mn Atoms. Since the diffusion of the incorporated Mn atoms not only is an indispensable process for the formation of the surface alloy but also is directly correlated with the incorporation process, as will be discussed in the coming paragraphs, we shall first discuss the atomic mechanisms for the diffusion of incorporated Mn atoms. In contrast to the well-developed hierarchies of diffusion mechanisms for single adatoms on a substrate surface, along island edges and across steps, the diffusion of an atom *incorporated into* (rather than adsorbed on) *the substrate surface* has received much less attention. Therefore this issue deserves a more extended discussion. We shall first list all the possible mechanisms which could lead to the mobility of an incorporated atom. By estimating the diffusion coefficient for each possible mechanism and comparing it with the experimental result, we shall finally conclude that *the motion of incorporated Mn atoms is mediated by diffusing vacancies.*

We shall initially restrict the discussion to small coverages around and below 0.1 ML Mn. A simple mechanism which could enable the diffusion of the incorporated Mn atoms could be an exchange process that occurs horizontally on the surface. The most favorable process is the ring mechanism, where four atoms rotate in the surface (see Fig. 7.47). However, such a process is rather unlikely in a close-packed arrangement of atoms, since the resulting activation barrier would be extremely high. Therefore, we do not believe that this mechanism is a reasonable candidate for the diffusion of incorporated Mn on Cu(100). Similar arguments have also been put forward for the bulk diffusion of impurities in Si (see, for example, [77]).

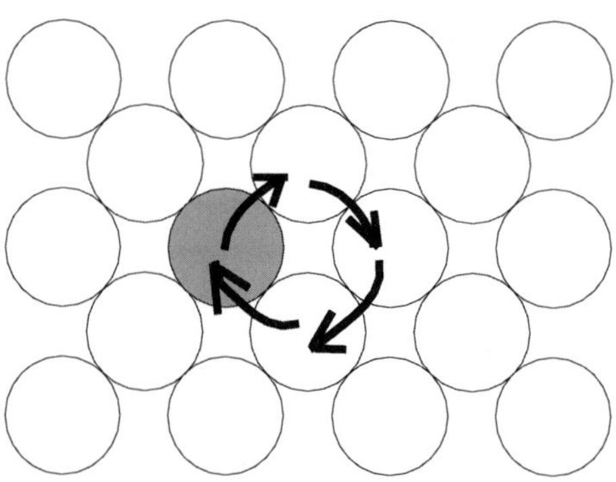

Fig. 7.47. Schematic illustration of the ring mechanism which allows an incorporated atom to diffuse in the surface. The *gray circle* denotes the incorporated atom. Its motion is brought about by a cooperative rotation of four atoms as indicated by the *arrows*

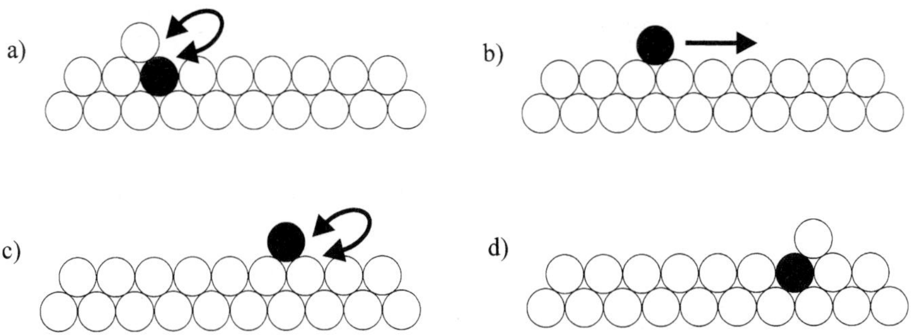

Fig. 7.48. Diffusion of incorporated Mn by a process of ejection and subsequent surface diffusion. An incorporated Mn atom is ejected onto the surface, for example by a Cu/Mn place exchange (**a**). The ejected Mn atom starts a diffusion process (**b**) and is finally incorporated by means of a Mn/Cu place exchange (*c*, *d*) *White circles* denote Cu atoms, and *black circles* Mn atoms. From [74], used with permission

Another conceivable scenario is the following: in the first step, an incorporated Mn atom is ejected onto the terrace, possibly via a Cu/Mn place exchange. Then this ejected Mn adatom diffuses on the surface until it is reincorporated through a Mn/Cu place exchange. This is illustrated in Fig. 7.48. For the Mn/Cu exchange process, we can use an Arrhenius "ansatz". The activation barrier for such an exchange can be estimated from the onset temperature for alloy formation, which is around 260 K on Cu(100). This leads to a considerable lifetime of a Mn adatom on the surface. On the other hand, such a Mn adatom can move rapidly on the surface. A lower limit on the diffusion coefficient of Mn adatoms on Cu(100) can be derived from the observation that the $c(8 \times 2)$ film structure can be formed at temperatures as low as 110 K. This gives an upper limit on the onset temperature of diffusion of 110 K. Then, from the Einstein relation, the average distance between the ejection and incorporation sites can be obtained. This results in a distance of several thousand angstroms, which is at least two to three orders of magnitude higher than the experimentally observed distance. Instead of a Mn/Cu place exchange, an alternative way in which the last step could occur in the above scenario could be incorporation by vacancy annihilation. In other words, the ejected Mn adatom diffuses on the terrace until it meets a vacancy and is then incorporated, thereby annihilating the vacancy. However, in this case also, owing to the extremely small equilibrium concentration of vacancies at room temperature, which is estimated in the next paragraph, the mean diffusion length of incorporated Mn would be between two and three orders of magnitude higher than the experimentally observed value. Hence we have to exclude the above scenario, in which the ejection of the incorporated atoms and their subsequent diffusion on the surface as adsorbed atoms are involved.

Having excluded the ring mechanism and the mechanism involving ejection, let us consider the vacancy-mediated mechanism, which is schemat-

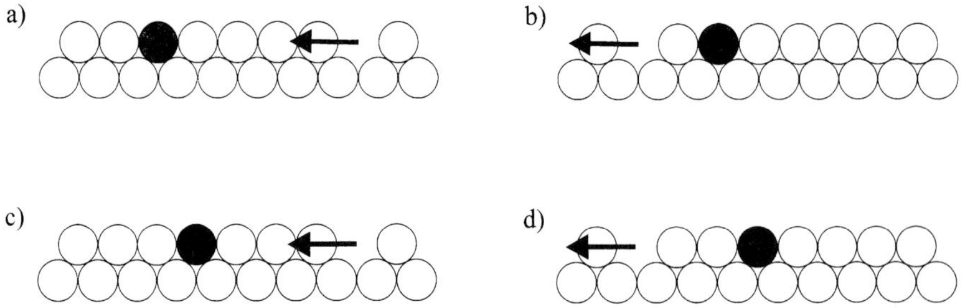

Fig. 7.49. Diffusion of incorporated Mn caused by diffusing vacancies. The vacancies move from the *right* to the *left*. As a consequence, the incorporated Mn atom jumps from the *left* to the *right*. *White circles* denote Cu atoms, and *black circles* Mn atoms. From [74], used with permission

ically depicted in Fig. 7.49. The experimental results can be successfully explained if diffusing vacancies cause the mobility of incorporated Mn. For such a mechanism, the total jump frequency of incorporated Mn is given by $\Gamma_{tot} = z \times c_V \times \Gamma_V$. Here, z is the number of nearest neighbors, Γ_V is the vacancy jump rate and c_V is the concentration of vacancies. Using a vacancy formation energy of $E_V = 0.47\,\text{eV}$, as calculated by Stoltze for Cu(100) [78], the equilibrium vacancy concentration can be estimated from $c_V = \exp(-E_V/kT)$ to be about 10^{-8}. The jump rate of the vacancies is $\Gamma_V = \nu \times \exp(-E_M/kT)$, where ν is the attempt frequency and E_M the vacancy migration energy. The migration energy of a vacancy on Cu(100) has been calculated to be 0.44 eV [78]. A similar value of 0.42 eV has been obtained in [79]. If we assume that the diffusion of vacancies is not affected by the presence of Mn, the migration energy of a vacancy on Cu(100) can be used. Using a typical attempt frequency of $10^{13\pm1}\ \text{s}^{-1}$, a total jump frequency of $\Gamma_{tot} = 10^{-2\pm1}\ \text{s}^{-1}$ is derived. This is in good agreement with the experimentally observed upper and lower limits on the jump rate, which lies between 10^{-2} and $5 \times 10^{-1}\ \text{s}^{-1}$ for 0.1 ML Mn.

We have made three assumptions in obtaining our theoretical estimate of the mobility of Mn (i) The equilibrium concentration of vacancies is not affected by incorporated Mn atoms. (ii) The jump rate of vacancies is not affected by incorporated Mn atoms. (iii) The concentration of vacancies corresponds to the equilibrium concentration, i.e., we are considering a stage where the rates of vacancy creation and vacancy annihilation are in equilibrium. The last assumption is justified, since we observe Mn mobility long after Mn incorporation. However, the first two assumptions hold only in the limit of low Mn coverage. Therefore the experimental data for extremely small Mn coverage are more meaningful for justifing the vacancy-mediated mechanism. The vacancy-mediated diffusion mechanism for incorporated Mn atoms is in fact also supported by our experimentally measured diffusion coefficient at 0.0033 ± 0.0005 ML Mn. From an evaluation of the Einstein relation, this diffusion coefficient has been determined to be $(5.3 \pm 0.3) \times 10^{-18}$ $cm^2\ s^{-1}$. For a vacancy-mediated motion, a relation between the diffusion coefficient D and the total jump frequency has been derived [74], which reads $\Gamma_{tot} = zD/fa^2$, where $f = 0.4669$ on an fcc (100) surface in the limit of small vacancy concentration. Thus the total jump frequency can be determined to be $(6.9 \pm 0.4) \times 10^{-2}\ s^{-1}$. This, again, is in good agreement with the theoretically estimated jump frequency for vacancy-mediated diffusion of Mn.

We can thus conclude that only a vacancy-driven motion of incorporated Mn can account for the observed mobility. Vacancy diffusion as a mechanism for diffusion of incorporated atoms has also been discussed in another, similar context [80]. The authors of [80] studied the adsorption and incorporation of In on Ag(100) and Ag(111) by perturbed-angular-correlation (PAC) spectroscopy. They attributed the incorporation to adatom diffusion and subsequent attachment to a step, followed by incorporation into the upper terrace through diffusing vacancies. On the basis of our STM observations of the incorporated Mn atoms in the Cu(100) surface [74, 75], we have studied the mechanisms of vacancy generation at steps and the role of the vacancies in forming the Mn–Cu surface alloy and a method has been proposed for studying step fluctuations via decoration of the vacancies [81]. The vacancy-mediated surface diffusion has been confirmed by a recent STM study, in which the motion of incorporated indium atoms in the Cu(100) surface was analyzed [82, 83]. Numerical calculations reveal that the diffusive motion of the indium atoms is caused by a low density of extremely mobile vacancies in the first layer of the Cu(100) surface.

For higher coverages, around 0.3 ML Mn and above, the discussion of the diffusion mechanism becomes more difficult. However, the assumption of a vacancy-driven process is supported by the fact that such a model also offers a plausible path for the ordering of the system at higher coverages. This is illustrated in Fig. 7.50. In terms of the vacancy model, the pronounced en-

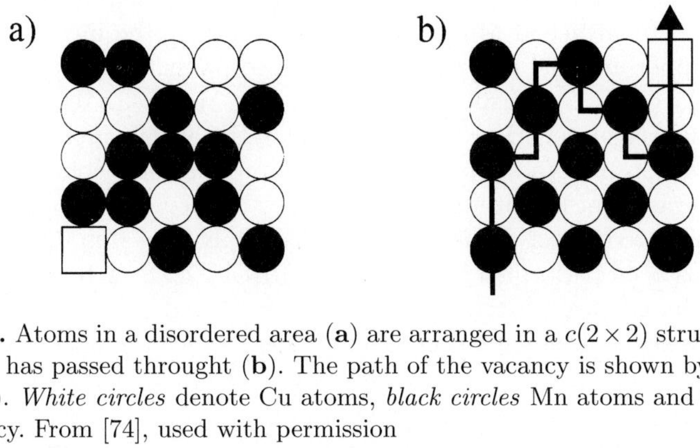

Fig. 7.50. Atoms in a disordered area (**a**) are arranged in a $c(2 \times 2)$ structure after a vacancy has passed throught (**b**). The path of the vacancy is shown by the *black line* in (b). *White circles* denote Cu atoms, *black circles* Mn atoms and the *square* the vacancy. From [74], used with permission

hancement of the mobility of incorporated Mn around 0.3 ML can be formally explained either by a higher jump rate of the vacancies, which corresponds to a decrease of the vacancy migration energy E_M, or by a larger equilibrium concentration of vacancies, i.e., a decrease of the vacancy formation energy E_V. Using $D = fa^2\Gamma_{tot}/Z$, a lower limit on the diffusion coefficient $D \geq 1.7 \times 10^{-16}$ cm^2 s^{-1} can be derived from the jump frequency Γ_{tot} (see Fig. 7.40). A similar estimate is given by Noh et al. [84], where $D \geq 1.5 \times 10^{-16}$ cm^2 s^{-1} for a comparable coverage [84]. These values are larger than that for small coverage by two orders of magnitude, requiring a decrease of the sum $E_F + E_M$ by approximately 0.12 eV. Such a decrease seems possible within the framework of a coverage dependence of E_F and/or E_M. Support for a pronounced composition dependence of E_F comes from studies of the bulk diffusion in binary alloys. These investigations have shown that in many ordered alloys of the DO_3 type, the vacancy concentration is up to a few orders of magnitude higher than in pure metals, resulting in fast bulk diffusion [85, 86]. An example is Fe_3Si. Such a scenario could also account for the high mobility that we observe for Mn incorporated into the Cu(100) substrate surface at higher coverage.

A similar coverage-dependent mobility has been observed for incorporated Pb atoms in the Cu(100) surface. Pb also forms ordered surface alloys on Cu(100) [87, 88]. The mobility of incorporated Pb atoms shows a pronounced maximum in an intermediate-coverage regime and is strongly reduced at higher and lower coverages [87, 89, 90]. It is remarkable that for both Pb and Mn, the increase in the mobility of the incorporated atoms is clearly correlated with an ordering process. We attribute this coverage dependence to the interactions between the incorporated atoms. Assuming a repulsive nearest-neighbor interaction, a Monte Carlo study by Uebing and Gomer [91] has qualitatively reproduced the enhancement of the diffusion coefficient in the intermediate-coverage regime (between 0 and 0.5 ML Mn).

Incorporation of Mn Atoms into the Upper Terrace of a Step. We have mentioned that Mn atoms can be attached to substrate steps after deposition. The attached Mn atoms can subsequently be incorporated into both the upper and the lower terrace. For incorporation into the lower terrace, interlayer mass transport mechanisms are required. We shall discuss these mechanisms later. Here we shall focus on the atomic processes involved in incorporation into the upper terrace. Two mechanisms could account for the incorporation of Mn atoms into the upper terrace: step fluctuations and vacancy-mediated direct diffusion. A detailed analysis of step fluctuations on Cu(100) has been performed by Giesen and Ibach [76]. Step fluctuations may be caused, for example, by adatoms, which are created at kinks, diffuse along the step edge and annihilate at another kink site. Such a process changes the shape of the step (see Fig. 7.51). The Mn atoms attached at a step could be surrounded by Cu atoms by step fluctuations and hence become incorporated. The contribution of the step fluctuations to Mn incorporation can be

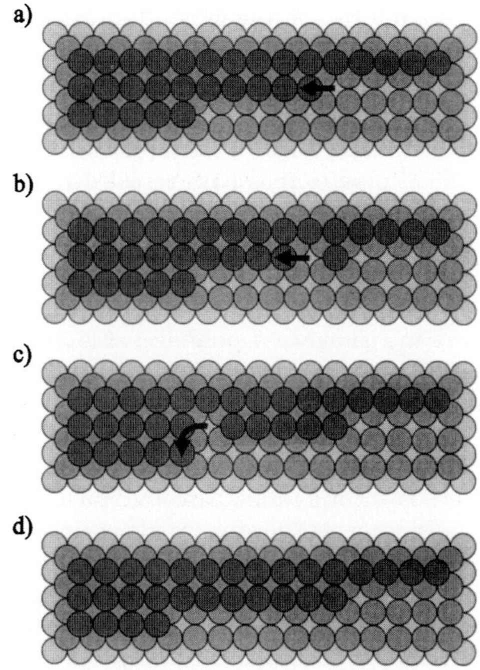

Fig. 7.51. Step fluctuations caused by diffusing vacancies: a vacancy is created at a kink site (**a**) and starts a diffusion process along the step (**b**), (**c**). Finally, the vacancy recombines at another kink (**d**). The shape of the step is changed by this process. From [75], used with permission

estimated from the mean square value $\langle(x(t) - x(t_0))^2\rangle = c(T) \times (t - t_0)^{1/4}$, where $c(T) \approx 0.2\ \mathrm{s}^{-1/4}$ for a [011] step on Cu(100) at room temperature [76]. Here, $x(t_0)$ and $x(t)$ are the positions of a step atom at the times t_0 and t, respectively. Compared with our deposition rate, $c(T)$ is too small for fast alloying. The attached Mn adatoms are not transferred into the upper-terrace step as fast as they arrive. As a consequence, a seam of essentially pure Mn will be formed by the attached Mn adatoms. This is in agreement with the observation of a Mn seam on the substrate steps of the vicinal Cu(1 1 141) surface [59]. The width of such a seam will depend upon the Mn coverage and the average terrace width. For small terrace widths, such as found on highly stepped surfaces, the seam will be rather thin at small Mn coverages. In this case, step fluctuations will be more important. For wider terraces, as present in the study described here, an almost pure Mn seam of considerable width will be formed. In this case, the step fluctuation mechanism does not work any more. Therefore vacancy diffusion, which has been described in detail above, should be the major contribution to the mixing at the Mn/Cu interface. Mediated by vacancy diffusion, the Mn atoms in the seam can afterwards diffuse into the upper terrace.

Atomic Mechanisms for Interlayer Mass Transport. Both step fluctuations and vacancy diffusion lead only to intralayer mass transport. However, to explain Mn incorporation into the lower terrace adjacent to a step, interlayer mass transport needs to occur as well. Interlayer mass transport is also needed to explain the incorporation of Mn atoms in the vicinity of island edges. In addition, STM observations have revealed that the incorporation of Mn depends strongly upon the step direction, while the attachment of Mn to a substrate step is independent of step orientation (see Fig. 7.42). Since different step orientations are associated with different densities of kinks, the kinks should play an important role in the mechanism of interlayer mass transport. In the following paragraphs, interlayer mass transport processes at island and step edges that could be relevant to alloy formation are discussed.

Two different classes of mechanism can produce interlayer mass transport. One class is based on a Mn/Cu place exchange, while the other includes vacancy generation and subsequent vacancy annihilation. Figure 7.52 shows possible pathways for Mn incorporation based on Mn/Cu place exchange. In the course of such a process, a Mn atom in the seam or, alternatively, as suggested in [81], a Mn adatom changes place with a Cu atom in the lower terrace. The Cu atom, after exchange, will be attached to the kink. These two processes are marked '(a)' and '(b)', respectively, in Fig. 7.52. A Mn/Cu place exchange could also occur on the terrace '(c)' in (Fig. 7.52). However, the processes at kinks should be energetically favorable, since the number of bonds that the ejected atom can form increases with exchange in the vicinity of a kink [81]. This would result in an inhomogeneous distribution of incorporated Mn, as is indeed observed. Since the activation energies necessary for an exchange mechanism of Mn on Cu(100) at the various different sites

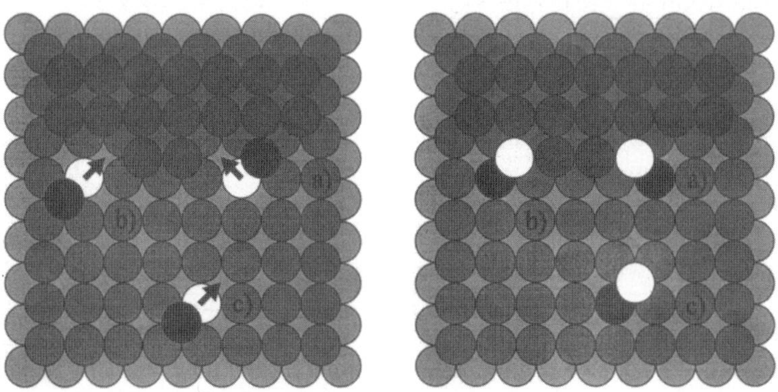

Fig. 7.52. Mn/Cu place exchange process at different sites on the surface: (a) in the vicinity of a kink, (b) near a step site and (c) on the open terrace. The initial arrangement is shown on the *left*, and the final configuration on the *right*. *Arrows* point in the directions of the jumps of the ejected Cu atoms. *White circles* denote Cu atoms, and *black circles* Mn atoms. From [75], used with permission

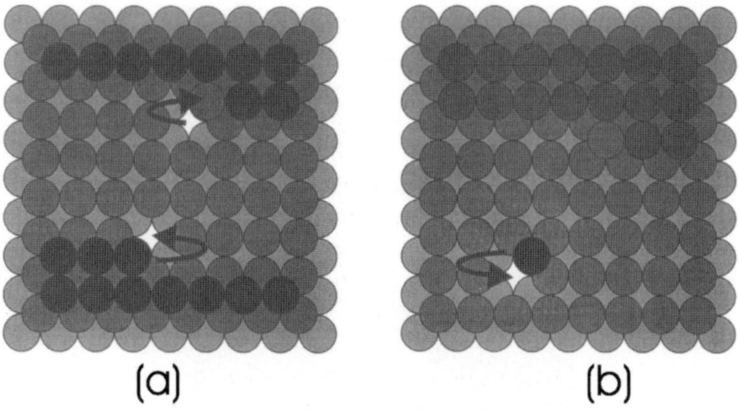

Fig. 7.53. Incorporation of Mn atoms via vacancy annihilation. (**a**) The generation of a vacancy at a kink site transfers a Cu atom from the lower terrace to the upper terrace (*upper part* of (**a**)). The annihilation of a vacancy at a kink site transports a Mn atom from the upper terrace to the lower terrace (*lower part* of (**a**)). Both processes support the mixing process, since mainly Mn atoms are situated in the upper terrace and mainly Cu atoms are situated in the lower terrace. (**b**) The annihilation of a vacancy transfers a Mn adatom to the lower terrace. The process is not restricted to island and step edges and can also occur on the open terrace. *Gray circles* denote Cu atoms, and *black circles* Mn atoms. From [75], used with permission

are not known, it is difficult to evaluate whether such an exchange process at a kink site is realistic and sufficient to explain the dominance of kinks in Mn incorporation.

In the second scenario for kink-induced Mn incorporation into the lower terrace, the incorporation is realized via vacancy generation and annihilation (see Fig. 7.53a). Vacancy generation at kinks allows the transfer of a Cu atom from the lower to the upper terrace, while vacancy annihilation at a kink site leads to transport of a Mn adatom into the lower terrace.

Since the equilibrium concentration of vacancies is, with $c_V \sim 10^{-8}$ on Cu(100), very low at room temperature, efficient Mn incorporation is only plausible if the vacancy generation rate is not too low. To justify the vacancy-mediated interlayer mass transport mechanism, we shall first estimate the vacancy generation rate. Vacancy creation can in principle take place at various sites on the surface. This is illustrated in Fig. 7.54. The rate of vacancy formation can be approximated by $g = c_S \times \nu \times \exp(-E_A/kT)$. Here, c_S is the concentration of sources where vacancies are created, and E_A is the activation energy to create a vacancy. The activation energy is composed of two contributions: the formation energy of a vacancy E_G and the migration energy of a vacancy away from the generation site E_M; hence $E_A = E_M + E_G$. Assuming that E_M is independent of the vacancy generation site, we can use the migration energy of a vacancy on Cu(100) (0.44 eV) in all cases [78]. However,

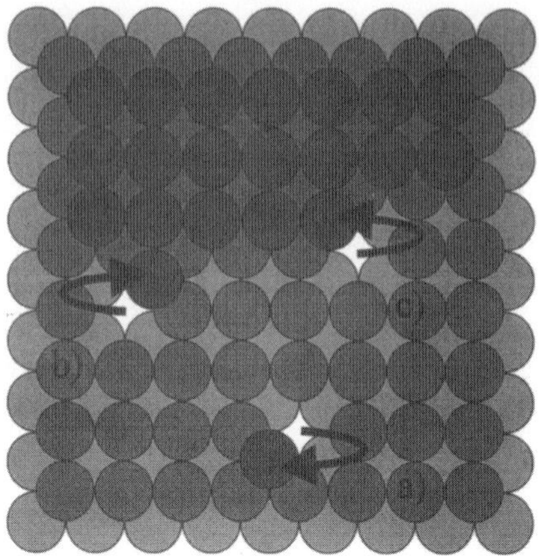

Fig. 7.54. Schematic illustration of the generation of a vacancy on a terrace (a), near a step (b) and in the vicinity of a kink site (c). From [75], used with permission

both c_S and E_G and therefore the rate of vacancy generation g have different values for each of the processes sketched in Fig. 7.54. On the open terrace, each atom is a potential source of vacancies, and therefore $c_S \sim 1$. Since the creation of a vacancy is accompanied by the generation of an adatom as well, E_G equals the formation energy of a vacancy plus that of an adatom in this case. Stoltze has determined this total energy E_G to be 0.98 eV [78]. Therefore, an activation barrier as large as 1.42 eV is encountered for the generation of a vacancy on an open terrace. This large activation energy means that the generation of a vacancy on an open terrace is quite difficult.

Less energy is necessary to form a vacancy close to a step. The energetically most favorable situation is the generation of a vacancy at a kink site. Here, the adatom is attached to the kink and therefore does not contribute to E_G. Therefore the total formation energy is reduced by one-half. Tréglia et al. [79] have calculated a formation energy $E_G = 0.47$ eV for the creation of a vacancy at a kink site. This result is in good agreement with the estimates presented above. Therefore the total activation energy is 0.91 eV for the generation of a vacancy at a kink. The kink concentration of a [011] step on Cu(100) at room temperature has been determined to be 2% [76]. In this case, the average source concentration for vacancy creation can be calculated to be $c_S = 5 \times 10^{-5}$ using the average terrace width of 1000 Å of our sample. This will be a lower limit for c_S, since only [011] steps are assumed and island formation has so far not been considered. This leads to a ratio of the generation rates at kink and terrace sites at room temperature $g_\mathrm{k}/g_\mathrm{t} \sim 10^4$. Therefore, kinks are the only effective source of vacancy generation on Cu(100) at room temperature. Monte Carlo simulations [81] confirm the role of kinks in vacancy production. Previous studies of Mn incorporation in Cu(1 1 141) [15, 59] and in Cu(1 1 17), [93] also show a correlation of the step density with the onset temperature of Mn incorporation. For Cu(1 1 17), the Mn incorporation starts around 170 K, while for the Cu(1 1 141) surface, with its lower step density, the onset temperature is above 200 K.

With the activation barrier and the source concentration known, the generation rate g_k can be calculated to be $2 \times 10^{-7\pm1}$ s^{-1} for a typical attempt frequency $\nu = 10^{13\pm1}$ s^{-1}. This is a rather low rate of generation of vacancies. However, this number merely represents a lower limit on the kink concentration, since only [011] steps have been considered. Other step directions have higher kink concentrations owing to forced kinks. In addition, the kink concentration increases dramatically during deposition because of island formation. At 0.1 ML Mn, from the shape, size and density of the islands, we estimate an enhancement of the kink concentration by at least two orders of magnitude. Therefore, in the typical time interval of 2 h between Mn deposition and the recording of the first STM images, the number of generated vacancies $g_k \times \Delta t$ is sufficiently large to account for the incorporated Mn atoms. Furthermore, an acceleration of vacancy formation at higher Mn coverages is most probably caused by a decrease of the activation energy for vacancy creation; we have discussed this decrease earlier. It should be kept in mind that the activation energies for vacancy generation have been estimated for a pure Cu(100) surface. Mn atoms attached to steps and kinks will probably reduce the corresponding activation barrier. Hannon and Ibach [92] have recently studied low-temperature incorporation of Mn on the Cu(1 1 13) surface and found that strong Mn–Cu bonding leads to a further reduction of the barriers at the step edge compared with those at a clean [110] step. Effective medium theory calculations [92] performed for Pd on Cu(1 1 13) imply that Pd reduces the activation barriers for vacancy formation at kinks and thus enhances Pd incorporation.

In the preceding paragraphs, we have focused on kink-assisted vacancy generation and annihilation (Fig. 7.53a). These processes are restricted to the close vicinity of island and step edges. Vacancy annihilation and hence Mn incorporation can also proceed by processes involving Mn adatoms (Fig. 7.53b). This is especially important at small Mn coverages, at which Mn islands have not yet formed or have a very low density. In this case, the mean lifetime and therefore the diffusion length of a vacancy depend on the concentration of the adatom gas. At small coverages, vacancies created at kink sites can diffuse on the open terrace before they are annihilated, leading to a more homogeneous distribution of incorporated Mn. Indeed, experiments at a coverage of 0.0033 ± 0.0005 ML Mn reveal that the incorporated Mn is more homogeneously distributed. In summary, the atomic mechanisms involved in forming an ordered Cu(100)–$c(2\times2)$-Mn surface alloy have been investigated using STM. Incorporation and hence alloy formation occur immediately upon the initial Mn deposition. The experiments reveal that Mn is preferentially incorporated in the vicinity of island edges and substrate steps. On the lower terrace, the rate of incorporation is correlated with the kink density and takes place either by Mn/Cu place exchange processes or by annihilation of vacancies generated at kinks. Vacancy annihilation by Mn adatoms can account qualitatively for both the observed coverage dependence of Mn incorporation and the enhanced incorporation close to kinks. The alloying of the upper terrace is driven by the attachment of Mn adatoms. Subsequent step fluctuations and, presumably more importantly, vacancy-mediated diffusion of incorporated Mn atoms lead to incorporation into the upper terrace. Subsequent vacancy diffusion produces a homogeneous Mn distribution with time.

7.7 Electronic and Magnetic Properties of $c(2 \times 2)$ MnCu and MnNi Surface Alloys

The magnetic properties of the $c(2 \times 2)$ MnCu and MnNi surface alloys have attracted a lot of interest, since they are an experimental realization of a two-dimensional alloy containing the magnetic Mn atom, which has the maximum atomic moment among 3d elements. A complete understanding of the electronic and magnetic properties of these surface phases would not only contribute to the knowledge of low-dimensional magnetism, but also provide new pathways to controlling the spin-dependent transport through interfaces and surfaces. Theoretically, the spin-resolved local density of states (LDOS) has been calculated by density functional theory [36, 41, 94, 95]. From the LDOS, the atomic moment can be determined. The magnetic order can be predicted by total-energy calculations assuming various different magnetic structures (ferromagnetic, antiferromagnetic and disordered structures). Experimentally, the electronic band structure of these surface alloys has been investigated by photoemission spectroscopy (for occupied states) and inverse photoemission spectroscopy (for empty states), mainly using synchrotron light [94, 95, 96]. X-ray absorption spectroscopy and x-ray magnetic dichroism have been used to determine the atomic moment and the magnetic order, respectively [97, 98, 99, 100].

Rader et al. have calculated the electronic band structure for MnCu and MnNi surface alloys on the basis of density functional theory in the local-spin-density approximation [94]. In these calculations, a structural optimization was conducted by total-energy minimization prior to the LDOS calculations, resulting in a corrugated structure for the topmost layer. A ferromagnetic ground-state configuration was assumed for the calculations. Figure 7.55 shows the calculated spin-resolved local density of states at the Mn sites[3]

[3] Mn states were identified as states that lead to a concentration of their charge density in muffin-tin spheres of Mn atoms. The criterion was 60% of the charge density at the Mn atoms for $c(2\times2)$ MnCu/Cu(100). For $c(2\times2)$ MnNi/Ni(100), the corresponding percentage was 50% for minority-spin states and 30% for majority-spin states. This distinction is necessary because the spin-dependent

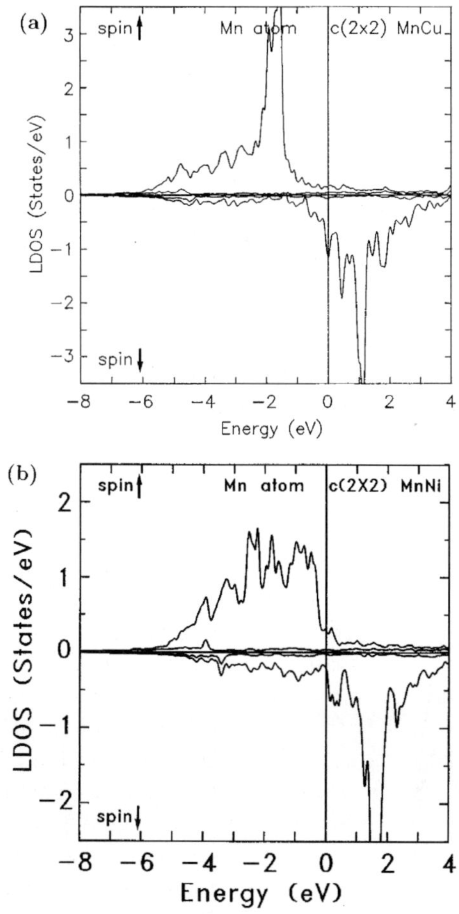

Fig. 7.55. Calculated spin-resolved local density of states at the Mn sites for (**a**) Cu(100)-$c(2\times2)$-Mn and (**b**) Ni(100)–$c(2\times2)$-Mn surface alloys. The majority band and the minority band are completely split. From [94], used with permission

for the MnCu and MnNi surface alloy phases. One can see a complete spin splitting between the majority and minority bands. This is in agreement with the total-energy calculation [36] which predict a large Mn magnetic moment for the $c(2\times2)$ MnCu/Cu(100) surface alloy. The calculations determine an exchange splitting of 2.7 eV for MnCu and 3.4 eV for MnNi at the center of the surface Brillouin zone ($\overline{\Gamma}$ point) [94]. The magnetic moment of the Mn atom is calculated to be $3.75\mu_B$ for $c(2\times2)$ MnCu/Cu(100) and $3.5\mu_B$ for $c(2\times2)$ MnNi/Ni(100). The local magnetic moment and exchange splitting have been calculated by several groups for MnCu and MnNi surface alloys.

Mn–3d-substrate–3d hybridization affects the majority states much more than the minority states.

For example, Eder et al. have predicted a magnetic moment of 4.1 μ_B and an exchange splitting of 4.4 eV at the $\overline{X}$ point [41]. These parameters are calculated to be 3.83μ_B and 3.5 eV by Schiller et al. [95]. Spišák and Hafner have predicted a local moment of 3.86μ_B for Mn and 0.41μ_B for Ni for the $c(2\times2)$ MnNi/Ni(100) surface alloy [101]. From Fig. 7.55a, both the majority band and the minority band of the MnCu surface alloy are very narrow, indicating a high localization of the Mn 3d electrons. However, for the MnNi surface alloy phase, a broad Mn majority band together with a narrow minority band are observed. This could be explained by a strong hybridization of Mn 3d minority states with Ni 3d states and a relatively weak hybridization with Cu 3d states.

Rader et al. have studied the electronic structure of $c(2\times2)$ MnCu/Cu(100) and $c(2\times2)$ MnNi/Ni(100) by photoemission and inverse photoemission spectroscopy [94]. The spectra at zero parallel momentum are shown in Figs. 7.56 and 7.57. The spectra were recorded for clean Cu(100) and Ni(100)

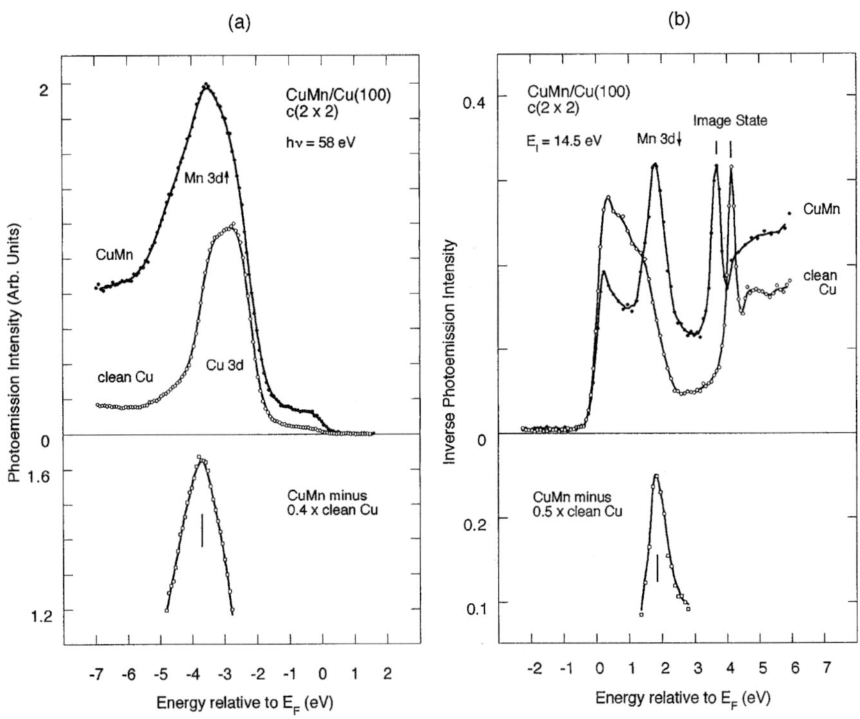

Fig. 7.56. Photoemission (**a**) and inverse photoemission (**b**) spectra for the $c(2\times2)$ MnCu/Cu(100) surface alloy. The *bottom panels* show difference spectra after subtracting the substrate signal. The majority- and minority-spin Mn-3d states show peaks at −3.7 eV and 1.85 eV, respectively, indicating an exchange splitting of 5.6 eV. From [94], used with permission

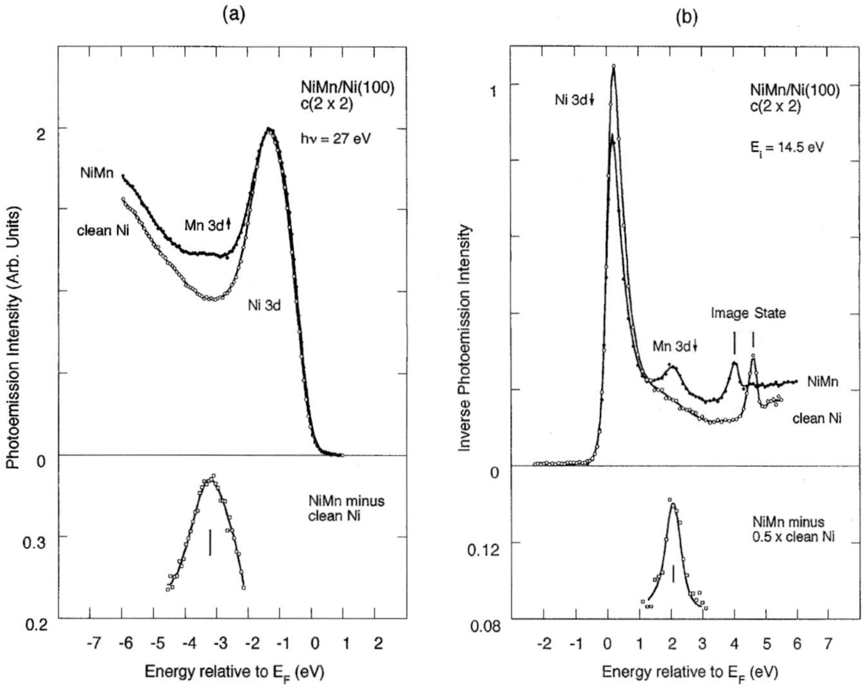

Fig. 7.57. Photoemission (**a**) and inverse photoemission (**b**) spectra for the $c(2{\times}2)$ MnNi/Ni(100) surface alloy. The *bottom panels* show difference spectra after subtracting the substrate signal. The majority- and minority-spin Mn 3d states show peaks at −3.2 eV and 2.05 eV, respectively, indicating an exchange splitting of 5.3 eV. From [94], used with permission

substrates and for the $c(2{\times}2)$ MnCu/Cu(100) and ($c(2{\times}2)$ MnNi/Ni(100) surfaces. The Mn 3d states shown in the bottom panels were obtained by subtracting the spectra of the substrate from the those of the surface alloys. The Mn 3d states show a peak well below the Fermi level in the photoemission spectrum and a peak well above the Fermi level in inverse photoemission, indicating a complete separation of the majority band and the minority band. However, the exchange splitting, which is 5.6 eV for MnCu and 5.3 eV for MnNi, is larger than the theoretical prediction. For both the MnCu and the MnNi surface alloy phase, the Mn 3d majority bands are found to be much broader than the minority bands, indicating a weaker hybridization for the minority band. Hayden et al. have performed an inverse photoemission study of the $c(2{\times}2)$ MnCu/Cu(100) surface alloy phase [102]. They find a well-resolved unoccupied state at 2.0 eV above the Fermi level, which is associated with the Mn adsorbate. This energy position is in good agreement with the 1.85 eV of the Mn 3d minority band obtained by Rader. Using angle-resolved photoemission spectra, the Fermi surface of $c(2\times2)$ MnCu/Cu(100) has been mapped by Schiller et al. [95].

The local magnetic moment of Mn in the $c(2\times2)$ MnCu/Cu(100) surface alloy has been estimated by O'Brien et al. using multiple techniques [97, 98, 103]. It has been concluded that the Mn atoms are in a high-spin ground state, on the basis of the following experimental results: (1) the branching ratio in the Mn $L_{2,3}$ soft x-ray absorption (SXA) spectrum, defined as $I(L_3)/[I(L_2 + I(L_3)]$ (see also Sect. 4.3.2), is larger for the surface alloy than for a Mn thin film on Cu(100) (see Fig. 7.58); (2) the Mn SXA line shape for the c(2×2) MnCu/Cu(100) surface alloy resembles the theoretical spectrum for the high-spin Hund's rule ground state ($S = 5/2$); and (3) the Mn 3s photoemission spectrum shows an exchange-split doublet [97]. The splitting of the 3s core-level could be caused by the interaction of the 3s hole and the spin-polarized 3d electrons. However, to our knowledge, no quantitative experimental value has been reported for the magnetic moment of the Mn atom in the $c(2\times2)$ MnCu/Cu(100) surface alloy.

In contrast to the consistent conclusion drawn from theoretical and experimental studies that the Mn atoms are in the high-spin state, the determination of the long-range order of the MnCu surface alloy seems to be a more difficult task. Though Wuttig et al. have predicted a large Mn local magnetic moment for the MnCu surface alloy and have excluded the paramagnetic phase as the ground-state magnetic order by total-energy calculations

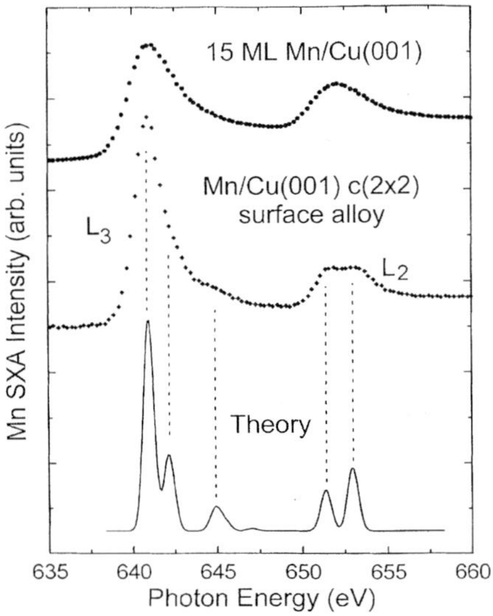

Fig. 7.58. Mn $L_{2,3}$ x-ray absorption spectrum for 15 ML of Mn on Cu(100) (*top*) and for the $c(2\times2)$ MnCu/Cu(100) surface alloy (*middle*), and the theoretical spectrum for the high-spin Hund's rule ground state ($S = 5/2$) of d^5 Mn^{2+}. Linearly polarized light was used in the measurement. From [98], used with permission

[36], their theoretical results do not tell us whether the MnCu surface alloy has a ferromagnetic or an antiferromagnetic ground state. However, their LEED observations rule out the possibility that an antiferromagnetic transition occurs above 100 K [36], since there are no extra LEED spots present in this temperature range. A recent ab initio density-functional study suggests that the MnCu surface alloy has a ferromagnetic ground state [41]. The total energy per unit cell of the ferromagnetic phase is lower than that of the antiferromagneic phase by more than 2 eV. O'Brien et al. have employed x-ray magnetic circular dichroism to detect the long-range order [97]. However, they did not capture any XMCD signal at room temperature, suggesting either no long-range order or an antiferromagnetic coupling in the MnCu surface alloy at room temperature. Rader et al. have extended the search for ferromagnetic order in the MnCu surface alloy down to 80 K [94] and derived the same conclusion. The first experimental evidence for ferromagnetic long-range order has recently been reported by Huttel et al. [100]. These authors have extended the XMCD measurements below 10 K and observed a clear XMCD signal at low temperature. The Curie temperature was estimated to be around 50 K (see Fig. 7.59). However, the Mn magnetic moment at 10 K estimated from the XMCD signals is extremely small ($0.67\mu_B$) compared with the theoretical predictions. The small magnetization has been ascribed to the surface contamination. Indeed, a decay of the XMCD signal with time has been observed [100]. This raises the question of whether the

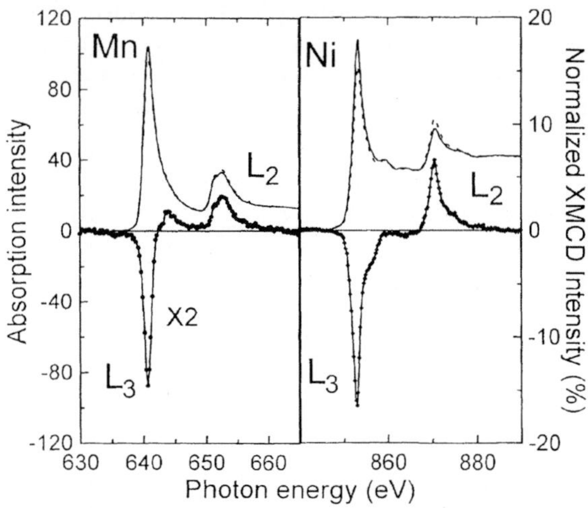

Fig. 7.59. Mn and Ni $L_{2,3}$ x-ray absorption spectra for the $c(2\times2)$ MnNi surface alloy with the photon spin parallel (σ_+ *dashed lines*) and antiparallel (σ_- *solid lines*) to the sample magnetization. The normalized Mn and Ni XMCD spectra ($2[\sigma_+ - \sigma_-]/[\sigma_+ + \sigma_-]$) are shown by *lines* with *solid symbols* on them. The nonzero XMCD intensity indicates a ferromagnetic long-range order for both Mn and Ni. From [98], used with permission

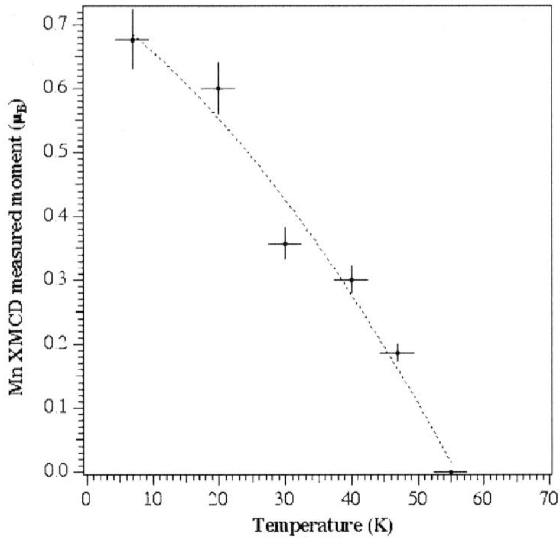

Fig. 7.60. Temperature dependence of the magnetization contributed by a Mn atom in the $c(2\times2)$ MnCu/Cu(100) surface alloy. The magnetic moment was derived from the integrated L_3 XMCD signal. From [100], used with permission

contamination also affects the Curie temperature. If it does, what is the ferromagnetic transition temperature for the clean MnCu surface alloy? These issues, therefore, still need further clarification.

It is not surprising that a considerable Mn XMCD signal is observed at room temperature for the $c(2\times2)$ MnNi/Ni(100) surface alloy (see Fig. 7.60) [98], since the substrate is ferromagnetic at room temperature. A comparison of the Mn and Ni XMCD signals reveals that the Mn magnetic moment is aligned ferromagnetically with respect to the Ni substrate [98]. The ferromagnetic order of the Mn atoms in the $c(2\times2)$ MnNi/Ni(100) surface alloy is believed to result from this ferromagnetic coupling between Mn and the substrate [99]. This is evidenced by the fact that the Mn always shows the same ordering temperature as the substrate when the MnNi surface alloy is prepared on an ultrathin Ni(100) film so that the Curie temperature of the Ni film changes with its thickness [99].

References

1. V. L. Moruzzi, P. M. Marcus and J. Kübler: Phys. Rev. B **39**, 6957(1989)
2. Landolt-Börnstein: *Structure Data of Elements and Intermetallic Phases III/6* (Springer Verlag, New York, 1971)
3. G. Fuster, N. E. Brener, J. Callaway, J. L. Fry, Y. Z. Zhao and D. A. Papaconstantopoupos, Phys. Rev. B **38**, 423(1988)
4. J. Kübler: J. Magn. Magn. Mat. **20**, 107(1980)
5. D. Tian, S. C. Wu, F. Jona and P. M. Marcus: Solid State Commun. **70**, 199(1989)

6. B. T. Jonker and J. J. Krebs, G. A. Prinz: Phys. Rev. B **39**, 1399(1989)
7. Y. U. Idzerda, B. T. Jonker, W. T. Elam and G. A. Prinz: J. Appl. Phys. **67**, 5385(1990)
8. S. T. Purcell, M. T. Johnson, N. W. E. McGee, R. Coehoorn and W. Hoving: Phys. Rev. B **45**, 13064(1992)
9. S. K. Kim, Y. Tian, M. Montesano, F. Jona and P. M. Marcus: Phys. Rev. B **54**, 5081(1996)
10. C. Binns and C. Norris: Surf. Sci. **116**, 338(1982)
11. W. F. Egelhoff, I. Jacob, J. M. Rudd, J. F. Cochran and B. Heinrich: J. Vac. Sci. Technol. A **8**, 1582(1990)
12. S. Blügel, B. Drittle,R. Zeller and P. H. Dederichs: Appl. Phys. A **49**, 547(1989)
13. E. Tamura, S. Blügel and R. Feder: Solid State Comm. 65, 1255(1988)
14. B. Heinrich, A. S. Arrott, C. Liu and S. T. Purcell: J. Vac. Sci. Technol. A **5**, 1935(1987)
15. Y. Gauthier, M. Poensgen and M. Wuttig: Surf. Sci. **303**, 36(1994)
16. W. Moritz: J. Phys. C. **17**, 353(1984)
17. G. Kleinle, W. Moritz, D. L. Adams and G. Ertl: Surf. Sci. **219**, L647(1989)
18. G. Kleinle, W. Moritz and G. Ertl: Surf. Sci. **238**, 119(1990)
19. J. B. Pendry: J. Phys. C **13**, 937(1980)
20. E. Zanazzi and F. Jona: Surf. Sci. **62**, 61(1977)
21. T. Flores, M. Hansen and M. Wuttig: Surf. Sci. **279** 251(1992)
22. M. Ohring: *The Materials Science of Thin Films* (Academic Press, New York, 1992)
23. R. A. Barker and P. J. Estrup: Phys. Rev. Lett. **41**, 1307(1978)
24. D. A. King and G. Thomas: Surf. Sci. **92**, 201(1980)
25. W. Moritz and S. Pflanz: private communication
26. J. A. Meyer, Y. Kuk, P. J. Estrup and P. J. Silverman: Phys. Rev. B **44**, 9104(1991)
27. H. M. Kramer and E. Bauer: Surf. Sci. **93**, 407(1980)
28. W. R. Tyson and W. A. Miller: Surf. Sci. **62**, 267(1977)
29. M. A. Van Hove, R. J. Koestner, P. C. Stair, J. P. Biberian, L. L. Kesmodel, I. Bartos and G. A. Somorjai: Surf. Sci. **103**, 189(1981)
30. W. Moritz: habilitation thesis, Ludwig-Maximilians-Universität München (1983)
31. Y. Gauthier, R. Baudoing-Savois, J. M. Bugnard, U. Bardi and A. Atrei: Surf. Sci. **276**, 1(1992)
32. Y. Gauthier, R. Baudoing-Savois, J. Runggren, M. Hammar and M. Gothelid: Surf. Sci. **327**, 100 (1995)
33. P. W. Palmberg and T. N. Rhodin: J. Chem. Phys. **49**, 147(1968)
34. W. Daum: PhD Thesis, RWTH Aachen(1988)
35. E. Lang, K. Müller, K. Heinz, M. A. Van Hove, R. J. Koestner and G. A. Somorjai: Surf. Sci. **127**, 347(1983)
36. M. Wuttig, C. C. Knight, T. Flores and Y. Gauthier: Surf. Sci. **292**, 189(1993)
37. J. Philip and J. Rundgren: in *Determination of Surface Structure by LEED*, ed by P. Marcus and F. Jona, (Plenum, New York 1984)
38. R. Toomes, A. Theobald, R. Lindsay, T. Geibel, O. Schaff, R. Didszhun, D. P. Woodruff, A. M. Bradshaw and V. Fritzsche: J. Phys. Condens. Matter **8**, 10231(1996)

39. D. Brown, T. C. Q. Noakes, D. P. Woodruff, P. Bailey and Y. Le Doaziou: J. Phys. Condens. Matter **11**, 1889(1999)
40. W. Wuttig, Y. Gauthier and S. Blügel: Phys. Rev. Lett. **70**, 3619(1993)
41. M. Eder, J. Hafner and E. G. Moroni: Phys. Rev. B **61**, 11492(2000)
42. Handbook of Auger Electron Spectroscopy (Physical Electronics, Eden Prarie, MI, 1972)
43. For example, Y. Gauthier and R. Baudoing: in *Segregation and Related Phenomena*, ed by P. Dowben and A. Miller, (CRC Press, Boca Raton, FL, 1990)
44. B. Dötsch, A. Schmidt, L. Hammer, N. Kottcke, R. Döll, K. Heinz, K. Müller, H. L. Davis and D. M. Zehner: private communication
45. W. Oed, H. Lindner, U. Starke, K. Heinz, K. Müller and J. B. Pendry: Sci. Sci. **224**, 179(1989)
46. H. Ralph: *Selected Values of the Thermodynamic Properties of Binary Alloys*, (American Society for Metals, Metal Park, Ohio, 1986)
47. J. W. Matthews: in *Epitaxial Growth*, ed by J. W. Matthews, (Academic Press, New York 1985)
48. A. Schmalt, S. Aminprooz, L. Becker, J. Haase, J. Neugebauer, M. Scheffler, D. R. Barcheler, D. L. Adams and E. Bϕgh: Phys. Rev. Lett. **67**, 2163(1991)
49. C. Stampfl. M. Scheffler, H. Over, J. Burchardt, M. Nielsen, D. L. Adams and W. Moritz: Phys. Rev. Lett. **69**, 1532(1992)
50. Z. Q. Wang, Y. S. Li, C. K. C. Lok, J. Quinn and F. Jona: Solid State Commun. **62**, 181(1987)
51. S. C. Wu, S. H. Lu, Z. Q. Wang, C. K. C. Lok, J. Quinn, Y. S. Li, D. Tian and F. Jona: Phys. Rev. B **38**, 5363(1988)
52. T. D. Pope, G. W. Anderson, K. Griffiths, P. R. Norton and G. W. Graham: Phys. Rev. B **44**, 11518(1991)
53. G. L. Kellog and P. J. Feibelmann: Phys. Rev. Lett. **64**, 3143(1990)
54. C. Chen and T. T. Tsong: Phys. Rev. Lett. **64**, 3147(1990)
55. S. C. Wang and G. Ehrlich: Phys. Rev. Lett. **67**, 469(1991); ibid, 2509(1991)
56. L. Hansen, P. Stoltze, K. W. Jacobsen, I. K. Nørskov: Phys. Rev. B **44**, 6523(1991)
57. D. D. Chambliss and S. Chiang: Surf. Sci. **264**, L187(1992)
58. J. F. Wolf: Ph. D. thesis, RWTH Aachen(1990)
59. M. Poensgen: Ph. D. thesis, RWTH Aachen(1993)
60. W. K. Burton, N. Cabrera and F. C. Frank: Trans. Roy. Soc. A **243**, 299(1951)
61. G. Zinsmeister: Thin Solid films **2**, 497(1968)
62. J. A. Venables: Phil. Mag. **27**, 693(1973)
63. S. M. Foiles: Surf. Sci. **191**, 329(1987)
64. U. von Barth, and L. Hedin: J. Phys. C **4**, 2064(1971)
65. E. Wimmer, H. Krakauer, M. Weinert and A. J. Freeman: Phys. Rev. B **24**, 864(1981)
66. S. Takizawa, PhD thesis, University of Tokyo, unpubished (1990)
67. A. Oswald, R. Zeller, P. J. Braspenning and P. H. Dederichs: J. Phys. F **15**, 193(1985)
68. U. Klemradt, B. Drittler, T. Hoshino, R. Zeller, P. H. Dederichs and H. Stefanou: Phys. Rev. B **43**, 9487(1991)
69. S. Blügel, B. Drittle, R. Zeller and P. H. Dederichs: Appl. Phys. A **49**, 547(1988)
70. D. Tian, H. Li, S. C. Wu, F. Jona, and P. M. Marcus: Phys. Rev. B **45**, 3749(1992)

71. M. Wuttig, S. Junghans, T. Flores and Blügel: Phys. Rev. B **53**, 7551(1996)
72. J. Tersoff and D. Hamann: Phys. Rev. Lett. **50**, 1998(1983)
73. D. Wortmann, S. Heinze, G. Bihlmayer and S. Blügel: Phys. Rev. B **62**, 2862(2000)
74. T. Flores, S. Junghans and M. Wuttig: Surf. Sci. **371**, 1(1997)
75. T. Flores, S. Junghans and M. Wuttig: Surf. Sci. **371**, 14(1997)
76. M. Giesen-Seibert and H. Ibach: Surf. Sci. **316**, 205(1994)
77. For example, S. K. Ghandhi: *VLSI Fabrication Principles* (Wiley, New York, 1995)
78. P. Stoltze: J. Phys.: Condens. Matter **6**, 9495(1994)
79. G. Tréglia, B. Legrand, A. Saúl, T. Flores and M. Wuttig: Surf. Sci. **352-354**, 552(1996)
80. R. Fink, R. Wesche, T. Klas, G. Krausch, R. Platzer, J. Voigt, U. WOhrmann and G. Schatz: Surf. Sci. **225**, 331(1990)
81. H. Ibach, M. Giesen, T. Flores, M. Wuttig and G. Treglia: Surf. Sci. **364**, 453(1996)
82. R. van Gastel, E. Somfai, W. van Saarloos and J. W. M. Frenken: Nature **408**, 665(2000)
83. R. van Gastel, E. Somfai, S. B. van Albada, W. van Saarloos and J. W. M. Frenken: Phys. Rev. Lett. **86**, 1562(2001)
84. H.P. Noh, T. Hashizume, D. Jeon, Y. Kuk, H.W. Picketing and T. Sakurai: Phys. Rev. B **50**, 2735(1994)
85. G. Vogl and W. Petry: Physikalische Blätter **10**, 925(1994)
86. T. Heumann, H. Meiners and H. Sttier: Z. Naturforsch. **25a**, 1883(1970)
87. C. Nagl, E. Platzgummer, O. Haller, M. Schmid and P. Varga: Surf. Sci. **331-333**, 831(1995)
88. Y. Gauthier, W. Moritz and W. Hösler: Surf. Sci. **345**, 53(1996)
89. Y. Girard, C. Cohen, J. Moulin and D. Schmaus: Nucl. Instrum. Methods B **64**, 73(1992)
90. C. Cohen, Y. Girard, P. Leroux-Hugon, A. L'Hoir, J. Moulin and D. Schmaus: Europhys. Lett. **24**, 767(1993)
91. C. Uebing and R. Gomer: J. Chem. Phys. **95**, 7626(1992)
92. J. B. Hannon, H. Ibach: Surf. Sci. **371**, 30(1997)
93. R. Niebuhr: Diploma Thesis, RWTH Aachen, 1993
94. O. Rader, W. Gudat, C. Carbone, E. Vescovo, S. Blügel, R. Kläsges, W. Eberhardt, M. Wuttig, J. Redinger and F. J. Himpsel: Phys. Rev. B **55**, 5404(1997)
95. F. Schiller, S. V. Halilov and C. Lauschat: Phys. Rev. B 65, 184430(2002); ibid **67**, 214431(2003)
96. Y. Huttel, F. Schiller, J. Avila and M. C. Asensio: Phys. Rev. B **61**, 4948(2000)
97. W. L. O'Brien, J. Zhang and B. P. Tonner: J. Phys.: Condens. Matter **5**, L515(1993)
98. W. L. O'Brien and B. P. Tonner: Phys. Rev. B **51**, 617(1995)
99. D. Schmitz, O. Rader, C. Carbone and W. Eberhardt: Phys. Rev. B **54**, 15352(1996)
100. Y. Huttel, C. M. Toedorescu, F. Bertran and G. Krill: Phys. Rev. B **64**, 94405(2001)
101. D. Spišák and J. Hafner: J. Phys.: Condens. Matter **11**, 6359(1999)
102. A. B. Hayden, P. Pervant and D. P. Woodruff: J. Phys.: Condsns. Matter **7**, 1139(1995)
103. W. L. O'Brien and B. P. Tonner: J. Appl. Phys. **76**, 6468(1994)

8 Summary and Outlook

It has been the aim of this book to illustrate the potential of epitaxially grown ultrathin films. In particular, metal-on-metal heteroepitaxy has been investigated. After a summary of the growth, structure and magnetism of 3d ultrathin films on various metallic substrates, considerable space has been devoted to a few prototype systems, i.e., Fe films on Cu(100) and Mn films on Cu(100) and Ni(100), with the goal of establishing a close correlation between the structure, growth and magnetism of ultrathin metallic films.

Our investigation of Fe films grown on Cu(100) demonstrates the intimate relationship between these properties, and the importance of their in-situ characterization. As already summarized in Chap. 6, after room-temperature deposition, three different regions of magnetic behavior are observed. In the first region (up to around 4 ML), the films are homogeneously magnetized, leading to a linear increase of the Kerr ellipticity with film thickness. Between 5 and 11 ML, a constant, small Kerr ellipticity is observed. This is a clear sign that only a small, constant number of layers contributes to the ferromagnetic signal. Postdeposition of CO reveals that it is the surface and not the interface that is magnetic. Up to film thicknesses of 11 ML, the magnetization is perpendicular to the surface. Above 11 ML, the magnetization switches from perpendicular to in-plane. A homogeneous magnetization of the whole film is found once again. The changes in magnetic behavior are direct consequences of a change in the growth mode and structure of the overlayer. The reorientation of the magnetization axis is connected with a structural transformation from fcc Fe to bcc Fe. An unusual structure characterizes the region where magnetic "live" surface layers are encountered. The first interlayer spacing has a surprisingly large expansion. It is this expansion and the resulting enlarged atomic volume that stabilize the magnetization of the surface atoms. Deeper bulk layers that do not contribute to the ferromagnetic signal have the interatomic spacings predicted for strained fcc films. The relationship between increased atomic spacing and enhanced magnetism is also reflected in the initial stages of growth. Here, a (5×1) structure with enlarged interatomic spacings throughout the whole film is magnetic.

An equally pronounced influence of the magnetic moments on the surface structure, or vice versa, of surface structure on the magnetic moments is exemplified by several different Mn phases on Cu(100) and Ni(100). The quasi-hexagonal Mn films grown at low temperature have an increased in-plane atomic spacing that is possibly a manifestation of an increased magnetic moment of these atoms. For the first manganese layer on Cu(100), this spacing is even larger than that on Ni, in line with calculations for Mn impurities in Cu and Ni which predict larger moments in Cu than in Ni. This correlation between the moments and the atomic arrangement is also encountered for ordered MnCu and MnNi surface alloys formed above 270 K. Theoretical calculations have made it clear that it is the gain in magnetic energy that stabilizes the Cu(100)–$c(2\times2)$ Mn surface alloy despite the positive heat of formation of Mn–Cu bonds. Mn atoms with enhanced magnetic moments occupy a large volume. The buckling movement of Mn is an efficient means of increasing the available space. Since atomic volumes and magnetic moments are tightly interwoven, the magnitude of the corrugation should be a measure of the size of the magnetic moment. Moreover, without the magnetization, this phase would be unstable against interdiffusion of Mn into the bulk.

To acquire a deeper understanding of this fascinating relationship between structure and magnetism, it would be most desirable to measure magnetic moments in an experimental setup that also enabled structure determinations. Several techniques, including photoemission, neutron diffraction, SQUID magnetometry and Mössbauer spectroscopy, allow either an indirect or a direct determination of magnetic moments. Accumulated theoretical evidence indicates that these moments can be drastically different on the surface or in a thin film compared with the bulk.

At this point, therefore, it is interesting to note that Co and Ni films on Cu(100) show behavior less exotic than that of Fe and Mn films on Cu(100). The Co/Cu(100) system has been studied extensively [1, 2]. Co films have an in-plane magnetization independent of film thickness. Their interatomic spacings, which have been determined by a LEED analysis [2], show no deviation from the expected behavior, i.e., the interlayer distance is dictated by the strain in the surface plane. The explanation for this more conventional behavior of Co and Ni films lies in their filled spin-up d bands. A change in magnetic moment is more difficult for these strong ferromagnets and cannot be accomplished even in the very different environment found at the surface (see Chap. 4). Consequently, weak magnets such as Cr, Mn and Fe are more interesting candidates when the goal is to learn how magnetic moments in thin films can be modified.

To strengthen our understanding of thin-film magnetization, it would be beneficial to examine the structure–property relationships for films of 4f-magnets, such as gadolinium and terbium. These metals, which approach more closely the Heisenberg model of ferromagnetism, have not yet been studied in comparable detail [3, 4, 5, 6]. In particular, structural investigations of ultrathin films are severely lacking. A systematic understanding of the correlation between structure and magnetism should go a long way in offering guidance as to how the magnetic properties of thin films can be manipulated and exploited to the advantage of technology and, more importantly, mankind.

Up to now, most ultrathin metal films have been prepared on metallic substrates. The growth, structure and physical properties of ultrathin metal films on nonmetallic substrates, especially on ionic crystals, have been much less studied [7]. First-principles density-functional calculations have predicted charge transfer between the metal film and the oxide substrate, as well as lateral polarization in the metal overlayer [8]. These effects can be neglected in ordinary thin films but could have important consequences for the film properties when the film is ultrathin. It has been already found that an oxide support strongly affects the catalytic behavior of transition metal clusters on it [9, 10]. Therefore, one has reason to believe that other physical properties, including the growth and magnetism, would also be modified if an ultrathin metal film was prepared on an ionic substrate.

For single-component films, epitaxy-stabilized metastable structures can be prepared only on those substrates which happen to have lattice constants close to those of the metastable phase. This limitation can be partially lifted by preparing alloy films. By appropriate alloying to reduce the strain energy, a greater number of metastable structures and forced structures with a considerable thickness could be stabilized by epitaxial growth. This provides new opportunities to observe novel magnetism in metals. Ultrathin alloy films are hence attracting more and more attention.

Ordered surface alloys, such as the Cu(100)–$c(2\times2)$ Mn and Ni(100)–$c(2\times2)$ Mn phases, not only are interesting candidates for the investigation of two-dimensional magnetism, but also might possibly act as templates for the preparation of ordered molecular monolayers. Nowadays, the preparation and characterization of self-assembled monolayers (SAMs) is an active research field. SAMs are expected to play an important role in future electronic devices. Usually, these molecular monolayers, such as thiols on a Au(111) substrate [11], are self-organized by the competing interactions between the molecules and the substrate and between the molecules. Since the dimensions of molecules are larger than those of atoms, site-selective adsorption of molecules on a surface alloy could produce a kind of epitaxial molecular monolayer. This is just an extreme example of applying templates to control the lateral arrangement of atoms, molecules or clusters. A reduction from two dimensions, i.e., ultrathin films, to one- and zero-dimensional objects should bring about new physics, and the first experiments to prove this have already been done. Exciting areas include the investigation of one-dimensional magnetic wires and of magnetic dots. The reduced dimensionality will lead not only to novel properties but possibly also to new products and applications.

As illustrated above, the research opportunities in thin-film growth and processing are exciting and numerous. Advances in the production of ultrathin films have created a route to the preparation of materials with altered lattice constants and the stabilization of crystal phases under conditions where they have no bulk analogue. In other words, novel materials with modified electronic and magnetic characteristics have been created. As a consequence of the coupling of these achievements to advanced theories of ultrathin films, innovative, new research directions have emerged.

References

1. J. J. de Miguel, A. Cebollada, J. M. Gallego, R. Miranda, C. M. Schneider, P. Schuster and J. Kirschner: J. Mag. Mag. Mat. **93**, 1(1993)
2. J. R. Cerda, P. L. de Andres, A. Cebollada, R. Miranda, E. Navas, P. Schuster and J. Kirschner: J. Phys. C 5, 2055(1993)
3. H. Li, D. Tian, J. Quinn, Y. S. Li, S. C. Wu and F. Jona: Phys. Rev. B **45**, 3853(1992)
4. J. B. Goedkoop, M. Grioni and J. C. Fuggle: Phys. Rev. B. **43**, 1179(1991)
5. R. J. H. Kappert, J. Vogel, M. Sacchi and J. C. Fuggle: Phys. Rev. B **48**, 2711(1993)
6. E. D. Tober, R. X. Ynzunza, C. Westphal and C. S. Fadley: Phys. Rev. B **53**, 5444(1996)
7. T. Kado: Surf. Sci. 454, 783(2000)
8. A. Bogicevic and D. R. Jennison: Phys. Rev. Lett. **82**, 4050(1999)
9. N. López and F. Illas: J. Phys. Chem. B **102**, 1430(1998)
10. N. López and J. K. Nørskov: Surf. Sci. 515 **515**, 175(2002)
11. Y. Yourdshahyana and A. M. Rappe: J. Chem. Phys. **117**, 825(2002)

Index

Springer Tracts in Modern Physics

167 **Photoelectric Properties and Applications of Low-Mobility Semiconductors**
By R. Könenkamp 2000. 57 figs. VIII, 100 pages

168 **Deep Inelastic Positron-Proton Scattering in the High-Momentum-Transfer Regime of HERA**
By U.F. Katz 2000. 96 figs. VIII, 237 pages

169 **Semiconductor Cavity Quantum Electrodynamics**
By Y. Yamamoto, T. Tassone, H. Cao 2000. 67 figs. VIII, 154 pages

170 **d–d Excitations in Transition-Metal Oxides**
A Spin-Polarized Electron Energy-Loss Spectroscopy (SPEELS) Study
By B. Fromme 2001. 53 figs. XII, 143 pages

171 **High-T_c Superconductors for Magnet and Energy Technology**
By B. R. Lehndorff 2001. 139 figs. XII, 209 pages

172 **Dissipative Quantum Chaos and Decoherence**
By D. Braun 2001. 22 figs. XI, 132 pages

173 **Quantum Information**
An Introduction to Basic Theoretical Concepts and Experiments
By G. Alber, T. Beth, M. Horodecki, P. Horodecki, R. Horodecki, M. Rötteler, H. Weinfurter, R. Werner, and A. Zeilinger 2001. 60 figs. XI, 216 pages

174 **Superconductor/Semiconductor Junctions**
By Thomas Schäpers 2001. 91 figs. IX, 145 pages

175 **Ion-Induced Electron Emission from Crystalline Solids**
By Hiroshi Kudo 2002. 85 figs. IX, 161 pages

176 **Infrared Spectroscopy of Molecular Clusters**
An Introduction to Intermolecular Forces
By Martina Havenith 2002. 33 figs. VIII, 120 pages

177 **Applied Asymptotic Expansions in Momenta and Masses**
By Vladimir A. Smirnov 2002. 52 figs. IX, 263 pages

178 **Capillary Surfaces**
Shape – Stability – Dynamics, in Particular Under Weightlessness
By Dieter Langbein 2002. 182 figs. XVIII, 364 pages

179 **Anomalous X-ray Scattering for Materials Characterization**
Atomic-Scale Structure Determination
By Yoshio Waseda 2002. 132 figs. XIV, 214 pages

180 **Coverings of Discrete Quasiperiodic Sets**
Theory and Applications to Quasicrystals
Edited by P. Kramer and Z. Papadopolos 2002. 128 figs., XIV, 274 pages

181 **Emulsion Science**
Basic Principles. An Overview
By J. Bibette, F. Leal-Calderon, V. Schmitt, and P. Poulin 2002. 50 figs., IX, 140 pages

182 **Transmission Electron Microscopy of Semiconductor Nanostructures**
An Analysis of Composition and Strain State
By A. Rosenauer 2003. 136 figs., XII, 238 pages

183 **Transverse Patterns in Nonlinear Optical Resonators**
By K. Staliūnas, V. J. Sánchez-Morcillo 2003. 132 figs., XII, 226 pages

184 **Statistical Physics and Economics**
Concepts, Tools and Applications
By M. Schulz 2003. 54 figs., XII, 244 pages

185 **Electronic Defect States in Alkali Halides**
Effects of Interaction with Molecular Ions
By V. Dierolf 2003. 80 figs., XII, 196 pages

Springer Tracts in Modern Physics

186 **Electron-Beam Interactions with Solids**
Application of the Monte Carlo Method to Electron Scattering Problems
By M. Dapor 2003. 27 figs., X, 110 pages

187 **High-Field Transport in Semiconductor Superlattices**
By K. Leo 2003. 164 figs.,XIV, 240 pages

188 **Transverse Pattern Formation in Photorefractive Optics**
By C. Denz, M. Schwab, and C. Weilnau 2003. 143 figs., XVIII, 331 pages

189 **Spatio-Temporal Dynamics and Quantum Fluctuations in Semiconductor Lasers**
By O. Hess, E. Gehrig 2003. 91 figs., XIV, 232 pages

190 **Neutrino Mass**
Edited by G. Altarelli, K. Winter 2003. 118 figs., XII, 248 pages

191 **Spin-orbit Coupling Effects in Two-dimensional Electron and Hole Systems**
By R. Winkler 2003. 66 figs., XII, 224 pages

192 **Electronic Quantum Transport in Mesoscopic Semiconductor Structures**
By T. Ihn 2003. 90 figs., XII, 280 pages

193 **Spinning Particles – Semiclassics and Spectral Statistics**
By S. Keppeler 2003. 15 figs., X, 190 pages

194 **Light Emitting Silicon for Microphotonics**
By S. Ossicini, L. Pavesi, and F. Priolo 2003. 206 figs., XII, 284 pages

195 **Uncovering *CP* Violation**
Experimental Clarification in the Neutral K Meson and B Meson Systems
By K. Kleinknecht 2003. 67 figs., XII, 144 pages

196 **Ising-type Antiferromagnets**
Model Systems in Statistical Physics and in the Magnetism of Exchange Bias
By C. Binek 2003. 52 figs., X, 120 pages

197 **Electroweak Processes in External Electromagnetic Fields**
By A. Kuznetsov and N. Mikheev 2003. 24 figs., XII, 136 pages

198 **Electroweak Symmetry Breaking**
The Bottom-Up Approach
By W. Kilian 2003. 25 figs., X, 128 pages

199 **X-Ray Diffuse Scattering from Self-Organized Mesoscopic Semiconductor Structures**
By M. Schmidbauer 2003. 102 figs., X, 204 pages

200 **Compton Scattering**
Investigating the Structure of the Nucleon with Real Photons
By F. Wissmann 2003. 68 figs., VIII, 142 pages

201 **Heavy Quark Effective Theory**
By A. Grozin 2004. 72 figs., X, 213 pages

202 **Theory of Unconventional Superconductors**
By D. Manske 2004. 84 figs., XII, 228 pages

203 **Effective Field Theories in Flavour Physics**
By T. Mannel 2004. 29 figs., VIII, 175 pages

204 **Stopping of Heavy Ions**
By P. Sigmund 2004. 43 figs., XIV, 157 pages

205 **Three-Dimensional X-Ray Diffraction Microscopy**
Mapping Polycrystals and Their Dynamics
By H. Poulsen 2004. 49 figs., XI, 154 pages

206 **Ultrathin Metal Films**
Magnetic and Structural Properties
By M. Wuttig and X. Liu 2004. 234 figs., XII, 375 pages

Printing: Saladruck, Berlin
Binding: Stein+Lehmann, Berlin